Nanoscale Energy Transport

Emerging phenomena, methods and applications

Nanoscale Energy Transport

Emerging phenomena, methods and applications

Edited by
Bolin Liao
Department of Mechanical Engineering, University of California, Santa Barbara, USA

IOP Publishing, Bristol, UK

ISBN 978-0-7503-1738-2 (ebook)
ISBN 978-0-7503-1736-8 (print)
ISBN 978-0-7503-1767-2 (myPrint)
ISBN 978-0-7503-1737-5 (mobi)

DOI 10.1088/978-0-7503-1738-2

Version: 20200301

IOP ebooks

British Library Cataloguing-in-Publication Data: A catalogue record for this book is available from the British Library.

Published by IOP Publishing, wholly owned by The Institute of Physics, London

IOP Publishing, Temple Circus, Temple Way, Bristol, BS1 6HG, UK

US Office: IOP Publishing, Inc., 190 North Independence Mall West, Suite 601, Philadelphia, PA 19106, USA

Contents

Part II Measurements and applications

Preface

Nanoscale energy transport is a fast-developing research field that studies the transport processes of fundamental energy carriers, including electrons, phonons, photons, magnons, etc, in devices and material structures with characteristic sizes in the nanometer range. Fundamentally, new physical phenomena emerge at the nanoscale due to the classical and quantum confinement effects of the energy carriers, leading to the breakdown of macroscopic constitutive laws, such as Fourier's law of heat conduction and Planck's law of blackbody radiation. Practically, as the advancement of nanotechnology has enabled routine fabrication of devices and materials at the nanoscale, a fundamental understanding of energy transport in these systems is crucial for achieving better efficiency and performance. Indeed, the improved understanding of nanoscale energy transport in the past two decades has led to better thermal management for microelectronic devices, more efficient thermoelectric modules and new strategies to efficiently harvest the full spectrum of solar power, to name a few examples. Therefore, nanoscale energy transport is a field of both fundamental interest and practical relevance. In this light, this multi-contributor volume aims to cover new developments in both the scientific basis and the practical relevance of nanoscale energy transport, with a particular emphasis on the emerging effects at the nanoscale that qualitatively differ from those at the macroscopic scale.

Excellent texts and monographs on nanoscale energy transport are available, for example by Chen [1], Zhang [2], Fisher [3], Volz [4] and others, where the fundamentals and the research developments at the time of publication are clearly elaborated. However, as this is an active field of research, new effects, experimental and computational methods, and applications are emerging at a fast pace. Complementary to these existing books, the goal of this volume is to cover recent developments in the theory, methods and applications of nanoscale energy transport from the past few years, and help researchers in this field obtain an overview of the current frontiers. To this end, I have invited active researchers in nanoscale energy transport to contribute chapters on their specialty topics and offer their expert perspectives on the important advancements in the past decade as well as future directions. In the end, 17 chapters were selected for this multi-contributor volume that cover a broad range of topics. In terms of microscopic energy carriers, the transport of phonons, electrons, photons and magnons in the nanoscale are discussed in various chapters. In terms of methods, state-of-the-art computational and experimental approaches are reviewed, including a chapter on the emerging material informatics method (chapter 5). In terms of material systems, a broad range from interfaces and molecular junctions to nanostructured bulk materials is included. While I believe this volume is a comprehensive survey of the state of the art of nanoscale energy transport, by no means does this book cover all significant new developments in the field. Notable omissions include spin caloritronics, where the coupling effects of phonons and magnons are investigated, and the energy transport in two-dimensional materials. For these topics, the interested reader is referred to excellent recent review articles such as [5] and [6].

The chapters are organized into two parts. Part I focuses on emerging theory and computational methods. Chapter 1 by Sangyeop Lee's group at the University of Pittsburgh discusses hydrodynamic phonon transport, particularly in two-dimensional materials, where normal phonon–phonon scatterings dominate Umklapp scatterings and the phonon thermal transport mimics fluid flow. Chapter 2, by Tianli Feng from Oak Ridge National Laboratory and Xiulin Ruan from Purdue University, reviews the recent development of calculating higher order phonon scattering rates, e.g. four-phonon processes, and its relevance to thermal transport in technologically important materials. Chapter 3, by Tengfei Luo's group at the University of Notre Dame, provides a detailed account of how bulk phonon scattering events affect interfacial thermal transport from first-principles and molecular dynamics simulations. Chapter 4, by Zhiting Tian's group at Cornell University, gives an introduction to the state-of-the-art atomistic Green's function (AGF) method for calculating interfacial phonon transport properties and interfacial thermal resistance. Chapter 5, by Junichiro Shiomi's group at the University of Tokyo, elaborates on using material informatics methods, in particular Bayesian optimization, for nanoscale thermal transport problems. This is an emerging front in computational materials science that has attracted intense interest recently due to the fast advancement of data science and machine learning methods. Chapter 6, by Chengyun Hua at Oak Ridge National Laboratory, reviews the current status of resolving the phonon mean free path distribution in real materials from both the computational and experimental perspectives, which is essential for engineering materials at the nanoscale to achieve desirable thermal transport properties. Chapter 7, by Keivan Esfarjani's group at the University of Virginia, provides a historic view of incorporating the lattice anharmonicity into first-principles phonon calculations at finite temperature, which is necessary to correctly describe phonon softening and phase transitions.

Part II focuses on the developments of experimental techniques and practical applications enabled by fundamental advancements. Chapter 8, by Professors Edgar Meyhofer and Pramod Reddy's group at the University of Michigan, details state-of-the-art measurement techniques to resolve thermal and thermoelectric transport across atomic and molecular junctions with extreme sensitivity. Chapter 9, by Xiaojia Wang's group at the University of Minnesota, reviews the recent applications of emerging time-resolved magneto-optical Kerr effect (TR-MOKE) spectroscopy to characterize both phonon and magnetization dynamics. Chapter 10, by Keshav Dani's group at Okinawa Institute of Science and Technology in Japan, introduces a class of powerful tools—ultrafast electron microscopy, in particular time-resolved photoemission electron microscopy (TR-PEEM)—and their applications in nanoscale energy transport. Chapter 11, by Chen Li's group at the University of California, Riverside summarizes recent results utilizing the inelastic neutron scattering (INS) technique to understand phonons and magnons in energy materials. Chapter 12, by Renkun Chen's group at the University of California, San Diego, reviews the historic development and recent applications of suspended micro-devices to characterize the thermal transport properties of nanomaterials. Chapters 13 and 15, by Nenad Miljkovic's group at the University of Illinois, Urbana–Champaign, discuss how micro- and nanostructured surfaces can significantly enhance condensation heat transfer and help combat frosting and icing in practical energy systems. Chapter 14, by Mona Zebarjadi's group at the University of Virginia, introduces the mechanisms and current frontiers of using thermionic emission to convert thermal energy into electricity as an alternative to thermoelectric conversion. Chapter 16, by Zhifeng Ren's group at the University of Houston, addresses a critical problem at the current development stage of thermoelectric materials—how to reliably measure the thermoelectric transport properties and energy conversion efficiency of thermoelectric materials. Chapter 17, by Andrej Lenert's group at the University of Michigan, reviews the recent development of thermophotovoltaic energy conversion, including both the material and the device aspects, and particularly the opportunities offered by the recent advancement of nanophotonics.

I wish to acknowledge all the authors for their valuable input and hard work, which have made this volume possible. I also want to thank the Institute of Physics Publishing for providing me the opportunity to work on this project, in particular Michael Slaughter and John Navas who initiated this project, and Caroline Mitchell, Daniel Heatley and Robert Trevelyan for their generous help (and patience) during the editing and production process. Last, but not least, I want to thank the support for research provided by the US Department of Energy, National Science Foundation and US Army Research Office.

Bolin Liao
University of California, Santa Barbara
November 2019

References

[1] Chen G 2005 *Nanoscale Energy Transport and Conversion: A Parallel Treatment of Electrons, Molecules, Phonons, and Photons* 1st edn (Oxford: Oxford University Press)

[2] Zhang Z 2007 *Nano/Microscale Heat Transfer* (New York: McGraw-Hill)

[3] Fisher T S 2013 *Thermal Energy At The Nanoscale* (Hackensack, NJ: WSPC)

[4] Volz S (ed) 2007 *Microscale and Nanoscale Heat Transfer* (Berlin: Springer)

[5] Boona S R, Myers R C and Heremans J P 2014 Spin caloritronics *Energy Environ. Sci.* **7** 885–910

[6] Gu X, Wei Y, Yin X, Li B and Yang R 2018 Colloquium: phononic thermal properties of two-dimensional materials *Rev. Mod. Phys.* **90** 041002

Editor biography

Bolin Liao

Bolin Liao is currently an assistant professor in the Department of Mechanical Engineering at the University of California, Santa Barbara. His main research interest is in nanoscale energy transport and its application to sustainable energy technologies. Specifically, his research aims to understand the transport and interaction processes of fundamental energy carriers, such as electrons, phonons, photons and magnons, at the smallest length and time scales, and then use this knowledge to develop more efficient clean energy devices, e.g. thermoelectric modules and photovoltaic cells. Current projects include first-principles and multiscale simulation of energy carrier transport, visualization of photophysics in space and time with scanning ultrafast electron microscopy, ultrafast optical techniques for thermal and thermoelectric characterization, and applied clean energy devices and systems.

Bolin obtained his PhD in mechanical engineering from MIT in March 2016, advised by Gang Chen. He was a Kavli Postdoctoral Fellow at the California Institute of Technology from May 2016 to June 2017, hosted by the late Ahmed Zewail, where he worked on scanning ultrafast electron microscopy.

Contributors

Xun Li

Xun Li received his BS degree in energy engineering from Zhejiang University, the People's Republic of China, in 2015. He is currently a PhD candidate in the Department of Mechanical Engineering and Materials Science at the University of Pittsburgh, PA. His research focuses on the computational study of first-principles based phonon transport in graphitic materials and interfacial phonon transport in semiconductors.

Sangyeop Lee

Sangyeop Lee is an Assistant Professor in the Department of Mechanical Engineering and Materials Science at the University of Pittsburgh, PA. He received his PhD from the Department of Mechanical Engineering at Massachusetts Institute of Technology in 2015. His current research interests are first-principles based simulation of hydrodynamic phonon transport and thermal transport in partially/fully disordered phases.

Tianli Feng

Tianli Feng is a postdoctoral fellow at Oak Ridge National Laboratory, TN. He received his BS in physics from the University of Science and Technology of China in 2011. He received his MS and PhD in mechanical engineering from Purdue University, IN in 2013 and 2017, respectively. His research interests include developing new atomistic-scale predictive simulation methods for thermal energy transport, and using these methods to guide the development of high-performance materials.

Xiulin Ruan

Xiulin Ruan is a Professor in the School of Mechanical Engineering and the Birck Nanotechnology Center at Purdue University, IN. He received his BS and MS in engineering thermophysics from Tsinghua University in 2000 and 2002, respectively. He received an MS in electrical engineering and a PhD in mechanical engineering from the University of Michigan at Ann Arbor in 2006 and 2007, respectively. His research interests include multiscale multiphysics simulations and experiments of phonon, electron, and photon transport and interactions, for various emerging applications. Ruan has been recognized with many awards, including the NSF CAREER Award (2012), the ASME Heat Transfer Division Best Paper Award (2015) and the Air Force Summer Faculty Fellowship (2010, 2011 and 2013). He currently serves as an associate editor for the ASME *Journal of Heat Transfer*.

Tengfei Luo

Tengfei Luo is the Dorini Family Collegiate Chair and Associate Professor in the Department of Aerospace and Mechanical Engineering (AME) at the University of Notre Dame (UND), IN. Before joining UND, he was a postdoctoral associate at Massachusetts Institute of Technology from 2009 to 2011, after obtaining his PhD from Michigan State University in 2009. At UND, Luo leads an interdisciplinary group focusing on nanoscale thermal transport, electronics thermal management, novel material design and manufacturing, and water treatment.

Eungkyu Lee

Eungkyu Lee is the Research Assistant Professor in the Department of Aerospace and Mechanical Engineering (AME) at the University of Notre Dame (UND), IN. He was a postdoctoral associate at the Department of AME at UND. He obtained his PhD from Seoul National University in 2015. At UND, Lee studies light–matter interaction involving nanophotonics and multiphase thermofluids and interfacial thermal transport.

Ruiyang Li

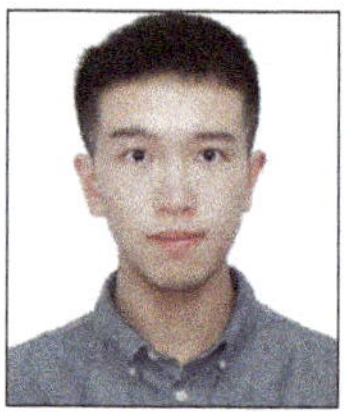

Ruiyang Li is currently pursuing a PhD degree at the University of Notre Dame, IN under the supervision of Tengfei Luo. He received his BS degree in energy and power engineering from Huazhong University of Science and Technology in 2018. His research interests include nanoscale thermal transport at semiconductor material interfaces, and the prediction of thermal properties using machine learning based techniques.

Zhiting Tian

Zhiting Tian is an Assistant Professor and a Eugene A Leinroth Sesquicentennial Faculty Fellow in the Sibley School of Mechanical and Aerospace Engineering at Cornell University, NY. Between 2014 and 2018 she was an Assistant Professor of Mechanical Engineering at Virginia Tech. Zhiting obtained her PhD in mechanical engineering from MIT in 2014. Zhiting's research focuses on the fundamental understanding of nanoscale thermal transport and energy conversion using experimental and computational tools. Zhiting's recent awards include the Office of Naval Research (ONR) Young Investigator Award, NSF CAREER Award, ACS Petroleum Research Fund Doctoral New Investigator Award and 3M Non-Tenured Faculty Award.

Jinghang Dai

Jinghang Dai is a PhD student in the Sibley School of Mechanical and Aerospace Engineering at Cornell University, NY. His research interests focus on the modeling of transport processes at interfaces and thermoelectric materials and experimental characterization of thermal transport properties.

Renjiu Hu

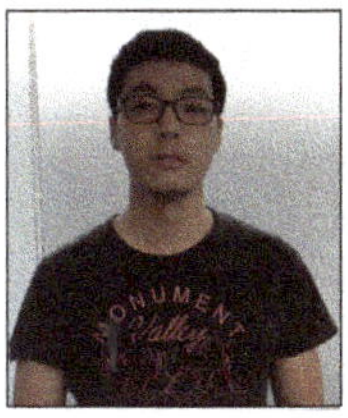

Renjiu Hu is a PhD student in the Sibley School of Mechanical and Aerospace Engineering at Cornell University, NY. His research focuses on interfacial thermal transport and Anderson localization.

Jiang Guo

Jiang Guo is currently a PhD student in the Department of Mechanical Engineering at the University of Tokyo, working on tailoring thermal radiative properties for energy applications via optimizing metamaterials using machine learning methods.

Shenghong Ju

Shenghong Ju received his PhD in 2014 from Tsinghua University, PRC. He is now an associated professor at the China–UK Low Carbon College, Shanghai Jiao Tong University, and a visiting scholar at the University of Tokyo. He is working on developing new energy materials, fundamental nanoscale heat transfer studies and materials informatics.

Junichiro Shiomi

Junichiro Shiomi received his PhD in 2004 from the Royal Institute of Technology (KTH), Sweden. He is currently a Professor in the Department of Mechanical Engineering, The University of Tokyo. His research interests include heat conduction of nanomaterials, polymer composites, thermoelectrics, phase change and fluidics in the nanoscale, interfacial thermofluid dynamics, thermal convections and materials informatics.

Chengyun Hua

Chengyun Hua received her BS in engineering physics from the University of Michigan in 2011, and her PhD in mechanical engineering from the California Institute of Technology in 2016. She joined the Oak Ridge National Laboratory, TN, in 2016 as a Liane B Russell Fellow and is currently working as an R&D associate. Her research focuses on gaining a comprehensive picture of how heat is transported in a solid at the scales of heat carriers, i.e. electrons and phonons (quantized vibrations in the lattice). Using both computation and experiments, she is currently working on using structures at the nanoscale as a tunable physical parameter to control and manipulate heat.

Keivan Esfarjani

Keivan Esfarjani obtained his general-engineering degree at the Ecole Centrale de Paris in France. He then pursued his studies in theoretical solid-state physics with a Master's (DEA) from the University of Paris followed by a PhD at the University of Delaware and a postdoc at Washington University in Saint Louis, MO. During his career, Esfarjani has held various positions at the Institute for Materials Research of Tohoku University, Sharif University of Technology, University of California Santa Cruz, MIT and Rutgers University. Currently he is an Associate Professor at the Departments of Mechanical Engineering, Materials Science and Physics at the University of Virginia. His research interests include the electronic structure calculation of materials from first-principles, lattice dynamics and phase change, electrical and thermal transport, and finally the storage and conversion of energy and information.

Yuan Liang

Yuan Liang received his BSc in applied physics from the University of Science and Technology of China, in Hefei, Anhui, in 2015. He was a member of the YAN Ji-Ci Elite Program in Physics, School of Gifted Young, working on the super-hydrophobicity of nano-membranes. He is currently working on his PhD in physics at the University of Virginia, Charlottesville. His current research interests include free energy computation and phase transition studies based on self-consistent phonon theory, computational schemes to extract anharmonic force constants in finite temperature—specifically stochastically designing a sampling method of lattice snapshots— and material properties analysis by applications of deep learning neural network packages.

Pramod Reddy

Pramod Reddy received a BTech and MTech in mechanical engineering from IIT-Bombay in 2002, and a PhD in applied science and technology from the University of California, Berkeley in 2007. He is currently a Professor in the Departments of Mechanical Engineering and Materials Science and Engineering at the University of Michigan, Ann Arbor.

Edgar Meyhofer

Edgar Meyhofer received a BS in biology from the University of Hannover in Germany, and an MS from Northeastern University in Boston, MA. In 1991 he earned a PhD from the University of Washington, Seattle. Since 2001 he has been a Professor of Mechanical Engineering and Biomedical Engineering at the University of Michigan in Ann Arbor. His current interests include nanoscale radiative heat transfer, energy transport in molecular junctions and nanoscale energy conversion.

Longji Cui

Longji Cui joined the faculty of the Mechanical Engineering and Materials Science and Engineering program of CU Boulder, CO, as an Assistant Professor in January 2020. Cui received his PhD in mechanical engineering in 2018 from the University of Michigan, Ann Arbor, under the supervision of Pramod Reddy and Edgar Meyhofer. From August 2018, he served as a Visiting Assistant Professor in the Mechanical Engineering department of CU and has been the J Evans Attwell-Welch Fellow of Rice University, and performed postdoctoral research in the Smalley-Curl Institute and Department of Physics and Astronomy (Natelson Research Group) at Rice University. Cui is a recipient of the

Robert M Caddell Memorial Award, the Richard and Eleanor Towner Prize for Outstanding PhD Research at U-M, the Chinese Government Award for Outstanding Student Abroad and the Material Research Society (MRS) Graduate Student Gold Medal.

Dustin Lattery

Dustin Lattery is currently a PhD student in mechanical engineering at the University of Minnesota, Twin Cities. He received his Bachelor's and Master's degrees in mechanical engineering from the University of Minnesota, Twin Cities in 2014 and 2017, respectively. His research focuses on magnetization dynamics in thin films. During his research, Lattery has optimized the signal from time-resolved magneto-optical Kerr effect measurements of perpendicular magnetic thin films, successfully combining both experimental and numerical methods. He has applied this method to a range of technologically important magnetic materials, for extracting information about their magnetic anisotropy and Gilbert damping.

Jie Zhu

Jie Zhu is currently an Associate Professor in the School of Energy and Power Engineering at Dalian University of Technology in China. He was previously a research associate in the Department of Mechanical Engineering at the University of Minnesota, Twin Cities. He received his BS in thermal science and energy engineering in 2004 from the University of Science and Technology of China, and a PhD from the Institute of Engineering Thermophysics, Chinese Academy of Sciences in 2011. Zhu's research interests include ultrafast non-equilibrium heat transfer, thermal transport across interfaces, the thermophysical properties of novel materials and the development of related experimental methods.

Dingbin Huang

Dingbin Huang is currently a PhD student in mechanical engineering at the University of Minnesota, Twin Cities. Mr Huang received his Master's degree in mechanical engineering from Shanghai Jiao Tong University in 2018, and a Bachelor's degree from Xi'an Jiaotong University in 2015. Mr Huang's research interests are the time-resolved study of magnetization dynamics and spin-heat coupling of novel materials for spintronic and data storage applications.

Xiaojia Wang

Professor Xiaojia Wang is an assistant professor in the Department of Mechanical Engineering at the University of Minnesota, Twin Cities. Prior to this, she was a postdoctoral research associate in the Department of Materials Science and Engineering at the University of Illinois at Urbana–Champaign. She received her PhD in mechanical engineering from the Georgia Institute of Technology in 2011. She received her MS in 2007 and BS in 2004 from Xi'an Jiaotong University, China, studying mechanical engineering. Her current research explores thermal and magnetic transport in functional materials and across material interfaces, using ultrafast optical techniques. These investigative efforts have a wide range of applications, including solid-state energy conversion and harvesting, data storage and spintronic devices.

Rebecca Wong

Rebecca Wong obtained her Bachelor's degree in physics and environmental science from Grinnell College. She has worked on various energy-related research projects at universities and national laboratories in the United States and abroad, and has experience in optics research at Osaka University and more recently the Okinawa Institute of Science and Technology Graduate University. Her research interests include novel semiconductor materials and devices, in particular for solar and renewable energy.

Michael Man

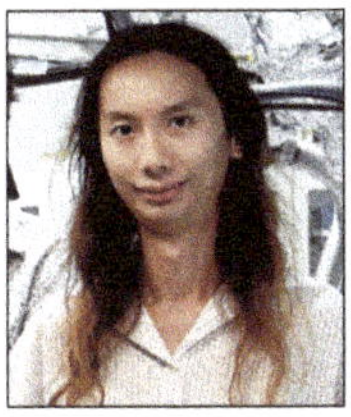

Michael Man received his Masters and PhD from Hong Kong University of Science and Technology. He moved to Okinawa and joined the Okinawa Institute of Science and Technology (OIST) as a postdoctoral researcher in 2012. Currently, he is a staff scientist in the Femtosecond Spectroscopy Unit at OIST. His research focuses on the development of ultrafast techniques in photoemission electron microscopy and in the study of electron dynamics and ultrafast phenomena in two-dimensional materials. He also has expertise in the field of surface science, covering phase transition, growth and magnetism in ultrathin films and two-dimensional materials.

Keshav Dani

Keshav Dani is currently an Associate Professor at the Okinawa Institute of Science and Technology (OIST), Graduate University in Okinawa, Japan. He joined OIST in November 2011 as a tenure-track Assistant Professor after completing a Director's Postdoctoral Fellowship at the Center for Integrated Nanotechnologies at Los Alamos National Laboratory, NM. Keshav graduated from UC Berkeley in 2006 with a PhD in physics, where he explored the nonlinear optical response of the quantum Hall system under the supervision of Daniel Chemla at LBNL. Prior to his PhD, he obtained a BS from Caltech in mathematics with a senior thesis in quantum information theory under John Preskill and Hideo Mabuchi. His current research interests lie in the use of ultrafast techniques to study the electron dynamics of two-dimensional materials and energy materials, develop optoelectronic applications in the terahertz regimes, and pursue interdisciplinary projects with OIST colleagues in neuroscience and art conservation.

Chen Li

Chen Li joined the Department of Mechanical Engineering and Materials Science and Engineering Program at the University of California, Riverside as an assistant professor in July 2016. Prior to joining UCR, Li worked as a research scientist for EFree (Energy Frontier Research in Extreme Environment Center), a DOE Energy Frontier Research Center (EFRC) centered at the Geophysical Laboratory of the Carnegie Institute of Washington and a joint faculty at Spallation Neutron Source (SNS) at Oak Ridge National Laboratory, TN.

Li obtained a BSc in physics from the Department of Physics, Peking University, and a PhD in materials science from the Department of Applied Physics and Materials Science, California Institute of Technology. After graduation, he worked as a postdoc for the Scattering and Thermophysics Group, Materials Science and Technology Division, at Oak Ridge National Laboratory.

Qiyang Sun

Qiyang Sun is a PhD student in the Mechanical Engineering Department, University of California, Riverside. Before joining UCR, Qiyang Sun obtained his BSc in engineering from the Energy and Power Engineering department, Huazhong University of Science and Technology, PRC.

Sunmi Shin

Sunmi Shin joined the Department of Mechanical Engineering at the National University of Singapore as a Research Assistant Professor in August 2019 and will be an Assistant Professor from July 2020. She received her PhD degree in materials science and engineering from the University of California, San Diego in 2019. She specializes in the experimental and theoretical investigation of fundamental nanoscale heat transport for thermal management and the development of personalized thermoregulators and energy harvesting devices using thermoelectric energy conversion. Her research interests include multidisciplinary approaches for efficient and active thermal energy technologies.

Renkun Chen

Renkun Chen is an Associate Professor in the Department of Mechanical and Aerospace Engineering at the University of California, San Diego (UCSD). He received his BS in thermophysics from Tsinghua University, Beijing in 2004, and his PhD in mechanical engineering from the University of California, Berkeley in 2008. He was a postdoctoral researcher at Berkeley prior to joining UCSD in 2009. His research group at UCSD is interested in the fundamentals and applications of thermal energy transport and conversion, including nanoscale energy transport phenomena, thermoelectric and solar–thermal energy conversion, phase-change heat transfer and thermal insulation technologies.

Hyeongyun Cha

Hyeongyun Cha received BS and MS degrees in mechanical engineering from the University of Illinois at Urbana–Champaign in 2014 and 2016, respectively, where he is currently pursuing a PhD degree in mechanical engineering. His current research focuses on the study of functional coating degradation using scanning probe microscopy techniques.

Soumyadip Sett

Soumyadip Sett received his BE degree in power engineering from Jadavpur University, India in 2011, and a PhD degree in mechanical engineering from the University of Illinois at Chicago (UIC) in 2016. He is currently a postdoctoral researcher with the University of Illinois at Urbana–Champaign (UIUC). His research interests intersect the multidisciplinary fields of thermofluid science, interfacial phenomena and renewable energy. His current work focuses on the phase-change heat transfer performance of micro/nanostructured surfaces, in particular involving low surface tension fluids and refrigerants. He was the recipient of the Deans Graduate Fellowship from UIC in 2015 for outstanding graduate research during his PhD, and in 2014 received the Chicago Consular Corps award for academic achievements as an international student.

Patrick Birbarah

Patrick Birbarah received his BE degree (with a high distinction) in mechanical engineering from the American University of Beirut, Lebanon in 2014. He received a PhD degree in mechanical science and engineering from the University of Illinois at Urbana–Champaign in 2019, where his work focused on phase-change heat transfer enhancement. He is currently working at Trane as a thermal systems engineer.

Tarek Gebrael

Tarek Gebrael received a BE degree (with a high distinction) in mechanical engineering with a minor in applied mathematics from the American University of Beirut, Lebanon in 2017. He is currently pursuing a PhD degree in mechanical engineering at the University of Illinois at Urbana–Champaign. His current research interests include electric field actuated droplet jumping during condensation and advanced phase-change thermal management techniques for next-generation high power density electronics.

Junho Oh

Junho Oh received BS and MS degrees in mechanical engineering from Sungkyunkwan University, Seoul, South Korea in 2013 and 2015, respectively, and a PhD degree in mechanical engineering from the University of Illinois at Urbana–Champaign. He is currently a Research Fellow at University College London, UK. His current research interests include nanoengineering, fluid mechanics and interfacial sciences for enhancing phase-change heat transfer and biomedical applications.

Nenad Miljkovic

Nenad Miljkovic received a BASc degree in mechanical engineering from the University of Waterloo, ON, Canada, in 2009, and MS and PhD degrees in mechanical engineering from the Massachusetts Institute of Technology in 2011 and 2013, respectively. He is currently an Associate Professor of mechanical science and engineering with the University of Illinois at Urbana–Champaign

(UIUC) where he leads the Energy Transport Research Laboratory. He has courtesy appointments in Electrical and Computer Engineering and the Materials Research Laboratory. His group's research intersects the multidisciplinary fields of thermo-fluid science, interfacial phenomena and renewable energy. Miljkovic was a recipient of the National Science Foundation CAREER Award, the American Chemical Society Petroleum Research Fund Doctoral New Investigator Award, the Office of Naval Research Young Investigator Award, a Distinguished Visiting Fellowship from the United Kingdom Royal Academy of Engineering, a US National Academy of Sciences Arab American Frontiers Fellowship, the ASME ICNMM Young Faculty Award, the ASME Pi Tau Sigma Gold Medal, the CERL Research and Development Technical Achievement Award, and the UIUC Dean's Award for Excellence in Research. He is the associate director of the Air Conditioning and Refrigeration Center, which is an NSF-founded I/UCRC at UIUC supported by 30 industrial partners.

Mona Zebarjadi

Mona Zebarjadi is a Joint Professor of the Electrical and Computer Engineering and Materials Science and Engineering Departments at the University of Virginia, Charlottesville, where she is leading the Energy Science and Nanotechnology Laboratory (ESNL). Prior to her current appointment she was a Professor at the Mechanical Engineering Department at Rutgers University. She received her Bachelor's and Master's degree in physics from Sharif University and her PhD in electrical engineering from University of California, Santa Cruz in 2009, after which she spent three years at MIT as a postdoctoral fellow working jointly with the Electrical and Mechanical Engineering Departments. Her current research interests include electron and phonon transport in thermoelectric, thermionic and thermomagnetic materials, and devices with a focus on two-dimensional structures.

Golam Rosul

Md Golam Rosul obtained his Bachelor of Science degree in electrical and electronic engineering from the Bangladesh University of Engineering and Technology (BUET) in 2015. He joined the PhD program in electrical engineering at the University of Virginia in the fall of 2017. His current research interest is thermionic transport in two-dimensional materials.

Sabbir Akhanda

Md Sabbir Akhanda obtained BS and MS degrees in applied physics, electronics and communication engineering and in electrical and electronic engineering, respectively, from the University of Dhaka, Bangladesh, in 2013 and 2015, respectively. He is currently a PhD student with the University of Virginia, Charlottesville. His current research interest is focused on thermoelectric and thermomagnetic materials.

Shreyas Chavan

Shreyas Chavan received his Bachelor of Technology degree in mechanical engineering from the Indian Institute of Technology Bombay in 2014, and MS and PhD degrees from the University of Illinois at Urbana–Champaign in 2016 and 2019, respectively.

During his PhD he worked with Nenad Miljkovic in the Energy Transport Research Laboratory. His primary research involved understanding the underlying fundamental physics of phase-change processes on structured superhydrophobic surfaces. He focused on the investigation of frosting and defrosting on superhydrophobic and biphilic surfaces.

Kalyan Boyina

Kalyan Boyina received a BS degree (Hons) in mechanical engineering from Oklahoma State University in 2014, and an MS degree from the University of Illinois, Urbana–Champaign in 2016. He is currently a doctoral student at the University of Illinois, Urbana–Champaign, where he works with Nenad Miljkovic in the Energy Transport Research Laboratory. His primary research interests include the large scale fabrication and performance characterization of anti-frosting superhydrophobic heat exchangers under a wide range of ambient conditions, optimization of superhydrophobic coating technologies, advanced defrosting techniques and brazed joint strength enhancement through surface modification.

Longnan Li

Longnan Li received a PhD degree in mechanical engineering from Sogang University, Seoul, South Korea in 2017. He was an engineer in the Haier group and Midea group before pursuing his PhD degree from 2007 to 2011. He is currently a Postdoctoral Research Associate in the Department of Mechanical Science and Engineering, University of Illinois Urbana–Champaign. His current research interests include developing durable and scalable coatings for high efficiency phase-change applications, micro/nanoscale fluid mechanics, electronics cooling and energy harvesting devices.

Qing Zhu

Qing Zhu is currently a PhD candidate in materials science and engineering at the University of Houston, TX. He received his Bachelor's degree in materials science and engineering from Northwestern Polytechnical University in China. His current research is focused on the measurement of conversion efficiency in thermoelectric materials.

Zhifeng Ren

Zhifeng Ren is the M D Anderson Chair Professor of Physics at the University of Houston, TX, and the director of the Texas Center for Superconductivity at the University of Houston (TcSUH). He received his BS degree from Xihua University in 1984, his MS degree from Huazhong University of Science and Technology in 1987, and his PhD degree from the Institute of Physics, Chinese Academy of Sciences, in 1990. He has published 510 peer-reviewed journal papers and has been awarded 55 patents. His research interests include high-performance thermoelectrics, ultrahigh thermal conductivity, amphiphilic Janus nanosheets for enhanced oil recovery, efficient catalysts for water splitting, including seawater, aligned carbon nanotubes, superconductivity, transparent flexible electrodes, etc.

Tobias Burger

Tobias Burger is a PhD candidate in chemical engineering at the University of Michigan. He received his MSE in chemical engineering from the University of Michigan and his BS in chemical engineering from the New Mexico Institute of Mining and Technology. He is working to develop thin-film PV devices for more affordable, high-performance thermophotovoltaic generators.

Caroline Sempere

Caroline Sempere is pursuing a BSE in chemical engineering at the University of Michigan. Her research experience includes the modeling of novel self-doped semiconductors and spectral control methods for thermophotovoltaic systems.

Andrej Lenert

Andrej Lenert is an Assistant Professor in the Department of Chemical Engineering at the University of Michigan. He completed his PhD at MIT in 2014, under the supervision of Evelyn Wang. He was a postdoctoral fellow at the University of Michigan, working with the Nanoscale Transport Laboratory and the Center for Photonic and Multiscale Nanomaterials. In 2016, he was named on the Forbes 30 under 30 list in Science for his contributions to the field of thermophotovoltaic energy conversion.

Part I

Theory and Computation

Chapter 1

Hydrodynamic phonon transport: past, present and prospects

Sangyeop Lee and Xun Li

Hydrodynamic phonon transport began to be studied several decades ago to verify the quantum theory of lattice thermal transport. The recent prediction of significant hydrodynamic phonon transport in graphitic materials shows its practical importance for high thermal conductivity materials and has brought this field renewed attention. As the study of this topic had been inactive to some extent for several decades, we aim to provide a brief overview of earlier studies as well as very recent studies. The topics we discuss in this chapter include the collective motion of phonons, several approaches to solving the Peierls–Boltzmann transport equation for hydrodynamic phonon transport, the role of normal scattering for thermal resistance and the propagation of second sound. Then, we close this chapter with our perspectives for future studies and the practical implications of hydrodynamic phonon transport.

1.1 Introduction

The transport of phonons, a major heat carrier in non-metallic solids, has usually been described using the diffusive limit, since Fourier's law was suggested 200 years ago. Fourier's law has a simple form that correlates thermodynamic driving force (i.e. temperature gradient, $-\nabla T$) and the resulting heat flux (q''):

$$\frac{1}{\kappa}q'' = -\nabla T. \tag{1.1}$$

This empirical law shows that there is always a damping coefficient, $1/\kappa$, involved in transport phenomena. $1/\kappa$ is the thermal resistance which determines the extent of damping in heat flow and the resulting heat flux at a given temperature gradient. However, such a damping effect is not observed in fluid flow, although both phonons and molecules are described well by the same Boltzmann transport theory.

doi:10.1088/978-0-7503-1738-2ch1

Also, they have similar thermodynamic driving forces—molecules are driven by pressure gradient-like phonons that are driven by a temperature gradient. Assuming an infinitely large domain to exclude any effect from the boundary, the molecular flow at the macroscale can be described by Euler's equation,

$$\frac{\mathrm{D}(\rho \mathbf{u})}{\mathrm{D}t} = -\nabla p, \tag{1.2}$$

where ρ and $\mathbf{u}$ are the density and velocity of the fluid element, respectively. With the Lagrangian coordinate, equation (1.2) shows the acceleration of molecules under the pressure gradient ($-\nabla p$) without any damping effect. This is the thermodynamic limit where the entropy generation is zero.

Now, one might ask a question—why does phonon flow as described by Fourier's law exhibit a damping effect while molecular flow does not? Interestingly, Nernst speculated a century ago that heat in high thermal conductivity materials may have inertia like a fluid [1]. The different behaviors of damping in molecular and phonon flows can be associated with the difference in the scattering processes of those two particles in terms of momentum conservation. For molecular flow, the total momentum of the molecules is always conserved upon molecule–molecule scattering. Therefore, inter-molecular scattering itself cannot cease the given molecular flow. For phonon flow, however, the total momentum of phonons is not always conserved upon phonon–phonon scattering. There are two different scattering mechanisms regarding the momentum conservation: normal and Umklapp scattering (hereafter N-scattering and U-scattering, respectively), suggested by Peierls [2]. As shown in figure 1.1(a), N-scattering involves phonon states with small wavevectors and the total momentum of phonon particles is conserved ($\mathbf{q}_1 + \mathbf{q}_2 = \mathbf{q}_3$), as in the inter-molecular scattering case. However, for U-scattering, the total momentum of phonon particles is not conserved. As a result, the phonon propagation direction is reversed upon U-scattering, thus directly causing thermal resistance. Phonon scattering by impurities, as shown in figure 1.1(c), also directly causes thermal resistance as it does not conserve total momentum. Hereafter, R-scattering refers to combined U- and impurity-scattering. In most materials at room temperature, N-scattering is weak compared to R-scattering, leading to the large damping effect of heat flow in solid materials.

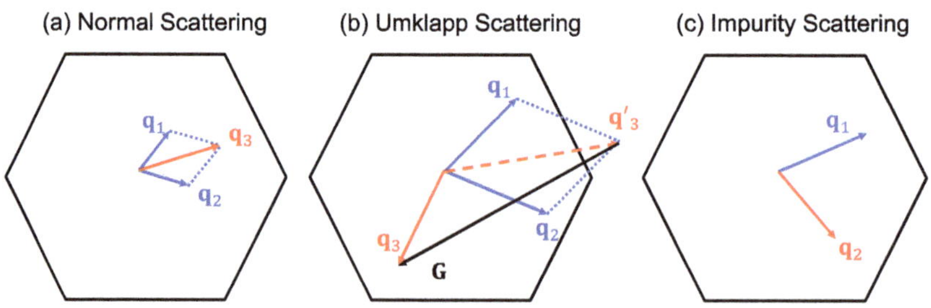

Figure 1.1. (a) Schematic of N-scattering, (b) U-scattering and (c) impurity scattering in the reciprocal space. The hexagon represents the first Brillouin zone.

We have assumed an infinitely large domain to compare the intrinsic damping of phonon flow and molecular flow. However, all solid materials have finite size, introducing phonon–boundary scattering. In most cases, the phonon–boundary scattering is diffuse boundary scattering rather than specular boundary scattering. The three types of phonon scattering (i.e. diffuse boundary scattering, N-scattering and R-scattering) influence phonon transport in different ways and thus there exist three regimes of phonon transport—the ballistic, hydrodynamic and diffusive regimes schematically shown in figure 1.2—depending on the dominant type of scattering mechanism. These three regimes occur in different ranges of temperature. The ballistic regime occurs at a low temperature where internal phonon scattering is much weaker than phonon–boundary scattering. Therefore, the phonon transport is limited by the diffuse boundary scattering and the thermal resistance is determined by the size and shape of the samples. As temperature increases, the internal phonon scattering starts to play a role in the transport process. At sufficiently low temperature, the majority of internal phonon scattering is N-scattering as phonon states with large wavevectors cannot be occupied. Because of the momentum conserving nature of N-scattering, the resulting phonon transport is similar to fluid flow and thus is called hydrodynamic phonon transport. Figure 1.2(b) shows the schematic of the heat flux profile, similar to the molecular Poiseuille flow. When temperature increases further, U-scattering becomes significant and the thermal resistance is due to the direct momentum destruction by U-scattering. As U-scattering occurs in any location, the heat flux has a spatially uniform profile as shown in figure 1.2(c).

The N- and U-scattering for phonons, since suggested by Peierls around a century ago [2], have been a foundation for the quantum theory of thermal transport in solids. Although the concept of N- and U-scattering was well accepted, the direct confirmation of N-scattering was still lacking. This led to the theoretical [4–10] and experimental efforts [11–15] for the prediction and observation of hydrodynamic phonon transport, namely, phonon Poiseuille flow and second sound, which will be discussed in more detail below. The phonon Poiseuille flow was first measured in solid He in the temperature range of 0.6–1.0 K [11]. Second sound was measured in solid ^{3}He at 0.5 K [12], in NaF at around 15 K [13, 14] and in Bi at 2 K [15]. Those experimental observations combined with the theoretical studies directly confirm the N-scattering for phonons and show remarkably different effects of N- and U-scattering on thermal transport. This was regarded as 'one of the great triumphs of the theory of lattice vibrations' [16].

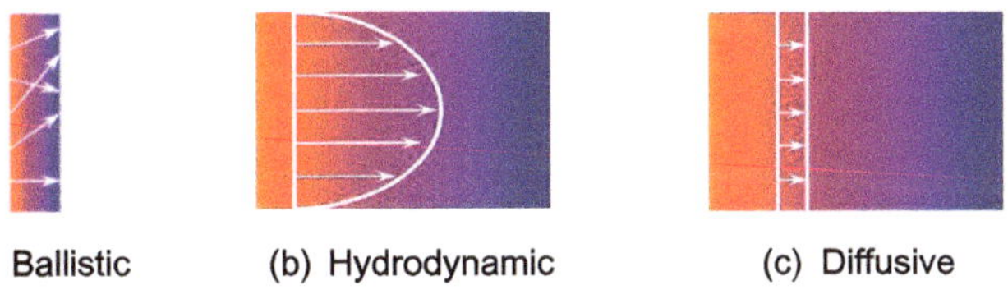

Figure 1.2. Schematic of phonon flux profile: (a) ballistic, (b) hydrodynamic and (c) diffusive. Reproduced with permission from [3]. Copyright 2015 Macmillan Publishers Ltd.

Despite the confirmation of hydrodynamic phonon transport, the study of this topic was inactive for several decades. As can be seen from the previous measurements, hydrodynamic phonon transport was observed in very low and narrow temperature ranges and thus was considered not relevant to practical applications. The conditions for hydrodynamic phonon transport are stringent because it is rare to satisfy the conditions of weak U-scattering and strong N-scattering at the same time. The U-scattering can easily be suppressed if the temperature is much lower than the Debye temperature so as to limit the phonon population to small wavevector states. However, if the temperature is lowered, there are not enough N-scattering events and the transport easily becomes ballistic. Thus, for hydrodynamic phonon transport to be significant, a material should exhibit a high Debye temperature and large anharmonicity at the same time. This is not common; a material with a high Debye temperature such as diamond usually exhibits small anharmonicity. The quality of the sample is another issue as impurity scattering is momentum destroying scattering and weakens the hydrodynamic features. It is interesting to note that NaF was chosen for the second sound experiments [13, 14, 17] because Na and F are naturally monoisotopic elements and thus at least isotope impurities do not exist.

Hydrodynamic phonon transport has recently received renewed attention after first-principles based studies predicted significant hydrodynamic phonon transport in graphitic materials including single-wall carbon nanotubes (SWCNTs) [18], graphene [3, 19] and graphite [20]. Interestingly, those graphitic materials exhibit a high Debye temperature and large anharmonicity at the same time, leading to strong N-scattering, shown in figure 1.3, and significant hydrodynamic phonon transport [3]. The light atomic mass of carbon and strong sp^2 bonding result in the high Debye temperature and weak U-scattering. Also, the flexural phonon modes from its layered atomistic structure are largely anharmonic for small wavevector states [21], leading to strong N-scattering.

The primary objective of this chapter is to provide a brief overview of basic concepts and recent studies of hydrodynamic phonon transport for those who have previously worked on ballistic and diffusive phonon transport. Other comprehensive

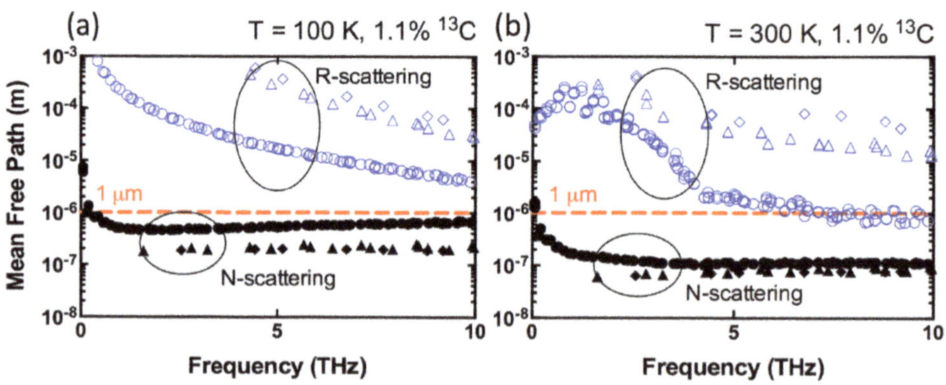

Figure 1.3. The mean free paths of N- and R-scattering in suspended graphene at 100 and 300 K from first-principles calculation.

review articles are available for advanced theoretical aspects [22, 23] and the macroscopic governing equations of the heatwave which is related to second sound [24]. This chapter is organized as follows. Section 1.2 discusses the displaced Bose–Einstein distribution as an equilibrium distribution under N-scattering and collective hydrodynamic phonon flow. Section 1.3 summarizes the methods to solve the Peierls–Boltzmann transport equation for hydrodynamic and quasi-hydrodynamic phonon transport. Section 1.4 provides our current understanding of the role of N-scattering for thermal resistance for various cases. Section 1.5 will review the theoretical and experimental studies of second sound. We then briefly discuss the future perspectives of phonon hydrodynamics in section 1.6.

We would also like to mention that the term 'hydrodynamic phonon transport' has been used in a different context in recent publications [25–31]. Those studies used phonon hydrodynamic equations that were derived assuming strong N-scattering compared to U-scattering and thus have a term similar to the viscous term of the Navier–Stokes equation [7, 8, 32]. However, to avoid any confusion, the phenomena studied in those studies are quasi-ballistic phonon transport and do not require strong N-scattering—the hydrodynamic equations were used to phenomenologically describe the quasi-ballistic transport. In this chapter, we focus on the hydrodynamic phenomena of phonon transport due to strong N-scattering and do not discuss the phenomenological hydrodynamic description of quasi-ballistic phonon transport. For the reader who is interested in the latter topic, a recent review article can provide a comprehensive summary [33].

1.2 Collective phonon flow

One unique feature of hydrodynamic transport that can be distinguished from the ballistic and diffusive regimes is the collective motion of particles. The term 'collective' is often used to describe different phenomena in solid-state physics. Here, we call the transport of particles collective when the flux of particles can be represented by a single value of velocity regardless of their quantum states. As an example, let us assume that we are able to track the movements of all molecules in a small fluid element. Assuming strong molecule–molecule scattering and small pressure gradient for the well-defined local equilibrium condition, the molecules then follow the displaced Boltzmann distribution:

$$f_{\mathrm{B}}^{\mathrm{disp}} = \left(\frac{m}{2\pi k_{\mathrm{B}}T}\right)^{3/2} \exp\left(-\frac{m\,|\mathbf{v}-\mathbf{u}|^2}{2k_{\mathrm{B}}T}\right), \tag{1.3}$$

where m, k_{B} and T represent the mass of a molecule, the Boltzmann constant and the temperature, respectively, $\mathbf{v}$ is the actual velocity of a molecule and $\mathbf{u}$ is the drift velocity. Note that the drift velocity is the same for all molecules regardless of their quantum states. Usually, the actual velocity is much larger than the drift velocity, making the movement of each molecule look random. However, the small drift velocity causes a net flow of molecules. As a result, the fluid element containing many molecules that seemingly move along random directions can move with the drift velocity as a whole. Thus, we call the molecular transport collective in this case.

Likewise, phonon particles show collective motion when the transport is hydrodynamic. The equilibrium distribution of phonons with N-scattering is the displaced Bose–Einstein distribution:

$$f^{\mathrm{disp}} = \left[\exp\left(\frac{\hbar(\omega - \mathbf{q}\cdot\mathbf{u})}{k_{\mathrm{B}}T}\right) - 1\right]^{-1}, \tag{1.4}$$

where $\mathbf{q}$ and $\mathbf{u}$ are the phonon wavevector and drift velocity (or displacement), respectively. In most cases where the transport is non-hydrodynamic, $\mathbf{u}$ differs for each phonon mode. However, in the hydrodynamic regime, $\mathbf{u}$ is a constant for all phonon modes. The displaced distribution function can be linearized assuming a small displacement, i.e. $\mathbf{q}\cdot\mathbf{u} \ll \omega$,

$$f^{\mathrm{disp}} \approx f^0 + \frac{\hbar}{k_{\mathrm{B}}T} f^0(f^0 + 1)\mathbf{q}\cdot\mathbf{u}. \tag{1.5}$$

The fact that the displaced Bose–Einstein distribution function is the equilibrium distribution upon N-scattering can be shown with Boltzmann's H-theorem [34]. For example, the rate of entropy generation upon coalescence three-phonon scattering is

$$\dot{S}_{\mathrm{scatt}} \sim \sum_{ijk}(\phi_i + \phi_j - \phi_k)^2 P_{i,j}^k, \tag{1.6}$$

where $P_{i,j}^k$ is the equilibrium transition rate of the coalescence process where the phonon particles at states i and j are merged to state k. A similar expression can be written for the decay process. ϕ_i represents the deviation of the distribution function from the stationary Bose–Einstein distribution f_i^0 (i.e. displaced Bose–Einstein distribution with zero displacement) and is defined as $\phi_i = (f_i - f_i^0)/(f_i^0(f_i^0 + 1))$. If the three phonon states exhibit the displaced Bose–Einstein distribution,

$$\phi_i + \phi_j - \phi_k = (\mathbf{q}_i + \mathbf{q}_j - \mathbf{q}_k)\cdot\mathbf{u}. \tag{1.7}$$

Considering the momentum conservation of N-scattering, $\mathbf{q}_i + \mathbf{q}_j = \mathbf{q}_k$, the entropy generation in this case is zero, verifying that the displaced Bose–Einstein distribution is an equilibrium distribution under N-scattering. From equation (1.7), even U-scattering ($\mathbf{q}_i + \mathbf{q}_j = \mathbf{q}_k \pm \mathbf{G}_m$) does not generate any entropy if the reciprocal lattice vector $\mathbf{G}_m$ is orthogonal to $\mathbf{u}$. This was also shown through the simulation of second sound in a recent study [35].

Whether a certain scattering process is N- or U-scattering depends on the choice of the Brillouin zone, which may lead to confusion or misunderstanding about the role of N- and U-scattering on phonon transport. We would like to emphasize that the concept of momentum conservation for understanding the phonon transport is valid only when the crystal momentum is defined with the first Brillouin zone, which is the Wigner–Seitz unit cell in reciprocal space. Otherwise, the displaced distribution function in equation (1.4) is incorrect and distinguishing N- and U-scattering based on the non-Wigner–Seitz unit cell is not meaningful.

With the linearized form of the displaced distribution function in equation (1.5), it is straightforward to show that the phonon particle flux $n_x^{''}$ can be described by the single value of **u**:

$$n_x^{''} = \frac{1}{NV}\sum_i v_x f^{\text{disp}} = \left(\frac{1}{NV}\sum_i v_x \frac{\hbar}{k_B T} f^0(f^0 + 1)\mathbf{q}\right) \cdot \mathbf{u}, \tag{1.8}$$

where N and V are the number of atoms and the volume of the unit cell, respectively. Similarly, heat flux $q_x^{''}$ is

$$q_x^{''} = \frac{1}{NV}\sum_i \hbar\omega v_x f^{\text{disp}} = \left(\frac{1}{NV}\sum_i \hbar\omega v_x \frac{\hbar}{k_B T} f^0(f^0 + 1)\mathbf{q}\right) \cdot \mathbf{u}. \tag{1.9}$$

It is worth noting that both particle flux and heat flux are linearly proportional to the local drift velocity **u**, representing the collective motion of phonon particles. The coefficients in the parentheses are constants determined by phonon dispersion and temperature. The fact that single value **u** can describe the transport of all phonon particles is the basis for the macroscopic transport equation about **u** which will be discussed in section 1.3.

As U-scattering cannot be completely avoided, the actual phonon distribution deviates from the displaced Bose–Einstein distribution to some extent:

$$f_i \approx f_i^0 + \frac{\hbar}{k_B T} f_i^0 \left(f_i^0 + 1\right)(\mathbf{q}_i \cdot \mathbf{u} + \delta_i), \tag{1.10}$$

where δ represents the deviation from the displaced Bose–Einstein distribution. It would be interesting to see how close the actual phonon distribution is to the displaced Bose–Einstein distribution in real materials in which hydrodynamic phonon transport is expected to be significant. In figure 1.4, we show the distribution function of phonon particles along the armchair direction in graphene at 100 K from the Peierls–Boltzmann transport equation (PBE) in an infinitely large sample case, which will be discussed in section 1.3. In most cases where the transport is not hydrodynamic, δ_i in equation (1.10) is larger compared to the collective part $\mathbf{q}_i \cdot \mathbf{u}$ and thus $(f_i - f_i^0)/(f_i^0(f_i^0 + 1))$ is not linear to $q_{i,x}$. However, in graphene, $(f_i - f_i^0)/(f_i^0(f_i^0 + 1))$ is nearly linear to $q_{i,x}$ with a constant slope, representing the collective motion of phonon particles with the same displacement regardless of the phonon mode. Figure 1.5 shows the contribution of the collective motion of phonon particles to total heat flux in (20,20) SWCNTs. At low temperature below 100 K, most of the heat is carried by the collective motion of phonon particles and the contribution of collective motion gradually decreases with temperature due to U-scattering.

1.3 Peierls–Boltzmann transport equation

The phonon distribution is described by the PBE:

$$\frac{\partial f_i(t, \mathbf{x})}{\partial t} + \mathbf{v}_i \cdot \nabla f_i(t, \mathbf{x}) = \sum_j G_{ij} f_j^{\text{d}}, \tag{1.11}$$

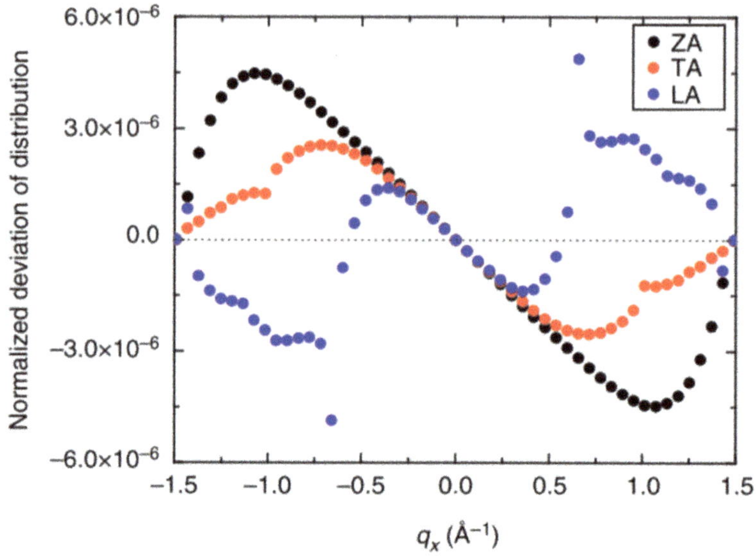

Figure 1.4. Normalized deviational distribution, $(f_i - f_i^0)/(f_i^0(f_i^0 + 1))$, in an infinitely large graphene at 100 K. Reproduced with permission from [3]. Copyright 2015 Macmillan Publishers Ltd.

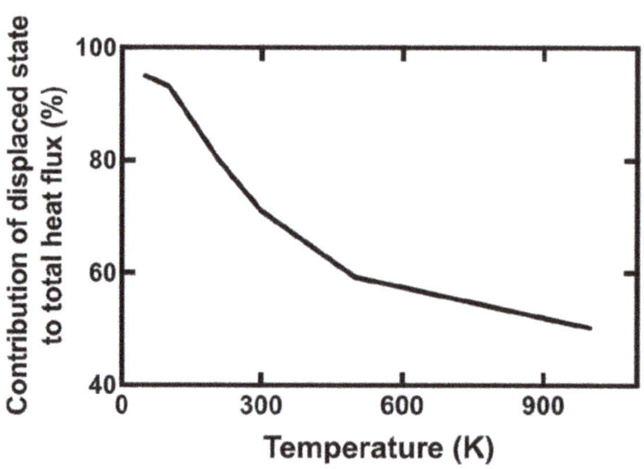

Figure 1.5. Contribution of collective motion of phonon particles to total heat flux in (20,20) SWCNT with naturally occurring ^{13}C isotope content (1.1%). Reproduced with permission from [18]. Copyright 2017 the American Physical Society.

where f_j^{d} is the deviational distribution function defined as $f_j - f_j^0$ and $\mathbf{G}$ is the scattering matrix. The original form of the PBE is known to be difficult to solve. The advection and scattering terms are in differential and integral forms and the unknown, $f_i(t, \mathbf{x})$, is a function in many dimensions including time, real space and reciprocal space domains. The equation has been often simplified assuming a steady state, a constant temperature gradient in an infinitely large sample, and very small deviation from the equilibrium distribution:

$$\mathbf{v}_i \cdot \nabla T \frac{\mathrm{d}f_i^0}{\mathrm{d}T} = \sum_j G_{ij} f_f^{\mathrm{d}}. \tag{1.12}$$

The differential advection term in equation (1.11) was replaced with the spatially homogeneous term, $\mathbf{v}_i \cdot \nabla T(\mathrm{d}f_i^0/\mathrm{d}T)$, by assuming the constant temperature gradient in an infinitely large sample. With these assumptions, the phonon distribution function is spatially homogeneous except for the change due to the temperature gradient. Then, the PBE could be simplified from the integro-differential equation to the homogeneous integral equation which is relatively easier to solve. The recently developed *ab initio* framework of lattice dynamics made it possible to calculate the scattering matrix, **G**, from first-principles [36, 37]. Also, several numerical techniques such as the full iterative method [38, 39] and the variational method [40] were developed to solve equation (1.12). Solving equation (1.12) with *ab initio* phonon dispersion and scattering matrix showed an excellent predictive power for the thermal conductivity of bulk samples [41].

For the hydrodynamic regime, however, the assumption of a spatially homogeneous distribution function is not valid. As schematically shown in figure 1.2(b), the heat flux and phonon distribution largely depend on the location in real space, and the advection term, $\mathbf{v}_i \cdot \nabla f_i$, in equation (1.11) cannot be homogeneous. Also, second sound is the temporal and spatial fluctuation of the temperature field which requires a description under an unsteady condition. Therefore, we would need to solve the PBE as an original form containing both differential and integral terms. We briefly review the past approaches used several decades ago to solve the PBE with several assumptions and also introduce recent approaches with minimal assumptions from first-principles.

One of the most challenging parts of solving the PBE is how to handle the integral scattering term. In the PBE, all phonon states are coupled to each other through the integral scattering term. Callaway suggested a simple form of scattering model from the fact that N-scattering and U-scattering tend to relax a phonon system to displaced and stationary Bose–Einstein distributions, respectively [42]. Although Callaway's scattering model was from intuition without rigorous theoretical considerations, it was later shown that the model can be formally derived by ignoring the off-diagonal terms of the N- and U-scattering matrices [43].

Early theoretical studies of phonon hydrodynamics derived macroscopic transport equations, such as the Navier–Stokes equation of fluid flow [4, 5, 7, 8, 32]. The work by Sussmann and Thellung [4] solved the PBE to the first order assuming no U-scattering and constructed momentum and energy balance equations. Some of Krumhansl group's work extended the transport equations to the case where U-scattering exists [5, 6]. The notable work by Guyer and Krumhansl [7, 8] solved the PBE in the eigenstate space of scattering operator which led to the concept of relaxon that will be discussed later. The derivation of these early studies was carefully examined and compared later by Hardy [32]. Although the details of derivation in the early studies are slightly different, they share the same basic idea. The idea is similar to how the Navier–Stokes equation is derived from the

Boltzmann transport equation with the BGK scattering model, which is analogous to the N-scattering term of Callaway's scattering model. We briefly discuss Sussmann and Thellung's derivation here.

The momentum and energy balance equations can be simply derived from the PBE by taking momentum ($\hbar\mathbf{q}$) and energy ($\hbar\omega$) as a moment of the PBE:

$$\frac{\partial E}{\partial t} + \nabla_\alpha Q_\alpha = 0 \tag{1.13}$$

$$\frac{\partial P_\alpha}{\partial t} + \nabla_\beta p_{\alpha\beta} = 0, \tag{1.14}$$

where

$$E = \frac{1}{NV}\sum_i \hbar\omega_i f_i \tag{1.15}$$

$$Q_\alpha = \frac{1}{NV}\sum_i \hbar\omega_i v_{\alpha,i} f_i \tag{1.16}$$

$$P_\alpha = \frac{1}{NV}\sum_i \hbar q_{\alpha,i} f_i \tag{1.17}$$

$$p_{\alpha\beta} = \frac{1}{NV}\sum_i \hbar q_{\alpha,i} v_{\beta,i} f_i\,. \tag{1.18}$$

E and Q_α are the energy density and heat flux along the α-direction. P_α and $p_{\alpha\beta}$ are the α-direction momentum density and the momentum flux along the β-direction. Note that the right-hand sides of equations (1.13) and (1.14) are zero because total momentum and energy are conserved upon N-scattering. If U-scattering is considered, the momentum destroying term by U-scattering would appear in the momentum balance equation. In order to complete those momentum and energy balance equations, the phonon distribution function is required. The phonon distribution can be found by solving the PBE with the N-scattering term of Callaway's scattering model:

$$\frac{\partial f}{\partial t} + \mathbf{v}\cdot\nabla f = -\frac{f - f^{\mathrm{disp}}}{\tau_{\mathrm{N}}}. \tag{1.19}$$

Equation (1.19) can be further simplified if we assume $\dot{f} \approx \dot{f}^{\mathrm{disp}}$ and $\nabla f \approx \nabla f^{\mathrm{disp}}$. This assumption is analogous to the Chapman–Enskog expansion to the first order and is valid when N-scattering is strong [44]. To be more specific, N-scattering is considered strong when the relaxation time and mean free path of N-scattering are much smaller than the characteristic time and size of the system (e.g. the time period of temperature fluctuation for second sound and the sample size for steady-state heat flow). With such assumptions, it is straightforward to solve equation (1.19). Based on the phonon distribution function from equation (1.19) being plugged into equations (1.13) and (1.14), the following macroscopic governing equations can be derived:

$$\dot{T}' - \frac{1}{3}v_g^2\nabla^2 T' + \frac{1}{3}\nabla \cdot \mathbf{u} = 0 \tag{1.20}$$

$$\dot{u}_\alpha + v_g^2\nabla_\alpha T' - v_g^2\tau_N\left(\frac{2}{5}\nabla_\alpha\nabla \cdot \mathbf{u} + \frac{1}{5}\nabla^2 u_\alpha\right) = 0, \tag{1.21}$$

where v_g is the group velocity. T' is the dimensionless deviational temperature defined as $(T - T_0)/T_0$, where T_0 is an equilibrium temperature.

Although early theoretical studies [4, 5, 7, 8, 32] are slightly different from the details of derivation, they are based on the same assumptions: (i) N-scattering being much stronger than U-scattering such that f is closer to f^{disp} than f^0 and (ii) N-scattering being strong enough that $\dot{f} \approx \dot{f}^{\mathrm{disp}}$ and $\nabla f \approx \nabla f^{\mathrm{disp}}$. Because of these assumptions, the hydrodynamic equations derived in the early studies have several limitations. The macroscopic hydrodynamic equations may not accurately describe the following cases: (i) the characteristic size of a system being comparable to the mean free path of N-scattering, namely, phonon transport somewhere between the ballistic and hydrodynamic limits, and (ii) N-scattering being not much stronger than U-scattering, namely, phonon transport somewhere between the diffusive and hydrodynamic limits. In addition, the validity of Callaway's scattering model is questionable for quantitative purposes [45, 46].

As the full scattering matrix can now be calculated from first-principles and the hydrodynamic phonon transport has gained renewed attention, there are two recently developed methods to solve the PBE with the full scattering matrix in both real and reciprocal spaces without the assumption of strong N-scattering. Both approaches provide a solution of the PBE without any significant assumptions and thus can be useful to study complex transport phenomena where features of all three regimes exist to some extent [47].

The first approach is based on the eigenstates of the scattering matrix. The scattering matrix can be symmetrized by multiplying a factor, $2\sinh(X_i/2)$, where $X_i = \hbar\omega_i/k_B T$, to equation (1.11) such that the scattering matrix has an orthogonal set of eigenstates [43]:

$$\left(2\sinh\frac{1}{2}X_i\right)\mathbf{v}_i \cdot \nabla f_i = \sum_j G_{ij}^* f_j^{\mathrm{d}^*}, \tag{1.22}$$

where $f_j^{\mathrm{d}^*}$ is $(2\sinh\frac{1}{2}X_j)f_j^{\mathrm{d}}$ and the scattering matrix, $\mathbf{G}^*$, is

$$G_{ij}^* = \left(\frac{2\sinh\frac{1}{2}X_i}{2\sinh\frac{1}{2}X_j}\right)G_{ij}. \tag{1.23}$$

The orthogonal eigenstates of $\mathbf{G}^*$ were later called relaxons [48]. The solution of the PBE, f^{d^*}, can then be expressed as a linear combination of relaxons and the equation for the coefficient (population) of each relaxon state can be derived from the PBE [48]. An advantage of the relaxon framework is that the relaxon has now a well-defined relaxation length and thus the thermal transport can be described with a simple kinetic description of relaxon particles. Phonons, if they experience strong N-scattering, do not have a well-defined relaxation length due to the complex interplay between N- and U-scattering processes and also its collective nature of motions.

The second approach employs the Monte Carlo (MC) method to solve the PBE with the full scattering matrix [49, 50]. The MC method was previously developed to solve the PBE with the single-mode relaxation time approximation (SMRT) for studying quasi-ballistic phonon transport [51–53]. The MC method with the SMRT stochastically determines the occurrence of scattering based on the probability of scattering. With the full scattering matrix, the MC method stochastically determines whether a certain scattering process occurs or not and the final state of phonon particles if the scattering is determined to occur. The energy-based PBE is chosen over the regular PBE due to its advantage of strict energy conservation:

$$\mathbf{v}_i \cdot \nabla(\omega f)_i = \sum_j B_{ij}\left(\omega_j f_j^{\mathrm{d}}\right), \tag{1.24}$$

where B_{ij} is the scattering matrix of the energy-based PBE, defined as $(\omega_i/\omega_j)G_{ij}$. The energy exchange upon scattering is described as

$$\omega_i f_i^{\mathrm{d}}(t + \Delta t) = \sum_j Z_{ij}(\Delta t)\omega_j f_j^{\mathrm{d}}(t), \tag{1.25}$$

where the energy propagator matrix $\mathbf{Z}$ can be found as

$$\mathbf{Z}(\Delta t) = \mathrm{e}^{\mathbf{B}\Delta t}. \tag{1.26}$$

If the off-diagonal terms of matrix $\mathbf{B}$ are ignored and only diagonal terms are considered, equation (1.25) is recovered to the exponential decay of energy which is equivalent to the SMRT:

$$\omega_i f_i^{\mathrm{d}}(t + \Delta t) = \exp(B_{ii}\Delta t)\omega_i f_i^{\mathrm{d}}(t). \tag{1.27}$$

Note B_{ii} is the same as $-\tau_i^{-1}$ from equation (1.24). The scattering in equation (1.25) describes the transfer of energy from phonon state j to i. In an MC simulation, the destination state i can be stochastically determined and its detailed MC algorithm can be found in the literature [49, 50].

1.4 Steady-state phonon hydrodynamics

The N-scattering itself does not directly cause thermal resistance because of its momentum conserving nature. However, the N-scattering can affect thermal resistance when combined with momentum destroying scattering (R-scattering or diffuse boundary scattering) or thermal reservoirs that emit phonons for which the distribution deviates from the displaced Bose–Einstein distribution. These situations are common in practical systems. We discuss the role of N-scattering for thermal resistance in three different cases: (i) an infinitely large sample, (ii) a sample with an infinite length but a finite width where diffuse boundary scattering destroys the phonon momentum along the flow direction and (iii) a sample with an infinite width but a finite length contacting hot and cold reservoirs that emit phonons with the stationary Bose–Einstein distribution.

1.4.1 Infinitely large sample

It is well known that the thermal conductivity is infinitely large when the N-scattering is the only scattering mechanism and the sample is infinitely large. Assuming that a local temperature gradient is applied and phonon flow is initiated, the phonons subsequently establish the displaced Bose–Einstein distribution through many N-scattering events. Then, the N-scattering does not further alter the displaced Bose–Einstein distribution and the phonons can continue to flow even without any temperature gradient, resulting in the infinite thermal conductivity. This leads to the simple statement that N-scattering itself does not cause thermal resistance. This simple statement, however, is true only when the distribution function is homogeneous in space as in the infinitely large sample. If there is a significant spatial variation of the distribution function, the N-scattering can cause thermal resistance. This will be discussed in section 1.4.3.

Even when the distribution function is homogeneous in space, N-scattering contributes to thermal resistance if U-scattering also exists. In general, phonon states with a small wavevector have very weak U-scattering. However, the small wavevector phonons can be scattered into larger wavevector states through N-scattering and then can be seen by U-scattering. A recent study on the thermal transport in SWCNTs [54] quantitatively shows the effect of N-scattering on thermal conductivity. The thermal conductivity of (10,10) SWCNT is 10 000 W m^{-1} K^{-1} when only U-scattering is considered, but it is significantly reduced to 2000 W m^{-1} K^{-1} when both N- and U-scattering processes are included.

1.4.2 Sample with an infinite length and a finite width

We consider a sample with an infinite length and a finite width to discuss the thermal resistance when N-scattering is combined with diffuse boundary scattering. As shown in figure 1.6, we consider a constant temperature gradient along the length direction, which drives the phonon flow.

The major mechanisms of thermal resistance in the diffusive and ballistic regimes are U-scattering and diffuse boundary scattering, respectively. In the hydrodynamic regime, we have a different mechanism for thermal resistance—the viscous damping effect which is a result of combined N- and diffuse boundary scattering. The drift velocity near boundaries is smaller than that in the middle of a sample due to diffuse boundary scattering. Thus, the drift velocity exhibits a gradient along the transverse direction (the y-direction in figure 1.6). Due to the drift velocity gradient, phonon momentum is transferred from the middle of the sample to the boundaries through

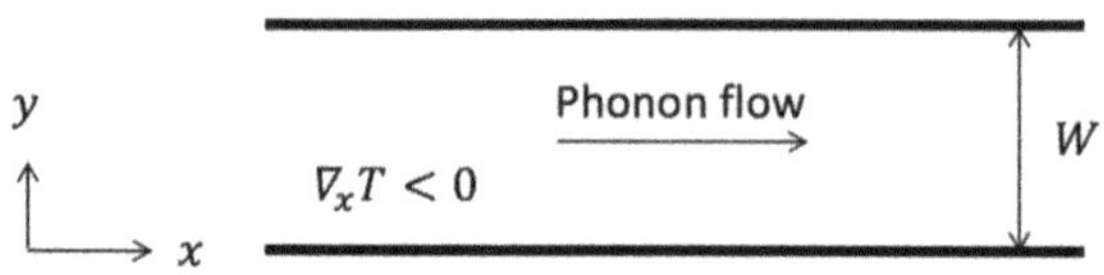

Figure 1.6. Schematic of phonon flow in an infinitely long sample with a finite width.

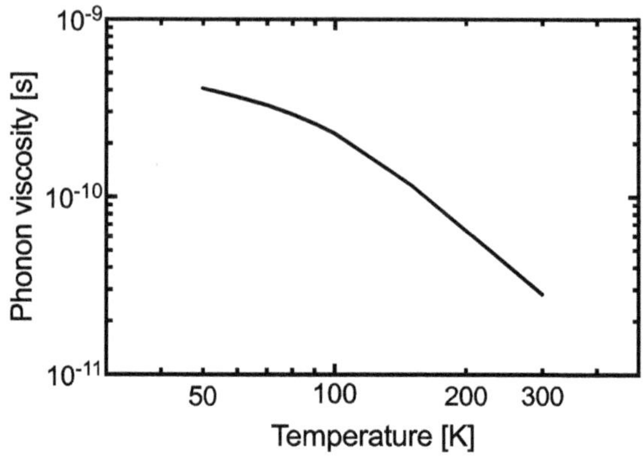

Figure 1.7. Temperature dependence of the phonon hydrodynamic viscosity of suspended graphene calculated with phonon dispersion and scattering rates from first-principles calculation. Reproduced with permission from [50]. Copyright 2018 the American Physical Society.

N-scattering processes and then finally is destroyed by the diffuse boundary scattering. The viscous damping term can be seen in the second-order derivative term in equation (1.21).

Based on the momentum balance equation from the PBE with its first-order solution discussed in section 1.3, an expression for the phonon hydrodynamic viscosity (μ_{ph}) can be derived [50]:

$$\mu_{\mathrm{ph}} = \frac{\sum_i q_x^2 v_{y,i}^2 f_i^0 \left(f_i^0 + 1\right) \tau_{\mathrm{N},i}}{\sum_i q_x v_{x,i} f_i^0 \left(f_i^0 + 1\right) \omega_i}. \tag{1.28}$$

A notable difference between the ballistic and hydrodynamic regimes is that the momentum transfer to the boundary in the hydrodynamic regime is impeded by N-scattering. As the N-scattering rate is increased, the rate of momentum transfer to the boundary, which determines the extent of viscous damping, is decreased. This can be seen in the phonon hydrodynamic viscosity as a function of temperature in figure 1.7. As temperature increases, the N-scattering rate is increased, resulting in the lower hydrodynamic phonon viscosity. The extent of viscous damping also depends on the width of the sample as indicated in the second-order derivative term in equation (1.21). The rate of momentum transfer in the hydrodynamic regime is proportional to $1/W^2$, where W is the width of a sample, while the rate in the ballistic regime is proportional to $1/W$.

The viscous damping effect of the hydrodynamic regime causes peculiar dependences of thermal conductivity on temperature and sample width, which are distinguished from the ballistic and diffusive cases. In the ballistic regime, the thermal conductivity is linearly proportional to the sample width. The thermal conductivity of the diffusive regime is constant regardless of sample width. However, the thermal conductivity of the hydrodynamic regime superlinearly increases with the sample width due to the viscous damping that decreases as W^2. In addition, the thermal conductivity of the hydrodynamic regime increases with temperature much faster than that of the ballistic regime as the viscous damping is weakened as temperature increases. The peculiar dependence of thermal conductivity on temperature was observed in solid He at a low temperature, verifying the existence of phonon Poiseuille flow [11]. Recently, these dependences have been predicted at a much higher temperature in graphene [3, 50, 55] and graphite [20], and experimentally observed in $SrTiO_3$ [56].

The peculiar dependences of thermal conductivity on temperature and sample width can be observed only when the actual transport phenomena are close to those in the ideal hydrodynamic regime without U-scattering. The thermal transport in graphitic materials at an intermediate temperature of above 100 K can exhibit all three different mechanisms of thermal resistance: U-scattering, direct diffuse boundary scattering, combined diffuse boundary and N-scattering. The significance of each mechanism can be evaluated using the momentum balance. The temperature gradient in figure 1.6 drives phonon flow and generates excess phonon momentum ($\Phi_{\nabla T}$). This momentum is balanced by momentum destructions by three different mechanisms: diffuse boundary scattering without internal phonon scattering (i.e. ballistic effect, Φ_{B}), diffuse boundary scattering combined with N-scattering (i.e. viscous damping or hydrodynamic effect, Φ_{H}) and direct momentum destruction by U-scattering (i.e. diffusive effect, Φ_{D}). The momentum balance can be expressed as

$$\Phi_{\nabla T} = \Phi_{\mathrm{B}} + \Phi_{\mathrm{H}} + \Phi_{\mathrm{D}}. \tag{1.29}$$

Figure 1.8(a) shows that the thermal conductivity of an infinitely long graphene has a temperature dependence of $T^{2.03}$ when the temperature is below 90 K, much larger than that of the ballistic case $T^{1.68}$. This temperature range agrees with the momentum balance analysis of the same sample, as shown in figure 1.8(b). It is clear that below 90 K, Φ_{H} is the major mechanism of the momentum destruction, indicating that viscous damping is significant at this condition [47].

1.4.3 Sample with an infinite width and a finite length contacting hot and cold reservoirs

When an infinitely wide sample contacts hot and cold reservoirs, as in figure 1.9, the phonons emitted from the reservoirs do not follow the displaced Bose–Einstein distribution. They follow a Bose–Einstein distribution distorted by a spectral transmission function at the interface between the sample and the reservoir. The N-scattering processes change this non-displaced Bose–Einstein distribution (i.e. non-collective) to the displaced Bose–Einstein distribution (i.e. collective).

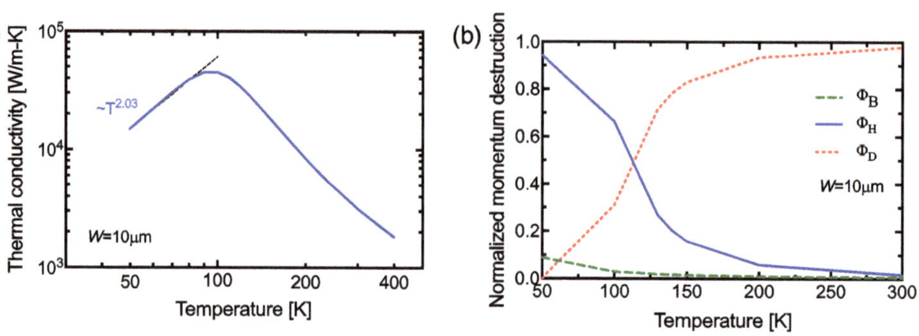

Figure 1.8. Temperature dependence of (a) thermal conductivity and (b) the momentum balance in an infinitely long graphene sample with the width of 10 μm from the MC solution of the PBE with the *ab initio* full three-phonon scattering matrix.

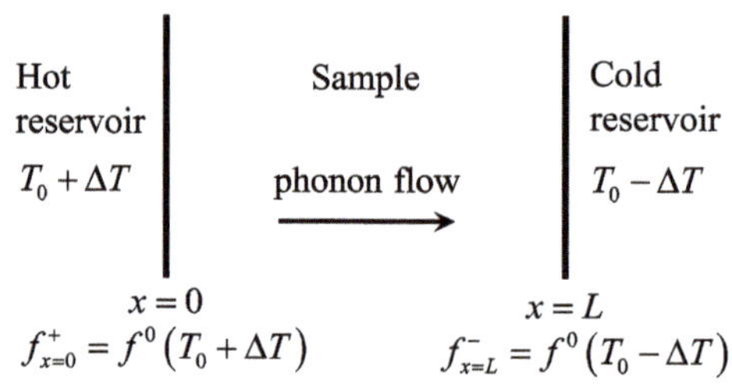

Figure 1.9. Schematic of sample geometry contacting hot and cold reservoirs.

As entropy is always generated when the distribution function is changed by scattering processes, as shown in equation (1.6), N-scattering causes thermal resistance near the interface between the sample and the reservoir where the emitted phonon flow becomes collective. The region where the thermal resistance occurs is within the order of the mean free path of N-scattering from the boundary. Figure 1.10 shows the formation of collective phonon flow at the cost of a temperature drop near the boundaries, resulting in the thermal resistance by N-scattering. Assuming N-scattering is the only scattering mechanism, the N-scattering far from the boundaries does not cause any temperature drop as the distribution function is already the displaced Bose–Einstein distribution.

The thermal resistance due to the transition between non-collective and collective phonon flows depends on materials. Figure 1.11 compares three-dimensional Debye phonon dispersion and graphite in terms of the reduction of phonon heat flux by N-scattering from the purely ballistic case. For the 3D Debye case, the reduction of heat flux is relatively small; the heat flux reduction by N-scattering is only around 5% for all three temperatures, 100, 200 and 300 K. However, for graphite, the reduction of heat flux is substantial; the heat flux is reduced by 20%, 30% and 40% at 100, 200 and 300 K, respectively. Other graphitic materials such as SWCNTs and graphene show a similar reduction of heat flux.

The large thermal resistance by N-scattering for graphitic materials can be explained with their nonlinear phonon dispersion with many phonon branches.

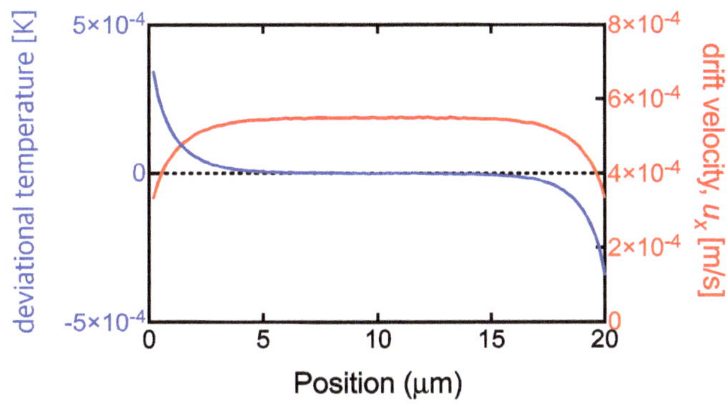

Figure 1.10. The profile of deviational temperature, defined as the difference between local temperature and global equilibrium temperature, and the drift velocity. The sample is (20,20) SWCNT contacting hot and cold reservoirs that have the deviational temperature of 0.001 and −0.001 K, respectively. The profile is calculated by the MC method of the PBE assuming Callaway's scattering model. The rate of N-scattering is assumed to be 10^{10} s^{-1} and U-scattering is ignored. Reprinted with permission from [57]. Copyright 2019 Taylor and Francis.

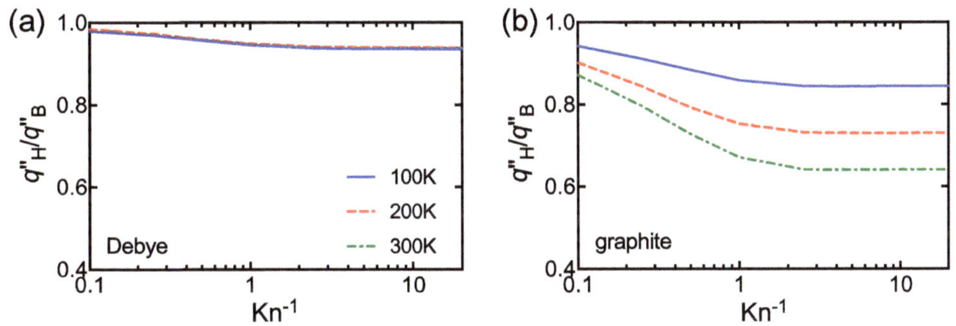

Figure 1.11. The ratio between heat flux with N-scattering (q''_{H}) and without any internal scattering (q''_{B}) as a function of inverse Knudsen number in (a) three-dimensional Debye model and (b) graphite. The heat flux is calculated with the MC solution of the PBE with Callaway's scattering model. The rate of N-scattering is assumed to be 10^{10} s^{-1} and U-scattering is ignored. Reproduced with permission from [57]. Copyright 2019 Taylor and Francis.

The rate of entropy generation due to scattering, equation (1.6), can be written as follows, assuming Callaway's scattering model and the stationary Bose–Einstein distribution for phonons emitted from the reservoirs:

$$\dot{S}_{\mathrm{scatt}} = \left(\frac{\Delta T}{T}\right)^2 \frac{\hbar^2}{\tau_{\mathrm{N}} k_{\mathrm{B}} T^2 N V} \sum_i f_i^0 \left(f_i^0 + 1\right) \omega_i |q_{x,i}| \left(v_{x,i}^* - u_x'\right), \tag{1.30}$$

where $v_{x,i}^*$ is $\omega_i/|q_{x,i}|$ and u_x' is the drift velocity per temperature difference and can be found from the momentum conservation:

$$u_x' = \frac{\sum_i |q_{x,i}| \omega_i f_i^0 \left(f_i^0 + 1\right)}{\sum_i q_{x,i}^2 f_i^0 \left(f_i^0 + 1\right)}. \tag{1.31}$$

Equation (1.30) shows that the temperature difference of two reservoirs drives the phonon flow with the displacement of $v_{x,i}^*$ which may vary depending on phonon modes while the drift velocity u_x' is the same for all phonon modes. If $v_{x,i}^*$ is a constant for all phonon modes (e.g. one-dimensional Debye phonon dispersion), $v_{x,i}^*$ is the same as u_x' by equation (1.31) and the entropy generation would be zero. However, if $v_{x,i}^*$ significantly varies with phonon states, the entropy generation is expected to be large. For the three-dimensional Debye model, $v_{x,i}^*$ varies with the direction of phonon wavevector and thus causes small thermal resistance, as can be seen in figure 1.11(a). The ratio $q_{\mathrm{H}}''/q_{\mathrm{B}}''$ in this case does not change with temperature as the variance of $v_{x,i}^*$ is associated with the direction of phonon wavevector only. For graphitic materials, however, the variance of $v_{x,i}^*$ is significant compared to the Debye model as a result of non-linear dispersion with many branches, resulting in the large thermal resistance that depends on temperature in figure 1.11(b). This indicates that the resistance due to the transition between non-collective and collective phonon flows is determined by the shape of phonon dispersion.

1.5 Unsteady phonon hydrodynamics (second sound)

The fundamental difference between N- and U-scattering in terms of momentum conservation leads to a different response upon temporal perturbation to a phonon system. One simple form of the perturbation is a heat pulse being applied to one end of a sample, as shown in figure 1.12. The heat pulse causes the increased local phonon density and the response of the phonon system is largely different depending on the transport regime. For the diffusive regime, the energy balance equation with Fourier's law indicates that the peak position of the heat pulse cannot move forward and remains at its original location. Then, the thermal energy of the heat pulse diffuses into the sample and finally the sample reaches an equilibrium with a slightly elevated temperature for the entire region. For the ballistic regime, the heat pulse can propagate through the sample as there is no phonon scattering. However, the shape of the heat pulse can spread out in space unless all phonon modes have the same group velocity along the heat-pulse propagation direction. This is expected to be particularly significant in graphitic materials where flexural phonon modes with a quadratic dispersion are important for thermal energy transport. For the hydrodynamic regime, the heat pulse leads to the local fluctuation of the temperature field which can propagate as a wave through the sample. An analogous

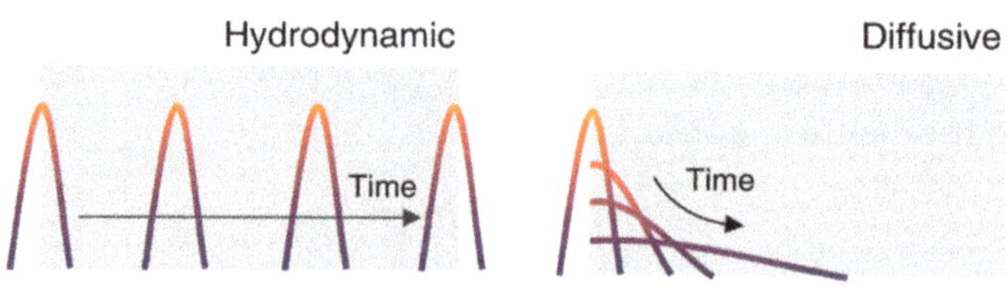

Figure 1.12. Propagation of a heat pulse in the diffusive and hydrodynamic regimes. Reproduced with permission from [3]. Copyright 2015 Macmillan Publishers Ltd.

phenomenon in the fluid system is the propagation of a pressure pulse in space, which is an acoustic sound. From the similarity of the two phenomena, the temperature pulse propagation in the form of a wave in the hydrodynamic regime is called second sound.

The second sound was first studied with superfluid He II in which the phonon is an elemental excitation of the system. The speed of second sound in liquid He was predicted by Landau using the two-fluid theory of rotons and phonons [58] and later confirmed by an experiment [59]. The predicted temperature wave was named second sound by Landau in order to distinguish it from the first sound, which is ordinary acoustic sound (i.e. pressure wave propagation). Later, it was shown that the same speed of second sound can be directly derived by using the phonon gas model without rotons [60, 61], which motivated the study of second sound in crystalline solids.

The second sound in solids can be observed with two different methods: a heat-pulse experiment [12–15, 62] and a light scattering method [10, 63–68]. In the heat pulse experiment, a heat pulse was applied to one end of a few millimeter long samples and the temperature to the opposite end was recorded as a function of time. At sufficiently low temperature such that internal phonon–phonon scattering is negligibly weak, two peaks of temperature pulse were observed, each of which represents the ballistic transport of transverse and longitudinal phonons. No significant dispersion of the temperature peak was observed because three-dimensional bulk materials where long wavelength phonons have a linear dispersion relation were used. With slightly increased temperature (around 15 K for NaF [14]), another peak in addition to those two peaks was observed. The delay time of the new peak agrees well with the predicted speed of second sound and the third peak was considered second sound. As temperature is further increased, the third peak disappears, indicating that U-scattering becomes significant. The light scattering method measures the inelastic light scattering by a local change of dielectric constants due to the second sound wave. A challenge lies in very weak coupling between light and thermal fluctuation at low temperatures. To solve this problem, a relatively strong thermal fluctuation field was induced by an optical grating method and the second sound in NaF was successfully measured [66]. The measured speed of second sound agrees well with that from the previous heat pulse experiments. Later, the light scattering measurements were carried out without inducing a thermal fluctuation field for $SrTiO_3$. $SrTiO_3$ has soft transverse optical phonons with small wavevector that are strongly anharmonic and thus cause strong N-scattering [67, 68]. The measured spectrum at around 30 K exhibits a doublet with a frequency shift (~20 GHz) that is comparable to the expected frequency of second sound in this temperature range.

As the conditions for the clear observation of second sound are narrow in the variable space, second sound measurements critically require *a priori* knowledge of the wavelength and frequency of the second sound as well as the speed of the second sound. The wavelength and frequency of second sound are determined by the mean free path and scattering rate of N- and U-scattering processes. If the pulse duration is much longer than the rate of U-scattering, the pulse can be destroyed by the

U-scattering and thus cannot propagate as a second sound. If the pulse duration is much shorter than the rate of N-scattering, phonons will travel with their own group velocity and do not have a chance to establish collective motion because of the lack of N-scattering. In this case, the pulse also cannot maintain its original shape and the thermal energy smears out.

The speed, frequency and wavelength of second sound were theoretically studied by calculating the dispersion relation of second sound. The speed of second sound was derived for the simplest case where Debye phonon dispersion is assumed and there is no U-scattering, giving the well-known relation for the speed of second sound, $v_{II} = v_g/\sqrt{3}$, where v_{II} is the speed of second sound [60, 61]. The speed of second sound was also derived for more realistic phonon dispersion consisting of one longitudinal and two degenerate transverse acoustic branches, all having Debye-type dispersion [4]. Later, theoretical studies considered the possible mechanisms for the attenuation of second sound and predicted the possible second sound frequency ranges [5–7]. In the literature, two different types of second sound called drifting and driftless second sounds were discussed [22, 69]. The driftless second sound differs from the second sound we discuss and does not require strong N-scattering; it occurs when all eigenstates of scattering operator have a similar relaxation time such that collective-looking thermal transport can occur. To our best knowledge, there was no experimental observation of the driftless second sound.

The dispersion relation of second sound can be derived from the momentum and energy balance equations in equations (1.13) and (1.14). If U-scattering is considered, the momentum destruction by U-scattering needs to be added to the right-hand side of equation (1.14). An example dispersion relation of second sound in (20,20) SWCNT is shown in figure 1.13(a). The real and imaginary frequencies represent the propagation and attenuation of a pulse, respectively. The imaginary frequency in the limit of a small wavevector (i.e. long wavelength) is mostly determined by the rate of U-scattering. As the wavevector is increased (i.e. wavelength becomes shorter), the viscous damping effect by N-scattering becomes strong, causing the significant attenuation of second sound. Figure 1.13(b) shows the required length of a sample.

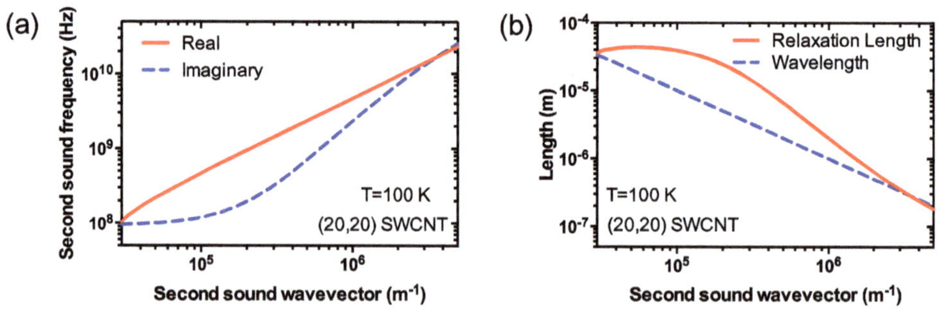

Figure 1.13. The propagation and attenuation of second sound in (20,20) SWCNT. (a) The dispersion relation of second sound showing propagation (real) and attenuation (imaginary) of second sound. (b) The comparison between relaxation length and wavelength of second sound. Reproduced with permission from [18]. Copyright 2017 the American Physical Society.

For the second sound to propagate, the sample length should be larger than the wavelength of the second sound but smaller than its relaxation length defined as $v_{\text{II}}\text{Im}(\Omega)$, where v_{II} and $\text{Im}(\Omega)$ are the speed of the second sound and the imaginary part of the second sound frequency, respectively.

1.6 Summary and future perspectives

In this chapter, we briefly reviewed past and recent studies on hydrodynamic phonon transport. We first discussed the displaced Bose–Einstein distribution representing the collective motion of phonon particles as an equilibrium state under N-scattering. Then, we introduced several approaches to solve the Peierls–Boltzmann transport equation for the case where N-scattering is significant. Based on the solution of the Peierls–Boltzmann transport equation, we then showed how N-scattering affects thermal phonon transport in both the steady-state and transient cases. For the steady-state cases, we discuss three scenarios: when N-scattering is combined with (i) U-scattering, (ii) diffuse boundary scattering and (iii) thermal reservoirs that emit phonons following non-displaced Bose–Einstein distribution functions. In all cases, N-scattering affects thermal transport indirectly. For the first case where N-scattering is combined with U-scattering, it transfers energy from small wavevector states where U-scattering is relatively weak to large wavevector states where U-scattering is strong, thereby contributing to thermal resistance. For the second case, the N-scattering impedes the momentum transfer to the boundaries which act as a momentum sink by diffuse boundary scattering. We discussed that stronger N-scattering leads to a lesser viscous damping effect and larger thermal conductivity. For the last case, the N-scattering itself causes thermal resistance when the distribution function is not homogeneous in space due to thermal reservoirs emitting phonons with a non-displaced distribution. The thermal resistance occurs while those non-collective phonon flows become collective through N-scattering processes. The thermal resistance by the transition between collective and non-collective phonon flows depends on the shape of phonon dispersion; while the thermal resistance due to this effect is small for Debye phonon dispersion, it can be significant in graphitic materials because of their highly nonlinear phonon dispersion with many branches. The second sound was discussed as a representative phenomenon of phonon hydrodynamics in the transient case. The N-scattering causes the damping of second sound even without U-scattering. If the fluctuation of the temperature field is fast in the time and space domains, for example, the frequency and wavelength of second sound are shorter than the rate and mean free path of N-scattering, respectively, the fluctuation can be largely damped. Thus, the N-scattering imposes the limit of frequency and wavelength of second sound for its propagation.

Although the recently developed *ab initio* framework for phonon transport has been proved for its high accuracy and predictive power [41], the significant hydrodynamic phonon transport in graphitic materials needs to be experimentally confirmed. The significant contributions from the flexural phonon modes to thermal transport were experimentally shown in [70], but its strong N-scattering due to extremely large anharmonicity for small wavevector states has not been verified. The explicit observation of hydrodynamic phonon transport has several challenges. First, the measurements need to be done in a much smaller length and time scale compared to the previous studies performed several decades ago. The characteristic length and time scale of hydrodynamic phonon transport scales with the mean free paths and rate of internal phonon scattering. As those previous studies measured the hydrodynamic phonon transport at extremely low temperatures below 15 K, the internal phonon scattering was weak; therefore, the Poiseuille flow was measured with a several millimeter sized sample [11] and the second sound propagation was measured with a time scale of microseconds [14]. However, as the hydrodynamic phonon transport in graphitic materials is expected to occur at a much higher temperature, the internal phonon scattering is accordingly strong. Thus, the experiments need to be performed with sub-millimeter sized samples and nanosecond temporal resolution. Recent advancements on the microscale platform for the measurement of thermal conductivity [71, 72] as well as the ultrafast spectroscopy technique [73–75] are perhaps well suited for the measurement of hydrodynamic phonon transport in graphitic materials. Second, we would need a large sample with minimal defects. The *ab initio* simulation shows that the sample size should be at least 10 μm for measuring phonon Poiseuille flow and second sound at 100 K [50], but typical graphitic material samples with this sample size contain many defects. Interestingly, the observation of second sound was reported very recently using a highly oriented pyrolytic graphite (HOPG) sample [76]. This study used the transient grating method to generate the standing wave of second sound and could measure the fluctuation of temperature which is associated with the second sound. As the second sound is in standing-wave form in this study, it does not need to propagate throughout the entire sample and could be measured with less significant damping. The observation of phonon Poiseuille flow is expected to be more challenging compared to the second sound case. The theoretical prediction of phonon Poiseuille flow assumed infinitely long samples for the condition of fully developed phonon flow [50]. If a sample has a finite length, there would be the so-called entrance effect which is due to the transition from spatially uniform phonon flow to parabolic phonon flow near the entrance. This would require a sample with the length being much larger than the width. In the previous study, the phonon Poiseuille flow was predicted with the width of 10 μm, thus the length should be much longer than this value.

The recent prediction of significant hydrodynamic phonon transport indicates that the hydrodynamic regime is practically important for high thermal conductivity materials where N-scattering is often strong and cannot be ignored. Although the clear observation of hydrodynamic phonon transport is expected at sub-room temperatures, the hydrodynamic phonon transport is still important for understanding the thermal transport. As shown in figure 1.3, the mean free path of N-scattering and U-scattering has a large gap in the length ranges from sub-micrometer to micrometer for 300 K. If the sample size lies in this gap which is common in the practical applications of high thermal conductivity materials for thermal management, the diffusive–ballistic phonon transport may not correctly describe the thermal transport phenomena; it ignores the thermal resistance due to the momentum transfer and the formation of collective phonon flow by N-scattering. Therefore, the hydrodynamic regime needs to be considered another limit of thermal transport in addition to the ballistic and diffusive limits which were extensively studied in the past [77, 78]. The detailed mechanisms of how N-scattering contributes to thermal resistance when combined with other scattering processes has not been rigorously discussed in the past. We think this is partly because of the lack of available numerical tools; it has been very challenging to solve the Peierls–Boltzmann transport equation in both real and reciprocal spaces with minimal assumptions. With the recently developed *ab initio* frameworks for solving the Peierls–Boltzmann transport equation in both real and reciprocal spaces [48, 50], it is now possible to quantitatively study the influence of N-scattering on the overall thermal transport process when it is combined with other scattering processes. This would complete the understanding of phonon transport in high thermal conductivity materials and lead to the better design of thermal devices using those high thermal conductivity materials.

Acknowledgments

We acknowledge support from the National Science Foundation (Award Nos 1705756 and 1709307).

References

[1] Nernst W 1917 *Die Theoretischen Grundlagen des Neuen Wärmesatzes* (Halle: Knapp)
[2] Peierls R 1929 Zur kinetischen Theorie der Wärmeleitung in Kristallen *Ann. Phys.* **395** 1055–101
[3] Lee S, Broido D, Esfarjani K and Chen G 2015 Hydrodynamic phonon transport in suspended graphene *Nat. Commun.* **6** 6290
[4] Sussmann J A and Thellung A 1963 Thermal conductivity of perfect dielectric crystals in the absence of Umklapp processes *Proc. Phys. Soc.* **81** 1122
[5] Prohofsky E W and Krumhansl J A 1964 Second-sound propagation in dielectric solids *Phys. Rev.* **133** A1403–10
[6] Guyer R A and Krumhansl J A 1964 Dispersion relation for second sound in solids *Phys. Rev.* **133** A1411–7
[7] Guyer R A and Krumhansl J A 1966 Thermal conductivity, second sound, and phonon hydrodynamic phenomena in nonmetallic crystals *Phys. Rev.* **148** 778–88
[8] Guyer R A and Krumhansl J A 1966 Solution of the linearized phonon Boltzmann equation *Phys. Rev.* **148** 766–78
[9] Gurzhi R 1964 Thermal conductivity of dielectrics and ferrodielectrics at low temperatures *J. Exp. Theor. Phys.* **46** 719–24
[10] Griffin A 1965 On the detection of second sound in crystals by light scattering *Phys. Lett.* **17** 208–10
[11] Mezhov-Deglin L 1965 Measurement of the thermal conductivity of crystalline He^4 *J. Exp. Theor. Phys.* **49** 66–79
[12] Ackerman C C and Overton W C 1969 Second sound in solid helium-3 *Phys. Rev. Lett.* **22** 764–6
[13] McNelly T F *et al* 1970 Heat pulses in NaF: onset of second sound *Phys. Rev. Lett.* **24** 100–2

[14] Jackson H E, Walker C T and McNelly T F 1970 Second sound in NaF *Phys. Rev. Lett.* **25** 26–8
[15] Narayanamurti V and Dynes R C 1972 Observation of second sound in bismuth *Phys. Rev. Lett.* **28** 1461–65
[16] Ashcroft N and Mermin N 1976 *Solid State Physics* (Philadelphia, PA: Saunders)
[17] Jackson H E and Walker C T 1971 Thermal conductivity, second sound, and phonon–phonon interactions in NaF *Phys. Rev.* B **3** 1428–39
[18] Lee S and Lindsay L 2017 Hydrodynamic phonon drift and second sound in a (20,20) single-wall carbon nanotube *Phys. Rev.* B **95** 184304
[19] Cepellotti A, Fugallo G, Paulatto L, Lazzeri M, Mauri F and Marzari N 2015 Phonon hydrodynamics in two-dimensional materials *Nat. Commun.* **6** 6400
[20] Ding Z, Zhou J, Song B, Chiloyan V, Li M, Liu T-H and Chen G 2018 Phonon hydrodynamic heat conduction and Knudsen minimum in graphite *Nano Lett.* **18** 638–49
[21] Schelling P K and Keblinski P 2003 Thermal expansion of carbon structures *Phys. Rev.* B **68** 035425
[22] Beck H, Meier P F and Thellung A 1974 Phonon hydrodynamics in solids *Phys. Status Solidi* a **24** 11–63
[23] Gurevich V L v 1986 *Transport in Phonon Systems Modern Problems in Condensed Matter Sciences* (Amsterdam: Elsevier)
[24] Joseph D D and Preziosi L 1989 Heat waves *Rev. Mod. Phys.* **61** 41–73
[25] Alvarez F X, Jou D and Sellitto A 2009 Phonon hydrodynamics and phonon–boundary scattering in nanosystems *J. Appl. Phys.* **105** 014317
[26] Ziabari A *et al* 2018 Full-field thermal imaging of quasiballistic crosstalk reduction in nanoscale devices *Nat. Commun.* **9** 255
[27] Torres P *et al* 2018 Emergence of hydrodynamic heat transport in semiconductors at the nanoscale *Phys. Rev. Mater.* **2** 076001
[28] Guo Y and Wang M 2018 Phonon hydrodynamics for nanoscale heat transport at ordinary temperatures *Phys. Rev.* B **97** 035421
[29] Dong Y, Cao B-Y and Guo Z-Y 2011 Generalized heat conduction laws based on thermomass theory and phonon hydrodynamics *J. Appl. Phys.* **110** 063504
[30] Cimmelli V A, Sellitto A and Jou D 2010 Nonlinear evolution and stability of the heat flow in nanosystems: beyond linear phonon hydrodynamics *Phys. Rev.* B **82** 184302
[31] Sellitto A, Alvarez F X and Jou D 2010 Temperature dependence of boundary conditions in phonon hydrodynamics of smooth and rough nanowires *J. Appl. Phys.* **107** 114312
[32] Hardy R J and Albers D L 1974 Hydrodynamic approximation to the phonon Boltzmann equation *Phys. Rev.* B **10** 3546–51
[33] Guo Y and Wang M 2015 Phonon hydrodynamics and its applications in nanoscale heat transport *Phys. Rep.* **595** 1–44
[34] Ziman J M 1960 *Electrons and Phonons: The Theory of Transport Phenomena in Solids* (Oxford: Oxford University Press)
[35] Ding Z, Zhou J, Song B, Li M, Liu T-H and Chen G 2018 Umklapp scattering is not necessarily resistive *Phys. Rev.* B **98** 180302
[36] Broido D A, Malorny M, Birner G, Mingo N and Stewart D A 2007 Intrinsic lattice thermal conductivity of semiconductors from first principles *Appl. Phys. Lett.* **91** 231922
[37] Esfarjani K, Chen G and Stokes H T 2011 Heat transport in silicon from first-principles calculations *Phys. Rev.* B **84** 085204

[38] Omini M and Sparavigna A 1996 Beyond the isotropic-model approximation in the theory of thermal conductivity *Phys. Rev.* B **53** 9064–73
[39] Omini M and Sparavigna A 1997 Heat transport in dielectric solids with diamond structure *Nuovo Cimento* D **19** 1537
[40] Fugallo G, Lazzeri M, Paulatto L and Mauri F 2013 *Ab initio* variational approach for evaluating lattice thermal conductivity *Phys. Rev.* B **88** 045430
[41] Lindsay L, Hua C, Ruan X L and Lee S 2018 Survey of *ab initio* phonon thermal transport *Mater. Today Phys.* **7** 106–20
[42] Callaway J 1959 Model for lattice thermal conductivity at low temperatures *Phys. Rev.* **113** 1046
[43] Krumhansl J A 1965 Thermal conductivity of insulating crystals in the presence of normal processes *Proc. Phys. Soc.* **85** 921
[44] Vincenti W G and Kruger C H 1965 *Introduction to Physical Gas Dynamics* (New York: Wiley)
[45] Allen P B 2013 Improved Callaway model for lattice thermal conductivity *Phys. Rev.* B **88** 144302
[46] Ma J, Li W and Luo X 2014 Examining the Callaway model for lattice thermal conductivity *Phys. Rev.* B **90** 035203
[47] Li X and Lee S 2019 Crossover of ballistic, hydrodynamic, and diffusive phonon transport in suspended graphene *Phys. Rev.* B **99** 085202
[48] Cepellotti A and Marzari N 2016 Thermal transport in crystals as a kinetic theory of relaxons *Phys. Rev.* **6** 041013
[49] Landon C D and Hadjiconstantinou N G 2014 Deviational simulation of phonon transport in graphene ribbons with *ab initio* scattering *J. Appl. Phys.* **116** 163502
[50] Li X and Lee S 2018 Role of hydrodynamic viscosity on phonon transport in suspended graphene *Phys. Rev.* B **97** 094309
[51] Majumdar A 1993 Microscale heat conduction in dielectric thin films *J. Heat Transfer* **115** 7–16
[52] Péraud J-P M and Hadjiconstantinou N G 2011 Efficient simulation of multidimensional phonon transport using energy-based variance-reduced Monte Carlo formulations *Phys. Rev.* B **84** 205331
[53] Hao Q, Chen G and Jeng M-S 2009 Frequency-dependent Monte Carlo simulations of phonon transport in two-dimensional porous silicon with aligned pores *J. Appl. Phys.* **106** 114321
[54] Lindsay L, Broido D and Mingo N 2009 Lattice thermal conductivity of single-walled carbon nanotubes: beyond the relaxation time approximation and phonon–phonon scattering selection rules *Phys. Rev.* B **80** 125407
[55] Guo Y and Wang M 2017 Heat transport in two-dimensional materials by directly solving the phonon Boltzmann equation under Callaway's dual relaxation model *Phys. Rev.* B **96** 134312
[56] Martelli V, Jiménez J L, Continentino M, Baggio-Saitovitch E and Behnia K 2018 Thermal transport and phonon hydrodynamics in strontium titanate *Phys. Rev. Lett.* **120** 125901
[57] Lee S, Li X and Guo R 2019 Thermal resistance by transition between collective and non-collective phonon flows in graphitic materials *Nanoscale Microscale Thermophys. Eng.* **2019** 1–12
[58] Landau L 1941 Theory of the superfluidity of helium II *Phys. Rev.* **60** 356–8

[59] Peshkov V 1946 Determination of the velocity of propagation of the second sound in helium II *J. Phys. USSR* **10** 389–96
[60] Ward J and Wilks J 1951 The velocity of second sound in liquid helium near the absolute zero *Philos. Mag.* **42** 314–6
[61] Ward J C and Wilks J III 1952 Second sound and the thermo-mechanical effect at very low temperatures *Philos. Mag.* **43** 48–50
[62] Rogers S J 1971 Transport of heat and approach to second sound in some isotopically pure alkali-halide crystals *Phys. Rev.* B **3** 1440–57
[63] Guyer R A 1965 Light scattering detection of thermal waves *Phys. Lett.* **19** 261
[64] Griffin A 1968 Brillouin light scattering from crystals in the hydrodynamic region *Rev. Mod. Phys.* **40** 167–205
[65] Wehner R K and Klein R 1972 Scattering of light by entropy fluctuations in dielectric crystals *Physica* **62** 161–97
[66] Pohl D W and Irniger V 1976 Observation of second sound in NaF by means of light scattering *Phys. Rev. Lett.* **36** 480–3
[67] Hehlen B, Pérou A-L, Courtens E and Vacher R 1995 Observation of a doublet in the quasielastic central peak of quantum-paraelectric $SrTiO_3$ *Phys. Rev. Lett.* **75** 2416–9
[68] Koreeda A, Takano R and Saikan S 2007 Second sound in $SrTiO_3$ *Phys. Rev. Lett.* **99** 265502
[69] Hardy R J 1970 Phonon Boltzmann equation and second sound in solids *Phys. Rev.* B **2** 1193–207
[70] Seol J H *et al* 2010 Two-dimensional phonon transport in supported graphene *Science* **328** 213–6
[71] Shi L *et al* 2003 Measuring thermal and thermoelectric properties of one-dimensional nanostructures using a microfabricated device *J. Heat Transfer* **125** 881–8
[72] Ou E, Li X, Lee S, Watanabe K and Taniguchi T 2019 Four-probe measurement of thermal transport in suspended few-layer graphene with polymer residue *J. Heat Transfer* **141** 061601–5
[73] Cahill D G 2004 Analysis of heat flow in layered structures for time-domain thermoreflectance *Rev. Sci. Instrum.* **75** 025901
[74] Johnson J A *et al* 2013 Direct measurement of room-temperature nondiffusive thermal transport over micron distances in a silicon membrane *Phys. Rev. Lett.* **110** 131–4
[75] Salcedo J R, Siegman A E, Dlott D D and Fayer M D 1978 Dynamics of energy transport in molecular crystals: the picosecond transient-grating method *Phys. Rev. Lett.* **41** 131–4
[76] Huberman S *et al* 2019 Observation of second sound in graphite at temperatures above 100 K *Science* **364** 375–9
[77] Cahill D G *et al* 2014 Nanoscale thermal transport. II. 2003–2012 *Appl. Phys. Rev.* **1** 011305
[78] Cahill D G *et al* 2003 Nanoscale thermal transport *J. Appl. Phys.* **93** 793–818

IOP Publishing

Nanoscale Energy Transport
Emerging phenomena, methods and applications
Bolin Liao

Chapter 2

Higher-order phonon scattering: advancing the quantum theory of phonon linewidth, thermal conductivity and thermal radiative properties

Tianli Feng and Xiulin Ruan

Phonon scattering plays a central role in the quantum theory of phonon linewidth, which in turn governs important properties including infrared spectra, Raman spectra, lattice thermal conductivity, thermal radiative properties, and also significantly affects other important processes such as hot electron relaxation. Since Maradudin and Fein's classic work in 1962, three-phonon scattering had been considered as the dominant intrinsic phonon scattering mechanism and has seen tremendous advances. However, the role of the higher-order four-phonon scattering had been persistently unclear and so was ignored. The tremendous complexity of the formalism and computational challenges stood in the way, prohibiting the direct and quantitative treatment of four-phonon scattering. In 2016, a rigorous four-phonon scattering formalism was developed, and the prediction was realized using empirical potentials. In 2017, the method was extended using first-principles calculated force constants, and the thermal conductivities of boron arsenides (BAs), Si and diamond were predicted. The predictions for BAs were later confirmed by several independent experiments. Four-phonon scattering has since been investigated in a range of materials and established as an important intrinsic scattering mechanism for thermal transport and radiative properties. Specifically, four-phonon scattering is important when the fourth-order scattering potential or phase space becomes relatively large. The former scenario includes: (i) nearly all materials when the temperature is high; (ii) strongly anharmonic (low thermal conductivity) materials, including most rocksalt compounds, halides, hydrides, chalcogenides and oxides. The latter scenario includes: (iii) materials with large acoustic–optical phonon band gaps, such as XY compounds with a large atomic mass ratio between X and Y; (iv) two-dimensional materials with reflection symmetry, such as single-layer graphene, single-layer boron nitride and carbon nanotubes; and (v) phonons with a large

doi:10.1088/978-0-7503-1738-2ch2

density of states, such as optical phonons, which are important for Raman, infrared and thermal radiative properties. Four-phonon scattering is expected to gain broad interest in various technologically important materials for thermoelectrics, thermal barrier coatings, thermal energy storage, phase change, nuclear power, ultra-high temperature ceramics, infrared spectra, Raman spectra, radiative transport, hot electron relaxation and radiative cooling. Four-phonon scattering has been, and will continue to be, established as an important intrinsic phonon scattering mechanism beyond three-phonon scattering. The prediction of four-phonon scattering will transition from a breakthrough to a new routine in the next decade.

2.1 Overview

Phonon scattering plays a central role in the quantum theory of phonon linewidth, thermal conductivity and thermal radiative properties. For over half a century, three-phonon scattering has been considered as the dominant intrinsic phonon scattering mechanism. Starting from the third-order anharmonic Hamiltonian and Fermi's golden rule (FGR), Maradudin and Flinn [1], Maradudin and Fein [2], and Maradudin *et al* [3] derived an anharmonic lattice dynamics (ALD) method to predict intrinsic three-phonon scattering rates in solids. Debernardi *et al* [4] combined ALD and first-principles methods based on the density functional theory (DFT) to predict three-phonon scattering rates and linewidths for carbon, silicon and germanium, and the results agreed well with Raman spectra. This work was followed by the first-principles prediction of the phonon linewidths of a variety of materials [5–8]. More recently, Broido *et al* combined first-principles calculations of three-phonon scattering rates and the phonon Boltzmann transport equation (BTE) and enabled first-principles prediction of thermal conductivity [9]. Many studies have since been conducted on the thermal transport based on three-phonon scattering, and the calculated thermal conductivity (κ) has found incredible agreement with measured κ values for a variety of systems [9–15]. First-principles calculations of three-phonon scattering rates of zone-center optical phonons have also been combined with the Lorentz oscillator model to predict the thermal radiative properties of polar materials [16].

However, a persistent fundamental question remained: what is the role of four-phonon and higher-order scattering? The observations of a series of experiments deviated from the three-phonon scattering theory, but no accepted explanation had emerged. In the early years, Joshi *et al* found in experiments that the thermal conductivity of silicon at a high temperature decreases more rapidly than the $1/T$ trend as would be given by the three-phonon scattering theory, and they assumed the existence of four-phonon scattering to be responsible for this behavior [17]. However, Ecsedy and Klemens made calculations and concluded that this trend could not be due to four-phonon scattering [18]. The optical phonon linewidths based on three-phonon theory often show large underestimation for materials with significant infrared applications such as cubic BN (*c*-BN), 3*C*-SiC, GaN, GaP, GaAs, InAs and AlAs at high temperature or even room temperature (RT) [4, 19, 20]. Moreover, first-principles methods overestimated the measured thermal conductivities of a number of materials [10, 11, 21–23]. For example, while some predictions gave reasonable accuracy with measured data at low temperature, they over-predicted significantly at higher temperature [10], diminishing the predictive power for applications such as thermal barrier coatings and high temperature thermoelectrics. Even at RT, such deviations could become quite large for some technologically important materials [11, 21–23]. Such deviations had often been attributed to defects and impurities in the materials, and the role of four-phonon scattering was unclear and ignored, largely due to the lack of theoretical formalism and computational power needed to treat four-phonon scattering. Recent explorations of four-phonon scattering included checking the phase space [24] and examining its significance from molecular dynamics [25]. Direct and quantitative prediction of four-phonon scattering rates was greatly desired to uncover the physics but was not available.

In 2016, Feng and Ruan [26] developed the four-phonon scattering formalism and mitigated the challenges in computation, and rigorously predicted the four-phonon scattering rates for several benchmark materials such as diamond, silicon, germanium and solid argon. Based on empirical interatomic potentials, they predicted strong four-phonon scattering rates in silicon and germanium at high temperatures and in argon even at low temperatures. Their four-phonon scattering results have explained well the discrepancy in phonon scattering rates between the perturbation theory of three-phonon scattering and molecular dynamics (MD) using the same empirical potentials, because MD naturally includes all the orders of anharmonicity. The inclusion of four-phonon scattering makes the ALD and MD consistent with each other, and pushes a step forward towards the 'unification' of these simulation methods of phonon and thermal transport.

In 2017, Feng *et al* [27] extended the method by calculating the fourth-order force constants from DFT instead of empirical interatomic potentials and predicted a significant impact of four-phonon scattering on the thermal conductivity of diamond and silicon at high temperatures and BAs even at RT. Their predictions for Si agree well with early experiments, and those for BAs were later verified by three independent experimental works published in 2018 that directly measured the thermal conductivity of high-quality single-crystal BAs [28–30]. Since then, the prediction of the large impact of four-phonon scattering has been accepted, and more investigations are going on. For example, its significance was recently found in two-dimensional material graphene [31] and other strongly anharmonic materials such as PbTe [32] and NaCl [33].

Compared to three-phonon scattering, the most apparent complexity of four-phonon scattering is its large phase space. As illustrated in figure 2.1, in a three-phonon process, a phonon mode λ can either split into two other modes ($\lambda \rightarrow \lambda_1 + \lambda_2$) or combine with one other mode to a new one ($\lambda + \lambda_1 \rightarrow \lambda_2$). In a four-phonon process, it can split into three other modes ($\lambda \rightarrow \lambda_1 + \lambda_2 + \lambda_3$), generate two new modes by absorbing one ($\lambda + \lambda_1 \rightarrow \lambda_2 + \lambda_3$), or convert to a new mode by absorbing two ($\lambda + \lambda_1 + \lambda_2 \rightarrow \lambda_3$). All these processes must obey the conservation law of energy and quasi-momentum. The number of possible combinations of four-phonon modes that satisfy the conservation laws is usually several orders larger than

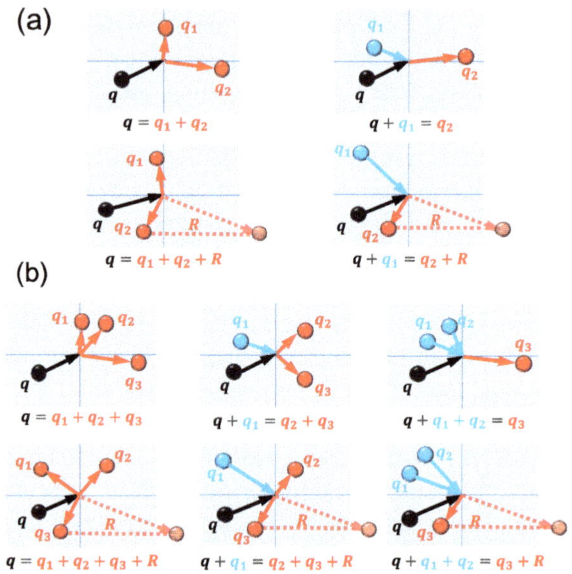

Figure 2.1. Three- and four-phonon scattering diagrams. (a) Three-phonon splitting and combination processes. (b) Four-phonon splitting, redistribution and combination processes. The shaded rectangles represent the first Brillouin zone (BZ). The phonon momentum is $\hbar\mathbf{q}$. The processes with momentum conserved are normal processes. The others with momentum non-conserved are Umklapp processes, in which the resulting phonons are folded back by reciprocal lattice vectors **R**. Reproduced with permission from [27]. Copyright 2017 the American Physical Society.

that of three-phonon processes. For example, in silicon, by using a $16 \times 16 \times 16$ **q**-mesh, a phonon mode can find $\sim 10^3$ possible combinations with other modes for three-phonon processes, while it can find 10^7–10^8 possible combinations for four-phonon scattering. Therefore, four-phonon scattering is less dependent on the dispersive nature of phonon frequencies compared to three-phonon scattering.

Feng *et al*'s works [26, 27, 31] have established that four-phonon scattering is non-negligible with two origins: strong scattering potential and large scattering phase space. The former presents as strong anharmonicity, as found in strongly anharmonic materials, i.e. most low thermal conductivity materials such as solid argon [26], PbTe [32] and NaCl [33]. Even for weakly anharmonic materials with high thermal conductivity, such as diamond and silicon, the anharmonicity could become strong when the temperature is high. For example, four-phonon scattering reduces the thermal conductivity of silicon by 30% at 1000 K [26, 27]. Quantitatively, four-phonon scattering rates scale with temperature quadratically ($\sim T^2$), which is one order faster than three-phonon scattering. Regarding the scattering phase space, the four-phonon process generally has a several orders larger phase space than three-phonon scattering since the conservation laws can easily be satisfied. This effect is reflected most significantly in higher-frequency phonons including, in particular, optical phonons, which have a large density of states and thus four-phonon scattering phase space. In addition, the four-phonon scattering can be exceptionally important in systems where three-phonon processes have a suppressed phase space, either due to a large acoustic–optical phonon band gap or the reflection symmetry in 2D materials. One example of the former is BAs, a quite harmonic crystal, for which neglecting four-phonon scattering leads to 57% over-prediction in thermal conductivity at room temperature [27]. The optical phonon relaxation times in these materials with large acoustic–optic phonon band gaps are exceptionally suppressed by four-phonon scattering [27]. Single-layer graphene is an example of the latter, and the four-phonon scattering for the flexural acoustic (ZA) mode was predicted to be suppressed less than the three-phonon scattering by reflection symmetry [31], although the prediction still needs experimental validation.

Beyond the single-mode relaxation time approximation (SMRTA or RTA), Feng and Ruan [31] derived the exact solution to phonon BTE that incorporates the four-phonon scattering's phase space into the iteration in the calculation of thermal conductivity. Due to the large phase space of four-phonon scattering, the iteration is extremely computationally expensive. Fortunately, they found that four-phonon scattering is usually dominated by the Umklapp process even in the materials where three-phonon scattering is dominated by the normal process [26, 27], indicating that in order to save time it is not necessary to include the four-phonon's phase space in the iterative scheme in these materials. However, for some materials such as graphene where the four-phonon scattering is dominated by the normal process, the iterative scheme involving four-phonon phase space is crucial to the thermal conductivity prediction [31].

The remainder of this chapter is organized as follows. In section 2.2, the four-phonon scattering formalism is derived in the context of solving phonon BTE. For generality, multiple scattering mechanisms including three-phonon, four-phonon, phonon–impurity and phonon–boundary scatterings are included in the solution since they are coupled together with each other in the exact solution that involves the iteration of the phase spaces of these scattering processes. The SMRTA solution is presented at the zeroth iteration. In section 2.3, the significance of four-phonon scattering induced by large scattering potential is presented. For weakly anharmonic materials, the large four-phonon scattering potential could be induced by raising the temperature. For strongly anharmonic materials, the strong four-phonon scattering potential is caused by the intrinsic strongly anharmonic interatomic bonding, even at low temperatures. In section 2.4, the significance of four-phonon scattering induced by large scattering phase space is presented. This is either induced by restricted three-phonon scattering phase space or a large density of states (DOS). The former includes certain groups of materials such as those with large acoustic–optical phonon band gaps and two-dimensional materials with reflection symmetry. The latter is remarkably represented in optical phonon modes, which often show high four-phonon scattering rates due to large DOS. The prediction of zone-center optical phonon linewidth and thermal radiative properties is extensively presented here. In section 2.5, we supplement some discussion of a few important issues related to four-phonon scattering such as the frequency scaling law, the Umklapp scattering and the three-phonon scattering to the second order. In section 2.6, a brief summary is provided and an outlook of future research directions is presented.

2.2 Formalism of four-phonon scattering

In this section, we derive the solution to the phonon BTE that includes three-phonon, four-phonon, phonon–impurity and phonon–boundary scattering. The derivation is extensively presented in [26, 31], and it is summarized here again. The phonon BTE [34–36]

$$\mathbf{v}_\lambda \cdot \nabla n_\lambda = \left.\frac{\partial n_\lambda}{\partial t}\right|_s \tag{2.1}$$

describes the balance of the phonon population between diffusive drift and collision. λ labels the phonon mode ($\mathbf{q}$, ν), with $\mathbf{q}$ representing the wave vector and ν representing the dispersion branch, $\mathbf{v}_\lambda$ is the group velocity, and n_λ is the phonon occupation number. Due to a small temperature gradient, n_λ has a small derivation n_λ' from its equilibrium Bose–Einstein distribution $n_\lambda^0 = [\exp(\hbar\omega_\lambda/k_\mathrm{B}T) - 1]^{-1}$ so that $n_\lambda = n_\lambda^0 + n_\lambda'$. By assuming that n_λ' is independent of temperature [35], $(\partial n_\lambda/\partial T) \simeq (\partial n_\lambda^0/\partial T)$, we have

$$\mathbf{v}_\lambda \cdot \nabla T \frac{\partial n_\lambda^0}{\partial T} = \left.\frac{\partial n_\lambda'}{\partial t}\right|_s \tag{2.2}$$

considering $\nabla n_\lambda = (\partial n_\lambda/\partial T)\nabla T$.

The scattering term $(\partial n_\lambda'/\partial t)|_s$ is the decay rate of the perturbation n_λ' due to the scattering processes of the mode λ, including the three-phonon processes $\lambda \to \lambda_1 + \lambda_2$ and $\lambda + \lambda_1 \to \lambda_2$, the four-phonon processes $\lambda \to \lambda_1 + \lambda_2 + \lambda_3$, $\lambda + \lambda_1 \to \lambda_2 + \lambda_3$ and $\lambda + \lambda_1 + \lambda_2 \to \lambda_3$, and the isotope and boundary scattering processes $\lambda \to \lambda_1$. The scattering rates of these processes are given by the scattering probabilities, P, which are determined by FGR:

$$P_{i\to f} = \frac{2\pi}{\hbar} |\langle f|\hat{H}|i\rangle|^2\, \delta(E_i - E_f), \tag{2.3}$$

where $|i\rangle$ and $|f\rangle$ are the initial and final quantum states, respectively. The net transition rate from $|i\rangle$ to $|f\rangle$ is, therefore, written as

$$P_{i\to f} - P_{f\to i} = \frac{2\pi}{\hbar}\left(|\langle f|\hat{H}|i\rangle|^2 - |\langle i|\hat{H}|f\rangle|^2\right)\delta(E_i - E_f). \tag{2.4}$$

The initial and final quantum states depend on scattering processes. For example, for the three-phonon process $\lambda \to \lambda_1 + \lambda_2$, the initial and final quantum states are $|i\rangle = |n_\lambda + 1, n_{\lambda_1}, n_{\lambda_2}\rangle$ and $|f\rangle = |n_\lambda, n_{\lambda_1} + 1, n_{\lambda_2} + 1\rangle$, respectively. Similarly, for the four-phonon process $\lambda \to \lambda_1 + \lambda_2 + \lambda_3$, the initial and final quantum states are $|i\rangle = |n_\lambda + 1, n_{\lambda_1}, n_{\lambda_2}, n_{\lambda_3}\rangle$ and $|f\rangle = |n_\lambda, n_{\lambda_1} + 1, n_{\lambda_2} + 1, n_{\lambda_3} + 1\rangle$, respectively. The other processes can be analogized.

The transition rate in equation (2.4) is determined by the lattice Hamiltonian $\hat{H}$ [2, 37]:

$$\hat{H} = \hat{H}_0 + \hat{H}_3 + \hat{H}_4 + \cdots + \hat{H}_\mathrm{iso} + \cdots, \tag{2.5}$$

which includes the harmonic part

$$\hat{H}_0 = \sum_{\lambda} \hbar\omega_\lambda \left(a_\lambda^\dagger a_\lambda + 1/2\right), \tag{2.6}$$

the first-order anharmonic part

$$\hat{H}_3 = \sum_{\lambda\lambda_1\lambda_2} H^{(3)}_{\lambda\lambda_1\lambda_2}\left(a^\dagger_{-\lambda} + a_\lambda\right)\left(a^\dagger_{-\lambda_1} + a_{\lambda_1}\right)\left(a^\dagger_{-\lambda_2} + a_{\lambda_2}\right), \tag{2.7}$$

with

$$H^{(3)}_{\lambda\lambda_1\lambda_2} = \frac{\hbar^{3/2}}{2^{3/2} \times 6N_{\mathbf{q}}^{1/2}} \Delta_{\mathbf{q}+\mathbf{q}_1+\mathbf{q}_2,\mathbf{R}} \frac{V^{(3)}_{\lambda\lambda_1\lambda_2}}{\sqrt{\omega_\lambda \omega_{\lambda_1} \omega_{\lambda_2}}}, \tag{2.8}$$

$$V^{(3)}_{\lambda\lambda_1\lambda_2} = \sum_{b,l_1b_1,l_2b_2} \sum_{\alpha\alpha_1\alpha_2} \Phi^{\alpha\alpha_1\alpha_2}_{0b,l_1b_1,l_2b_2} \frac{\mathrm{e}^{\lambda}_{\alpha b} \mathrm{e}^{\lambda_1}_{\alpha_1 b_1} \mathrm{e}^{\lambda_2}_{\alpha_2 b_2}}{\sqrt{\bar{m}_b \bar{m}_{b_1} \bar{m}_{b_2}}} \mathrm{e}^{\mathrm{i}\mathbf{q}_1\cdot\mathbf{r}_{l_1} + \mathrm{i}\mathbf{q}_2\cdot\mathbf{r}_{l_2}}, \tag{2.9}$$

the second-order anharmonic part

$$\hat{H}_4 = \sum_{\lambda\lambda_1\lambda_2\lambda_3} H^{(4)}_{\lambda\lambda_1\lambda_2\lambda_3}\left(a^\dagger_{-\lambda} + a_\lambda\right)\left(a^\dagger_{-\lambda_1} + a_{\lambda_1}\right)\left(a^\dagger_{-\lambda_2} + a_{\lambda_2}\right)\left(a^\dagger_{-\lambda_3} + a_{\lambda_3}\right) \tag{2.10}$$

with

$$H^{(4)}_{\lambda\lambda_1\lambda_2\lambda_3} = \frac{\hbar^2}{2^2 \times 24N_{\mathbf{q}}} \Delta_{\mathbf{q}+\mathbf{q}_1+\mathbf{q}_2+\mathbf{q}_3,\mathbf{R}} \frac{V^{(4)}_{\lambda\lambda_1\lambda_2\lambda_3}}{\sqrt{\omega_\lambda \omega_{\lambda_1} \omega_{\lambda_2} \omega_{\lambda_3}}} \tag{2.11}$$

$$\begin{aligned} V^{(4)}_{\lambda\lambda_1\lambda_2\lambda_3} = &\sum_{b,l_1b_1,l_2b_2,l_3b_3} \sum_{\alpha\alpha_1\alpha_2\alpha_3} \Phi^{\alpha\alpha_1\alpha_2\alpha_3}_{0b,l_1b_1,l_2b_2,l_3b_3} \\ &\times \frac{\mathrm{e}^{\lambda}_{\alpha b} \mathrm{e}^{\lambda_1}_{\alpha_1 b_1} \mathrm{e}^{\lambda_2}_{\alpha_2 b_2} \mathrm{e}^{\lambda_3}_{\alpha_3 b_3}}{\sqrt{\bar{m}_b \bar{m}_{b_1} \bar{m}_{b_2} \bar{m}_{b_3}}} \mathrm{e}^{\mathrm{i}\mathbf{q}_1\cdot\mathbf{r}_{l_1} + \mathrm{i}\mathbf{q}_2\cdot\mathbf{r}_{l_2} + \mathrm{i}\mathbf{q}_3\cdot\mathbf{r}_{l_3}}, \end{aligned} \tag{2.12}$$

and the extrinsic perturbations such as the isotopes

$$\hat{H}_{\text{iso}} = \sum_{\lambda\lambda_1} H^{(\text{iso})}_{\lambda\lambda_1}\left(a^\dagger_{-\lambda} + a_\lambda\right)\left(a^\dagger_{-\lambda_1} + a_{\lambda_1}\right) \tag{2.13}$$

with

$$H^{(\text{iso})}_{\lambda\lambda_1} = -\frac{1}{4N_{\mathbf{q}}} \sum_{l,b} \sum_{\mathbf{q}_l} \Delta m_{l,\,b} \sqrt{\omega_\lambda \omega_{\lambda_1}} \Delta_{\mathbf{q}+\mathbf{q}_1+\mathbf{q}_l,\,\mathbf{R}} \mathbf{e}^{\lambda}_b \cdot \mathbf{e}^{\lambda_1}_b \mathrm{e}^{-\mathrm{i}\mathbf{q}_l\cdot\mathbf{r}_l}, \tag{2.14}$$

where b, l and α label the indices of the basis atom, unit cell and direction, respectively. $N_{\mathbf{q}}$ is the total number of $\mathbf{q}$ points of a uniform mesh in the first BZ. The Kronecker delta $\Delta_{i,j}$ is 0 if $i \neq j$ or 1 if $i = j$. $\mathbf{e}$ is the phonon eigenvector. The summation of $\sum_{l,b}$ goes over all the unit cells in the domain and the summation over

$\mathbf{q}_{1,2,3,l}$ goes over all the $\mathbf{q}$ points in the first BZ. $a^{\dagger}$ and a are the phonon creation and annihilation operators, respectively. $\mathbf{R}$ represents any reciprocal lattice vector that can be decomposed into the superposition of integer reciprocal lattice basis vectors. Φ is the interatomic force constant (IFCs). $\mathbf{r}_l$ is the position vector of the lth unit cell. (Attention should be paid to the usage of $\mathbf{e}$ and $\mathbf{r}_l$ as discussed at the end of this section.)

Substituting equation (2.5) into equation (2.4), the right-hand side of (2.2) can be rewritten as [2, 13, 26, 34–40]

$$\begin{aligned}\left.\frac{\partial n'_\lambda}{\partial t}\right|_s = &-\sum_{\lambda_1\lambda_2}\left\{\frac{1}{2}[n_\lambda(1+n_{\lambda_1})(1+n_{\lambda_2})-(1+n_\lambda)n_{\lambda_1}n_{\lambda_2}]\mathcal{L}_-\right.\\&\left.+[n_\lambda n_{\lambda_1}(1+n_{\lambda_2})-(1+n_\lambda)(1+n_{\lambda_1})n_{\lambda_2}]\mathcal{L}_+\right\}\\&-\sum_{\lambda_1\lambda_2\lambda_3}\left\{\frac{1}{6}[n_\lambda(1+n_{\lambda_1})(1+n_{\lambda_2})(1+n_{\lambda_3})-(1+n_\lambda)n_{\lambda_1}n_{\lambda_2}n_{\lambda_3}]\mathcal{L}_{--}\right.\\&+\frac{1}{2}[n_\lambda n_{\lambda_1}(1+n_{\lambda_2})(1+n_{\lambda_3})-(1+n_\lambda)(1+n_{\lambda_1})n_{\lambda_2}n_{\lambda_3}]\mathcal{L}_{+-}\\&\left.+\frac{1}{2}[n_\lambda n_{\lambda_1}n_{\lambda_2}(1+n_{\lambda_3})-(1+n_\lambda)(1+n_{\lambda_1})(1+n_{\lambda_2})n_{\lambda_3}]\mathcal{L}_{++}\right\}\\&-\sum_{\lambda_1}(n_\lambda-n_{\lambda_1})\mathcal{L}_{\mathrm{iso}}-(n_\lambda-n_\lambda^0)\frac{1}{\tau_{b,\lambda}^0}.\end{aligned}\tag{2.15}$$

The first summation on the right-hand side represents the three-phonon scattering rate of the mode λ, with the first term accounting for the splitting process $\lambda \rightarrow \lambda_1 + \lambda_2$ and the second the combination process $\lambda + \lambda_1 \rightarrow \lambda_2$. The physical meaning of the first term is the difference between the transition rates of $\lambda \rightarrow \lambda_1 + \lambda_2$ and $\lambda \leftarrow \lambda_1 + \lambda_2$ and thus indicates the decay rate of n_λ due to the splitting process. Similarly, the second term illustrates the transition rate difference between $\lambda + \lambda_1 \rightarrow \lambda_2$ and $\lambda + \lambda_1 \leftarrow \lambda_2$, indicating the decay rate of n_λ due to the combination process. $\mathcal{L}_\pm$ contains the information of the intrinsic transition probability and the transition selection rules for energy and momentum, $\omega_\lambda \pm \omega_{\lambda_1} - \omega_{\lambda_2} = 0$ and $\mathbf{q} \pm \mathbf{q}_1 - \mathbf{q}_2 = \mathbf{R}$, where $\mathbf{R} = 0$ implies the normal (N) process and $\mathbf{R} \neq 0$ the Umklapp (U) process. The second summation accounts for the four-phonon scattering of mode λ, with the first parenthesis representing the process $\lambda \rightarrow \lambda_1 + \lambda_2 + \lambda_3$, the second the process $\lambda + \lambda_1 \rightarrow \lambda_2 + \lambda_3$ and the third $\lambda + \lambda_1 + \lambda_2 \rightarrow \lambda_3$. Similarly, $\mathcal{L}_{\pm\pm}$ accounts for the transition probabilities and the selection rules, i.e., $\omega_\lambda \pm \omega_{\lambda_1} \pm \omega_{\lambda_2} - \omega_{\lambda_3} = 0$ and $\mathbf{q} \pm \mathbf{q}_1 \pm \mathbf{q}_2 - \mathbf{q}_3 = \mathbf{R}$, for those processes. The third summation is the phonon–isotope scattering rate for $\lambda \rightarrow \lambda_1$ given by Tamura [37], with the selection rules $\omega_\lambda = \omega_{\lambda_1}$ and $\mathbf{q} \neq \mathbf{q}_1$. The last term on the right-hand side of equation (2.15) indicates the phonon–boundary scattering rate. The minus sign before each scattering term indicates that the perturbation n'_λ is decreasing with time, i.e. the phonon distribution tends to recover its equilibrium state due to the scattering. The expressions for $\mathcal{L}_\pm$, $\mathcal{L}_{\pm\pm}$ and $\mathcal{L}_{\mathrm{iso}}$ are given as

$$\mathcal{L}_{\pm} = \frac{\pi\hbar}{4N_{\mathbf{q}}} \left| V_{\pm}^{(3)} \right|^2 \Delta_{\pm} \frac{\delta(\omega_\lambda \pm \omega_{\lambda_1} - \omega_{\lambda_2})}{\omega_\lambda \omega_{\lambda_1} \omega_{\lambda_2}}, \tag{2.16}$$

$$\mathcal{L}_{\pm\pm} = \frac{\pi\hbar}{4N_{\mathbf{q}}} \frac{\hbar}{2N_{\mathbf{q}}} \left| V_{\pm\pm}^{(4)} \right|^2 \Delta_{\pm\pm} \frac{\delta(\omega_\lambda \pm \omega_{\lambda_1} \pm \omega_{\lambda_2} - \omega_{\lambda_3})}{\omega_\lambda \omega_{\lambda_1} \omega_{\lambda_2} \omega_{\lambda_3}}, \tag{2.17}$$

$$\mathcal{L}_{\text{iso}} = \frac{\pi}{2N_{\mathbf{q}}} \omega_\lambda \omega_{\lambda_1} \sum_b^n g_b \left| \mathbf{e}_\lambda^b \cdot \mathbf{e}_{\lambda_1}^{b*} \right|^2 \delta(\omega_\lambda - \omega_{\lambda_1}). \tag{2.18}$$

The Kronecker deltas $\Delta_{\pm} = \Delta_{\mathbf{q}\pm\mathbf{q}_1-\mathbf{q}_2,\mathbf{R}}$ and $\Delta_{\pm\pm} = \Delta_{\mathbf{q}\pm\mathbf{q}_1\pm\mathbf{q}_2-\mathbf{q}_3,\mathbf{R}}$ describe the momentum selection rule. The delta function $\delta(\Delta\omega)$ in the calculation of each $\mathcal{L}$ can be evaluated by the Lorentzian function $(1/\pi)(\zeta/((\Delta\omega)^2 + \zeta^2))$. In the isotope scattering formula, $g_b = \sum_i f_{ib}(1 - m_{ib}/\bar{m}_b)^2$ measures the mass disorder, where i indicates isotope types, f_{ib} is the fraction of isotope i in lattice sites of basis atom b, m_{ib} is the mass of isotope i and $\bar{m}_b$ is the average atom mass of basis b sites. The transition probability matrices $V_{\pm}^{(3)}$ and $V_{\pm\pm}^{(4)}$ are

$$V_{\pm}^{(3)} = \sum_{b,l_1b_1,l_2b_2} \sum_{\alpha\alpha_1\alpha_2} \Phi_{0b,\,l_1b_1,\,l_2b_2}^{\alpha\alpha_1\alpha_2} \frac{\mathbf{e}_{\alpha b}^{\lambda} \mathbf{e}_{\alpha_1 b_1}^{\pm\lambda_1} \mathbf{e}_{\alpha_2 b_2}^{-\lambda_2}}{\sqrt{\bar{m}_b \bar{m}_{b_1} \bar{m}_{b_2}}} e^{\pm i\mathbf{q}_1\cdot\mathbf{r}_{l_1} - i\mathbf{q}_2\cdot\mathbf{r}_{l_2}}, \tag{2.19}$$

$$V_{\pm\pm}^{(4)} = \sum_{b,l_1b_1,l_2b_2,l_3b_3} \sum_{\alpha\alpha_1\alpha_2\alpha_3} \Phi_{0b,\,l_1b_1,\,l_2b_2,\,l_3b_3}^{\alpha\alpha_1\alpha_2\alpha_3} \frac{\mathbf{e}_{\alpha b}^{\lambda} \mathbf{e}_{\alpha_1 b_1}^{\pm\lambda_1} \mathbf{e}_{\alpha_2 b_2}^{\pm\lambda_2} \mathbf{e}_{\alpha_3 b_3}^{-\lambda_3}}{\sqrt{\bar{m}_b \bar{m}_{b_1} \bar{m}_{b_2} \bar{m}_{b_3}}} e^{\pm i\mathbf{q}_1\cdot\mathbf{r}_{l_1} \pm i\mathbf{q}_2\cdot\mathbf{r}_{l_2} - i\mathbf{q}_3\cdot\mathbf{r}_{l_3}}, \tag{2.20}$$

$\Phi_{0b,l_1b_1,l_2b_2}^{\alpha\alpha_1\alpha_2}$ and $\Phi_{0b,l_1b_1,l_2b_2,l_3b_3}^{\alpha\alpha_1\alpha_2\alpha_3}$ are the third-order and fourth-order IFCs.

Assume a perturbation in all the phonon modes [34, 41–43], we have

$$n_\lambda = n_\lambda^0 + n_\lambda', \; n_\lambda' = -\Psi_\lambda \frac{\partial n_\lambda^0}{\partial(\hbar\omega_\lambda)} = \Psi_\lambda \cdot \frac{1}{k_{\text{B}}T} n_\lambda^0 \left(n_\lambda^0 + 1\right), \tag{2.21}$$

$$n_{\lambda_1} = n_{\lambda_1}^0 + n_{\lambda_1}', \; n_{\lambda_1}' = -\Psi_{\lambda_1} \frac{\partial n_{\lambda_1}^0}{\partial(\hbar\omega_{\lambda_1})} = \Psi_{\lambda_1} \cdot \frac{1}{k_{\text{B}}T} n_{\lambda_1}^0 \left(n_{\lambda_1}^0 + 1\right), \tag{2.22}$$

$$n_{\lambda_2} = n_{\lambda_2}^0 + n_{\lambda_2}', \; n_{\lambda_2}' = -\Psi_{\lambda_2} \frac{\partial n_{\lambda_2}^0}{\partial(\hbar\omega_{\lambda_2})} = \Psi_{\lambda_2} \cdot \frac{1}{k_{\text{B}}T} n_{\lambda_2}^0 \left(n_{\lambda_2}^0 + 1\right), \tag{2.23}$$

$$n_{\lambda_3} = n_{\lambda_3}^0 + n_{\lambda_3}', \; n_{\lambda_3}' = -\Psi_{\lambda_3} \frac{\partial n_{\lambda_3}^0}{\partial(\hbar\omega_{\lambda_3})} = \Psi_{\lambda_3} \cdot \frac{1}{k_{\text{B}}T} n_{\lambda_3}^0 \left(n_{\lambda_3}^0 + 1\right), \tag{2.24}$$

where Ψ measures the derivation in the phonon distribution from equilibrium, weighted with a factor that depends on the equilibrium distribution of that mode [34]. In the final step of each of equations (2.21)–(2.24), we used the fact that $\partial n^0/\partial(\hbar\omega) = -n^0(n^0 + 1)/k_{\text{B}}T$. By substituting equations (2.21)–(2.24) into equation

(2.15) and dropping the higher-order terms $O(\Psi^2)$ and $O(\Psi^3)$, the scattering term of the linearized phonon BTE is written as

$$\begin{aligned}
\left.\frac{\partial n_\lambda'}{\partial t}\right|_s = &-\sum_{\lambda_1\lambda_2}\frac{1}{k_\mathrm{B}T}\left\{\frac{1}{2}(\Psi_\lambda - \Psi_{\lambda_1} - \Psi_{\lambda_2})n_\lambda^0\left(1 + n_{\lambda_1}^0\right)\left(1 + n_{\lambda_2}^0\right)\mathcal{L}_-\right.\\
&\left.+ (\Psi_\lambda + \Psi_{\lambda_1} - \Psi_{\lambda_2})n_\lambda^0 n_{\lambda_1}^0\left(1 + n_{\lambda_2}^0\right)\mathcal{L}_+\right\}\\
&-\sum_{\lambda_1\lambda_2\lambda_3}\frac{1}{k_\mathrm{B}T}\left\{\frac{1}{6}(\Psi_\lambda - \Psi_{\lambda_1} - \Psi_{\lambda_2} - \Psi_{\lambda_3})\left(1 + n_\lambda^0\right)n_{\lambda_1}^0 n_{\lambda_2}^0 n_{\lambda_3}^0\mathcal{L}_{--}\right.\\
&+\frac{1}{2}(\Psi_\lambda + \Psi_{\lambda_1} - \Psi_{\lambda_2} - \Psi_{\lambda_3})\left(1 + n_\lambda^0\right)\left(1 + n_{\lambda_1}^0\right)n_{\lambda_2}^0 n_{\lambda_3}^0\mathcal{L}_{+-}\\
&\left.+\frac{1}{2}(\Psi_\lambda + \Psi_{\lambda_1} + \Psi_{\lambda_2} - \Psi_{\lambda_3})\left(1 + n_\lambda^0\right)\left(1 + n_{\lambda_1}^0\right)\left(1 + n_{\lambda_2}^0\right)n_{\lambda_3}^0\mathcal{L}_{++}\right\}\\
&-\sum_{\lambda_1}\frac{1}{k_\mathrm{B}T}(\Psi_\lambda - \Psi_{\lambda_1})n_\lambda^0\left(1 + n_\lambda^0\right)\mathcal{L}_\mathrm{iso} - \frac{1}{k_\mathrm{B}T}\Psi_\lambda n_\lambda^0\left(1 + n_\lambda^0\right)\frac{1}{\tau_{b,\lambda}^0}.
\end{aligned} \tag{2.25}$$

Here, we have taken advantage of the identical relations

$$\lambda \to \lambda_1 + \lambda_2 : \; n_\lambda^0\left(1 + n_{\lambda_1}^0\right)\left(1 + n_{\lambda_2}^0\right) - \left(1 + n_\lambda^0\right)n_{\lambda_1}^0 n_{\lambda_2}^0 = 0, \tag{2.26}$$

$$\lambda + \lambda_1 \to \lambda_2 : \; n_\lambda^0 n_{\lambda_1}^0\left(1 + n_{\lambda_2}^0\right) - \left(1 + n_\lambda^0\right)\left(1 + n_{\lambda_1}^0\right)n_{\lambda_2}^0 = 0, \tag{2.27}$$

$$\lambda \to \lambda_1 + \lambda_2 + \lambda_3 : \; n_\lambda^0\left(1 + n_{\lambda_1}^0\right)\left(1 + n_{\lambda_2}^0\right)\left(1 + n_{\lambda_3}^0\right) - \left(1 + n_\lambda^0\right)n_{\lambda_1}^0 n_{\lambda_2}^0 n_{\lambda_3}^0 = 0, \tag{2.28}$$

$$\lambda + \lambda_1 \to \lambda_2 + \lambda_3 : \; n_\lambda^0 n_{\lambda_1}^0\left(1 + n_{\lambda_2}^0\right)\left(1 + n_{\lambda_3}^0\right) - \left(1 + n_\lambda^0\right)\left(1 + n_{\lambda_1}^0\right)n_{\lambda_2}^0 n_{\lambda_3}^0 = 0, \tag{2.29}$$

$$\lambda + \lambda_1 + \lambda_2 \to \lambda_3 : \; n_\lambda^0 n_{\lambda_1}^0 n_{\lambda_2}^0\left(1 + n_{\lambda_3}^0\right) - \left(1 + n_\lambda^0\right)\left(1 + n_{\lambda_1}^0\right)\left(1 + n_{\lambda_2}^0\right)n_{\lambda_3}^0 = 0, \tag{2.30}$$

and

$$\begin{aligned}
\lambda \to \lambda_1 + \lambda_2 \text{: } \left(1 + n_{\lambda_1}^0\right)\left(1 + n_{\lambda_2}^0\right) - n_{\lambda_1}^0 n_{\lambda_2}^0 &= \frac{\left(1 + n_{\lambda_1}^0\right)\left(1 + n_{\lambda_2}^0\right)}{\left(1 + n_\lambda^0\right)}\\
&= \frac{n_{\lambda_1}^0 n_{\lambda_2}^0}{n_\lambda^0} = 1 + n_{\lambda_1}^0 + n_{\lambda_2}^0,
\end{aligned} \tag{2.31}$$

$$\lambda + \lambda_1 \rightarrow \lambda_2: n_{\lambda_1}^0\left(1 + n_{\lambda_2}^0\right) - \left(1 + n_{\lambda_1}^0\right)n_{\lambda_2}^0 = \frac{\left(1 + n_{\lambda_1}^0\right)n_{\lambda_2}^0}{n_\lambda^0}$$
$$= \frac{n_{\lambda_1}^0\left(1 + n_{\lambda_2}^0\right)}{1 + n_\lambda^0} = n_{\lambda_1}^0 - n_{\lambda_2}^0, \tag{2.32}$$

$$\lambda \rightarrow \lambda_1 + \lambda_2 + \lambda_3: \left(1 + n_{\lambda_1}^0\right)\left(1 + n_{\lambda_2}^0\right)\left(1 + n_{\lambda_3}^0\right) - n_{\lambda_1}^0 n_{\lambda_2}^0 n_{\lambda_3}^0 = \frac{n_{\lambda_1}^0 n_{\lambda_2}^0 n_{\lambda_3}^0}{n_\lambda^0}, \tag{2.33}$$

$$\lambda + \lambda_1 \rightarrow \lambda_2 + \lambda_3: n_{\lambda_1}^0\left(1 + n_{\lambda_2}^0\right)\left(1 + n_{\lambda_3}^0\right) - \left(1 + n_{\lambda_1}^0\right)n_{\lambda_2}^0 n_{\lambda_3}^0 = \frac{\left(1 + n_{\lambda_1}^0\right)n_{\lambda_2}^0 n_{\lambda_3}^0}{n_\lambda^0}, \tag{2.34}$$

$$\lambda + \lambda_1 + \lambda_2 \rightarrow \lambda_3: n_{\lambda_1}^0 n_{\lambda_2}^0\left(1 + n_{\lambda_3}^0\right) - \left(1 + n_{\lambda_1}^0\right)\left(1 + n_{\lambda_2}^0\right)n_{\lambda_3}^0$$
$$= \frac{\left(1 + n_{\lambda_1}^0\right)\left(1 + n_{\lambda_2}^0\right)n_{\lambda_3}^0}{n_\lambda^0}. \tag{2.35}$$

Equations (2.26–2.35) are obtained based on the energy conservation rule combined with the Bose–Einstein distribution. For example, equations (2.26) and (2.31) are derived by substituting ω of the Bose–Einstein distribution $e^{\hbar\omega/k_BT} = 1 + 1/n_\lambda^0$ into the energy conservation (selection rule) $\omega = \omega_1 + \omega_2$, giving the result $1 + 1/n_\lambda^0 = (1 + 1/n_{\lambda_1}^0)(1 + 1/n_{\lambda_2}^0)$, which directly deduces equations (2.26) and (2.31).

The final expression of the right-hand side of the original phonon BTE (equation (2.2)) is obtained by defining the form [43] of

$$\Psi = -\hbar\omega\tau\mathbf{v} \cdot \nabla T/T \tag{2.36}$$

and putting it into equation (2.25) for all the modes λ, λ_1, λ_2 and λ_3, while the left-hand side of equation (2.2) is transformed by the identical relation

$$\frac{\partial n_\lambda^0}{\partial T} = \frac{1}{T}\frac{\hbar\omega_\lambda}{k_BT}n_\lambda^0\left(n_\lambda^0 + 1\right). \tag{2.37}$$

Thus, the phonon BTE (equation (2.2)) is transformed as

$$
\begin{aligned}
1 = \sum_{\lambda_1\lambda_2}\Bigg\{ & \frac{1}{2}(\tau_\lambda - \tau_{\lambda_1}\xi_{\lambda\lambda_1} - \tau_{\lambda_2}\xi_{\lambda\lambda_2})\frac{\left(1+n_{\lambda_1}^0\right)\left(1+n_{\lambda_2}^0\right)}{1+n_\lambda^0}\mathcal{L}_- \\
& + (\tau_\lambda + \tau_{\lambda_1}\xi_{\lambda\lambda_1} - \tau_{\lambda_2}\xi_{\lambda\lambda_2})\frac{n_{\lambda_1}^0\left(1+n_{\lambda_2}^0\right)}{1+n_\lambda^0}\mathcal{L}_+\Bigg\} \\
& + \sum_{\lambda_1\lambda_2\lambda_3}\Bigg\{\frac{1}{6}(\tau_\lambda - \tau_{\lambda_1}\xi_{\lambda\lambda_1} - \tau_{\lambda_2}\xi_{\lambda\lambda_2} - \tau_{\lambda_3}\xi_{\lambda\lambda_3})\frac{n_{\lambda_1}^0 n_{\lambda_2}^0 n_{\lambda_3}^0}{n_\lambda^0}\mathcal{L}_{--} \\
& + \frac{1}{2}(\tau_\lambda + \tau_{\lambda_1}\xi_{\lambda\lambda_1} - \tau_{\lambda_2}\xi_{\lambda\lambda_2} - \tau_{\lambda_3}\xi_{\lambda\lambda_3})\frac{\left(1+n_{\lambda_1}^0\right)n_{\lambda_2}^0 n_{\lambda_3}^0}{n_\lambda^0}\mathcal{L}_{+-} \\
& + \frac{1}{2}(\tau_\lambda + \tau_{\lambda_1}\xi_{\lambda\lambda_1} + \tau_{\lambda_2}\xi_{\lambda\lambda_2} - \tau_{\lambda_3}\xi_{\lambda\lambda_3})\frac{\left(1+n_{\lambda_1}^0\right)\left(1+n_{\lambda_2}^0\right)n_{\lambda_3}^0}{n_\lambda^0}\mathcal{L}_{++}\Bigg\} \\
& + \sum_{\lambda_1}(\tau_\lambda - \tau_{\lambda_1}\xi_{\lambda\lambda_1})\mathcal{L}_{\rm iso} + \frac{\tau_\lambda}{\tau_{b,\lambda}^0},
\end{aligned}
$$

and further as

$$\tau_\lambda = \tau_\lambda^0(1 + \Xi_{3,\lambda} + \Xi_{4,\lambda} + \Xi_{{\rm iso},\lambda}), \tag{2.38}$$

with

$$\frac{1}{\tau_\lambda^0} = \frac{1}{\tau_{3,\lambda}^0} + \frac{1}{\tau_{4,\lambda}^0} + \frac{1}{\tau_{{\rm iso},\lambda}^0} + \frac{1}{\tau_{b,\lambda}^0}, \tag{2.39}$$

$$\frac{1}{\tau_{3,\lambda}^0} = \sum_{\lambda_1\lambda_2}\left\{\frac{1}{2}\left(1 + n_{\lambda_1}^0 + n_{\lambda_2}^0\right)\mathcal{L}_- + \left(n_{\lambda_1}^0 - n_{\lambda_2}^0\right)\mathcal{L}_+\right\}, \tag{2.40}$$

$$
\begin{aligned}
\frac{1}{\tau_{4,\lambda}^0} = \sum_{\lambda_1\lambda_2\lambda_3}\Bigg\{ & \frac{1}{6}\frac{n_{\lambda_1}^0 n_{\lambda_2}^0 n_{\lambda_3}^0}{n_\lambda^0}\mathcal{L}_{--} + \frac{1}{2}\frac{\left(1+n_{\lambda_1}^0\right)n_{\lambda_2}^0 n_{\lambda_3}^0}{n_\lambda^0}\mathcal{L}_{+-} \\
& + \frac{1}{2}\frac{\left(1+n_{\lambda_1}^0\right)\left(1+n_{\lambda_2}^0\right)n_{\lambda_3}^0}{n_\lambda^0}\mathcal{L}_{++}\Bigg\},
\end{aligned} \tag{2.41}
$$

$$\frac{1}{\tau_{{\rm iso},\lambda}^0} = \sum_{\lambda_1}\mathcal{L}_{\rm iso}, \tag{2.42}$$

$$\frac{1}{\tau_{b,\lambda}^0} = \frac{2|v_{\lambda,x}|}{L} + \frac{2|v_{\lambda,y}|}{W}\frac{1-p}{1+p}. \tag{2.43}$$

$$\Xi_{3,\lambda} = \sum_{\lambda_1\lambda_2}\left\{\frac{1}{2}(\tau_{\lambda_1}\xi_{\lambda\lambda_1} + \tau_{\lambda_2}\xi_{\lambda\lambda_2})\left(1 + n_{\lambda_1}^0 + n_{\lambda_2}^0\right)\mathcal{L}_- \right. \\ \left. + (\tau_{\lambda_2}\xi_{\lambda\lambda_2} - \tau_{\lambda_1}\xi_{\lambda\lambda_1})\left(n_{\lambda_1}^0 - n_{\lambda_2}^0\right)\mathcal{L}_+\right\}, \tag{2.44}$$

$$\Xi_{4,\lambda} = \sum_{\lambda_1\lambda_2\lambda_3}\left\{\frac{1}{6}(\tau_{\lambda_1}\xi_{\lambda\lambda_1} + \tau_{\lambda_2}\xi_{\lambda\lambda_2} + \tau_{\lambda_3}\xi_{\lambda\lambda_3})\frac{n_{\lambda_1}^0 n_{\lambda_2}^0 n_{\lambda_3}^0}{n_\lambda^0}\mathcal{L}_{--} \right. \\ + \frac{1}{2}(\tau_{\lambda_2}\xi_{\lambda\lambda_2} + \tau_{\lambda_3}\xi_{\lambda\lambda_3} - \tau_{\lambda_1}\xi_{\lambda\lambda_1})\frac{\left(1 + n_{\lambda_1}^0\right)n_{\lambda_2}^0 n_{\lambda_3}^0}{n_\lambda^0}\mathcal{L}_{+-} \\ \left. + \frac{1}{2}(\tau_{\lambda_3}\xi_{\lambda\lambda_3} - \tau_{\lambda_1}\xi_{\lambda\lambda_1} - \tau_{\lambda_2}\xi_{\lambda\lambda_2})\frac{\left(1 + n_{\lambda_1}^0\right)\left(1 + n_{\lambda_2}^0\right)n_{\lambda_3}^0}{n_\lambda^0}\mathcal{L}_{++}\right\}, \tag{2.45}$$

$$\Xi_{\text{iso},\lambda} = \sum_{\lambda_1}\tau_{\lambda_1}\xi_{\lambda\lambda_1}\mathcal{L}_{\text{iso}}, \tag{2.46}$$

$$\xi_{\lambda\lambda_1} \equiv \frac{\omega_{\lambda_1}\mathbf{v}_{\lambda_1}\cdot\nabla T}{\omega_\lambda\ \mathbf{v}_\lambda\cdot\nabla T} = \frac{\omega_{\lambda_1}v_{\lambda_1 x}}{\omega_\lambda v_{\lambda x}}, \tag{2.47}$$

$$\xi_{\lambda\lambda_2} \equiv \frac{\omega_{\lambda_2}\mathbf{v}_{\lambda_2}\cdot\nabla T}{\omega_\lambda\ \mathbf{v}_\lambda\cdot\nabla T} = \frac{\omega_{\lambda_2}v_{\lambda_2 x}}{\omega_\lambda v_{\lambda x}}, \tag{2.48}$$

$$\xi_{\lambda\lambda_3} \equiv \frac{\omega_{\lambda_3}\mathbf{v}_{\lambda_3}\cdot\nabla T}{\omega_\lambda\ \mathbf{v}_\lambda\cdot\nabla T} = \frac{\omega_{\lambda_3}v_{\lambda_3 x}}{\omega_\lambda v_{\lambda x}}. \tag{2.49}$$

In summary, τ_λ is obtained by solving equation (2.38), with equations (2.16)–(2.20) and (2.39)–(2.48). Since both the left- and right-hand sides contain the unknown τ_λ, equation (2.38) is solved iteratively and thus is also called the iterative scheme. τ_λ^0 in equation (2.39) is the phonon relaxation time based on SMRTA [26]. In the boundary scattering term, L and W represent the length (along the heat flow direction) and the width (perpendicular to the heat flow direction) of the material. $0 \leqslant p \leqslant 1$ is the specularity parameter with $p = 0$ indicating an extremely rough surface and $p = 1$ indicating a mirror-like surface.

Generally, four-phonon scattering is dominated by the Umklapp processes and, therefore, the SMRTA τ_λ^0 in equation (2.39) is accurate enough to account for the four-phonon scattering in general materials. In other words, to save time it is not necessary to take into account the iteration of $\Xi_{4,\lambda}$ in equation (2.38). However, in some materials such as graphene, the four-phonon scattering is dominated by normal processes, which lead to a collective behavior of phonons. In this case, $\Xi_{4,\lambda}$ cannot be neglected in the iteration in equation (2.38). Such a phenomenon results from the fact that N-scattering itself does not contribute to thermal resistance since it conserves momentum.

Attention should be paid to the usage of eigenvectors **e** and the phases in the exponential terms in equations (2.19) and (2.20). In these equations, we assume that the eigenvectors are obtained by solving the dynamical matrix

$$D_{\alpha\alpha_1}^{bb_1}(\mathbf{q}) = \frac{1}{\sqrt{m_b m_{b_1}}} \sum_{l_1} \Phi_{0b,l_1b_1}^{\alpha\alpha_1} e^{i\mathbf{q}\cdot\mathbf{r}_{l_1}}, \tag{2.50}$$

which uses the positions of cells instead of atoms. However, if the eigenvectors are obtained by solving the dynamical matrix using

$$D_{\alpha\alpha_1}^{bb_1}(\mathbf{q}) = \frac{1}{\sqrt{m_b m_{b_1}}} \sum_{l_1} \Phi_{0b,l_1b_1}^{\alpha\alpha_1} e^{i\mathbf{q}\cdot(\mathbf{r}_{l_1b_1} - \mathbf{r}_{0b})}, \tag{2.51}$$

as implemented in Phonopy [44], one should use the positions of atoms ($\mathbf{r}_{lb}$) instead of the positions of cells ($\mathbf{r}_l$) in the phases of the exponential terms in equations (2.19) and (2.20), which should read

$$e^{\pm i\mathbf{q}_1\cdot\mathbf{r}_{l_1b_1} - i\mathbf{q}_2\cdot\mathbf{r}_{l_2b_2}} \tag{2.52}$$

for three-phonon scattering and

$$e^{\pm i\mathbf{q}_1\cdot\mathbf{r}_{l_1b_1} \pm i\mathbf{q}_2\cdot\mathbf{r}_{l_2b_2} - i\mathbf{q}_3\cdot\mathbf{r}_{l_3b_3}} \tag{2.53}$$

for four-phonon scattering.

2.3 Strong four-phonon scattering potential

In this section, we demonstrate the significance of four-phonon scattering originating from strong scattering potential, i.e. strong anharmonicity. Figure 2.2 shows the sketches of the potential wells of weakly and strongly anharmonic materials. The exact potential energy is decomposed into the Tylér series to the second, third and fourth orders. The second-order expansion is harmonic and has large derivation from the exact potential. Since anharmonicity basically increases with increasing temperature, such derivation becomes large at high temperaures. The third-order (anharmonic) correction could alleviate such a derivation but is not adequate at high temperatures, while the fourth-order (anharmonic) correction brings the potential well much closer to exactly one. For weakly anharmonic materials, the fourth-order correction is negligible at low temperatures while it becomes significant at elevated temperatures. For strongly anharmonic materials, even starting from a low temperature, the fourth-order correction is large (while at high temperatures the fifth-order correction seems non-negligible). Note that 'high' or 'low' temperature is a relative quantity, depending on the Debye temperature (T_D) of a material. For example, 80 K is 'high' for solid argon ($T_D \approx 80$ K), while 300 K is 'low' for silicon ($T_D \approx 640$ K) and diamond ($T_D \approx 2220$ K).

Therefore, this section consists of two parts, demonstrating the two categories of materials with strong four-phonon scattering potential, i.e. weakly anharmonic

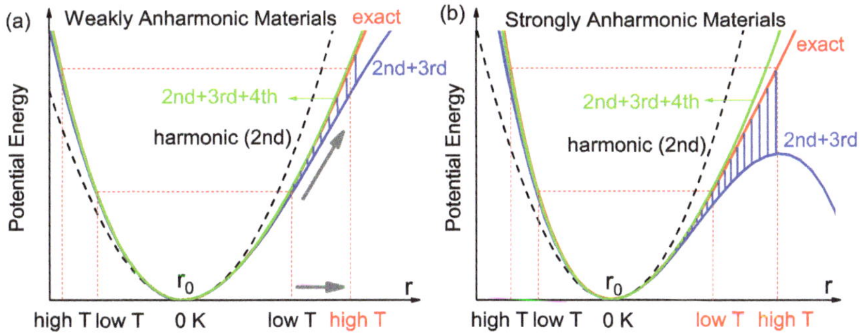

Figure 2.2. Sketches of the interatomic potentials of (a) weakly and (b) strongly anharmonic materials near the equilibrium position r_0. Shown are the exact potential (red solid curve), harmonic approximation (black dashed curve), third-order approximation (blue solid curve) and fourth-order approximation (green solid curve). The difference between the third-order approximation and the exact potential is marked by blue solid lines. The atomic vibration range is represented by temperature, i.e. atoms deviate from the equilibrium position more at higher temperatures. For weakly anharmonic materials, the fourth-order correction becomes important at elevated temperatures, while for strongly anharmonic materials, it is significant starting from low temperature.

materials at high temperatures (section 2.3.1) and strongly anharmonic materials even at low temperatures (section 2.3.2). For the first category, we take diamond, silicon and germanium as examples, which have high RT thermal conductivity. For the second category, we use solid argon, PbTe and NaCl as examples.

2.3.1 High temperature

The first category with strong four-phonon scattering potential is found in general solids at high temperatures, even with high RT thermal conductivity, such as diamond, silicon and germanium. As a starting point, classical potentials [45, 46] are used to calculate the fourth-order force constants as well as the four-phonon scattering rates [26]. The accuracy of four-phonon calculations is examined by comparing the thermal conductivities obtained from the three-/four-phonon scattering to those obtained from Green–Kubo MD simulations.

In figure 2.3, the three- and four-phonon scattering rates, $\tau_{3,\lambda}^{-1}$ and $\tau_{4,\lambda}^{-1}$, of diamond, Si and Ge are shown as a function of temperature. Far below their Debye temperatures, 2220, 640 and 374 K for diamond, Si and Ge, respectively, $\tau_{4,\lambda}^{-1}$ is generally negligible. However, $\tau_{4,\lambda}^{-1}$ increases faster than $\tau_{3,\lambda}^{-1}$ with increasing temperature and is no longer negligible when the temperature is high. Even for diamond, the most harmonic material with stiff bonds, it can be speculated from the trend that four-phonon scattering could be important when the temperature is close to its Debye temperature. We also noted that four-phonon scattering is relatively more important for optical phonons than acoustic phonons. This will be discussed in the following sections. The scaling laws $\tau_{3,\lambda}^{-1} \sim T$ and $\tau_{4,\lambda}^{-1} \sim T^2$ are found to be valid for both acoustic and optical phonons for all materials. These temperature dependences

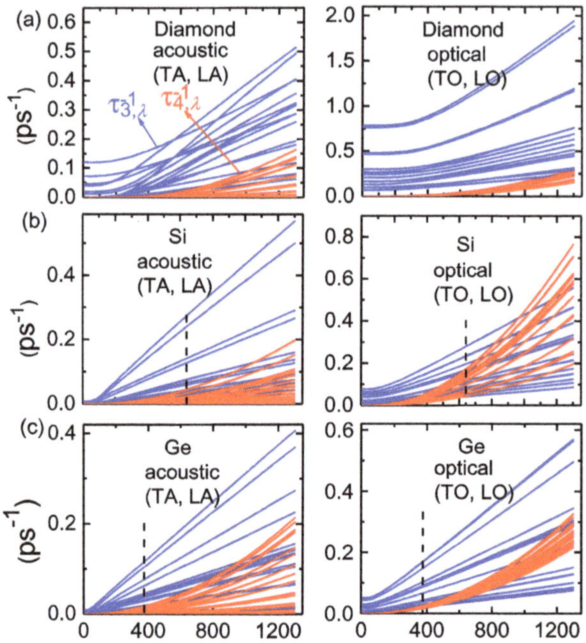

Figure 2.3. Temperature-dependent $\tau_{3,\lambda}^{-1}$ (blue) and $\tau_{4,\lambda}^{-1}$ (red) of the eight evenly sampled **q** points from Γ to X in diamond, silicon and germanium. Each curve represents an individual mode (a branch of a sampled **q** point). The dashed lines label Debye temperatures. The force constants are obtained from classical Tersoff potentials. Reproduced with permission from [26]. Copyright 2016 the American Physical Society.

result from equation (2.39), which roughly indicates $\tau_{3,\lambda}^{-1} \sim n^0$ and $\tau_{4,\lambda}^{-1} \sim (n^0)^2$, leading to $\tau_{3,\lambda}^{-1} \sim T$ and $\tau_{4,\lambda}^{-1} \sim T^2$ since n^0 is proportional to T at high temperatures.

The impact of four-phonon scattering on thermal conductivity is demonstrated by comparing

$$\kappa_{3,\mathrm{RTA},z} = \frac{1}{V}\sum_{\lambda} v_{z,\lambda}^2 c_\lambda \tau_{3,\lambda} \tag{2.54}$$

and

$$\kappa_{3+4,\mathrm{RTA},z} = \frac{1}{V}\sum_{\lambda} v_{z,\lambda}^2 c_\lambda \left(\tau_{3,\lambda}^{-1} + \tau_{4,\lambda}^{-1}\right)^{-1} \tag{2.55}$$

as shown in figures 2.4 and 2.8. For diamond, Si and Ge, κ_3 and κ_{3+4} match well with each other at low temperatures, indicating that four-phonon scattering is negligible. At room temperature, κ_{3+4} is lower than κ_3 by 1%, 8% and 15% for diamond, Si and Ge, respectively, as shown in the inset in figure 2.4. As the temperature increases to 1000 K, this discrepancy grows to 15%, 25% and 36%, respectively. Such results, again, indicate that even in weakly anharmonic materials, four-phonon scattering may play a critical role at high temperatures.

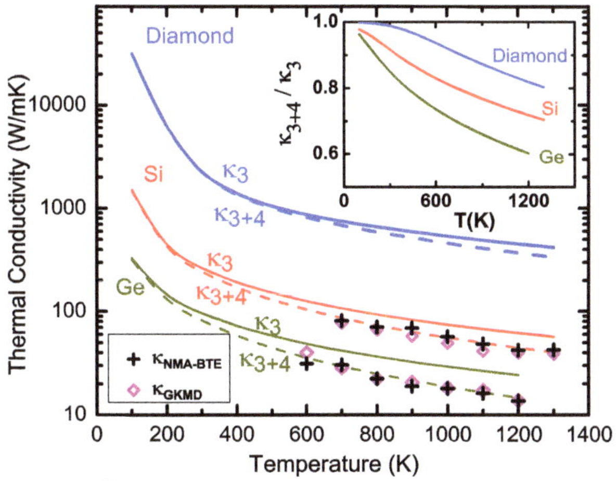

Figure 2.4. The comparison among $\kappa_{3,\mathrm{RTA}}$, $\kappa_{3+4,\mathrm{RTA}}$, $\kappa_{\mathrm{NMA,RTA}}$ and $\kappa_{\mathrm{Green-Kubo,RTA}}$ for diamond, silicon and germanium predicted from classical Tersoff potentials. The inset shows the ratio $\kappa_{3+4,\mathrm{RTA}}/\kappa_{3,\mathrm{RTA}}$. Reproduced with permission from [26]. Copyright 2016 the American Physical Society.

The accuracy of the four-phonon scattering calculation is examined by MD simulations, which naturally include all the orders of anharmonicities. κ_3 and κ_{3+4} are compared to κ_{NMA}, which is calculated by the BTE using the linewidth $\tau_{\mathrm{NMA},\lambda}^{-1}$ obtained from normal mode analysis (NMA) based on MD. They are also compared to $\kappa_{\mathrm{GK(MD)}}$ that is directly obtained from the Green–Kubo formalism based on MD. A good agreement between κ_{3+4} and κ_{NMA} as well as $\kappa_{\mathrm{GK(MD)}}$ is found for Si and Ge in figure 2.4. The comparison in diamond is not done since diamond has a high Debye temperature, below which κ_{3+4} obtained from quantum mechanics is not comparable to κ_{NMA} and $\kappa_{\mathrm{GK(MD)}}$ from classical MD. In contrast, κ_3 is considerably over-predicted, particularly at high temperatures. For clearer insight, we plot the ratio of κ_{3+4}/κ_3 as a function of temperature in the insets. Since we use empirical interatomic potentials that are approximations to the true atomic interactions, the numbers presented here should be understood with caution or on a semi-quantitative basis.

ALD and MD were regarded as two different methods. The phonon scattering rates calculated from the former and the phonon linewidths calculated from the latter were regarded as two separate quantities with their quantitative agreement remaining missing for a long time [25]. With the aid of four-phonon scattering, these two quantities match, and the two methods become consistent with each other. The four-phonon calculation pushes a step forward towards the 'unification' of the simulation methods of phonon and thermal transport.

To have a quantitative comparison between the four-phonon calculation and experiment, it is necessary to use DFT instead of empirical potentials for the calculation of force constants. The calculations were performed for diamond and silicon. While diamond will be shown in section 2.4.1 for special purposes, here we take silicon for demonstration. The three- and four-phonon scattering rates for all

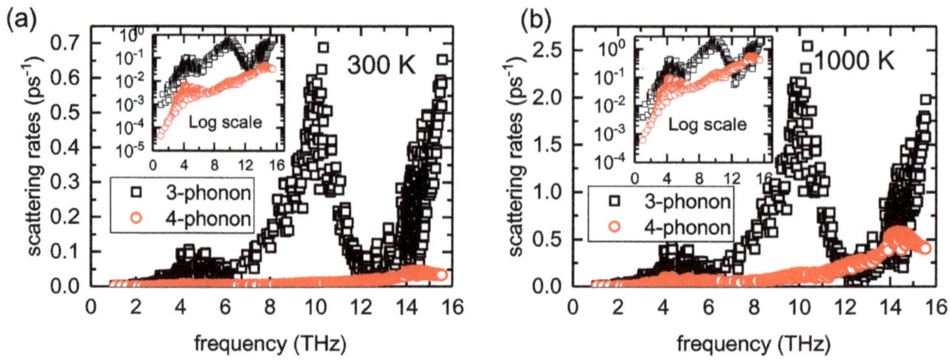

Figure 2.5. First-principles three-phonon (black squares) and four-phonon (red circles) scattering rates of Si at 300 and 1000 K with 16 × 16 × 16 phonon **q** point grids. The insets are in log–linear scales to give a better view of the low-frequency regions. Reproduced with permission from [27]. Copyright 2017 the American Physical Society.

modes throughout the Brillouin zone of Si calculated from DFT are shown in figure 2.5 at 300 and 1000 K. The insets show the low-frequency behavior. We note that the four-phonon scattering rates increase quadratically with temperature, while three-phonon scattering rates increase linearly, not shown here. The trend is consistent with the preceding discussions based on classical potentials. At 300 K, τ_4^{-1} is well below τ_3^{-1} in Si throughout the frequency domain. As T increases to 1000 K, the four-phonon rates of the low-frequency phonons remain insubstantial; however, higher-energy longitudinal acoustic (LA) modes and all the optical modes exhibit large τ_4^{-1}, comparable to τ_3^{-1}. The large τ_4^{-1} of the heat-carrying LA phonons will have a substantial effect on the thermal conductivity of these materials, and that of the optical modes can affect infrared optical properties [19].

The thermal conductivity of silicon is calculated by solving the iterative phonon BTE beyond the RTA. Due to the high computational cost, the four-phonon scattering rates are computed at the RTA level only and inserted into the iterative scheme that determines the nonequilibrium phonon distributions from mixing of the three-phonon processes. This is similar to employing phonon–isotope and phonon–boundary scattering terms in the full BTE solution [9, 11, 47–51]. We will show that such an approximation is likely valid as the four-phonon scattering is dominated by Umklapp processes in section 2.5.2. We also include phonon–isotope scattering [37] in these κ calculations of naturally occurring materials. The iterative solution to the BTE for κ of the naturally occurring Si is shown in figure 2.6. The three-phonon predictions agree well with measured data at low temperature (<600 K); however, significant deviations from experiment occur at high temperatures. For example, at 1000 K three-phonon resistance alone over-predicts the measured κ of silicon by 26%. In section 2.4.1, we will see that the over-prediction is also around 30% for diamond at 1000 K. Four-phonon scattering eliminates such discrepancies and brings the prediction to match well with experimental values. These examples demonstrate the significance of four-phonon scattering due to a strong scattering potential at high temperatures.

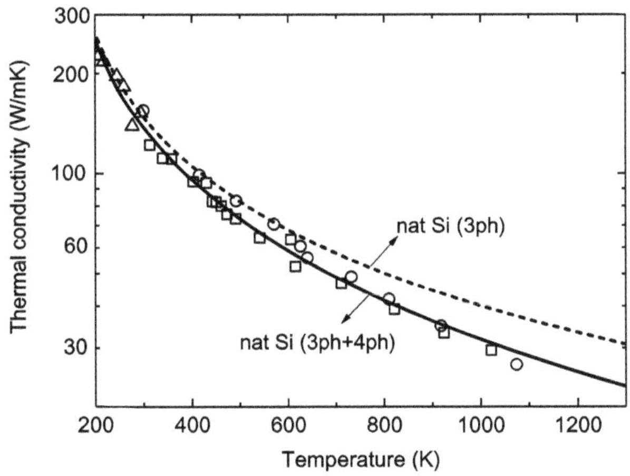

Figure 2.6. Thermal conductivities of naturally occurring Si. Dashed lines give first-principles calculated κ_3, while solid lines give κ_{3+4}. Calculation data are taken from [27]. Symbols represent measured data from: [52] triangles; [53] squares; and [54] circles.

2.3.2 Strongly anharmonic materials

For materials with low thermal conductivity, the four-phonon scattering potential is high even at ordinary temperature. We take Lennard-Jones argon as the benchmark material and then move onto more practically important materials, PbTe and NaCl. The scattering rates and thermal conductivity of argon are shown in figures 2.7 and 2.8, respectively. The summation of three- and four-phonon rates $\tau_{3,\lambda}^{-1} + \tau_{4,\lambda}^{-1}$ agrees well with MD results, $\tau_{\mathrm{NMA},\lambda}^{-1}$. It is seen that four-phonon is non-negligible even far below its Debye temperature and that it reduces the κ of argon by 35%–65% from 20 to 80 K.

PbTe and NaCl are two strongly anharmonic materials, in which the four-phonon scattering potential is high even at RT [32, 33]. It is also found that for those materials the phonon frequency shift at finite temperature is large and would affect the phonon scattering rates. Phonon modes at finite temperature are softened by anharmonicity via anharmonic three- and four-phonon scatterings as well as thermal expansion [55]. The phonon scattering, phonon frequency and thermal expansion are coupled together. Phonon scattering affects thermal expansion and changes phonon frequency, while thermal expansion also shifts phonon frequency, which in turn affects phonon scattering.

Strongly anharmonic materials generally have strong four-phonon scattering and, at the same time, their phonon frequencies at finite temperatures shift considerably away from those at 0 K due to the strong anharmonicity, which typically induces large thermal expansion as well as bond softening. The steps to reasonably include most anharmonic effects for thermal conductivity prediction with the current computational power are described in [32, 33] and are briefly summarized as follows. Step 1: Determine the correct temperature-dependent lattice constant, i.e. thermal expansion coefficient, by using quasiharmonic approximation (QHA)

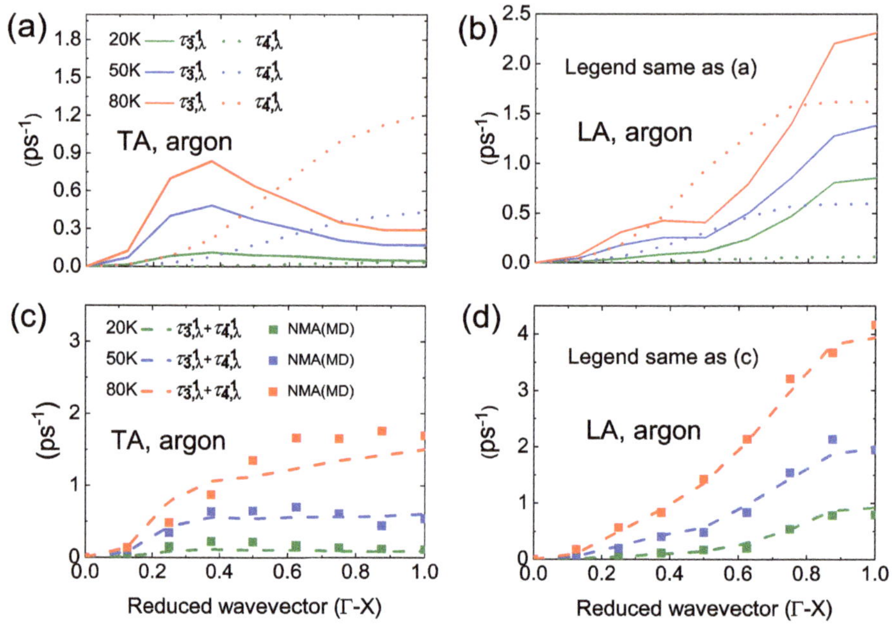

Figure 2.7. (a) and (b) A comparison between three- and four-phonon scattering rates for Lennard-Jones argon. (c) and (d) The total scattering rates ($\tau_{3,\lambda}^{-1} + \tau_{4,\lambda}^{-1}$) compared to the linewidths obtained from NMA based on MD simulations. Reproduced with permission from [26]. Copyright 2016 the American Physical Society.

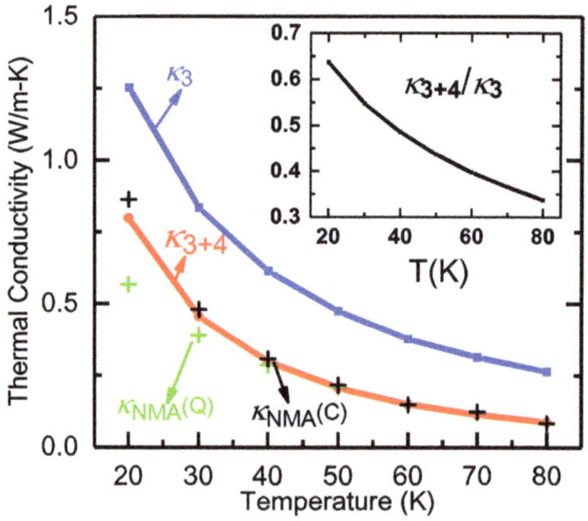

Figure 2.8. The κ values of Lennard-Jones argon predicted from $\tau_{3,\lambda}^{-1}$, $\tau_{3,\lambda}^{-1} + \tau_{4,\lambda}^{-1}$ and $\tau_{\mathrm{NMA},\lambda}^{-1}$ as a function of temperature, with the inset showing the ratio of κ_{3+4}/κ_3. $\kappa_{\mathrm{NMA}}(\mathrm{Q})$ and $\kappa_{\mathrm{NMA}}(\mathrm{C})$ represent that the specific heat is calculated by the quantum (Bose–Einstein) and classical (Boltzmann) phonon distributions, respectively. The phonon dispersion used in the calculation of κ_{NMA} is from lattice dynamics (LD) calculation at 0 K, to be consistent with the κ_3 and κ_{3+4} calculations. Reproduced with permission from [26]. Copyright 2016 the American Physical Society.

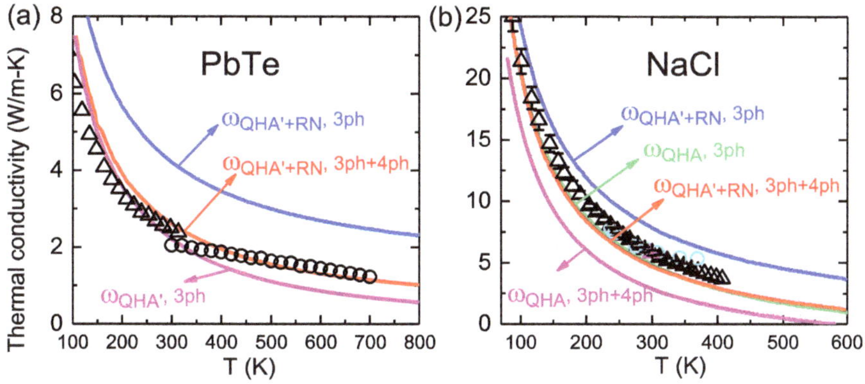

Figure 2.9. The thermal conductivities of PbTe and NaCl calculated from DFT (solid curves) in comparison to experimental data (symbols). First-principles data are taken from [32] for PbTe and [33] for NaCl. The experimental data are taken from [56, 57] for PbTe and [58–60] for NaCl.

together with phonon renormalization (RN), hereafter referred to as QHA′. Step 2: Determine the harmonic phonon frequency $\omega_{QHA'}$ at the temperature-dependent lattice constant for a given temperature. Step 3: Calculate the anharmonic frequency $\omega_{QHA'+RN}$ by taking into account phonon renormalization. Step 4: Calculate the three- and four-phonon scattering rates as well as thermal conductivity based on the anharmonic frequency $\omega_{QHA'+RN}$. Steps 3 and 4 could be performed at the same time by calculating the real and imaginary parts of the phonon self-energy [2, 26, 32, 55] to avoid the double counting of phonon anharmonicity.

Figure 2.9 shows the thermal conductivities of PbTe and NaCl calculated by considering different degrees of anharmonicity as compared to experimental values. It is found that the three-phonon scattering without phonon renormalization can somehow predict the thermal conductivities well. After the phonon renormalization is included, the predicted thermal conductivities appear far above experimental values. If four-phonon scattering is considered, it pulls back the thermal conductivity prediction to agree well with experiments. Therefore, the agreement between the prediction and experimental values achieved by three-phonon scattering calculations without phonon renormalization or four-phonon scattering is a coincidence. The reason is that phonon renormalization largely shrinks the phonon scattering phase space while the inclusion of four-phonon scattering increases the scattering phase space, and they cancel each other for PbTe and NaCl. When the phonon renormalization effect and the four-phonon scattering are both considered, the thermal conductivity prediction reaches the best accuracy. At this moment, it is still an open question whether this phenomenon appears in PbTe and NaCl only or in a broader scope of strongly anharmonic materials.

2.4 Large four-phonon or suppressed three-phonon phase space

Apart from a strong scattering potential, the other origin of strong four-phonon scattering is the large scattering phase space. In this section, we will discuss such groups of materials, including the materials with large acoustic–optical phonon band gaps and two-dimensional materials with reflection symmetry. We will also particularly discuss optical phonon modes, which usually have a large density of states and, therefore, large four-phonon scattering phase space.

2.4.1 Materials with large acoustic–optical phonon band gaps

When a material is composed of two elements with a large atomic mass ratio, the phonon vibration spectrum often presents a band gap. For example, the III–V or IV–IV binary zinc-blende BAs, SiC, AlAs, AlSb, GaN, GaP, InAs and InP all show a band gap, as shown in figure 2.10. Lindsay *et al* [11] found that phonon band gaps limited the three-phonon scattering phase space since the energy summation of two lower-branch phonons could hardly reach the high energy of optical phonon branches. As a result, the thermal conductivity predicted by three-phonon scattering

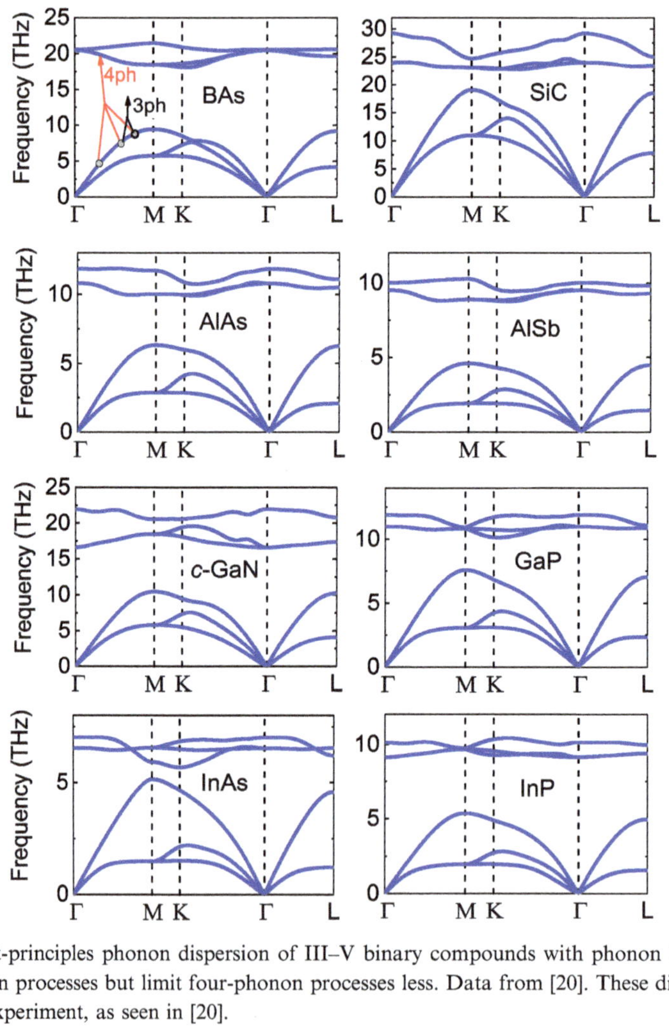

Figure 2.10. First-principles phonon dispersion of III–V binary compounds with phonon band gaps, which limit three-phonon processes but limit four-phonon processes less. Data from [20]. These dispersion relations agree well with experiment, as seen in [20].

for these materials is often very high. However, Feng *et al* [27] found that such band gaps do not restrict four-phonon scattering significantly since three lower-branch phonons are easily combined into a high-branch phonon mode, allowed by the energy conservation law, as shown in figure 2.10. As a consequence, four-phonon scattering in these materials is of significance to their thermal conductivity predictions.

A predominant example is BAs, which has been predicted using three-phonon scattering to own a thermal conductivity of 2 200 W $(m\ K)^{-1}$ at room temperature [11], being comparable to diamond, the highest in nature. Given such high thermal conductivity, it potentially opens an opportunity for thermal management of electronic devices.

However, Feng *et al* [27] found that four-phonon scattering is strong in BAs. In figure 2.11, the first-principles three- and four-phonon scattering rates of BAs are compared to those of diamond. The relative importance of four-phonon scattering in these two materials shows a clear difference. At room temperature, diamond shows negligible τ_4^{-1} throughout the whole spectrum, while BAs present non-negligible τ_4^{-1} for higher-branch acoustic phonons as well as all optical phonons. For example, three-phonon scattering rates have a deep valley at around 21 THz, i.e. for the optical phonon at the Γ point. These modes with high energy and small

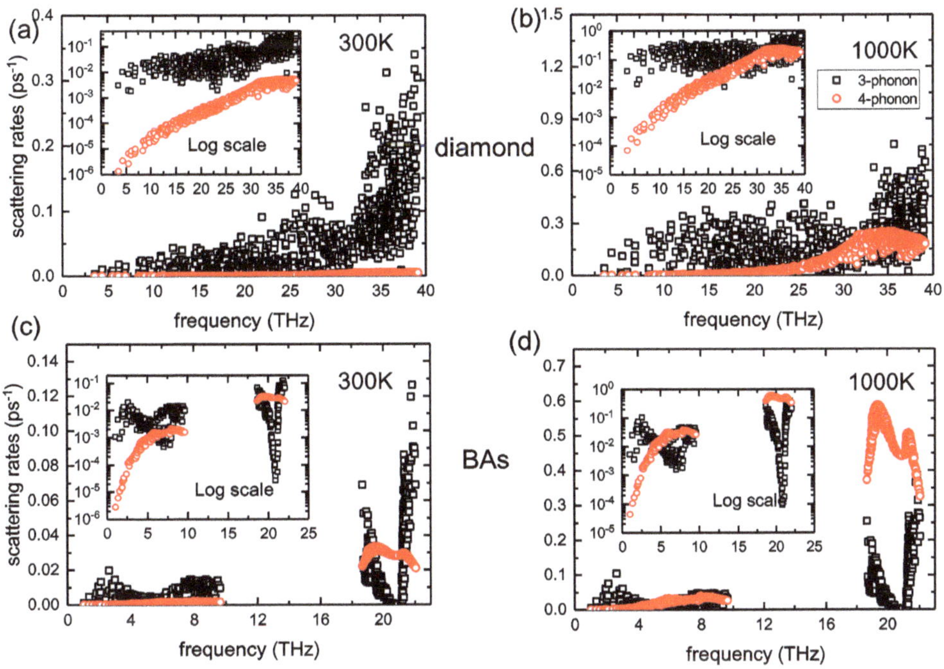

Figure 2.11. First-principles three-phonon (black squares) and four-phonon (red circles) scattering rates of diamond and BAs at 300 and 1000 K with 16 × 16 × 16 phonon **q** point grids. The insets are in log–linear scale to give a better view of the low-frequency regions. Reproduced with permission from [27]. Copyright 2017 the American Physical Society.

momentum can hardly find two other phonon modes that satisfy energy conservation and momentum conservation simultaneously. Such a large phonon band gap, however, does not forbid three acoustic phonons from combining into an optical mode. The predicted thermal conductivity of BAs after including four-phonon scattering at room temperature is reduced significantly, from ~2200 W(m K)$^{-1}$ to ~1400 W (m K)$^{-1}$, as shown in figure 2.12. Such reduction grows with increasing temperature, and the temperature-scaling trend at elevated T changes from $\kappa_3 \sim T^{-0.84}$ to $\kappa_{34} \sim T^{-1.64}$.

In 2018, several experimental works [28–30] grew high-purity BAs single crystals and verified the predicted thermal conductivity of BAs by including four-phonon scattering. As shown in figure 2.12, both the thermal conductivity values and temperature dependence measured from experiments agree well with the predictions. Without four-phonon scattering, one can fit the thermal conductivity to match the experiment at a certain temperature by adding phonon–defect scattering [29], which, however, cannot reproduce a correct temperature-scaling trend. Therefore, it is safe to conclude that the lower thermal conductivity of experimental samples compared to κ_3 is due to the strong four-phonon scattering, rather than defects. Later, the four-phonon scattering in boron phosphide (BP) was also verified in an experiment by Cahill *et al* [65].

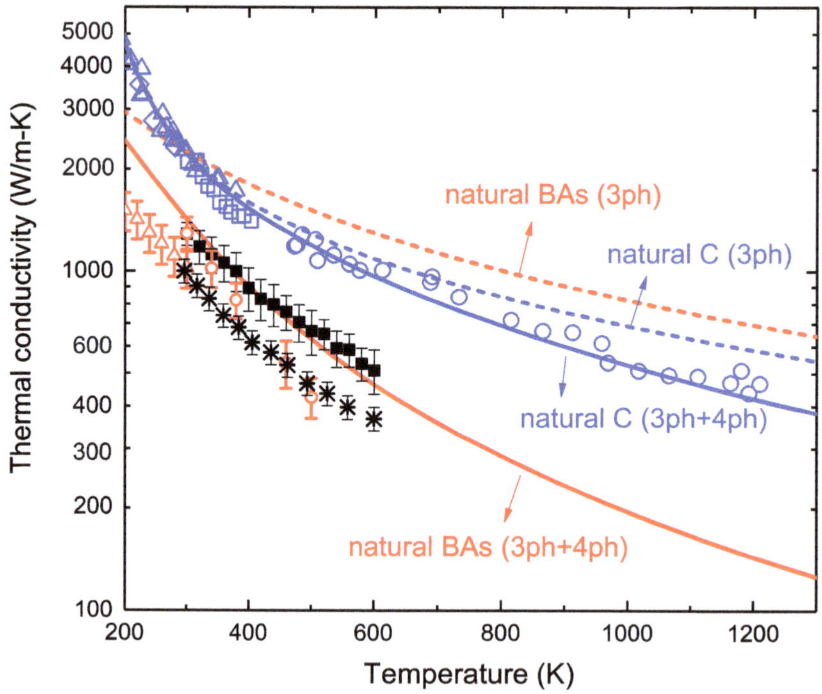

Figure 2.12. Thermal conductivities of naturally occurring diamond and BAs. Dashed lines give calculated κ_3, while solid lines give κ_{3+4}. First-principles data (curves) are taken from [27]. The symbols represent measured data. For diamond: blue triangles [61], blue squares [62], blue circles [63], blue squares [64]; for BAs: black squares [29], black stars [30], red circles [28] and red triangles [28].

In diamond, which does not present a phonon band gap, four-phonon scattering is not as strong as in BAs but is certainly not negligible at high temperatures. As T increases to 1000 K, four-phonon rates of the low-frequency phonons remain insubstantial, however, higher-energy LA modes and all the optical modes exhibit large τ_4^{-1}, comparable to τ_3^{-1} leading to a 23% reduction to the thermal conductivity.

The other materials with phonon band gaps may also show a large reduction of thermal conductivity with four-phonon scattering. Figure 2.13 shows another example, cubic GaN (*c*-GaN), in which the four-phonon scattering brings the thermal conductivity prediction down by a considerable amount to agree well with experiment.

Yang *et al* [66] found that for some materials in which optical branches have long three-phonon lifetimes, e.g. AlSb, four-phonon scattering is even more critical than three-phonon scattering as it diminishes optical phonon thermal transport, and therefore significantly reduces the thermal conductivities. Also, they showed that four-phonon scattering can play an extremely important role in weakening the isotope effect on κ. Specifically, four-phonon scattering reduces the room-temperature κ of the isotopically pure and naturally occurring AlSb by 70% and 50%, respectively (figure 2.14). The reduction for isotopically pure and naturally occurring *c*-GaN is about 34% and 27%, respectively. For isotopically pure wurtzite GaN (*w*-GaN), the reduction is about 13% at room temperature and 25% at 400 K. These results provided important guidance for experimentalists for achieving high thermal conductivities in III–V compounds for applications in semiconductor industry.

2.4.2 Optical phonons

The linewidths of infrared-active zone-center phonons are also important for the infrared dielectric functions of polar materials, which are key for applications in sensing, radiative cooling, energy harvesting, metamaterials, etc. For example, in polar compound semiconductors such as BAs, BN and SiC crystals, the zone-center longitudinal optical (LO) phonon lifetime plays an essential role in mediating the

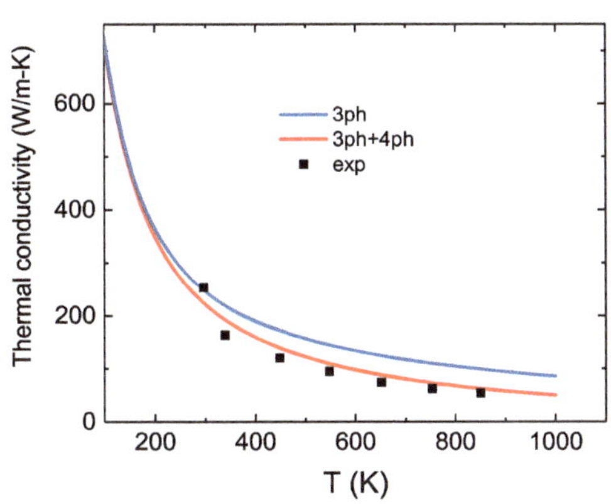

Figure 2.13. Thermal conductivity of *c*-GaN. First-principles data (solid curves) are taken from [66]. The experiment values (black squares) are taken from [67].

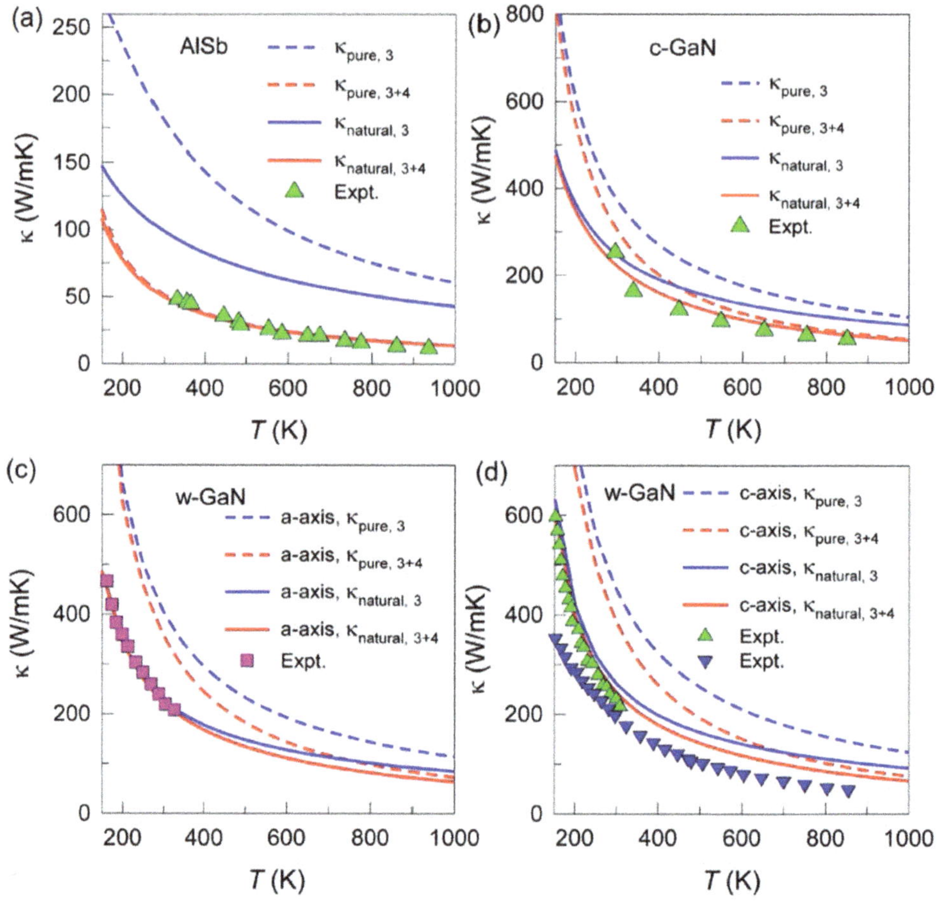

Figure 2.14. Lattice thermal conductivity as a function of temperature for (a) AlSb, (b) cubic GaN and wurtzite GaN along the (c) in-plane and (d) through-plane directions. Dashed lines represent the calculated isotopically pure κ_{pure} and solid lines represent the calculated naturally occurring κ_{natural}. The blue lines give the calculated κ with only three-phonon scattering and red lines give the results after including four-phonon scattering. All symbols represent experimental data for naturally occurring materials, which can be found in [66].

energy exchanges between the hot electrons and the lattice through Fröhlich interaction [68]. The materials such as GaN, GaAs and AlAs are promising candidates for use in optoelectronic devices and electronics. In addition, optical phonon scattering is also crucial to the thermal conductivity of certain materials that have a large number of optical phonon branches.

Optical phonons often exhibit relatively stronger four-phonon scattering compared to acoustic phonons as shown in the preceding sections. The reason is that optical phonons typically have a much higher density of states. Such a large amount of modes crowded within a narrow frequency range enabled large four-phonon scattering phase space for the process $\lambda_1 + \lambda_2 \rightarrow \lambda_3 + \lambda_4$ with $\lambda_{1,2,3,4}$ having similar

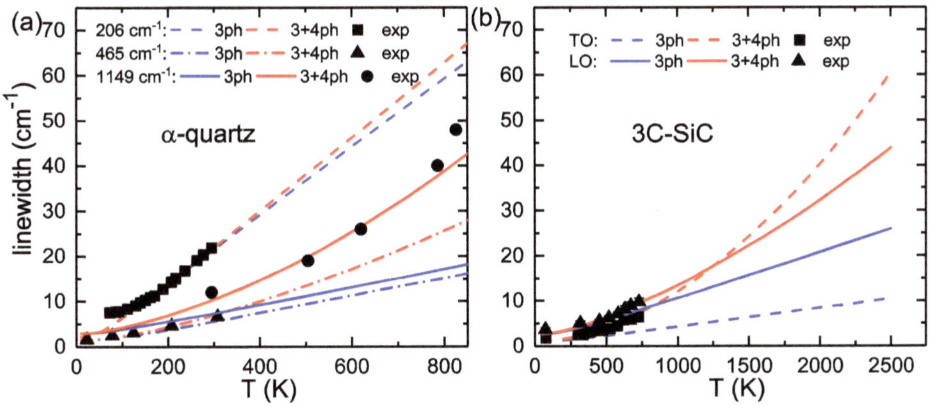

Figure 2.15. The optical phonon linewidth at the Γ point for α-quartz and 3*C*-SiC. First-principles data from [20]. Experimental data from [69, 70] for α-quartz and [16, 71] for SiC.

energy. More importantly, for the materials with phonon band gaps, four-phonon scattering is exceptionally important.

Figure 2.15 shows the zone-center optical phonon linewidths of α-quartz and 3*C*-SiC [20]. Figure 2.16 shows those of III–V compounds including *c*-BN, BAs, AlP, AlAs, AlSb, *c*-GaN, GaP, GaAs, GaSb, InP, InAs and InSb [20]. The first-principles calculated linewidths without four-phonon scattering are compared to available experimental values. The predicted optical phonon linewidths with only three-phonon scattering significantly disagree with the Raman measurements at mid and high temperatures. With four-phonon scattering included, reasonable agreements with available experimental data are achieved, demonstrating the significance of four-phonon processes in determining their infrared phonon linewidths.

For BAs in particular, it can be seen that the three-phonon processes have no contribution to optical phonon linewidth. This is due to the large acoustic–optical (a–o) gap as discussed in the preceding section. A similar case is also found in AlSb.

Note that for *c*-BN, BAs, InAs and InSb, even including four-phonon scattering, the prediction still has a significant discrepancy with experiments at higher temperatures, reflecting that five-phonon and higher-order phonon scattering may not be negligible in these particular materials.

To understand the high τ_4^{-1} of optical phonons, we show the contributions of different four-phonon processes of AlAs as an example in figure 2.17 (a). Clearly, it shows that the redistribution process $\lambda_1 + \lambda_2 \rightarrow \lambda_3 + \lambda_4$ dominates the four-phonon scattering of the optical phonon mode. This can be understood since the redistribution process is largely facilitated by the crowded branches of the optical phonon modes, as shown in figure 2.17 (b), among which the conservation law of energy and momentum can easily be satisfied by the four-phonon redistribution process. Actually, this is a general phenomenon since optical branches are bunched closely in energy in general materials, not limited to those with large acoustic–optical band gaps. This could also be the reason why optical phonons generally have strong four- and even higher-order phonon scattering.

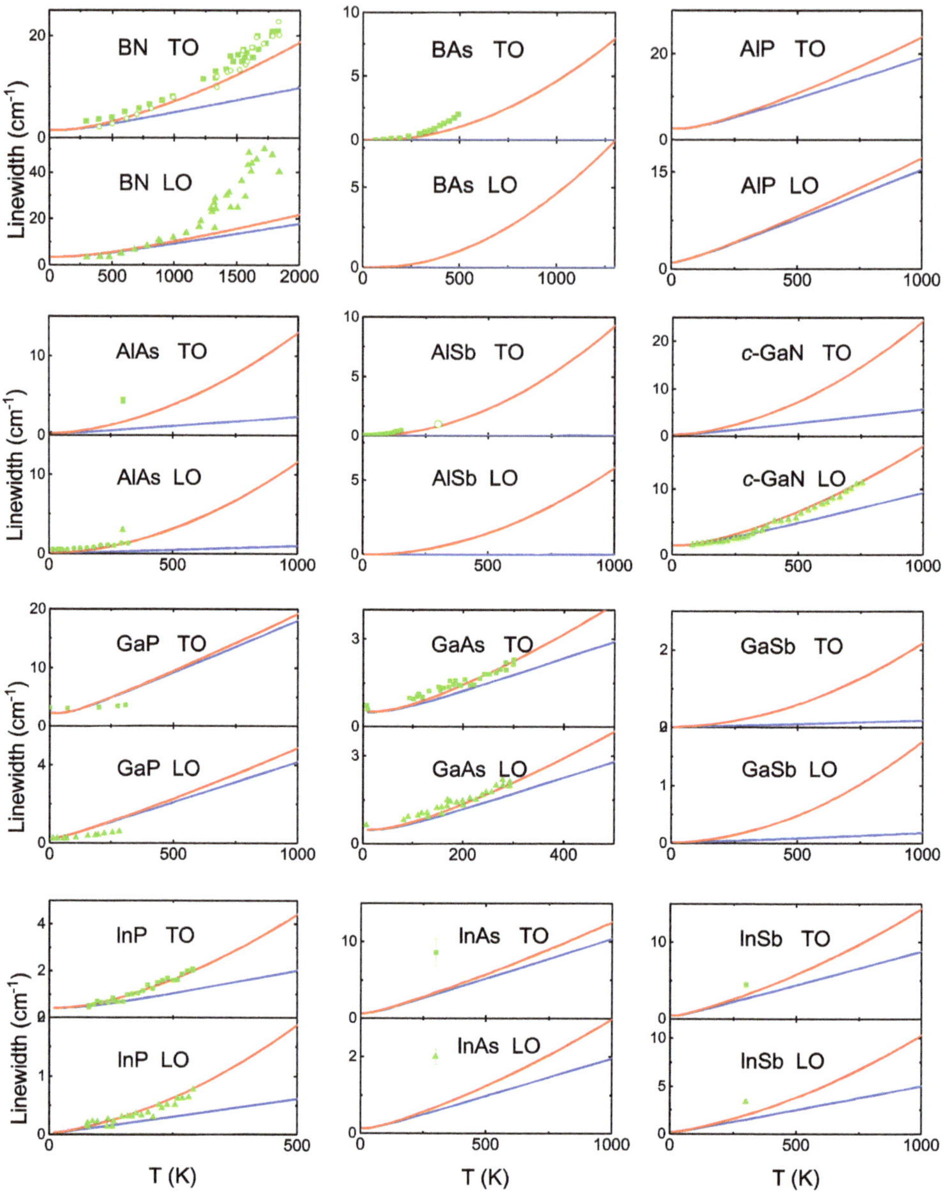

Figure 2.16. The optical phonon linewidth at the Γ point for group III–V zinc-blende compounds. First-principles data: τ_3^{-1} (blue curves) and $\tau_3^{-1} + \tau_4^{-1}$ (red curves) are taken from [20]. Note that the isotope or defect scattering is not included in the theoretical data here. Experimental values (green dots): *c*-BN [72], BAs [73], AlAs [74], AlSb [75, 76], *c*-GaN [77], GaP [6], GaAs [78], InP [78], InAs [79, 80] and InSb [75, 76].

The impact on optical phonon linewidth can also significantly affect thermal transport. For example, the three-phonon scattering predicts that the RT thermal conductivity of AlSb is about 98.8 W (m K)$^{-1}$, in which optical phonons contribute 49% [66]. However, after the four-phonon scattering is included, the RT κ is reduced

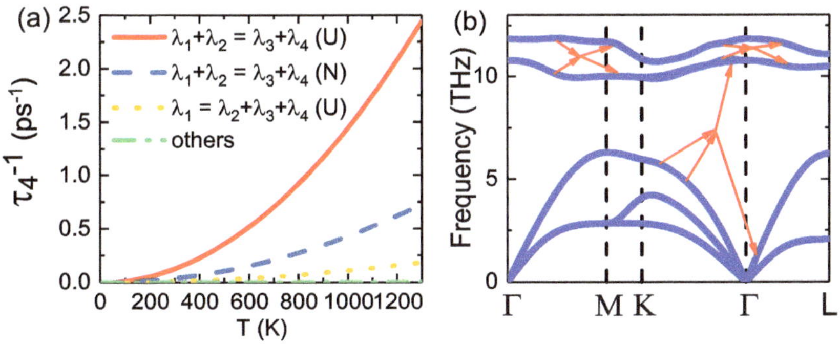

Figure 2.17. (a) First-principles scattering rates of the TO mode of AlAs from different four-phonon processes. (b) Sketch of the redistribution processes $\lambda_1 + \lambda_2 \rightarrow \lambda_3 + \lambda_4$. Data from [20].

to 39.5 W(m K)$^{-1}$ and the contribution of optical phonons is reduced to about 4%, that is, four-phonon scattering nearly kills the thermal transport of optical phonons in AlSb [66]. At 1000 K, the reduction is even larger, from 64% to 2%.

2.4.3 Two-dimensional materials with reflection symmetry

Two-dimensional materials with reflection symmetry are another example in which three-phonon scattering is largely limited [81, 82], leaving four-phonon scattering plenty of room to make a significant difference [31]. Graphene and hexagonal BN are two examples that have been demonstrated by Lindsay *et al* to have reflection symmetry [81, 82]. In this section, we take graphene as an example to demonstrate the possible role of four-phonon scattering.

Graphene has attracted intense interest for both fundamental research and practical applications due to its unique structure and extraordinary properties. The two-dimensional honeycomb structure, the zero band gap and the strong sp^2 bond endow graphene with unique electronic, thermal, optical and mechanical behaviors. The thermal transport in graphene has been quite intriguing since it was discovered that the scattering of flexural (out-of-plane) modes is largely forbidden by the reflection symmetry, leading to long relaxation times and high thermal conductivity. Three-phonon scattering theory predicts the room-temperature thermal conductivity of single-layer graphene as being around 3000 W (m K)$^{-1}$ [83, 84]. Different experimental methods, conditions and samples, however, showed quite different thermal conductivity values ranging from ~1500 to ~4000 W (m K)$^{-1}$ [85–89], and the widely used Raman technique has been questioned for use on graphene recently [90], leaving the thermal conductivity value of graphene even more mysterious.

Reflection symmetry in 2D materials forbids all the phonon–phonon scattering processes that involve an odd number of flexural modes [82]. Lindsay *et al* found numerically that in graphene the three-phonon scattering rates of the processes that involve 1 or 3 flexural modes are zero [82]. Feng and Ruan have verified numerically that the four-phonon scattering rates of the processes that involve 1 or 3 flexural modes are zero as well. Therefore, three-phonon processes can only involve 0 or 2

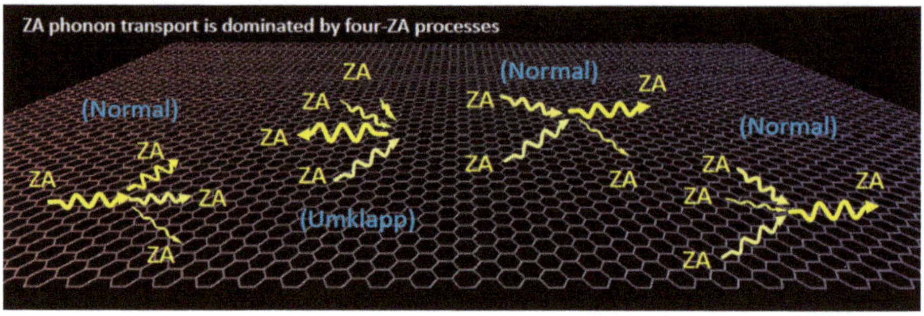

Figure 2.18. The 4-ZA processes in SLG. Reproduced with permission from [31]. Copyright 2018 the American Physical Society.

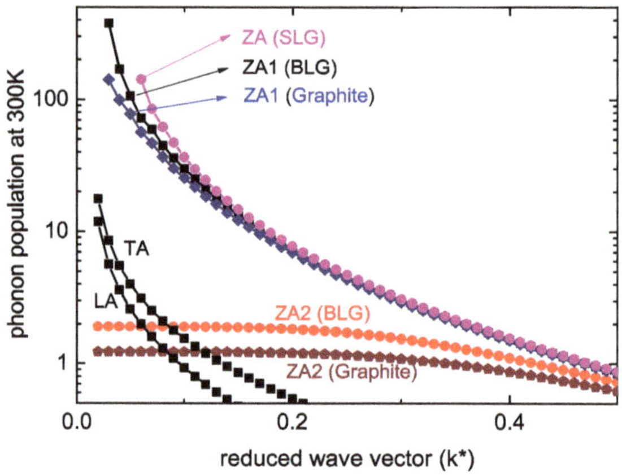

Figure 2.19. The phonon populations for different branches in SLG, bilayer graphene (BLG) and graphite. Data from [31].

flexural modes, while the four-phonon processes may involve 0, 2 or 4 flexural modes. Feng and Ruan found that most (60%–90%) of the three-phonon scattering processes of the ZA branch are forbidden by the reflection symmetry, while only about 40% of four-phonon scattering processes of the ZA branch are forbidden. Most importantly, four-phonon scattering allows two important processes, ZA + ZA → ZA + ZA and ZA ↔ ZA + ZA + ZA, which are called 4-ZA processes, shown in figure 2.18. Due to the quadratic dispersion relation, the ZA mode has quite a high phonon population near the Γ point as shown in figure 2.19. Therefore, these processes have ultra-high scattering rates since the four-phonon scattering is roughly proportional to the square of the phonon population.

The SMRTA-based three- and four-phonon scattering rates obtained from optimized Tersoff potential are compared in figure 2.20 [31]. Since the scattering rates follow the temperature dependence of $1/\tau_{3,\lambda}^0 \sim T$ and $1/\tau_{4,\lambda}^0 \sim T^2$, similarly to those in bulk materials [26, 27], we take the temperatures at 300 and 700 K as examples to show the amplitudes of $1/\tau_{3,\lambda}^0$ and $1/\tau_{4,\lambda}^0$ as a function of the reduced wave vector from Γ to M. We find that $1/\tau_{4,\lambda}^0$ is comparable to or even much higher than

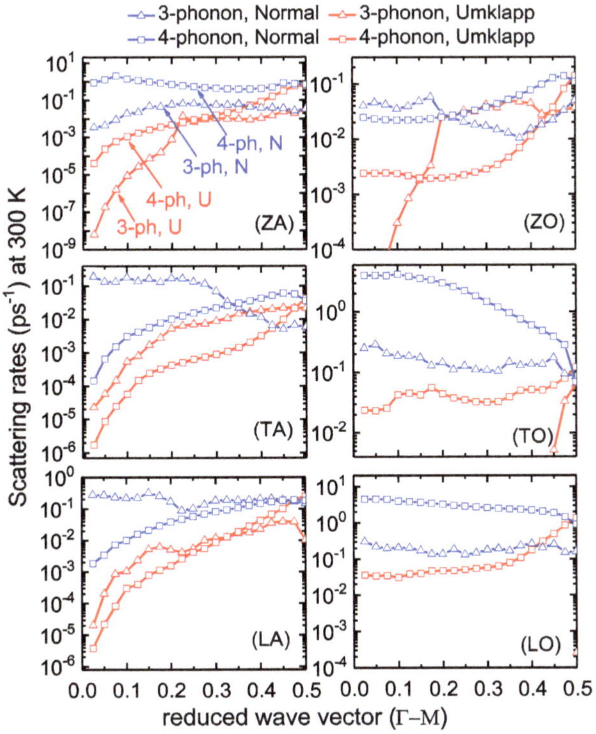

Figure 2.20. The three-phonon and four-phonon scattering rates, $\tau_{3(N)}^{-1}$, $\tau_{3(U)}^{-1}$, $\tau_{4(N)}^{-1}$ and $\tau_{4(U)}^{-1}$, of the six branches of SLG with respect to the reduced wave vector (Γ–M) at 300 K calculated by using the optimized Tersoff potential. Data from [31].

$1/\tau_{3,\lambda}^{0}$, even at room temperature, in particular for the ZA, TO and LO branches. For instance, $1/\tau_{3,\lambda}^{0}$ of the ZA branch at room temperature is typically below 0.08 ps^{-1} while the value of $1/\tau_{4,\lambda}^{0}$ is about 0.42-2 ps^{-1}, which indicates the relaxation time of ZA mode at room temperature is about 0.5–2 ps, far below expectations. At 700 K, the $1/\tau_{4,\lambda}^{0}$ of the ZA, TO and LO branches even reach above 10 ps^{-1}, being 2–3 orders higher than $1/\tau_{3,\lambda}^{0}$.

Feng and Ruan [31] found that the ZA four-phonon scattering is dominated by the normal (N) process, as shown in figure 2.20, indicating hydrodynamic behavior of phonon transport. Therefore, the SMRTA is not accurate to calculate the thermal conductivity, instead, the iterative solution to the BTE is needed. Feng and Ruan solved the iterative solution of BTE by including both three- and four-phonon scatterings, and they showed that the thermal conductivity of 9 μm graphene is reduced significantly from ~3383 to ~810 W(m K)$^{-1}$ after including the four-phonon scattering [31] (figure 2.21). Later in 2019, Gu *et al* refined the calculation of four-phonon scattering in single-layer graphene by using a smeared broadening factor as well as a temperature-dependent phonon dispersion relation [91]. They further confirmed the striking importance of four-phonon scattering in SLG and found that the thermal conductivity, using the optimized Tersoff potential, is about

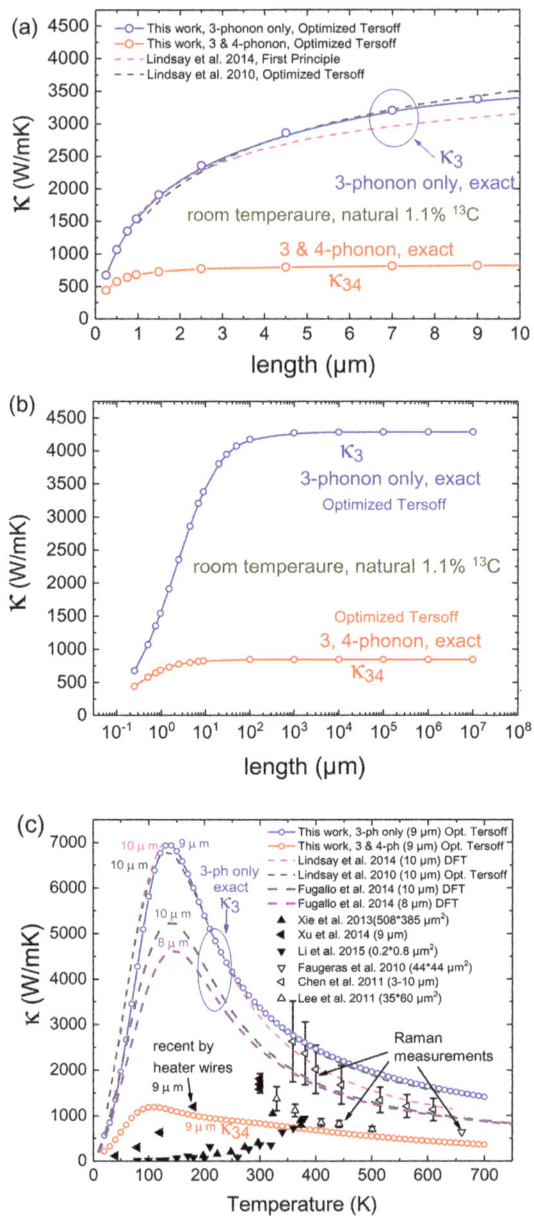

Figure 2.21. (a) Length-dependent thermal conductivity of single-layer graphene at 300 K. (b) Length convergence of thermal conductivity at 300 K. (c) Temperature-dependent lattice thermal conductivity. The dashed lines represent the theoretical predictions from the literature with three-phonon scattering only. The solid lines with open circles that are marked 'this work' are the predictions from [31]. In all the predictions, the natural 1.1% ^{13}C is included with the exact solution to the linearized BTE. The triangles show the experimental measured results. Data from Lindsay *et al* [82, 83], Fugallo *et al* [84], Xie *et al* [93], Xu *et al* [87], Li *et al* [94], Faugeras *et al* [85], Chen *et al* [95] and Lee *et al* [96]. Reproduced with permission from [31]. Copyright 2018 the American Physical Society.

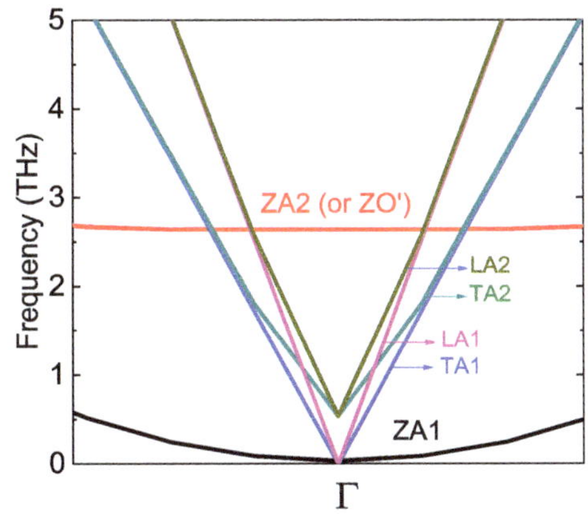

Figure 2.22. The phonon dispersion near the zone center along Γ–*M* in bilayer graphene calculated using an optimized Tersoff potential.

1900 W(m K)$^{-1}$. Since the fourth-order force constant of the classical interatomic potential [92] has not been validated against first-principles, the absolute values of the thermal conductivity after including four-phonon scattering should be interpreted qualitatively.

Since the four-phonon scattering rates in SLG are high, two natural questions are: (i) Does four-phonon scattering play an important role in multilayer graphene and graphite? (ii) Is five-phonon scattering important in SLG? The answers are negative for both questions. To address the first question, we plot the phonon dispersion of bilayer graphene in figure 2.22. Due to the interlayer van der Waals interaction, the ZA mode of SLG is split into the ZA and ZO′ modes in bilayer graphene or graphite. ZO′ represents a breathing mode between adjacent layers. We find that even such a small splitting can result in a large reduction of the phonon population, as shown in figure 2.19. Due to the splitting, the phase space of the four-ZA process becomes 1/16 of that in SLG, being unimportant. This explains the fact that the three-phonon thermal conductivity prediction of graphite agrees well with experiments [84, 97]. Regarding the second question, we need to refer to the reflection symmetry. The five-phonon process can at most involve four ZA modes, the same as four-phonon scattering. Without increasing the population, the higher order makes the five-phonon scattering negligible compared to four-phonon scattering.

2.5 Further discussion

2.5.1 Scaling with frequency

Due to the simplicity, scaling laws of phonon scattering rates are very useful in thermal nanoengineering. For example, the power laws of three-phonon scattering $\tau_3^{-1} \sim \omega^2 T$ and phonon–defect scattering $\tau_d^{-1} \sim \omega^4$ have been widely used for advanced thermoelectric materials in understanding experimental thermal conductivity [98–100]. Therefore, it is important to have a scaling law $\tau_4^{-1} \sim \omega^\beta T^2$ for four-phonon scattering. In the early literature [18, 101], β was taken as 2 by using drastic approximations. The fittings of diamond, Si and BAs in figure 2.23 show that the value of β varies from 2 to 4. In comparison, τ_3^{-1} has a more dispersed distribution with a frequency, and the scaling law $\tau_3^{-1} \sim \omega^2 T$ cannot fit all the data well simultaneously, while τ_4^{-1} has a more concentrated distribution and the scaling law can fit all the phonons well. This is because the selection rules in four-phonon processes can be more easily satisfied, that is, they become less restrictive and less dependent on the dispersive nature of the phonon frequency.

2.5.2 Strong Umklapp scattering

The difference between iterative and RTA thermal conductivities comes from subtle differences in the normal and Umklapp processes. Umklapp processes provide thermal resistance, degradation of a flowing distribution of phonons. Normal processes do not degrade the overall current but play the important role of redistributing thermal energy among various modes in the system. If normal processes dominate over Umklapp processes, the RTA solution does not accurately represent κ as it treats normal processes as purely resistive and underestimates κ [9]. As shown in figure 2.24, three-phonon scattering is dominated by normal processes in diamond and BAs, not so in Si. Thus, diamond and BAs require an exact solution for three-phonon scattering [9]. As for four-phonon scattering, all three materials show dominant Umklapp processes over normal processes. Thus, treating the four-phonon scattering at the RTA level for most materials within the iteration scheme is probably a good approximation.

2.5.3 Negligible three-phonon scattering to the second order

We note that two three-phonon processes, $\lambda + \lambda_1 \rightarrow \lambda'$ and $\lambda' \rightarrow \lambda_2 + \lambda_3$, may be combined to give the three-phonon scattering to the second order, which is another type of fourth-order process [34, 101], as shown in figure 2.25 (b). Here, λ' is an intermediate virtual state. The energy is conserved from the initial state $\lambda + \lambda_1$ to the final state $\lambda_2 + \lambda_3$, while the energy is not necessarily conserved in the first step or in the second step alone [34]. The energy denominators of three-phonon scattering

$$\frac{\langle i|\hat{H}_3|f\rangle}{|E_i - E_f|} \tag{2.56}$$

and four-phonon scattering

$$\frac{\langle i|\hat{H}_4|f\rangle}{|E_i - E_f|} \tag{2.57}$$

vanish due to the energy conservation law $E_i = E_f$. In contrast to equations (2.56) and (2.57), the transition matrix element in the combined three-phonon process is

$$\frac{\langle i|\hat{H}_3|\mathrm{vir}\rangle\langle \mathrm{vir}|\hat{H}_3|f\rangle}{|E_i - E_{\mathrm{vir}}|}. \tag{2.58}$$

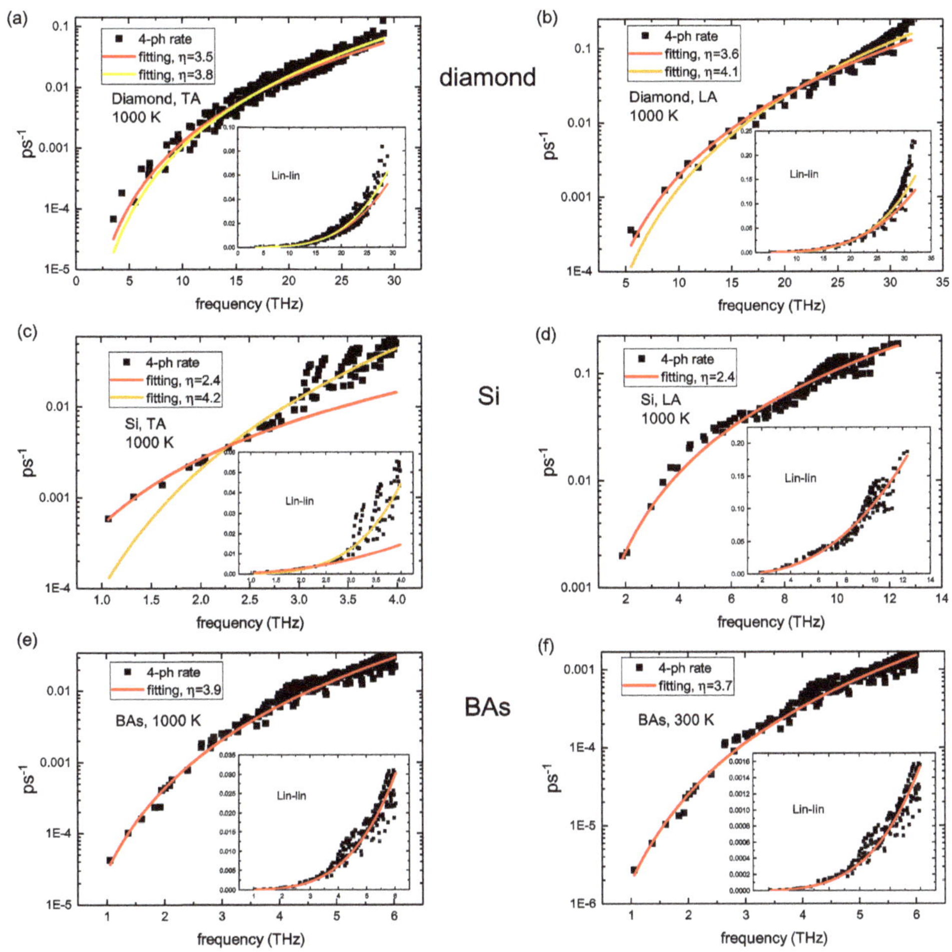

Figure 2.23. Power law fitting $\tau_4^{-1} = A\omega^\beta$ of the acoustic phonons in diamond, Si and BAs calculated from first principles. Each panel is plotted in a log–linear scale to give a clear view of the low-frequency behavior, while the inset is in linear–linear scale for a clearer view of the high-frequency behavior. (a) and (b) are the TA and LA modes of diamond at 1000 K, respectively. (c) and (d) are the TA and LA modes of Si at 1000 K, respectively. (e) and (f) are the acoustic modes of BAs at 1000 and 300 K, respectively. We note that four-phonon scattering is only important for diamond and Si at higher temperatures. For each of (a), (b) and (c), we have two fitting curves: the red curve (lower power) fits the low-frequency behavior better, while the yellow curve (higher power) fits better in the higher-frequency range. Reproduced with permission from [27]. Copyright 2017 the American Physical Society.

$|\text{vir}\rangle$ is the intermediate virtual state. The discussion of the denominator in equation (2.58) can be divided into two cases. In case 1, the energy is not conserved in the first or the second step [34]. The energy denominators for the transition are not small. Therefore, the transition rate is not considered to be large, as discussed in [102]. In case 2, the energy conservation condition for the first step is nearly satisfied or satisfied. This process was named 'resonance in three-phonon scattering' and is

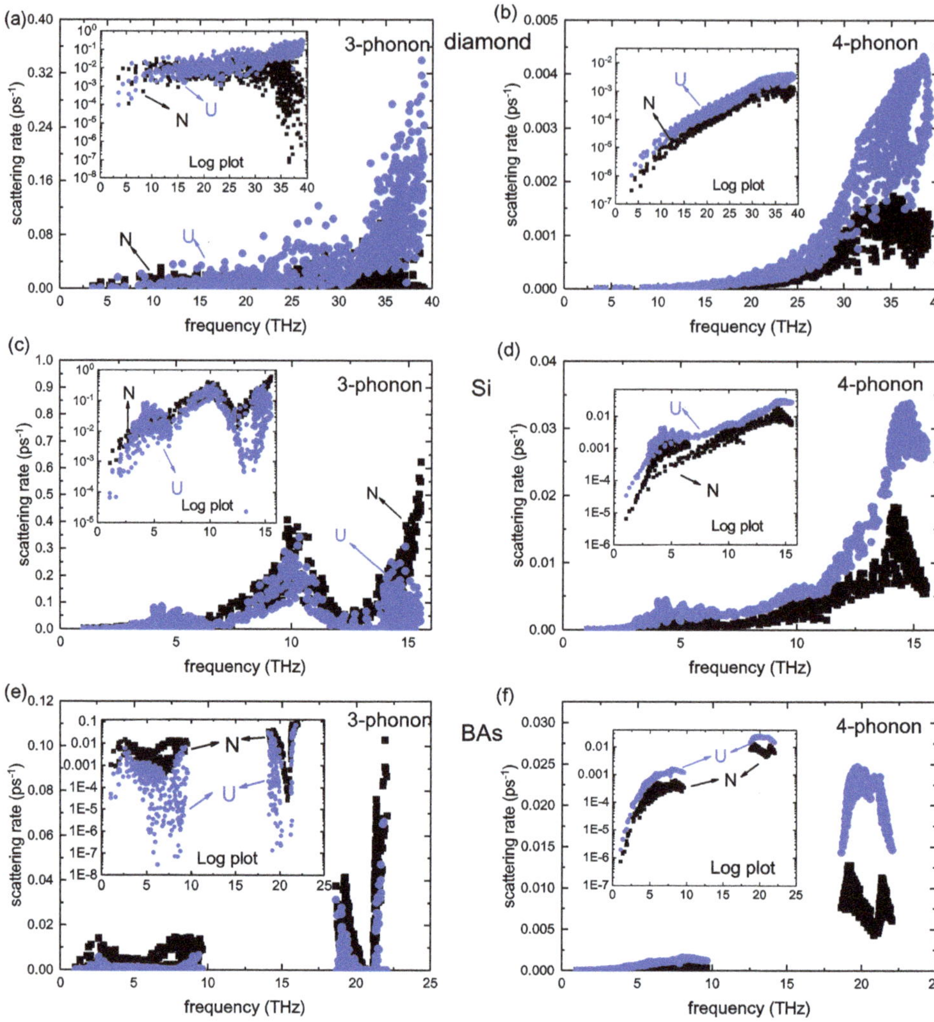

Figure 2.24. Comparison between normal and Umklapp scattering rates for diamond, Si and BAs at $T = 300$ K calculated from first-principles. Each panel is plotted in a linear–linear scale, while each inset is in a log–linear scale. Reproduced with permission from [27]. Copyright 2017 the American Physical Society.

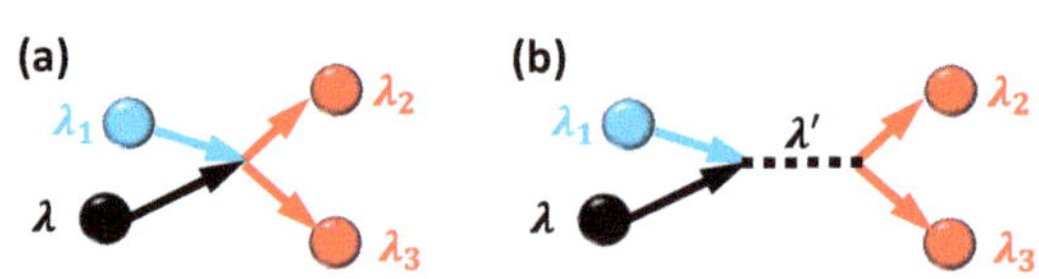

Figure 2.25. The diagram examples for the comparison between (a) the intrinsic four-phonon scattering and (b) the three-phonon scattering to the second order. Reproduced with permission from [26].

discussed by Carruthers [102]. In this case, although the scattering is in the same order as the intrinsic four-phonon scattering, the number of scattering events that satisfy the energy and momentum selection rule is only 10^{-3}–10^{-5} of that in the intrinsic four-phonon scattering in our study. This is because the resonant three-phonon scattering has a strong requirement that the intermediate state has to be an existing phonon mode in the **q**-mesh, while the intrinsic four-phonon scattering has no such requirement. For example, for Si with a $16 \times 16 \times 16$ **q**-mesh and the energy-conservation-tolerant range of 1.24 meV (0.3 THz), the TA mode at $\mathbf{q}^* = (0.5,0,0)$ has 4.6×10^7 intrinsic four-phonon events, and only 2.7×10^4 resonant three-phonon events. For the TA mode at $\mathbf{q}^* = (0.625,0,0)$, the number of intrinsic four-phonon events is similarly about 4.6×10^7, while the number of resonant three-phonon events is only 36. Therefore, the overall three-phonon to the second-order scattering rate is negligible compared to the intrinsic four-phonon scattering. We note that this conclusion is consistent with the conjecture in the literature [101].

2.6 Summary and outlook

In summary, predictive calculation of four-phonon scattering realized since 2016 has enabled the accurate prediction of thermal conductivity for a broader scope of materials over a much wider temperature range. Generally speaking, strong four-phonon scattering is originated from either a large scattering potential (e.g., a large phonon population or strong anharmonicity) or a large scattering phase space, or both. Based on these two origins, table 2.1 summarizes the categories and examples in which four-phonon scattering is strong.

The large scattering potential is induced by either high temperature or intrinsic strongly anharmonic interatomic bonding. The former is seen in general solids. Actually, raising temperature not only increases anharmonicity but also excites a greater phonon population that can boost four-phonon scattering. The latter could be found in many technically important materials, such as most inorganic thermoelectric materials. In addition, the rocksalt compounds generally have long-ranged resonant interaction that causes strong anharmonicity. Among them, PbTe and NaCl have been recently found to have strong four-phonon scattering at room temperature, while the others also need further investigation.

The large four-phonon scattering phase space is either caused by relatively restricted three-phonon scattering phase space or large DOS. The former is seen in the materials with a large acoustic–optical band gap and the two-dimensional materials with reflection symmetry. The latter is seen for high-frequency phonons, in particular optical phonons which are usually bunched together and allow a large $\lambda_1 + \lambda_2 \rightarrow \lambda_3 + \lambda_4$ scattering phase space.

Despite the advances of four-phonon scattering calculation in the past few years, its impact on a broader range of materials is still open to be examined. Some immediate examples are the materials showing ultra-low thermal conductivity, such as the halides (LiF, LiCl, LiBr, LiI, NaF, NaBr, NaI, KF, KCl, KBr, KI, RbF,

Table 2.1. The categories and examples in which four-phonon scattering is significant based on the two fundamental origins.

Origins	Categories	Applications	Examples examined	To be examined
Strong scattering potential (strong anharmonicity)	High temperature	Thermal barrier coating, nuclear materials, high-T thermoelectrics	In general	In general
	Strongly anharmonic (low-κ) materials	Thermoelectrics, thermal barrier coating, thermal energy storage, phase change materials	Ar, PbTe, NaCl	Rocksalt compounds, halides, hydrides, chalcogenides, oxides, others (Bi, Sb), etc
Large scattering phase space	With large acoustic–optical phonon band gap	Thermal management	BAs, AlP, AlAs, AlSb, c-GaN, GaP, GaAs, GaSb, InP, nAs, InSb, etc	I–VII, II–VI binary compounds with large mass ratio
	2D materials with reflection symmetry	Thermal management, surface plasmon	Single-layer graphene	Single-layer h-BN
	Large DOS: e.g. optical phonons	Infrared (sensing, radiative cooling, energy harvesting, metamaterials), hot electron relaxation, complex crystals (perovskites, MXenes, etc)	In general	In general

RbCl, RbBr, RbI, CsF, CsCl, CuCl, CuI, AgI, etc), hydrides (LiH, NaH, KH, RbH, CuH, etc), chalcogenides (CdSe, BaTe, CdTe, BaS, PbS, PbSe, Bi_2Te_3, SnS, SnSe, SnTe, GeTe, etc) and oxides (CdO, SrO, BaO, etc) [12]. Most of them have a rocksalt structure, and few are zinc-blende or wurtzite. The calculation of four-phonon scattering in these materials is straightforward since most crystal structures are simple, with only two basis atoms in a unit cell. Apart from these, it is also interesting to examine complex crystals such as perovskites and MXenes (transition metal carbides, nitrides, or carbonitrides), in which there are many optical branches. The impact of phonon renormalization on three- and four-phonon scattering urgently needs to be explored in more materials as they both are a representation of anharmonicity and coupled together with each other. Another interesting direction is to explore the interplay between four-phonon scattering and other factors such as pressure [103], interfaces and defects. In addition, we also look forward to seeing the exploration of other origins that could lead to strong four-phonon scattering and new thermal transport phenomena.

For ultra-low thermal conductivity materials, an open question is about the nature of thermal transport since four-phonon scattering could bring the phonon mean free path down below the interatomic distance, i.e., the Ioffe–Regel limit. Recently, it was found that for some single crystals such as Tl_3VSe_4, $YbFe_4Sb_{12}$, $CsSnI_3$, $CsPbI_3$ and $CsPbBr_3$, even three-phonon scattering could significantly underestimate the thermal conductivity at room temperature [104]. We suspect that the inclusion of the four-phonon scattering could bring the predicted thermal conductivity even lower. Therefore, the particle nature of phonon thermal transport still needs more investigation and better understanding when four-phonon scattering is included.

With the accurate prediction of phonon–phonon scattering rate enabled by adding four-phonon processes, a variety of processes that involve both phonon–phonon scattering and the scattering between phonons and other particles such as electrons, photons and polaritons could be re-investigated. As already discussed in this chapter, the prediction of infrared properties was significantly improved. We foresee that it will also generate an impact on the understanding and prediction of the laser heating process, hot electron relaxation, interfacial thermal transport, electrical transport, etc.

For many years, researchers have been devoted to the search for materials with extreme thermal properties, e.g., ultra-low or ultra-high thermal conductivity, due to the intriguing physics and promising cutting-edge applications. We anticipate that the generalization of four-phonon scattering calculation will create many opportunities towards this goal. With the rapid increase of the computational power, application of the three- and four-phonon scattering will generate significantly more impact on the prediction of thermal transport as well as other phonon-related applications.

References

[1] Maradudin A A and Flinn P A 1961 Anharmonic contributions to vibrational thermodynamic properties of solids: Part I *Ann. Phys.* **15** 337–59

[2] Maradudin A A and Fein A E 1962 Scattering of neutrons by an anharmonic crystal *Phys. Rev.* **128** 2589

[3] Maradudin A A, Fein A E and Vineyard G H 1962 On the evaluation of phonon widths and shifts *Phys. Status Solidi* B **2** 1479–92

[4] Debernardi A, Baroni S and Molinari E 1995 Anharmonic phonon lifetimes in semiconductors from density-functional perturbation theory *Phys. Rev. Lett.* **75** 1819–22

[5] Bechstedt F, Käckell P, Zywietz A, Karch K, Adolph B, Tenelsen K and Furthmüller J 1997 Polytypism and properties of silicon carbide *Phys. Status Solidi* B **202** 35–62

[6] Debernardi A 1998 Phonon linewidth in III–V semiconductors from density-functional perturbation theory *Phys. Rev.* B **57** 12847–58

[7] Lang G, Karch K, Schmitt M, Pavone P, Mayer A, Wehner R and Strauch D 1999 Anharmonic line shift and linewidth of the Raman mode in covalent semiconductors *Phys. Rev.* B **59** 6182–8

[8] Tang X and Fultz B 2011 First-principles study of phonon linewidths in noble metals *Phys. Rev.* B **84** 054303

[9] Broido D A, Malorny M, Birner G, Mingo N and Stewart D A 2007 Intrinsic lattice thermal conductivity of semiconductors from first principles *Appl. Phys. Lett.* **91** 231922

[10] Esfarjani K, Chen G and Stokes H T 2011 Heat transport in silicon from first-principles calculations *Phys. Rev.* B **84** 085204

[11] Lindsay L, Broido D A and Reinecke T L 2013 First-principles determination of ultrahigh thermal conductivity of boron arsenide: a competitor for diamond? *Phys. Rev. Lett.* **111** 025901

[12] Seko A, Togo A, Hayashi H, Tsuda K, Chaput L and Tanaka I 2015 Prediction of low-thermal-conductivity compounds with first-principles anharmonic lattice-dynamics calculations and Bayesian optimization *Phys. Rev. Lett.* **115** 205901

[13] Feng T and Ruan X 2014 Prediction of spectral phonon mean free path and thermal conductivity with applications to thermoelectrics and thermal management: a review *J. Nanomater.* **2014** 206370

[14] Bao H, Chen J, Gu X and Cao B 2018 A review of simulation methods in micro/nanoscale heat conduction *ES Energy Environ.* **1** 16

[15] Lindsay L, Hua C, Ruan X and Lee S 2018 Survey of *ab initio* phonon thermal transport *Mater. Today Phys.* **7** 106–20

[16] Tong Z, Liu L, Li L and Bao H 2018 Temperature-dependent infrared optical properties of 3c-, 4h- and 6h-SiC *Physica* B **537** 194–201

[17] Joshi Y P, Tiwari M D and Verma G S 1970 Role of four-phonon processes in the lattice thermal conductivity of silicon from 300 to 1300 K *Phys. Rev.* B **1** 642–6

[18] Ecsedy D and Klemens P 1977 Thermal resistivity of dielectric crystals due to four-phonon processes and optical modes *Phys. Rev.* B **15** 5957

[19] Bao H, Qiu B, Zhang Y and Ruan X 2012 A first-principles molecular dynamics approach for predicting optical phonon lifetimes and far-infrared reflectance of polar materials *J. Quant. Spectrosc. Radiat. Transf.* **113** 1683–8

[20] Yang X, Feng T, Kang J S, Hu Y, Li J and Ruan X 2019 Role of higher-order phonon scattering in the zone-center optical phonon linewidth and the Lorenz oscillator model (arXiv 1908.05121)
[21] Novikov M, Ositinskaya T, Shulzhenko O, Podoba O, Sokolov O and Petrusha I 1983 Heat-conductivity of cubic boron–nitride single-crystals *Dopov. Akad. Nauk Ukrain. RSR Seriya* A 72–5
[22] Zhao L-D, Lo S-H, Zhang Y, Sun H, Tan G, Uher C, Wolverton C, Dravid V P and Kanatzidis M G 2014 Ultralow thermal conductivity and high thermoelectric figure of merit in SnSe crystals *Nature* **508** 373–7
[23] Guo R, Wang X, Kuang Y and Huang B 2015 First-principles study of anisotropic thermoelectric transport properties of IV–VI semiconductor compounds SNSE and SNS *Phys. Rev.* B **92** 115202
[24] Lindsay L and Broido D A 2008 Three-phonon phase space and lattice thermal conductivity in semiconductors *J. Phys. Condens. Matter* **20** 165209
[25] Turney J, Landry E, McGaughey A and Amon C 2009 Predicting phonon properties and thermal conductivity from anharmonic lattice dynamics calculations and molecular dynamics simulations *Phys. Rev.* B **79** 064301
[26] Feng T and Ruan X 2016 Quantum mechanical prediction of four-phonon scattering rates and reduced thermal conductivity of solids *Phys. Rev.* B **93** 045202
[27] Feng T, Lindsay L and Ruan X 2017 Four-phonon scattering significantly reduces intrinsic thermal conductivity of solids *Phys. Rev.* B **96** 161201
[28] Kang J S, Li M, Wu H, Nguyen H and Hu Y 2018 Experimental observation of high thermal conductivity in boron arsenide *Science* **361** 575–8
[29] Li S, Zheng Q, Lv Y, Liu X, Wang X, Huang P Y, Cahill D G and Lv B 2018 High thermal conductivity in cubic boron arsenide crystals *Science* **361** 579–81
[30] Tian F *et al* 2018 Unusual high thermal conductivity in boron arsenide bulk crystals *Science* **361** 582–5
[31] Feng T and Ruan X 2018 Four-phonon scattering reduces intrinsic thermal conductivity of graphene and the contributions from flexural phonons *Phys. Rev.* B **97** 045202
[32] Xia Y 2018 Revisiting lattice thermal transport in PBTE: the crucial role of quartic anharmonicity *Appl. Phys. Lett.* **113** 073901
[33] Ravichandran N K and Broido D 2018 Unified first-principles theory of thermal properties of insulators *Phys. Rev.* B **98** 085205
[34] Ziman J M 1960 *Electrons and Phonons* (London: Oxford University Press)
[35] Kaviany M 2008 *Heat Transfer Physics* (New York: Cambridge University Press)
[36] Klemens P 1958 *Solid State Physics* vol 7 (New York: Academic)
[37] Tamura S-I 1983 Isotope scattering of dispersive phonons in Ge *Phys. Rev.* B **27** 858
[38] Casimir H B G 1938 Note on the conduction of heat in crystals *Physica* **5** 495–500
[39] Berman R, Simon F E and Ziman J M 1953 The thermal conductivity of diamond at low temperatures *Proc. R. Soc. Lond.* A **220** 171–83
[40] Berman R, Foster E L and Ziman J M 1955 Thermal conduction in artificial sapphire crystals at low temperatures. I. Nearly perfect crystals *Proc. R. Soc. Lond.* A **231** 130–44
[41] Omini M and Sparavigna A 1995 An iterative approach to the phonon Boltzmann equation in the theory of thermal conductivity *Physica* B **212** 101–12
[42] Omini M and Sparavigna A 1997 Heat transport in dielectric solids with diamond structure *Nuovo Cimento Soc. Ital. Fis.* D **19D** 1537–63

[43] Broido D A, Ward A and Mingo N 2005 Lattice thermal conductivity of silicon from empirical interatomic potentials *Phys. Rev.* B **72** 014308
[44] Togo A and Tanaka I 2015 First principles phonon calculations in materials science *Scr. Mater.* **108** 1–5
[45] Tersoff J 1989 Modeling solid-state chemistry: interatomic potentials for multicomponent systems *Phys. Rev.* B **39** 5566–8
[46] Tersoff J 1990 Erratum: Modeling solid-state chemistry: interatomic potentials for multicomponent systems *Phys. Rev.* B **41** 3248
[47] Li W, Mingo N, Lindsay L, Broido D A, Stewart D A and Katcho N A 2012 Thermal conductivity of diamond nanowires from first principles *Phys. Rev.* B **85** 195436
[48] Li W, Lindsay L, Broido D A, Stewart D A and Mingo N 2012 Thermal conductivity of bulk and nanowire $Mg_2Si_xSn_{1-x}$ alloys from first principles *Phys. Rev.* B **86** 174307
[49] Lindsay L, Broido D and Mingo N 2009 Lattice thermal conductivity of single-walled carbon nanotubes: beyond the relaxation time approximation and phonon–phonon scattering selection rules *Phys. Rev.* B **80** 125407
[50] Lindsay L and Broido D A 2012 Theory of thermal transport in multilayer hexagonal boron nitride and nanotubes *Phys. Rev.* B **85** 035436
[51] Lindsay L, Broido D A and Reinecke T L 2012 Thermal conductivity and large isotope effect in GaN from first principles *Phys. Rev. Letters* **109** 095901
[52] Ruf T, Henn R, Asen-Palmer M, Gmelin E, Cardona M, Pohl H-J, Devyatych G and Sennikov P 2000 Thermal conductivity of isotopically enriched silicon *Solid State Commun.* **115** 243–7
[53] Abeles B, Beers D, Cody G and Dismukes J 1962 Thermal conductivity of Ge–Si alloys at high temperatures *Phys. Rev.* **125** 44
[54] Glassbrenner C J and Slack G A 1964 Thermal conductivity of silicon and germanium from 3 K to the melting point *Phys. Rev.* **134** A1058
[55] Feng T, Yang X and Ruan X 2018 Phonon anharmonic frequency shift induced by four-phonon scattering calculated from first principles *J. Appl. Phys.* **124** 145101
[56] Morelli D T, Jovovic V and Heremans J P 2008 Intrinsically minimal thermal conductivity in cubic I–V–vi_2 semiconductors *Phys. Rev. Lett.* **101** 035901
[57] El-Sharkawy A, El-Azm A A, Kenawy M, Hillal A and Abu-Basha H 1983 Thermophysical properties of polycrystalline PBS, PBSE, and PBTE in the temperature range 300–700 K *Int. J. Thermophys.* **4** 261–9
[58] Håkansson B and Andersson P 1986 Thermal conductivity and heat capacity of solid NaCl and NaI under pressure *J. Phys. Chem. Solids* **47** 355–62
[59] McCarthy K A and Ballard S S 1960 Thermal conductivity of eight halide crystals in the temperature range 220 K to 390 K *J. Appl. Phys.* **31** 1410–12
[60] Yukutake H and Shimada M 1978 Thermal conductivity of NaCl, MgO, coesite and stishovite up to 40 kbar *Phys. Earth Planet. Inter.* **17** 193–200
[61] Wei L, Kuo P K, Thomas R L, Anthony T R and Banholzer W F 1993 Thermal conductivity of isotopically modified single crystal diamond *Phys. Rev. Lett.* **70** 3764–7
[62] Onn D G, Witek A, Qiu Y Z, Anthony T R and Banholzer W F 1992 Some aspects of the thermal conductivity of isotopically enriched diamond single crystals *Phys. Rev. Lett.* **68** 2806–9
[63] Olson J R, Pohl R O, Vandersande J W, Zoltan A, Anthony T R and Banholzer W F 1993 Thermal conductivity of diamond between 170 and 1200 K and the isotope effect *Phys. Rev.* B **47** 14850–6

[64] Berman R, Hudson P R W and Martinez M 1975 Nitrogen in diamond: evidence from thermal conductivity *J. Phys. C: Solid State Phys.* **8** L430

[65] Zheng Q, Li S, Li C, Lv Y, Liu X, Huang P Y, Broido D A, Lv B and Cahill D G 2018 High thermal conductivity in isotopically enriched cubic boron phosphide *Adv. Funct. Mater.* **28** 1805116

[66] Yang X, Feng T, Li J and Ruan X 2019 Stronger role of four-phonon scattering than three-phonon scattering in thermal conductivity of III-V semiconductors at room temperature *Phys. Rev.* B **100** 245203

[67] Jeżowski A, Stachowiak P, Plackowski T, Suski T, Krukowski S, Boćkowski M, Grzegory I, Danilchenko B and Paszkiewicz T 2003 Thermal conductivity of GaN crystals grown by high pressure method *Phys. Status Solidi* B **240** 447–50

[68] Hess S, Taylor R, O'Sullivan E, Ryan J, Cain N, Roberts V and Roberts J 1999 Hot carrier relaxation by extreme electron–LO phonon scattering in GaN *Phys. Status Solidi* B **216** 51–5

[69] Gervais F and Piriou B 1975 Temperature dependence of transverse and longitudinal optic modes in the α and β phases of quartz *Phys. Rev.* B **11** 3944–50

[70] Dean K, Sherman W and Wilkinson G 1982 Temperature and pressure dependence of the Raman active modes of vibration of α-quartz *Spectrochim. Acta* A **38** 1105–8

[71] Ulrich C, Debernardi A, Anastassakis E, Syassen K and Cardona M 1999 Raman linewidths of phonons in Si, Ge, and Sic under pressure *Phys. Status Solidi* B **211** 293–300

[72] Herchen H and Cappelli M A 1993 Temperature dependence of the cubic boron nitride Raman lines *Phys. Rev.* B **47** 14193–9

[73] Hadjiev V G, Iliev M N, Lv B, Ren Z F and Chu C W 2014 Anomalous vibrational properties of cubic boron arsenide *Phys. Rev.* B **89** 024308

[74] Lockwood D, Yu G and Rowell N 2005 Optical phonon frequencies and damping in AlAs, GaP, GaAs, InP, InAs and InSb studied by oblique incidence infrared spectroscopy *Solid State Commun.* **136** 404–9

[75] Turner W J and Reese W E 1962 Infrared lattice bands in AlSb *Phys. Rev.* **127** 126–31

[76] McCluskey M D, Haller E E and Becla P 2001 Carbon acceptors and carbon–hydrogen complexes in AlSb *Phys. Rev.* B **65** 045201

[77] Cuscó R, Domènech-Amador N, Novikov S, Foxon C T and Artús L 2015 Anharmonic phonon decay in cubic GaN *Phys. Rev.* B **92** 075206

[78] Irmer G, Wenzel M and Monecke J 1996 The temperature dependence of the LO(T) and TO(T) phonons in GaAs and InP *Phys. Status Solidi* B **195** 85–95

[79] Stimets R W and Lax B 1970 Reflection studies of coupled magnetoplasma–phonon modes *Phys. Rev.* B **1** 4720–35

[80] Hass M and Henvis B 1962 Infrared lattice reflection spectra of III–V compound semiconductors *J. Phys. Chem. Solids* **23** 1099–104

[81] Lindsay L and Broido D A 2011 Enhanced thermal conductivity and isotope effect in single-layer hexagonal boron nitride *Phys. Rev.* B **84** 155421

[82] Lindsay L, Broido D A and Mingo N 2010 Flexural phonons and thermal transport in graphene *Phys. Rev.* B **82** 115427

[83] Lindsay L, Li W, Carrete J, Mingo N, Broido D A and Reinecke T L 2014 Phonon thermal transport in strained and unstrained graphene from first principles *Phys. Rev.* B **89** 155426

[84] Fugallo G, Cepellotti A, Paulatto L, Lazzeri M, Marzari N and Mauri F 2014 Thermal conductivity of graphene and graphite: collective excitations and mean free paths *Nano Lett.* **14** 6109–14

[85] Faugeras C, Faugeras B, Orlita M, Potemski M, Nair R R and Geim A K 2010 Thermal conductivity of graphene in corbino membrane geometry *ACS Nano* **4** 1889–92

[86] Chen S, Wu Q, Mishra C, Kang J, Zhang H, Cho K, Cai W, Balandin A A and Ruoff R S 2012 Thermal conductivity of isotopically modified graphene *Nature Mater.* **11** 203–7

[87] Xu X *et al* 2014 Length-dependent thermal conductivity in suspended single-layer graphene *Nat. Commun.* **5** 3689

[88] Ghosh S, Calizo I, Teweldebrhan D, Pokatilov E P, Nika D L, Balandin A A, Bao W, Miao F and Lau C N 2008 Extremely high thermal conductivity of graphene: prospects for thermal management applications in nanoelectronic circuits *Appl. Phys. Lett.* **92** 151911

[89] Ghosh S, Bao W, Nika D L, Subrina S, Pokatilov E P, Lau C N and Balandin A A 2010 Dimensional crossover of thermal transport in few-layer graphene *Nat. Mater.* **9** 555–8

[90] Vallabhaneni A K, Singh D, Bao H, Murthy J and Ruan X 2016 Reliability of Raman measurements of thermal conductivity of single-layer graphene due to selective electron–phonon coupling: a first-principles study *Phys. Rev.* B **93** 125432

[91] Gu X, Fan Z, Bao H and Zhao C Y 2019 Revisiting phonon–phonon scattering in single-layer graphene *Phys. Rev.* B **100** 064306

[92] Lindsay L and Broido D A 2010 Optimized Tersoff and Brenner empirical potential parameters for lattice dynamics and phonon thermal transport in carbon nanotubes and graphene *Phys. Rev.* B **81** 205441

[93] Xie H, Chen L, Yu W and Wang B 2013 Temperature dependent thermal conductivity of a free-standing graphene nanoribbon *Appl. Phys. Lett.* **102** 111911

[94] Li Q-Y, Takahashi K, Ago H, Zhang X, Ikuta T, Nishiyama T and Kawahara K 2015 Temperature dependent thermal conductivity of a suspended submicron graphene ribbon *J. Appl. Phys.* **117** 065102

[95] Chen S *et al* 2011 Raman measurements of thermal transport in suspended monolayer graphene of variable sizes in vacuum and gaseous environments *ACS Nano* **5** 321–8

[96] Lee J-U, Yoon D, Kim H, Lee S W and Cheong H 2011 Thermal conductivity of suspended pristine graphene measured by Raman spectroscopy *Phys. Rev.* B **83** 081419

[97] Lindsay L, Broido D A and Mingo N 2011 Flexural phonons and thermal transport in multilayer graphene and graphite *Phys. Rev.* B **83** 235428

[98] Kim S I *et al* 2015 Dense dislocation arrays embedded in grain boundaries for high-performance bulk thermoelectrics *Science* **348** 109–14

[99] Hong M, Chasapis T C, Chen Z-G, Yang L, Kanatzidis M G, Snyder G J and Zou J 2016 *n*-type $Bi_2Te_{3-x}Se_x$ nanoplates with enhanced thermoelectric efficiency driven by wide-frequency phonon scatterings and synergistic carrier scatterings *ACS Nano* **10** 4719

[100] Xu B *et al* 2017 Nanocomposites from solution-synthesized PbTe–BiSbTe nanoheterostructure with unity figure of merit at low–medium temperatures (500–600 K) *Adv. Mater.* **29** 1605140

[101] Joshi Y P, Tiwari M D and Verma G S 1969 Role of four-phonon processes in the lattice thermal conductivity of silicon from 300 to 1300 K *Phys. Rev.* B **1** 642

[102] Carruthers P 1962 Resonance in phonon–phonon scattering *Phys. Rev.* **125** 123–5

[103] Ravichandran N K and Broido D 2019 Non-monotonic pressure dependence of the thermal conductivity of boron arsenide *Nat. Commun.* **10** 827

[104] Mukhopadhyay S, Parker D S, Sales B C, Puretzky A A, McGuire M A and Lindsay L 2018 Two-channel model for ultralow thermal conductivity of crystalline Tl_3VSe_4 *Science* **360** 1455–8

IOP Publishing

Nanoscale Energy Transport
Emerging phenomena, methods and applications
Bolin Liao

Chapter 3

Pre-interface scattering influenced interfacial thermal transport across solid interfaces

Ruiyang Li, Eungkyu Lee and Tengfei Luo

In this chapter, an overview of recent research progress on the effects of pre-interface scattering on interfacial thermal transport is provided. Interfacial thermal transport is of great importance to thermal management in electronics and the properties of thermoelectrics. Due to the unique scattering mechanism near the interface, much research effort has been focused on controlling the thermal boundary conductance (TBC) by engineering interfacial properties, such as interfacial bonding and geometry. Common wisdom tells us that scattering should impede thermal transport. However, it has been found that phonon scattering away from the interface may enhance TBC, which efficiently facilitates the energy communication among different phonon modes inside the material and redistributes the phonon energy into a more favorable distribution for interfacial thermal transport. Here, we briefly review the mechanism behind the pre-interface scattering effects and then discuss a few examples of such phenomena, where scatterings such as phonon–phonon, phonon–electron and phonon–isotope scattering in the pre-interface region can all serve the purpose of enhancing TBC. This may offer a unique and unconventional avenue for improving thermal transport across interfaces.

Interfaces play a critical role in the thermal management of next-generation high-power density electronics such as high-performance parallel computing or millimeter-wave communications for gigabit-data transmission [1, 2]. To develop advanced electronics, the dimensions of the unit devices (e.g. transistor) continue to decrease and their number density steadily increases. This inevitably leads to large densities of interfaces and presents unprecedented power densities in a single integrated circuit. However, the large thermal resistance across a large number of interfaces prevents effective power dissipation, leading to local hotspots, in particular in electrical channels, and strongly degrades the performance of devices, such as carrier-mobility drop, current level hysteresis, or gate-leakage current [3, 4].

doi:10.1088/978-0-7503-1738-2ch3

The thermal transport across solid interfaces can be characterized by the thermal boundary resistance (TBR) or its inverse, thermal boundary conductance (TBC) [5], which is a proportionality constant that relates the heat flux perpendicular to the interface to the temperature drop at the interface via

$$G = \frac{J}{A\Delta T}, \tag{3.1}$$

where J is the heat flux, A is the interfacial area and ΔT is the temperature difference between the two sides of the interface.

The interfacial thermal transport is determined by the transmission and scattering of energy carriers at the interfaces [6]. In most dielectrics and semiconductors, the major heat carrier is the phonon, which is the quantization of lattice vibration. In general, interfacial thermal resistance originates from phonon spectral mismatch of two different materials, which leads to the incomplete transmission of phonon energy. In the harmonic limit, where phonons scatter elastically at the interface and the transmitted phonon has the same frequency as the incident one, the interfacial thermal resistance is directly related to the relative discrepancies of the phonon density of states at a certain frequency. To estimate TBC without considering inelastic processes, the phonon has been traditionally interpreted either as a perfectly diffusive quasi-particle (e.g. the diffuse mismatch model (DMM) [6, 7]) or as a coherent wave (e.g. the acoustic mismatch model (AMM) [8]). The harmonic-limited mismatch models have been employed extensively to predict the TBC across simple solid–solid interfaces [9–11]. However, in many cases, these models cannot quantitatively reproduce experimentally measured TBR for various interfaces [12–14]. This inaccuracy of the harmonic-limited models is not unexpected because phonon–interface scattering can actually be a combination of elastic and inelastic processes, where the scattering processes change not only momentum but also frequency. Although many improvements have been made to consider the anharmonicity effect [15–18], neither of those improved models can accurately include the impact of atomic interfacial features (roughness, stress, imperfection, etc), and there is still no mismatch model that considers anharmonicity inside materials. Another powerful simulation tool is the atomistic Green's function (AGF) [19–21], which is, however, largely limited to small system sizes and harmonic phonon interactions. The AGF method was also improved to include the anharmonic effect in a one-dimensional system [22], but the effect should be dependent on many factors, such as material and temperature, which can significantly complicate the AGF calculation in the 3D case. In a recent work [14], the measured TBC of the GaN/ZnO interface considerably disagrees with the predictions using the DMM and AGF methods, suggesting that these models may be fundamentally flawed.

Molecular dynamics (MD) simulations, which fully include elastic and inelastic scattering and can incorporate the influence of disorders in the structure, have been extensively used to investigate thermal transport across different interfaces [23]. In MD simulations, atoms and molecules follow classical dynamics based on the time integration of Newton's equations of motion, while the interactions between atoms are decided by empirical interatomic potentials. The non-equilibrium molecular dynamics (NEMD) method, which imposes a temperature difference on the system, can provide a direct measure of the thermal resistance from the temperature difference at the interface. The shortcoming of MD simulation is that it cannot account for quantum mechanical phonon populations, which can be important below a material's Debye temperature. To study the transmission of individual phonon mode, the wave-packet (WP) method [24–26] has been employed to calculate a single phonon mode's transmission coefficient in an MD simulation. However, this method requires all other modes at zero temperature (not excited) and therefore is unable to include phonon–phonon interactions.

It is well acknowledged that TBC is closely related to many interfacial properties (geometry, bonding, disorder, etc) [27–31], which can lead to additional carrier scattering mechanisms that significantly affect interfacial thermal transport. According to traditional mismatch models, enhancing phonon scatterings at the interface will definitely impair thermal transport. However, with the power of the aforementioned simulation tools, many predictions draw questions on the validity of this conventional belief. Giri *et al* [32] have used MD simulations to demonstrate that the TBC across amorphous interfaces can be greater than that across the corresponding crystalline interfaces—an example where scattering is not necessarily always disadvantageous to interfacial thermal transport. Through MD simulations, Lee and Luo [33] predicted AlN, which can increase the coupling of some optical phonons, to be an optimal intermediate layer for GaN/SiC interfaces which leads to TBC enhancement. In addition, experimental techniques, such as time-domain thermoreflectance (TDTR) and frequency domain thermoreflectance (FDTR) [34–36], have been employed to investigate the thermal transport across engineered interfaces experimentally. Using the TDTR technique, Lee *et al* [37] measured the TBC of nanopatterned Si–metal interfaces and found that by introducing nano-pillars (50–100 nm) at the interface, the TBC can be increased by as much as 88%. This is somewhat counterintuitive since surface roughness is traditionally believed to scatter phonons and reduce TBC [38]. A recent study by Yates *et al* [39] showed that the conversion of an amorphous SiN layer between GaN and diamond to a disordered mixture of Si–C–N can significantly increase the TBC for GaN/diamond interfaces. Moreover, through the introduction of N_2 defects to amorphous SiOC:H/SiC:H interfaces, Giri *et al* [40] experimentally proved that interfacial defects can enhance the TBC.

All these findings suggest that the enhancement of the phonon scattering effect near the interface can potentially increase the TBC. Instead of contributing more to the TBC by not interacting with other phonon modes, phonons tend to transfer more energy across the interface by stronger interactions. Methods such as modal analyses were used to evaluate the phonon interactions in the vicinity of interfaces [41–46]. They found that the inelastic scattering can contribute significantly to the TBC, and that the interfacial region can bridge the phonons with significantly different energies. However, much less attention is paid to the scattering away from the interface, whose impact can even outweigh the impact of scattering at the interface. It is found that the pre-interface phonon scattering can not only facilitate the energy communication among different phonon modes but also optimize the energy distribution of phonons into a state that favors an overall more efficient interfacial energy transmission [47–49]. Based on this pre-interface scattering mechanism, several strategies were proposed to serve the purpose of energy redistribution. It has been proved that scattering from anharmonicity, isotopes and electrons can all help redistribute phonon energy and further improve the thermal transport across interfaces.

Using the NEMD method, Wu and Luo [47] provided a detailed study of the influence of anharmonicity on thermal transport across solid–solid interfaces. A schematic of the NEMD simulation set-up and an example temperature profile are shown in figure 3.1. The interface is formed by joining two model materials together. A typical simulation begins with equilibration of the whole system at a certain temperature with periodic boundary conditions in all three directions. Then, two thermostats with different temperatures are applied at each end to induce a heat flux across the system. After the steady state is reached, the production run is performed to extract the temperature profile. The temperature difference at the interface is then calculated by taking the difference of the linearly extrapolated temperature profiles of the two sides (figure 3.1(c)). Heat flux can be extracted by monitoring the energy change of the thermostats. The TBC is finally calculated according to equation (3.1).

Before digging into the simulation results, it is necessary to review all possible channels for heat transfer across interfaces in this case. Basically, there are two channels for interfacial thermal transport. One is the elastic channel, which means that the energy transfer occurs in the overlapped frequency region of two different materials. Another channel is attributed to the inelastic phonon transmission, which involves energy exchange among phonons with different frequencies. When we have a model interface consisting of two dissimilar lattices whose vibrational spectra are shown in figure 3.1(b), we can evaluate the anharmonic contribution because there is no overlap between the optical phonons of two lattices. It has been proved that inelastic channels can have contributions to the interfacial heat flux, but how exactly such inelastic channels enable frequency change of phonons is not clear.

As mentioned above, anharmonicity inside the materials is often ignored in many prediction models. Wu and Luo [47] found that the increased anharmonicity inside the materials can facilitate the energy communication of different phonon modes away from the interface, and that the pre-interface anharmonic phonon scattering can largely affect the interfacial thermal transport, while the scatterings right at the interface do not show an obvious impact on the TBC. In their study, the interatomic interactions are modeled using interatomic force constants (IFCs) up to the third order. The interatomic force is calculated as defined in equation (3.2),

$$\vec{F}_i = -\frac{\partial V}{\partial \vec{u}_i} = -\alpha_i - \sum_j \beta_{i,j}\vec{u}_j - \frac{1}{2!}\sum_{j,k} \gamma_{i,j,k}\vec{u}_j\vec{u}_k - \dots, \tag{3.2}$$

where α, β and γ are IFCs of the first, second and third orders. $\vec{u}_j$ is the atomic displacement, and the subscripts denote the atoms. By tuning the amplitude of the

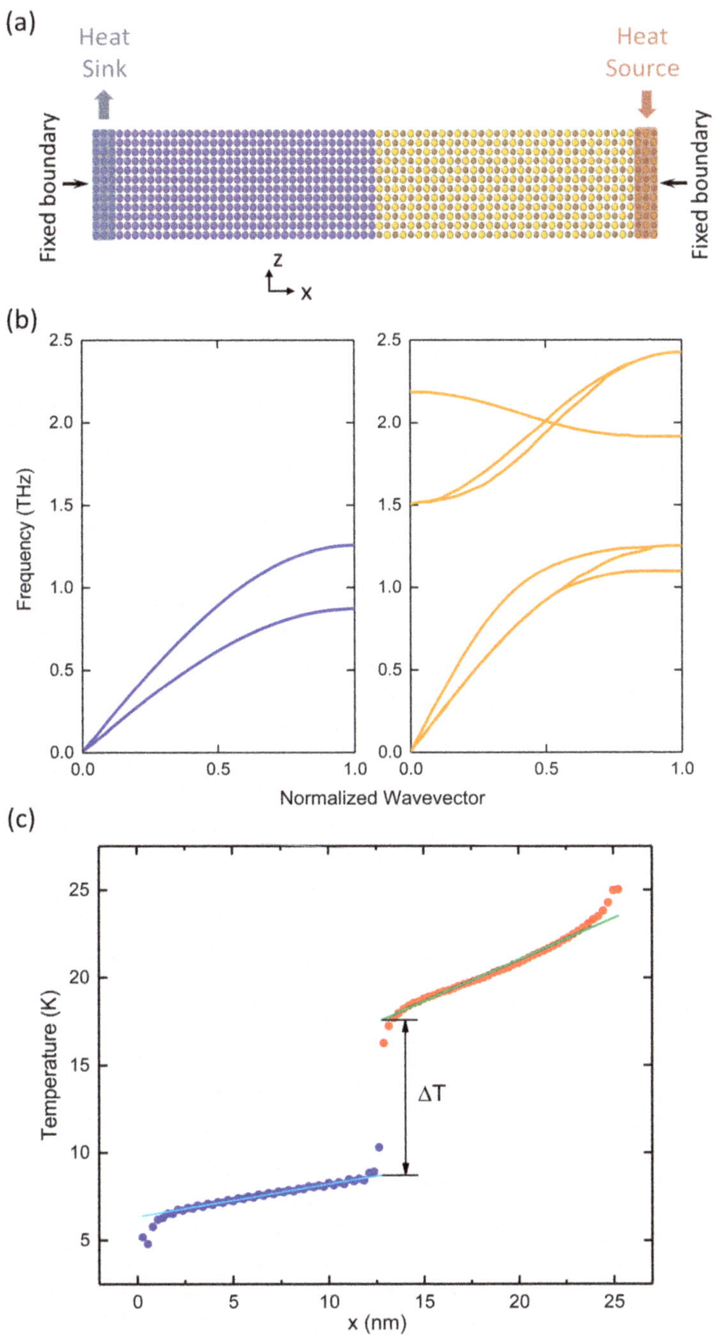

Figure 3.1. (a) An example NEMD simulation set-up of a solid–solid interface consisting of a monoatomic lattice (left) and a diatomic lattice (right). (b) Phonon dispersion of two lattices. (c) An example steady-state temperature profile in the simulation. Reproduced with permission from [47]. Copyright 2014 AIP Publishing.

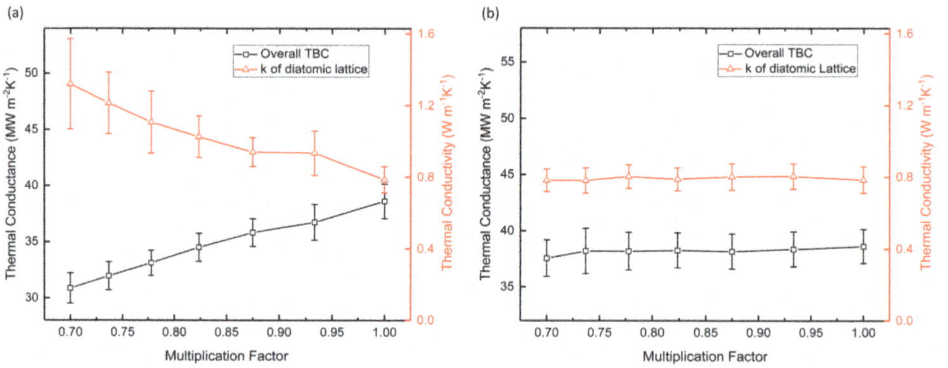

Figure 3.2. (a) TBC and thermal conductivity of the diatomic lattice when the cubic IFCs of the diatomic lattice are changed. The multiplication factor refers to the factor used to tune the cubic IFC. (b) TBC and thermal conductivity of the diatomic lattice when the cubic IFCs right at the interface are changed.

cubic IFCs (third-order), this model could isolate the anharmonic effect without changing the phonon spectra determined by harmonic IFCs. The two sides of the interface are made of monatomic FCC lattices and diatomic FCC lattices, respectively, and both lattices have the same IFCs. The monatomic lattice only has acoustic phonons, while the diatomic lattice has both acoustic and optical phonons (figure 3.1(b)). It should be noted that the masses of the atoms are chosen to ensure no direct spectral overlap between the optical phonons in the diatomic lattice and the acoustic phonons in the monatomic lattice (figure 3.1(b)), and thus it is possible to extract the unique behavior of the optical modes. As illustrated in figure 3.2, the overall TBC increases with the increased cubic IFCs inside the diatomic lattice (figure 3.2(a)), while it shows virtually no change with larger cubic IFCs right at the interface (figure 3.2(b)). This observation could not be explained by traditional mismatch models since the phonon spectra of two contacting materials are kept the same with the unchanged harmonic IFCs. It is also observed that the thermal conductivity of the diatomic lattice decreases as the increased anharmonicity inside the diatomic lattice enhances the phonon–phonon scattering. The thermal conductivity of the monoatomic lattice is not influenced. In the cases where the interfacial cubic IFCs are modified, there is no obvious change in the thermal conductivity of the two lattices.

The authors explained these results by proposing that the enhanced phonon scattering in the pre-interface region can increase the energy transferred from the optical phonons into the acoustic phonons that can easily carry energy across the interface. Spectral temperatures are calculated to verify this, which are related to the vibrational power spectrum of the non-equilibrium state and that of the equilibrium state at various positions [50–52]. The differences between spectral temperatures of different phonon groups can provide information about the efficiency of energy conversion among those groups. They divided the phonons into two groups: acoustic modes (0–1.3 THz) and optical modes (1.5–2.4 THz). Figures 3.3(a) and (b) illustrate the spectral temperatures for different cubic IFCs. There are smaller

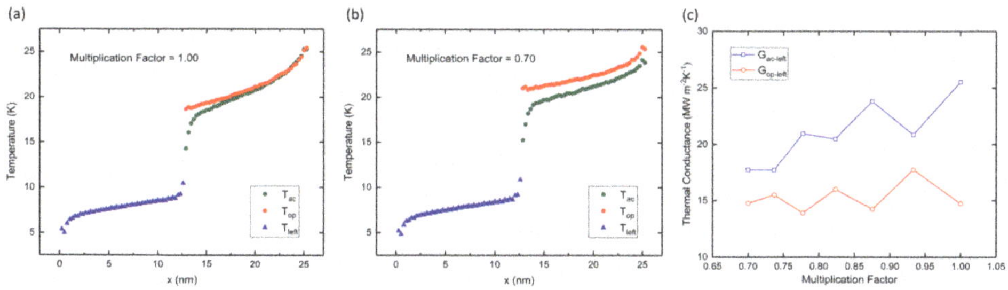

Figure 3.3. Spectral temperature profiles as a function of position in the x-direction, with the cubic force constants multiplied by (a) 1.00 and (b) 0.70. (c) TBC of different phonon groups ($G_{\text{ac-left}}$, $G_{\text{op-left}}$) with different anharmonicity.

differences between the spectral temperatures of two phonon groups in the diatomic lattice as the cubic IFCs increase, indicating that the enhanced anharmonicity improves the energy exchange between acoustic and optical modes inside the diatomic lattice due to stronger anharmonic phonon scattering processes.

The concept of the two-temperature model [10, 53] is borrowed to quantify the increase in the contribution of acoustic modes to thermal conductance when the anharmonicity is increased. This model can derive the contribution to the heat transfer of various phonon groups. $G_{\text{op-left}}$ ($G_{\text{ac-left}}$) describes the thermal conductance between the optical (acoustic) modes of the diatomic lattice and acoustic modes of the monoatomic lattice. As shown in figure 3.3(c), $G_{\text{ac-left}}$ is obviously larger than $G_{\text{op-left}}$, and $G_{\text{ac-left}}$ increases with the enhanced anharmonicity, but no significant change is present for $G_{\text{op-left}}$. This proves that the energy carried by acoustic phonons across the interface is enhanced as the pre-interface anharmonic scattering becomes stronger. Non-zero $G_{\text{op-left}}$ contradicts the traditional mismatch model, suggesting that the contribution of optical modes exists although there is no spectral overlap, and $G_{\text{op-left}}$ can also be viewed as the contribution of the inelastic channel for heat transfer. Specifically, the enhanced anharmonicity inside the material can facilitate the energy redistribution among phonon modes, where energy is transferred from optical modes to acoustic modes that carry most of the energy across the interface through coupling with phonons on the other side of the interface. In this case, phonon scattering away from the interface plays a key role in the increased TBC. A similar analysis was also conducted to investigate the effect of electron–phonon scattering on the phonon–phonon thermal transport at metal–nonmetal interfaces [48]. It was found that energy communications between different phonon modes inside the metal could be strengthened by electron–phonon interactions and further lead to larger phonon–phonon thermal conductance. While electron–phonon interaction usually cannot change the phonon frequency, it can alter the population of the scattered phonon modes and subsequently influence the phonon–phonon scattering where the phonon population is a factor in the scattering rate according to Fermi's golden rule [54].

In addition to scattering from anharmonicity and electrons, scattering from defects and isotopes may also help change the phonon energy distribution [55–57].

In a preliminary study using a similar model in figure 3.1(a), it was observed that doping isotopes with 10% higher mass randomly in the monoatomic lattice can increase the TBC (figure 3.4(a)). As the doping concentration of isotopes is enlarged, the TBC continues to increase. Inspired by that, Lee and Luo [49] used NEMD simulations to study the isotope effect on the thermal transport across the SiC/GaN interface, which is an important interface in power electronics. In their simulations, ^{15}N atoms are introduced as the isotopes to GaN, where ^{15}N randomly substitutes ^{14}N with different concentrations (f^{iso}). The calculated TBC ($G_{SiC/GaN}$) and thermal conductivity of GaN (k_{GaN}) as functions of f^{iso} are shown in figure 3.4(b). Apparently, isotope doping can enhance $G_{SiC/GaN}$ compared to the pristine case, and the increase is significant even with small concentrations. It can also be observed from the figure that k_{GaN} decreases drastically at low f^{iso} and then remains flat. Considering that the decreased thermal conductivity of doped GaN is related to the isotope–phonon scattering, the similar trend of enhanced $G_{SiC/GaN}$ should be the result of the same mechanism.

Since the isotope–phonon scattering can happen inside the material, further examination of the possible pre-interface scattering effect would be important. They also evaluate the effects of several geometrical features, such as the skin depth of the isotope region (L_{iso}) and its distance from the interface (L_D). Figure 3.5(a) depicts $G_{SiC/GaN}$ as a function of the skin depth of the isotope-doped region. All data show $G_{SiC/GaN}$ increases and then saturates at certain L_{iso}. It is notable that a thin doped layer is sufficient to induce obvious increases in TBC regardless of isotopic concentration. It is also seen that when the isotope-doped region is thin (<40), higher f^{iso} leads to larger TBC. These observations suggest that a larger number of isotopes will result in stronger isotope scattering, which should be responsible for an improved thermal transport across the interface.

Systems with isotope-doped regions placed at certain distances away from the interface are also studied in this work. The thickness of the doped layer is fixed to be

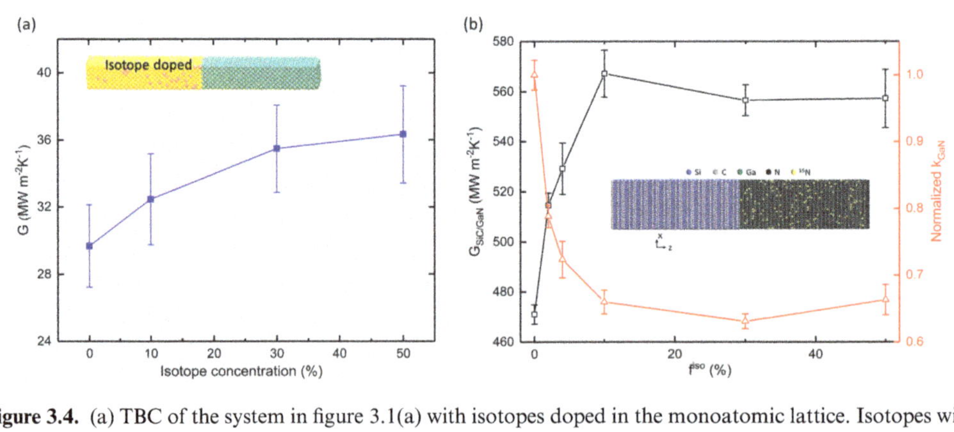

Figure 3.4. (a) TBC of the system in figure 3.1(a) with isotopes doped in the monoatomic lattice. Isotopes with 10% heavier mass are used. (b) TBC of isotope-doped SiC/GaN ($G_{SiC/GaN}$) and normalized thermal conductivity of GaN (k_{GaN}) as functions of isotope concentration (f^{iso}). The subplot is a schematic of the SiC/GaN interface structure with 10% ^{15}N isotope doped in the GaN layer.

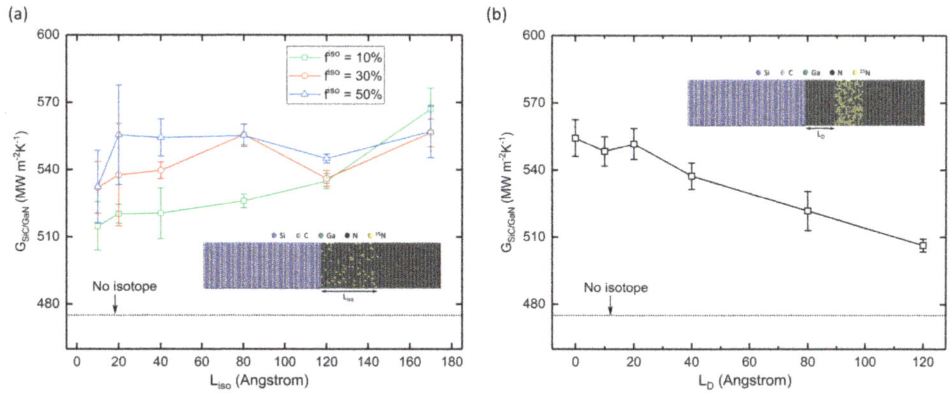

Figure 3.5. $G_{SiC/GaN}$ as a function of (a) the skin depth of the isotope-doped region (L_{iso}) and (b) the distance from the interface to the doped region (L_D), respectively. The insets are representative structures used in MD simulations. Reproduced with permission from [49]. Copyright 2018 AIP Publishing.

40 Å with f^{iso} = 50%. As shown in figure 3.5(b), the increase in $G_{SiC/GaN}$ drops as isotope scattering happens further away from the interface. However, all cases show higher thermal conductance than the pristine case, which confirms the effect of pre-interface isotope–phonon scattering. Similar to the study of electron–phonon scattering, the increased TBC can be explained in the way that the elastic isotope–phonon scattering helps optimize the phonon energy distribution so that acoustic–optical phonon interactions will be enhanced. Through these scattering processes, the energy carried by optical phonons is transferred to acoustic phonons, which play a pivotal role in transferring energy across the interface. But when the isotope-doped region is located away from the interface, phonons can be scattered again in the non-isotope region and this will offset the effect of isotope scattering before phonons reach the interface.

Spectral analysis can be used to quantify such a pre-interface isotope scattering effect. A system with 10% ^{15}N randomly distributed in the GaN layer was studied as an example. The phonon vibrational power spectrum (VPS) for this doped system shows that ^{15}N isotopes actually lead to a decrease in the VPS overlap between GaN and SiC, suggesting that the enhanced interfacial thermal transport cannot be explained by the mismatch models. Moreover, the spectral temperatures of two phonon groups for the pristine case and the doped case are depicted in figure 3.6. Although in the GaN layer the spectral temperatures of the low-frequency phonon group (0–20 THz) do not differ much from those of the high-frequency group (20–40 THz), an obvious difference is observed in the SiC layer, indicating that the inter-mode energy transfer is less efficient in SiC. After isotopes are introduced to the system, the spectral temperature difference between two phonon groups increases in SiC, which is mainly due to the increase in the spectral temperatures of low-frequency phonons. This means that the introduction of isotopes enables more energy from GaN to be received by low-frequency phonons in SiC. The source of the increased energy should correspond to low-frequency phonons in the GaN layer because the elastic process has the dominant impact in the thermal transport across

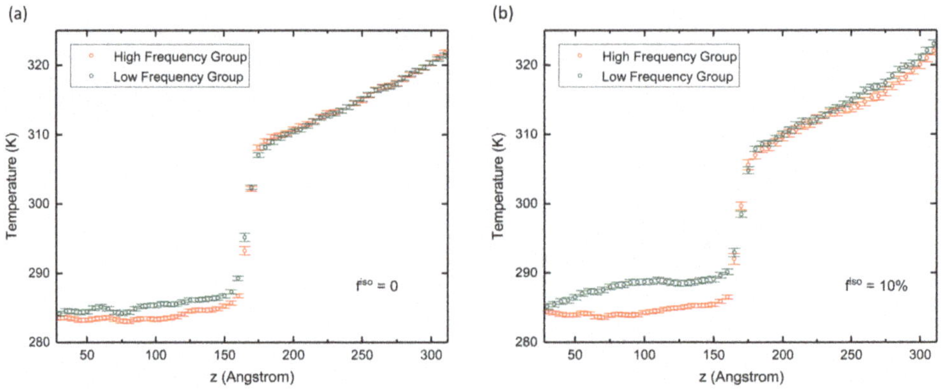

Figure 3.6. Spectral temperatures of high-frequency phonon group (0–20 THz) and low-frequency phonon group (20–40 THz) for the cases $f^{\mathrm{iso}} = 0$ and $f^{\mathrm{iso}} = 10\%$ in the GaN layer. Reproduced with permission from [49]. Copyright 2018 AIP Publishing.

interfaces. Therefore, the fact can be derived from the spectral analysis that the isotope–phonon interactions keep transferring energy from high-frequency phonons to low-frequency phonons, which can easily carry energy across the interface. In the meantime, the heavier ^{15}N isotope also helps reduce the frequency gap between two phonon groups and enhance the energy transfer among phonons.

These studies on pre-interface scattering indicate that scattering inside the materials should not be neglected when we try to understand interfacial thermal transport and predict TBC. This may provide new possibilities of improving TBC: we can engineer the materials rather than focus on the narrow interfacial regime. Additionally, useful guidance for interface design and material growth can be derived from these studies. For example, strategies such as doping isotopes can be achieved in conventional experiments, where growth techniques such as molecular beam epitaxy (MBE) are powerful tools to control the concentration of isotopes with high precision [58]. It is also noted that for real interface design in power electronics, the modification of interactions among heat carriers may lead to a variety of tradeoffs between thermal and electrical properties. Therefore, detailed assessments of possible tradeoffs are necessary when we are looking for optimal interface design in the development of future devices.

In summary, we reviewed the effects of pre-interface scattering on TBC across the solid–solid interface. Using various experimental and computational methods, many interfacial features were proven to have significant impacts on the TBC by modifying the phonon scatterings. Counterintuitively, the enhancement of phonon scattering effect near the interface can potentially increase the TBC. Moreover, it has been found that engineering some properties inside materials can improve the interfacial thermal transport as well, which is mainly attributed to the enhanced phonon scattering away from the interface. In general, controlled pre-interface scattering (phonon–phonon, phonon–electron and phonon–isotope scattering, etc) can effectively facilitate the energy communication among phonons and further improve the phonon energy distribution to favor better interfacial thermal transport.

Although there may be challenges still to be overcome to properly utilize these strategies in interface design without affecting other properties, these findings offer a new angle for improving thermal transport across interfaces.

References

[1] Pop E, Sinha S and Goodson K E 2006 Heat generation and transport in nanometer-scale transistors *Proc. IEEE* **94** 1587–601

[2] Pop E 2010 Energy dissipation and transport in nanoscale devices *Nano Res.* **3** 147–69

[3] Borkar S 2001 Low power design challenges for the decade *Proc. of the ASP-DAC 2001 Asia and South Pacific Design Automation Conf. 2001 (Cat. No. 01EX455) (IEEE),* pp 293–6

[4] Mahajan R, Nair R, Wakharkar V, Swan J, Tang J and Vandentop G 2002 Emerging directions for packaging technologies *Intel Technol. J.* **6** 62

[5] Cahill D G *et al* 2014 Nanoscale thermal transport. II. 2003–2012 *Appl. Phys. Rev.* **1** 011305

[6] Swartz E T and Pohl R O 1989 Thermal boundary resistance *Rev. Mod. Phys.* **61** 605–68

[7] Swartz E T and Pohl R O 1987 Thermal resistance at interfaces *Appl. Phys. Lett.* **51** 2200–2

[8] Little W A 1959 The transport of heat between dissimilar solids at low temperatures *Can. J. Phys.* **37** 334–49

[9] Cahill D G, Ford W K, Goodson K E, Mahan G D, Majumdar A, Maris H J, Merlin R and Phillpot S R 2003 Nanoscale thermal transport *J. Appl. Phys.* **93** 793–818

[10] Majumdar A and Reddy P 2004 Role of electron–phonon coupling in thermal conductance of metal–nonmetal interfaces *Appl. Phys. Lett.* **84** 4768–70

[11] Hopkins P E and Norris P M 2007 Effects of joint vibrational states on thermal boundary conductance *Nanoscale Microscale Thermophys. Eng.* **11** 247–57

[12] Lyeo H-K and Cahill D G 2006 Thermal conductance of interfaces between highly dissimilar materials *Phys. Rev.* B **73** 144301

[13] Hua C, Chen X, Ravichandran N K and Minnich A J 2017 Experimental metrology to obtain thermal phonon transmission coefficients at solid interfaces *Phys. Rev.* B **95** 205423

[14] Gaskins J T *et al* 2018 Thermal boundary conductance across heteroepitaxial ZnO/GaN interfaces: assessment of the phonon gas model *Nano Lett.* **18** 7469–77

[15] Reddy P, Castelino K and Majumdar A 2005 Diffuse mismatch model of thermal boundary conductance using exact phonon dispersion *Appl. Phys. Lett.* **87** 211908

[16] Prasher R 2009 Acoustic mismatch model for thermal contact resistance of van der Waals contacts *Appl. Phys. Lett.* **94** 041905

[17] Shin S, Kaviany M, Desai T and Bonner R 2010 Roles of atomic restructuring in interfacial phonon transport *Phys. Rev.* B **82** 081302

[18] Hopkins P E, Duda J C and Norris P M 2011 Anharmonic phonon interactions at interfaces and contributions to thermal boundary conductance *J. Heat Trans. ASME* **133** 062401

[19] Mingo N and Yang L 2003 Phonon transport in nanowires coated with an amorphous material: an atomistic Green's function approach *Phys. Rev.* B **68** 245406

[20] Tian Z, Esfarjani K and Chen G 2012 Enhancing phonon transmission across a Si/Ge interface by atomic roughness: first-principles study with the Green's function method *Phys. Rev.* B **86** 235304

[21] Zhang W, Fisher T S and Mingo N 2006 Simulation of interfacial phonon transport in Si–Ge heterostructures using an atomistic Green's function method *J. Heat Transfer* **129** 483–91

[22] Mingo N 2006 Anharmonic phonon flow through molecular-sized junctions *Phys. Rev.* B **74** 125402

[23] Luo T and Chen G 2013 Nanoscale heat transfer—from computation to experiment *Phys. Chem. Chem. Phys.* **15** 3389–412
[24] Baker C H, Jordan D A and Norris P M 2012 Application of the wavelet transform to nanoscale thermal transport *Phys. Rev.* B **86** 104306
[25] Schelling P K, Phillpot S R and Keblinski P 2002 Phonon wave-packet dynamics at semiconductor interfaces by molecular-dynamics simulation *Appl. Phys. Lett.* **80** 2484–6
[26] Schelling P K, Phillpot S R and Keblinski P 2004 Kapitza conductance and phonon scattering at grain boundaries by simulation *J. Appl. Phys.* **95** 6082–91
[27] Luo T and Lloyd J R 2012 Enhancement of thermal energy transport across graphene/graphite and polymer interfaces: a molecular dynamics study *Adv. Funct. Mater.* **22** 2495–502
[28] Jiang T, Zhang X, Vishwanath S, Mu X, Kanzyuba V, Sokolov D A, Ptasinska S, Go D B, Xing H G and Luo T 2016 Covalent bonding modulated graphene–metal interfacial thermal transport *Nanoscale* **8** 10993–1001
[29] Lee E, Zhang T, Hu M and Luo T 2016 Thermal boundary conductance enhancement using experimentally achievable nanostructured interfaces—analytical study combined with molecular dynamics simulation *Phys. Chem. Chem. Phys.* **18** 16794–801
[30] Zhang T, Gans-Forrest A R, Lee E, Zhang X, Qu C, Pang Y, Sun F and Luo T 2016 Role of hydrogen bonds in thermal transport across hard/soft material interfaces *ACS Appl. Mater. Interfaces* **8** 33326–34
[31] Lee E, Menumerov E, Hughes R A, Neretina S and Luo T 2018 Low-cost nanostructures from nanoparticle-assisted large-scale lithography significantly enhance thermal energy transport across solid interfaces *ACS Appl. Mater. Interfaces* **10** 34690–8
[32] Giri A, Hopkins P E, Wessel J G and Duda J C 2015 Kapitza resistance and the thermal conductivity of amorphous superlattices *J. Appl. Phys.* **118** 165303
[33] Lee E and Luo T 2017 The role of optical phonons in intermediate layer-mediated thermal transport across solid interfaces *Phys. Chem. Chem. Phys.* **19** 18407–15
[34] Cahill D G 2004 Analysis of heat flow in layered structures for time-domain thermoreflectance *Rev. Sci. Instrum.* **75** 5119–22
[35] Schmidt A J, Chen X and Chen G 2008 Pulse accumulation, radial heat conduction, and anisotropic thermal conductivity in pump–probe transient thermoreflectance *Rev. Sci. Instrum.* **79** 114902
[36] Schmidt A J, Cheaito R and Chiesa M 2009 A frequency-domain thermoreflectance method for the characterization of thermal properties *Rev. Sci. Instrum.* **80** 094901
[37] Lee E, Zhang T, Yoo T, Guo Z and Luo T 2016 Nanostructures significantly enhance thermal transport across solid interfaces *ACS Appl. Mater. Interfaces* **8** 35505–12
[38] Kazan M 2009 First principles calculation of the thermal conductance of GaN/Si and GaN/SiC interfaces as functions of the interface conditions *Appl. Phys. Lett.* **95** 141904
[39] Yates L *et al* 2018 Low thermal boundary resistance interfaces for GaN-on-diamond devices *ACS Appl. Mater. Interfaces* **10** 24302–9
[40] Giri A *et al* 2018 Interfacial defect vibrations enhance thermal transport in amorphous multilayers with ultrahigh thermal boundary conductance *Adv. Mater.* **30** e1804097
[41] Chalopin Y and Volz S 2013 A microscopic formulation of the phonon transmission at the nanoscale *Appl. Phys. Lett.* **103** 051602
[42] Murakami T, Hori T, Shiga T and Shiomi J 2014 Probing and tuning inelastic phonon conductance across finite-thickness interface *Appl. Phys. Exp.* **7** 121801

[43] Sääskilahti K, Oksanen J, Tulkki J and Volz S 2014 Role of anharmonic phonon scattering in the spectrally decomposed thermal conductance at planar interfaces *Phys. Rev.* B **90** 134312
[44] Gordiz K and Henry A 2015 A formalism for calculating the modal contributions to thermal interface conductance *New J. Phys.* **17** 103002
[45] Gordiz K and Henry A 2016 Phonon transport at interfaces: determining the correct modes of vibration *J. Appl. Phys.* **119** 015101
[46] Zhou Y and Hu M 2017 Full quantification of frequency-dependent interfacial thermal conductance contributed by two- and three-phonon scattering processes from nonequilibrium molecular dynamics simulations *Phys. Rev.* B **95** 115313
[47] Wu X and Luo T 2014 The importance of anharmonicity in thermal transport across solid–solid interfaces *J. Appl. Phys.* **115** 014901
[48] Wu X and Luo T 2015 Effect of electron–phonon coupling on thermal transport across metal–nonmetal interface—a second look *EPL (Europhys. Lett.)* **110** 67004
[49] Lee E and Luo T 2018 Thermal transport across solid–solid interfaces enhanced by pre-interface isotope–phonon scattering *Appl. Phys. Lett.* **112** 011603
[50] Carlborg C F, Shiomi J and Maruyama S 2008 Thermal boundary resistance between single-walled carbon nanotubes and surrounding matrices *Phys. Rev.* B **78** 205406
[51] Shenogina N, Keblinski P and Garde S 2008 Strong frequency dependence of dynamical coupling between protein and water *J. Chem. Phys.* **129** 155105
[52] Luo T and Lloyd J R 2010 Non-equilibrium molecular dynamics study of thermal energy transport in Au–SAM–Au junctions *Int. J. Heat Mass Transfer* **53** 1–11
[53] An M, Song Q, Yu X, Meng H, Ma D, Li R, Jin Z, Huang B and Yang N 2017 Generalized two-temperature model for coupled phonons in nanosized graphene *Nano Lett.* **17** 5805–10
[54] Luo T, Garg J, Shiomi J, Esfarjani K and Chen G 2013 Gallium arsenide thermal conductivity and optical phonon relaxation times from first-principles calculations *EPL (Europhys. Lett.)* **101** 16001
[55] Yang N, Zhang G and Li B 2008 Ultralow thermal conductivity of isotope-doped silicon nanowires *Nano Lett.* **8** 276–80
[56] Chen S, Wu Q, Mishra C, Kang J, Zhang H, Cho K, Cai W, Balandin A A and Ruoff R S 2012 Thermal conductivity of isotopically modified graphene *Nat. Mater.* **11** 203–7
[57] Lindsay L, Broido D A and Reinecke T L 2012 Thermal conductivity and large isotope effect in GaN from first principles *Phys. Rev. Lett.* **109** 095901
[58] Novikov S V, Morris R D, Kent A J, Geen H L and Foxon C T 2007 MBE growth of GaN using ^{15}N isotope for nuclear magnetic resonance applications *J. Cryst. Growth* **301–302** 417–9

Chapter 4

Introduction to the atomistic Green's function approach: application to nanoscale phonon transport

Renjiu Hu, Jinghang Dai and Zhiting Tian

The atomistic Green's function (AGF) approach is a powerful tool for modeling nanoscale phonon flows, in particular interface heat transport. A review of its development and comprehensive deduction of its formulas are presented in this chapter. The numerical implementation of AGF is illustrated in detail using a one-dimensional atom chain example. Recent advancements, such as the extension to three dimensions, mode decomposition and anharmonicity correction, are also included.

4.1 Introduction

The non-equilibrium Green's function was developed by Datta [1] to simulate electron transport. Mingo and Yang [2] extended it to study phonon transport and called it the atomistic Green's function (AGF). Although AGF was initially only applied to a one-dimensional system without translational periodicity, there has recently been significant progress in AGF development, including expansion in the three-dimensional model [3, 4], incorporation of first-principles force constants [5–7], mode-specific transmission calculation [8, 9] and anharmonicity correction [10, 11, 24], which make the AGF method a useful tool in thermal conductance calculations.

AGF is a very sophisticated method for phonon ballistic transport calculations which requires that a phonon propagates without any inelastic or phase-breaking scattering in the bulk. Therefore, when the scattering happens on the boundary or the interface is dominant, the scattering in the bulk can be neglected, and AGF can be a perfect tool for interfacial transport calculation.

The model we will use for interfacial transport is shown in figure 4.1.

doi:10.1088/978-0-7503-1738-2ch4

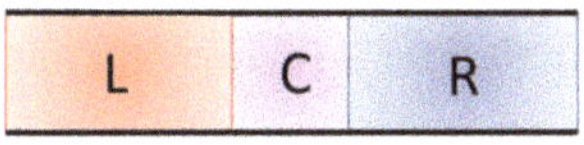

Figure 4.1. A general model of interfacial transmission.

The model is composed of two semi-infinite leads and the central region where the scattering might happen. Details about AGF will be introduced in the following sections.

4.2 Atomistic Green's function

4.2.1 Deduction of atomistic Green's functions

We start from a simple case. Consider an infinitely long one-dimensional atomic chain, which has uniform atoms with mass m and force constants K with neighboring atoms. Starting from the equation of motion, we could have

$$m\frac{\mathrm{d}^2u_n}{\mathrm{d}t^2} = -K(2u_n - u_{n-1} - u_{n+1}), \tag{4.1}$$

where u_n is the displacement of atom n. Meanwhile, we will also assume that the displacement function u_n has the form of plane waves (i.e. the plane wave assumption):

$$u_n(t) \sim \exp[\mathrm{i}(kna - \omega t)], \tag{4.2}$$

where k is the wavevector and ω is the angular frequency.

Substitute u_n into equation (4.1) with equation (4.2):

$$-\omega^2 u_n = -\frac{K}{m}(2u_n - u_{n-1} - u_{n+1}).$$

In matrix form

$$(\omega^2\mathbf{I} - \mathbf{D})\mathbf{u} = 0, \tag{4.3}$$

where the dynamical matrix $\mathbf{D}$ is

$$\mathbf{D} = \frac{1}{m}\begin{pmatrix} 2K & -K & & \\ -K & 2K & -K & \\ & -K & 2K & \ddots \\ & & \ddots & \ddots \end{pmatrix}.$$

In the general case, the elements in $\mathbf{D}$ are defined as [2]

$$\mathbf{D}_{ij} = \frac{1}{\sqrt{M_iM_j}}\begin{cases} \dfrac{\partial^2 U}{\partial u_i \partial u_j}, & i \neq j \\ -\displaystyle\sum_{i\neq j}\frac{\partial^2 U}{\partial u_i \partial u_j}, & i = j \end{cases},$$

where M_i is the mass of the ith atom.

Equation (4.3) could be regarded as a linear ordinary differential operator $\mathcal{L}$ times the displacement matrix $\mathbf{u}$,

$$\mathcal{L}\mathbf{u} = 0,$$

where $\mathcal{L} = (\omega^2\mathbf{I} - \mathbf{D})$. Therefore, we could solve this equation with the atomistic Green's function method.

Generally, suppose one would like to solve an ordinary differential equation

$$\mathcal{L}\mathbf{u}(x) = f(x),$$

where $f(x)$ is linearly independent of $\mathbf{u}(x)$. Then, the solution of $\mathbf{u}(x)$ can be written as

$$\mathbf{u}(x) = \int G(x, x')f(x)\mathrm{d}x',$$

where $G(x, x')$ is the atomistic Green's function that satisfies

$$\mathcal{L}G(x, x') = \delta(x - x'). \tag{4.4}$$

From quantum mechanics, the atomistic Green's function is the correlation function containing the information of the system under study, such as density of state (DOS). Therefore, we can use equation (4.4) to obtain the expression of the atomistic Green's function and subsequently the information of the system.

Solving equation (4.4) with $\mathcal{L} = (\omega^2\mathbf{I} - \mathbf{D})$, we could have

$$G = (\omega^2 - \mathbf{D})^{-1}.$$

However, this is the answer for a closed system, which means that there is no phonon exchange with the environment. For the transport problem we are discussing here, the two leads keep the constant temperatures, which indicates phonon exchange should happen between leads and the environment [13]. Therefore, to display the character of an open system, we add an infinitesimal complex $\mathrm{i}\eta(\eta \ll 1)$ into the expression

$$G = [(\omega + \mathrm{i}\eta)^2 - \mathbf{D}]^{-1}.$$

Here, $\mathrm{i}\eta$ indicates the leakage of phonons from leads [13].

Also, we notice that the system we are focusing on is a finite central region with two semi-infinite leads, which means that the size of $\mathbf{D}$ is infinite, making the inverse operation implausible on the computer. Therefore, an effective finite matrix representation of leads is necessary.

Suppose that when there is no coupling between the leads and the central region, the dynamical matrix of the whole system is D_0. The expression of D_0 should be

$$D_0 = \begin{pmatrix} D_\mathrm{L} & & \\ & D_\mathrm{c} & \\ & & D_\mathrm{R} \end{pmatrix}$$

which satisfies

$$(\omega^2 - D_0)G_0 = \mathbf{I}. \tag{4.5}$$

where G_0 is the corresponding atomistic Green's function. Here, the right-hand side of equation (4.5) is **I** because the matrix is under the site basis. Meanwhile, for the same system with coupling between leads and the central region, the dynamical matrix of the whole system D should be

$$D = \begin{pmatrix} D_{\mathrm{L}} & D_{\mathrm{LC}} & \\ D_{\mathrm{CL}} & D_{\mathrm{C}} & D_{\mathrm{CR}} \\ & D_{\mathrm{RC}} & D_{\mathrm{R}} \end{pmatrix}.$$

Here, we neglect the direct coupling between the two leads. It also satisfies

$$(\omega^2 - D)G = \mathbf{I}. \tag{4.6}$$

Here, G is also the corresponding atomistic Green's function of the dynamical matrix D.

The coupled dynamical matrix could also be presented by the uncoupled counterpart

$$D = D_0 + V,$$

where V is the coupling potential

$$V = \begin{pmatrix} & D_{\mathrm{LC}} & \\ D_{\mathrm{CL}} & & D_{\mathrm{CR}} \\ & D_{\mathrm{RC}} & \end{pmatrix}.$$

Substituting the dynamical matrix D in equation (4.6) with this expression and comparing it with equation (4.5), we could have the Dyson function:

$$G = G_0 + G_0VG = G_0 + GVG_0.$$

The Dyson function builds a bridge between the coupled and the uncoupled system. If the uncoupled atomistic Green's function and the coupling potential are simple for calculation, we can obtain the coupled counterpart rapidly via the Dyson function even if it is difficult to achieve directly.

Therefore, we could rewrite the atomistic Green's function in the central region as

$$G_{\mathrm{c}} = G_{\mathrm{c}}^0 + G_{\mathrm{c}}^0 D_{\mathrm{CL}} G_{\mathrm{LC}} + G_{\mathrm{c}}^0 D_{\mathrm{CR}} G_{\mathrm{RC}} \tag{4.7a}$$

$$G_{\mathrm{LC}} = G_{\mathrm{LC}}^0 + G_{\mathrm{L}}^0 D_{\mathrm{LC}} G_{\mathrm{C}} = G_{\mathrm{L}}^0 D_{\mathrm{LC}} G_{\mathrm{c}} \tag{4.7b}$$

$$G_{\mathrm{RC}} = G_{\mathrm{RC}}^0 + G_{\mathrm{R}}^0 D_{\mathrm{RC}} G_{\mathrm{C}} = G_{\mathrm{R}}^0 D_{\mathrm{RC}} G_{\mathrm{c}}. \tag{4.7c}$$

Here, we change the subscript 0 of the uncoupled system physical quantities to the superscript for compact. Substituting G_{LC} and G_{RC} in equation (4.7a) with equation (4.7b) and equation (4.7c), we could have

$$G_{\mathrm{c}} = G_{\mathrm{c}}^0 + G_{\mathrm{c}}^0 D_{\mathrm{CL}} G_{\mathrm{L}}^0 D_{\mathrm{LC}} G_{\mathrm{c}} + G_{\mathrm{c}}^0 D_{\mathrm{CR}} G_{\mathrm{R}}^0 D_{\mathrm{RC}} G_{\mathrm{c}} = G_{\mathrm{c}}^0 + G_{\mathrm{c}}^0 \Sigma G_{\mathrm{c}},$$

which has the same form as the Dyson function. Here, we define the 'self-energy' of the left lead as $\Sigma_{\mathrm{L}} = D_{\mathrm{CL}} G_{\mathrm{L}}^0 D_{\mathrm{LC}}$ and the 'self-energy' of the right lead as

$\Sigma_R = D_{CR}G_R^0 D_{RC}$. The 'self-energy' of leads $\Sigma = \Sigma_L + \Sigma_R$. Considering that the leads are semi-infinite, which has a continuous spectrum, self-energy cannot simulate this kind of spectrum itself. However, with the help of the infinitesimal complex $i\eta$ which can broaden the discrete spectrum to a continuous one, self-energy is enough to approximate the effect from semi-infinite [13]. The method to calculate the uncoupled atomistic Green's function of semi-infinite leads will be introduced in the next section.

The expression of the atomistic Green's function of the central region in the coupled system is

$$G_c = [(G_c^0)^{-1} - \Sigma]^{-1} = [(\omega)^2 - D_c - \Sigma]^{-1}.$$

As the atomistic Green's function contains the information of the whole system, we can obtain various thermal properties from it. First, we can calculate the density of states (DOS) of the phonon.

The definition of the DOS is

$$\rho(\omega) = \sum\nolimits_n \delta(\omega - \omega_n), \tag{4.8}$$

where ω_n is the square root of the eigenvalue that satisfies equation (4.3). Perform spectrum decomposition on the atomistic Green's function of the system:

$$G = [(\omega + i\eta)^2 - D]^{-1} = U \begin{pmatrix} [(\omega + i\eta)^2 - \omega_1^2]^{-1} & & \\ & \ddots & \\ & & [(\omega + i\eta)^2 - \omega_n^2]^{-1} \end{pmatrix} U^{-1},$$

where $U = (u_1 \ \cdots \ u_n)$ is the matrix consisting of eigenvectors. The matrix in the center, which we call Λ, is the corresponding eigenvalue matrix in diagonal form. Using the Cauchy relation $\lim\limits_{\eta \to 0} \frac{1}{\omega - \omega_0 \pm i\eta} = \mathcal{P}(\frac{1}{\omega - \omega_0}) \mp i\pi\delta(\omega - \omega_0)$, we could rewrite the eigenvalue matrix as

$$\Lambda = \begin{pmatrix} \mathcal{P}\left(\frac{1}{\omega^2 - \omega_1^2}\right) - i\frac{\pi}{2\omega}\delta(\omega - \omega_1) & & \\ & \ddots & \\ & & \mathcal{P}\left(\frac{1}{\omega^2 - \omega_n^2}\right) - i\frac{\pi}{2\omega}\delta(\omega - \omega_n) \end{pmatrix}.$$

Comparing to equation (4.8), DOS can be expressed as

$$\rho(\omega) = -\frac{2\omega}{\pi}\mathbf{Im}(\mathbf{Tr}(\Lambda)) = -\frac{2\omega}{\pi}\mathbf{Im}(\mathbf{Tr}(G)). \tag{4.9}$$

Here, we used the relation $\mathbf{Tr}(A) = \sum_n \lambda_n$, where λ_n is the corresponding eigenvalue of matrix A.

Second, we can calculate the transmission coefficient. The atomistic Green's function method is entirely equivalent to the Laudauer scattering approach [12]. Therefore, we can obtain the expression of transmission from the scattering view. Here, we present an intuitive deduction of the transmission coefficient. Recall equation (4.6):

$$\begin{pmatrix} D_L & D_{LC} & \\ D_{CL} & D_C & D_{CR} \\ & D_{RC} & D_R \end{pmatrix} \begin{pmatrix} u_L \\ u_C \\ u_R \end{pmatrix} = \omega^2 \begin{pmatrix} u_L \\ u_C \\ u_R \end{pmatrix}.$$

In the scattering view, we can regard $u_i(i = \mathrm{L, C, R})$ as the 'scattering state'. u_L can be split into two sub-states: $u_L = u_L^0 + u_L^1$, where u_L^0 is the eigenvector of D_L and can be considered as the 'incident wave'. u_L^1, therefore, is the 'reflected wave'. Using u_L^0 to present u_i, we can have

$$\begin{cases} u_L = (1 + G_L^0 D_{LC} G_C D_{CL}) u_L^0 \\ u_R = G_R^0 D_{RC} G_C D_{CR} u_L^0 \\ u_C = G_C D_{CL} u_L^0 \end{cases},$$

where $G_l^0(l = \mathrm{L, R})$ is the atomistic Green's function of uncoupled leads. The partial current from the lead to the central region is

$$j_{l=\mathrm{L,R}} = \mathrm{i}\hbar\omega(u_l^\dagger V_{lC} u_C - u_C^\dagger V_{Cl} u_l).$$

Therefore, the partial current from the central region to the right lead is

$$\begin{aligned} j_n &= -j_R = -\mathrm{i}\hbar\omega(u_R^\dagger V_{RC} u_C - u_C^\dagger V_{CR} u_R) \\ &= -\mathrm{i}\hbar\omega(u_L^{0\dagger} V_{LC} G_C^\dagger V_{RC} (G_R^{0\dagger} - G_R^0) V_{RC} G_C V_{CL} u_L^0) \\ &= \hbar\omega(u_L^{0\dagger} V_{LC} G_C^\dagger \Gamma_R G_C V_{CL} u_L^0), \end{aligned}$$

where we introduce the level-width function $\Gamma_{l=\mathrm{L\cdot R}} = \mathrm{i}(\Sigma_l - \Sigma_l^\dagger)$, which characterizes the level-broadening caused by coupling between leads and the central region.

The total current from the left lead to the central region should satisfy

$$J_L = \sum_n j_n = \hbar\omega(u_{Ln}^{0\dagger} V_{LC} G_C^\dagger \Gamma_R G_C V_{CL} u_{Ln}^0) f_L(\omega_n).$$

The Bose–Einstein distribution function $f_L(\omega_n)$ shows the population of states of the left lead. Compared to the expression under the Laudauer formula:

$$J_L = \int_0^\infty \hbar\omega f_L(\omega)\tau(\omega)\frac{\mathrm{d}\omega}{2\pi}.$$

We can have the transmission function

$$
\begin{aligned}
\Xi(\omega) &= 2\pi\sum_n \delta(\omega - \omega_n)(u_{Ln}^{0\dagger} V_{LC} G_C^\dagger \Gamma_R G_C V_{CL} u_{Ln}^0) \\
&= 2\pi\sum_n\sum_m \delta(\omega - \omega_n)(u_{Ln}^{0\dagger} V_{LC} u_m)(u_m^\dagger G_C^\dagger \Gamma_R G_C V_{CL} u_{Ln}^0) \\
&= \sum_m \left(u_m^\dagger G_C^\dagger \Gamma_R G_C V_{CL} \left(2\pi \sum_n \delta(\omega - \omega_n) u_{Ln}^0 u_{Ln}^{0\dagger} \right) V_{LC} u_m \right) \\
&= \mathrm{Tr}(\Gamma_L G_C^\dagger \Gamma_R G_C).
\end{aligned} \tag{4.10}
$$

Here, in equation (4.10), line 2 uses a complete orthogonal basis to change the calculation order, line 3 uses the spectrum decomposition to obtain the value of the expression of the inner brackets and line 4 uses the property that the value is invariant under cyclic permutation.

Finally, we obtain the expression of the transmission function. A rigorous deduction can be found in [13].

With the transmission function, we can compute some thermal properties of the system. For instance, using the Landauer formula, the net thermal current of the system is

$$
J = J_L - J_R = \int_0^\infty \hbar\omega (f_L(\omega) - f_R(\omega)) \Xi(\omega) \frac{d\omega}{2\pi}.
$$

When the temperature difference between the left and the right lead is small enough, we can obtain the linear term of temperature difference from equation (4.11) via the Taylor expansion:

$$
J \approx \int_0^\infty \hbar\omega \frac{\partial f(\omega, T)}{\partial T}|_T (T_L - T_R) \Xi(\omega) \frac{d\omega}{2\pi}.
$$

Here, the reference temperature T could be picked as $T = \frac{T_L - T_R}{2}$. Then, we could calculate the thermal conductance σ

$$
\sigma(T) = \frac{J}{T_L - T_R} = \int_0^\infty \hbar\omega \frac{\partial f(\omega, T)}{\partial T} \Xi(\omega) \frac{d\omega}{2\pi}. \tag{4.11}
$$

4.2.2 Self-energy and surface Green's function

In the previous section, we have defined the self-energy matrices

$$
\Sigma_L^r \equiv D_{CL} g_L^r D_{LC} \tag{4.12}
$$

$$
\Sigma_R^r \equiv D_{CR} g_R^r D_{RC}, \tag{4.13}
$$

where D_{CL}, D_{LC}, D_{CR} and D_{RC} are the dynamics matrices connecting the central region to the leads, and $g^r{}_{\mathrm{L}}$ and $g^r{}_{\mathrm{R}}$ are the surface Green's functions for the semi-infinite leads. In order to compute transmission and thermal conductance, self-energy matrices are essential since they quantify the effect of the leads onto the central region. From the previous section, we know the dimension of the dynamical matrix is center-length × ∞, which is impossible for us to compute directly. In this scenario, we divide the semi-infinite lead into infinitely many layers with identical size. Moreover, we assume that the interaction only exists between nearest-neighboring cells. After this simplification, supposing the degree of freedom of one such principle layer is n_{PL}, the dynamics matrices, take D_{CR} for example, will be in the form

$$D_{\mathrm{CR}} = (\tilde{D}_{\mathrm{CR}} \;\; 0_{\mathrm{dC}\times n_{\mathrm{PL}}} \;\; \cdots \;\; 0_{\mathrm{dC}\times n_{\mathrm{PL}}}). \tag{4.14}$$

Although the dynamics matrices are large, there is now only one non-zero submatrix. Additionally, in order to compute the central atomistic Green's function G_{C}, we only need to compute the product of $D_{\mathrm{CL}}g_{\mathrm{L}}D_{\mathrm{LC}}$ and $D_{\mathrm{CR}}g_{\mathrm{R}}D_{\mathrm{RC}}$, rather than $g^r{}_{\mathrm{L}}$ and $g^r{}_{\mathrm{R}}$ explicitly. Since D_{CL} and D_{CR} have only one non-zero submatrix, we only need to compute the last diagonal element of $g^r{}_{\mathrm{L}}$ or the first diagonal element of $g^r{}_{\mathrm{R}}$:

$$\Sigma^r_{\mathrm{R}} \equiv (\tilde{D}_{\mathrm{CR}} \;\; 0_{\mathrm{dC}\times n_{\mathrm{PL}}} \;\; \cdots) \begin{pmatrix} \left(g^r_{\mathrm{R}}\right)_{0,0} & \cdots \\ \vdots & \ddots \end{pmatrix} \begin{pmatrix} \tilde{D}_{\mathrm{CR}} \\ 0_{n_{\mathrm{PL}}\times \mathrm{dC}} \\ \vdots \end{pmatrix} = \tilde{D}_{\mathrm{CR}}\left(g^r_{\mathrm{R}}\right)_{0,0}\tilde{D}_{\mathrm{CR}}, \tag{4.15}$$

where the $(g^r{}_{\mathrm{R}})_{0,0}$ element is the projection of the atomistic Green's function of the left lead onto the central region and it is the surface Green's function we defined before. From such a projecting process, we can obtain the self-energy matrices by only computing the surface Green's function rather than dealing with the infinite-dimension matrices. Various projection techniques have been developed in past decades, however, all of them were initially designed for electronic structure problems. Nonetheless, in the actual computation for phonons, we can apply these methods in a straightforward manner.

One conventional technique is to use the Dyson equation. Another obvious method is to solve the surface Green's function in a recursive formalism, such as the forward iteration scheme [14]. However, such a converging process is usually very slow. A more efficient and popular method is that of Sancho *et al* [15], which is well suited to evaluate the surface Green's function of a semi-infinite crystal. It will be easier to understand these methods with a simple model. Therefore, we use the following section for a detailed demonstration.

4.2.3 Phonon transport in one-dimensional systems

In this section, we will study different harmonic systems to illustrate how the AGFs are implemented in practice.

We will study a model of the atomic junction with two semi-infinite leads linked to the center region, as an illustration, where the center region is regarded as the interface in this model. The leads are assumed to be in thermal equilibrium all the

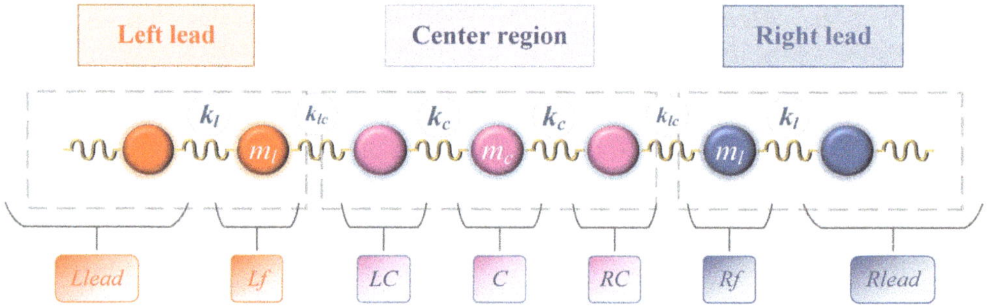

Figure 4.2. A one-dimensional toy model.

time. Their temperature difference will introduce a net phonon flow across the 'interface'. In order to obtain a quantitative description on the thermal transport, we first demonstrate how to derive the atomistic Green's functions of this system, in particular for obtaining different matrices such as D_{CL}, G_{C} and $(g^r{}_{\mathrm{R}})_{0,0}$.

As figure 4.2 shows, three different regions, the left lead region (Llead + Lf), the central region (LC + C + RC), and the right lead region (Rf + Rlead) are denoted. Each adjacent atom is connected by an elastic spring. For simplicity and generality, only three atoms are included in the central region and two of them are connected to two semi-infinite leads, respectively. The spacing is assumed to be identical.

In this scenario, the total dynamics matrix for the entire system is defined to be

$$\mathbf{D} = \begin{pmatrix} D_{\mathrm{L}} & D_{\mathrm{LC}} & 0 \\ D_{\mathrm{CL}} & D_{\mathrm{C}} & D_{\mathrm{CR}} \\ 0 & D_{\mathrm{RC}} & D_{\mathrm{R}} \end{pmatrix}$$

$$= \begin{pmatrix} \ddots & -\frac{k^l}{m^l} & & & & & \\ -\frac{k^l}{m^l} & \frac{k^{cl}+k^l}{m^l} & -\frac{k^{cl}}{\sqrt{m^c m^l}} & & & & \\ & -\frac{k^{cl}}{\sqrt{m^c m^l}} & \frac{k^c+k^l}{m^c} & -\frac{k^c}{m^c} & & & \\ & & -\frac{k^c}{m^c} & \frac{2k^c}{m^c} & -\frac{k^c}{m^c} & & \\ & & & -\frac{k^c}{m^c} & \frac{k^c+k^l}{m^c} & -\frac{k^{cl}}{\sqrt{m^c m^l}} & \\ & & & & -\frac{k^{cl}}{\sqrt{m^c m^l}} & \frac{k^{cl}+k^l}{m^l} & -\frac{k^l}{m^l} \\ & & & & & -\frac{k^l}{m^l} & \ddots \end{pmatrix}. \tag{4.16}$$

The sub-matrices will have the following forms:

$$D_{\mathrm{C}} = \begin{pmatrix} \frac{k^c + k^l}{m^c} & -\frac{k^c}{m^c} & 0 \\ -\frac{k^c}{m^c} & \frac{2k^c}{m^c} & -\frac{k^c}{m^c} \\ & -\frac{k^c}{m^c} & \frac{k^c + k^l}{m^c} \end{pmatrix}_{3\times 3} \tag{4.17a}$$

$$D_{\mathrm{L}} = \begin{pmatrix} \ddots & -\frac{k^l}{m^l} & 0 \\ -\frac{k^l}{m^l} & \frac{2k^l}{m^l} & -\frac{k^l}{m^l} \\ 0 & -\frac{k^l}{m^l} & \frac{k^{cl} + k^l}{m^l} \end{pmatrix}_{P\times P}, \tag{4.17b}$$

$$D_{\mathrm{R}} = \begin{pmatrix} \frac{k^{cl} + k^l}{m^l} & -\frac{k^l}{m^l} & 0 \\ -\frac{k^l}{m^l} & \frac{2k^l}{m^l} & -\frac{k^l}{m^l} \\ 0 & -\frac{k^l}{m^l} & \ddots \end{pmatrix}_{P\times P}, \tag{4.17c}$$

$$D_{\mathrm{CL}} = \begin{pmatrix} \cdots & 0 & \cdots & -\frac{k^{cl}}{\sqrt{m^c m^l}} \\ \cdots & 0 & \cdots & 0 \\ \cdots & 0 & \cdots & 0 \end{pmatrix}_{3\times P} \tag{4.17d}$$

$$D_{\mathrm{CR}} = \begin{pmatrix} -\frac{k^{cl}}{\sqrt{m^c m^l}} & \cdots & 0 & \cdots \\ 0 & \cdots & 0 & \cdots \\ 0 & \cdots & 0 & \cdots \end{pmatrix}_{3\times P}, \tag{4.17e}$$

where $D_{\mathrm{LC}} = D_{\mathrm{CL}}^{\dagger}$, $D_{\mathrm{RC}} = D_{\mathrm{CR}}^{\dagger}$.

The dimension of the central region is 3 for the only three atoms. The size of D_{L} and D_{R} will be $P \times P$, where P goes to ∞. The size of D_{LC} and D_{CR} is $3 \times P$. Just as we mentioned before, there is only one submatrix with size 3×3 that is non-zero. Before we apply the method in section 4.2.1 to calculate the transmission, we first introduce several frequently used approaches to compute the surface Green's function for this one-dimensional atom chain example.

(1) *Analytical formula of the surface Green's function.*

Although the matrices are large, just as we mentioned before, both D_{LC} and D_{CR} have only one non-zero element, provided that only one atom of the device is connected to the respective lead for the 1D atom chain. Consequently, to compute the atomistic Green's function G_{C}, we only need to compute the product of $D_{\mathrm{CL}}g_{\mathrm{L}}D_{\mathrm{LC}}$ and $D_{\mathrm{CR}}g_{\mathrm{R}}D_{\mathrm{RC}}$. Referring to the form of D_{CL} and D_{CR} in equations (4.17d) and (4.17d), it is evident that they have only one non-zero element. Hence, only the last diagonal element of g_{L} or the first diagonal element of g_{R} needs to be computed. The remaining elements in g_{L} and g_{R} are not necessary since they do not contribute to the product.

$g_{\mathrm{L}}(R, R)$ and $g_R(1, 1)$ correspond to the contact atoms that are bonded with the device. As illustrated in figure 4.2, the atoms in leads are split into two categories: Llead (Rlead) and Lf (Rf). The atoms in Llead (Rlead) correspond to the bulk of the contact which do not have any bonds with device atoms. Only the atoms in Lf (Rf) have bonds with central region atoms in the Llead (Rlead). On the other hand, the atoms in region C are the interior atoms of the device, which have no bonds with the leads. Based on this notation, the harmonic matrices of the contacts can be further split into components corresponding to the regions Llead (Rlead) and Llead (Rlead):

$$D_{\mathrm{L}} = \begin{pmatrix} D_{\mathrm{Llead}} & D^{\dagger}_{\mathrm{Llead,Lf}} \\ D_{\mathrm{Llead,Lf}} & D^{\mathrm{Lf}} \end{pmatrix}_{P\times P} \tag{4.18a}$$

$$D_{\mathrm{Llead}} = \begin{pmatrix} \ddots & -\dfrac{k^l}{m^l} & 0 \\ -\dfrac{k^l}{m^l} & \dfrac{2k^l}{m^l} & -\dfrac{k^l}{m^l} \\ 0 & -\dfrac{k^l}{m^l} & \dfrac{2k^l}{m^l} \end{pmatrix}_{(P-1)\times(P-1)} \tag{4.18b}$$

$$D_{\mathrm{Llead,Lf}} = \begin{pmatrix} \cdots & 0 & \cdots & 0 & -\dfrac{k^l}{m^l} \end{pmatrix}_{1\times(P-1)} \tag{4.18c}$$

$$D_{\mathrm{Lf}} = \left[\frac{k^c + k^l}{m^l}\right]_{1\times 1}. \tag{4.18d}$$

We only need the elements of the atomistic Green's function that correspond to atoms in the region Lf (Rf). Such elements are called the surface Green's functions. In this scenario, the self-energy matrices will be

$$\Sigma^r_{\mathrm{L}} \equiv D_{\mathrm{CL}}g^r_{\mathrm{L}}D_{\mathrm{LC}} = D_{\mathrm{Llead,Lf}}g^r_{\mathrm{Llead}}D^{\dagger}_{\mathrm{Llead,Lf}} \tag{4.19}$$

$$\Sigma_R^r \equiv D_{CL} g_L^r D_{LC} = D_{\text{Rlead,Rf}} g_{\text{Rlead}}^r D_{\text{Rlead,Rf}}^\dagger. \tag{4.20}$$

At the same time, we give a simple matrix inversion relation that is necessary to understand our method. The inversion of a 2 × 2 block matrix $\boldsymbol{M}$ can be reduced to an effective inversion focus on the last block [14]:

$$\boldsymbol{M} = \begin{pmatrix} A_{11} & A_{12} \\ A_{21} & A_{22} \end{pmatrix} \tag{4.21}$$

$$\boldsymbol{M}^{-1} = \begin{pmatrix} \ddots & \vdots \\ \cdots & [A_{22} - A_{21}A_{11}^{-1}A_{12}]^{-1} \end{pmatrix}. \tag{4.22}$$

The above relation can easily be verified. On the basis of this relation, we can present the surface Green's function in the form

$$g_L^{\text{surface}} = \left[(\omega + i\eta)^2 \boldsymbol{I} - D_{\text{Lf}} - D_{\text{Llead,Lf}}[(\omega + i\eta)^2 I - D^{\text{Llead}}]^{-1} D_{\text{Llead,Lf}}^\dagger\right]^{-1}, \tag{4.23}$$

where the submatrix in **A** should be

$$A_{11} = (\omega + i\eta)^2 \boldsymbol{I} - D_{\text{Llead}},\ A_{12} = D_{\text{Llead,Lf}}^\dagger,$$
$$A_{21} = D_{\text{Llead,Lf}},\ A_{22} = (\omega + i\eta)^2 I - D_{\text{Lf}}.$$

We can rewrite it to be

$$g_L^{\text{surface}} = \left[(\omega + i\eta)^2 \boldsymbol{I} - D_{\text{Lf}} - D_{\text{Llead,Lf}} g_{\text{Llead}}^{\text{surface}} D_{\text{Llead,Lf}}^\dagger\right]^{-1}, \tag{4.24}$$

where we have a new surface Green's function $g_{\text{Llead}}^{\text{surface}}$ with the size $(P-1) \times (P-1)$.

Similarly, we note that solving for $g_{\text{Llead}}^{\text{surface}}$ can use this inversion relation as well since $D_{\text{Llead,Lf}}$ still has only one non-zero element.

Obviously, such a process can be continued, through which we can find a way to evaluate the surface Green's function analytically. Suppose the corresponding lead contains only n layers, with the above method we can find the relationship between the surface Green's function of the nth layers' lead and the surface Green's function for the $(n-1)$th layers' lead with the frontier layer removed:

$$g_{\text{Llead},n}^{\text{surface}} = \left[(\omega + i\eta)^2 - \frac{2k^l}{m^l} - \frac{k^l}{m^l} g_{\text{Llead},n-1}^{\text{surface}} \frac{k^l}{m^l}\right]^{-1}. \tag{4.25}$$

To fulfill our previous assumption of the semi-infinite lead, we let n go to infinity. Now the surface Green's function for the nth layer and $(n-1)$th layer should be the same. Equation (4.24) will become

$$g_{\text{Llead}}^{\text{surface}} = \left[(\omega + i\eta)^2 - \frac{2k^l}{m^l} - \frac{k^l}{m^l} g_{\text{Llead}}^{\text{surface}} \frac{k^l}{m^l}\right]^{-1}. \tag{4.26}$$

Then we can have an analytical solution for the surface Green's function:

$$g_{\text{Llead}}^{\text{surface}} = \frac{\left[(\omega + \mathrm{i}\eta)^2 - \frac{2k^l}{m^l}\right] - \sqrt{\left[(\omega + \mathrm{i}\eta)^2 - \frac{2k^l}{m^l}\right]^2 - 4\left(\frac{k^l}{m^l}\right)^2}}{2\left(\frac{k^l}{m^l}\right)^2}. \tag{4.27}$$

(2) *Forward iteration method.*

As we discussed in the previous part, we usually compute the surface Green's function of an isolated, semi-infinite lead. For instance, the definition of the surface Green's function for the right lead is

$$[(\omega + \mathrm{i}\eta)^2 - D^{\mathrm{R}}]g_{\mathrm{R}}^r = \boldsymbol{I}, \tag{4.28}$$

where D^{R} is the dynamics matrix for the right lead. Generally, the form of the surface Green's function $g^r{}_{\mathrm{R}}$ will be

$$g_{\mathrm{R}}^r = \begin{pmatrix} g_{00} & g_{01} & \cdots & g_{0n} & \cdots \\ g_{10} & g_{11} & \cdots & g_{1n} & \cdots \\ \vdots & \vdots & \ddots & \cdots & \cdots \\ g_{n0} & g_{n1} & \vdots & g_{nn} & \cdots \\ \vdots & \vdots & \vdots & \vdots & \ddots \end{pmatrix}. \tag{4.29}$$

We only need the $(g^r{}_{\mathrm{R}})_{0,0}$ to calculate the self-energy. Following the definition of the right lead dynamics matrix in equation (4.17b) to expand equation (4.27), in the left-most column of $g^r{}_{\mathrm{R}}$, we have

$$\left[(\omega + \mathrm{i}\eta)^2 - \frac{(k^c + k^l)}{m^l}\right]g_{00} + \frac{k^l}{m^l}g_{10} = 1 \tag{4.30}$$

$$\frac{k^l}{m^l}g_{\xi-1,0} + \left[(\omega + \mathrm{i}\eta)^2 - \frac{2k^c}{m^l}\right]g_{\xi 0} + \frac{k^l}{m^l}g_{\xi+1,0} = 0,\ \xi = 1, 2, \ldots. \tag{4.31}$$

A forward iteration scheme gives a solution for $(g^r{}_{\mathrm{L}})_{0,0}$ by considering the original semi-infinite lead and the lead removed with one layer [16]. In the same way we obtain the formula of the central part of the atomistic Green's function, we can have

$$g_{00} = \left[(\omega + \mathrm{i}\eta)^2 I - \frac{k^c + k^l}{m^l} - \frac{k^l}{m^l}\hat{g}\frac{k^l}{m^l}\right]^{-1}, \tag{4.32}$$

where $\hat{g}$ is the surface Green's function for the lead with the layer nearest to the central region removed, i.e. $\hat{g}$ is the top-left element satisfying equation (4.28), with the corresponding column and row also being removed in matrix D^r. However, it is easy to see that such an operation will not change

D^{R} since our system is semi-infinite and translationally invariant. If we keep removing the layers, equation (4.32) will have the form

$$\hat{g} = \left[(\omega + \mathrm{i}\eta)^2 I - \frac{2k^l}{m^l} - \frac{k^l}{m^l}\hat{g}\frac{k^l}{m^l} \right]^{-1}. \quad (4.33)$$

This forward iteration can start with $\hat{g}$ equal to 0 until convergence.

However, such iterations converge slowly. There are various fast convergence methods for calculating the surface Green's function. We now describe one scheme which is suitable for most phonon computing cases.

(3) *Decimation scheme.*

The basic idea of the decimation scheme is similar to the previous forward iteration method. This method was developed by Sancho *et al*, and doubles the number of layers in each iteration and takes the effect of $2i$ layers in the ith iteration. We refer the reader to [15] for its derivation. A pseudo-code for this method is listed here:

$\mathrm{r} \leftarrow k_l/m_l$
$t \leftarrow 2k_l/m_l$
$\alpha \leftarrow -k_l/m_l$
$\beta \leftarrow \alpha^T$
$g \leftarrow [(\omega + \mathrm{i}\eta)^2 \mathcal{I} - t]^{-1}$
do
  $r' \leftarrow r + \alpha g \beta$
   if $|r' - r| < \epsilon$, exit
  $r \leftarrow r'$
  $t \leftarrow t + \alpha g \beta + \beta g \alpha$
  $\alpha \leftarrow \alpha g \alpha$
  $\beta \leftarrow \beta g \beta$
  $g \leftarrow [(\omega + \mathrm{i}\eta)^2 \mathcal{I} - t]^{-1}$
end
$g \leftarrow [(\omega + \mathrm{i}\eta)^2 \mathcal{I} - r]^{-1}$,

where the left arrow '$\leftarrow$' denotes assignment, ε is an error tolerance and $\mathrm{i}\eta$ indicates the leakage of phonons from leads.

If the spring constant and the mass are all set to be unity, we can obtain the real and imaginary parts of the surface Green's function, which is the same for both the left and right leads. We use the decimation scheme and the result is shown in figure 4.3.

We can also check this result using the analytical solution in equation (4.27), which should match the numerical solution.

After we obtain the dynamic matrix and the self-energy matrix of the leads, we can start to calculate the transmission function and thermal conductance for this toy model. We set a system with all the atoms being

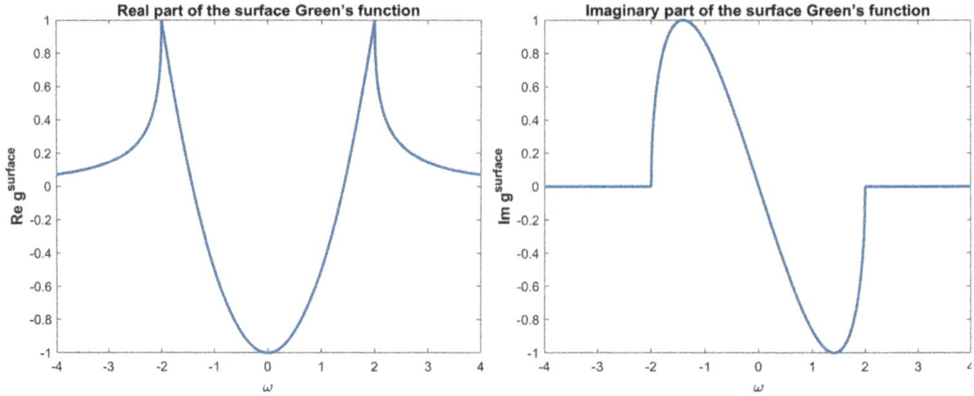

Figure 4.3. Plot of surface Green's function.

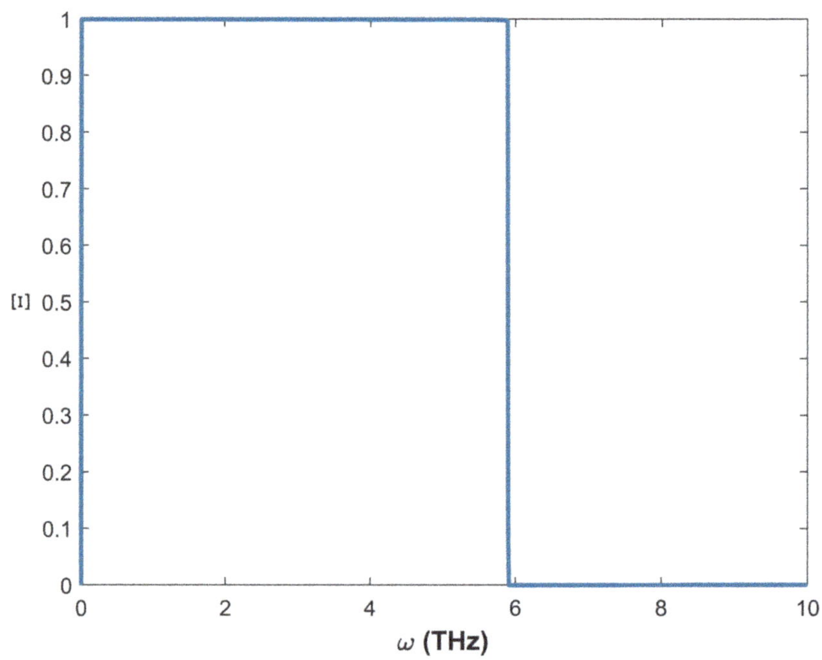

Figure 4.4. Transmission function for a one-dimensional silicon chain.

silicon atoms (mass set to 28.0850) and the spring constant is set to 1 eV $Å^{-2}$. Following the procedure in section 4.2.1, and using equation (4.10) to obtain the transmission (figure 4.4).

The heat flux and the thermal conductance can be computed with the transmission function. Figure 4.5 illustrates how thermal conductance changes concerning temperature for a one-dimensional atom chain. The

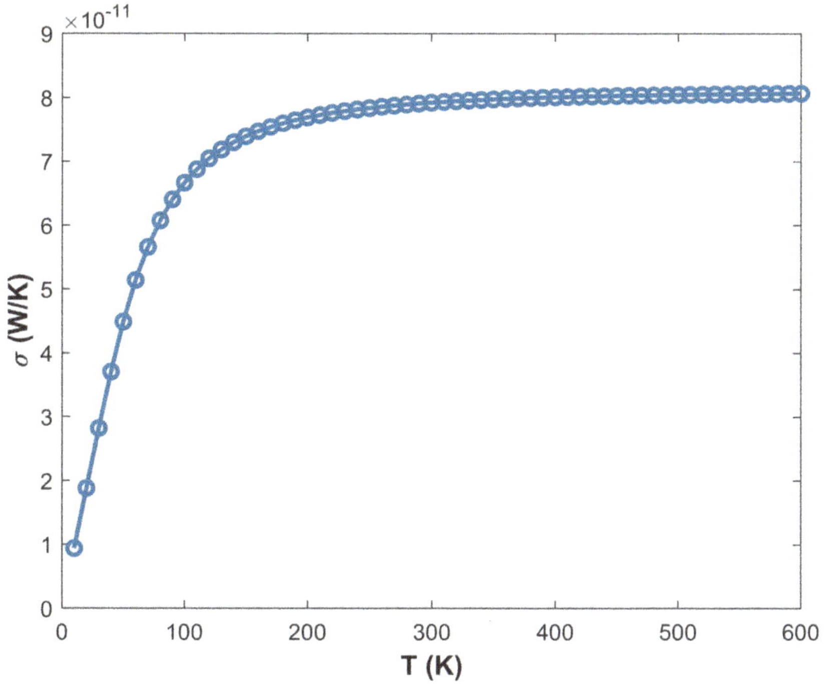

Figure 4.5. Thermal conductance for a one-dimensional silicon chain (W/K).

thermal conductance first increases and gradually converges when the temperature reaches a certain point. As the temperature increases, more phonon modes will be excited. However, the number of phonon modes is limited. After all modes are excited, thermal conductance cannot be raised further.

On the other hand, we can calculate the DOS for the system as shown in figure 4.6.

4.3 Recent progress

Although the traditional AGF method developed by Mingo and Yang [2] can give the total transmission coefficient of the interface in the central region, it cannot be applied widely. First, it can only work with finite sized systems in the transverse direction, which makes it fail in systems with infinite size in the transverse direction. Also, the transmission coefficient of specific polarization, which becomes more critical to the understanding and engineering of thermal transport, is not available. Finally, the anharmonic effect is be more evident as the temperature rises. In this case, the theoretical prediction would be biased.

Many people made have progress in enlarging its application range. Here, we introduce some extensions to traditional AGF.

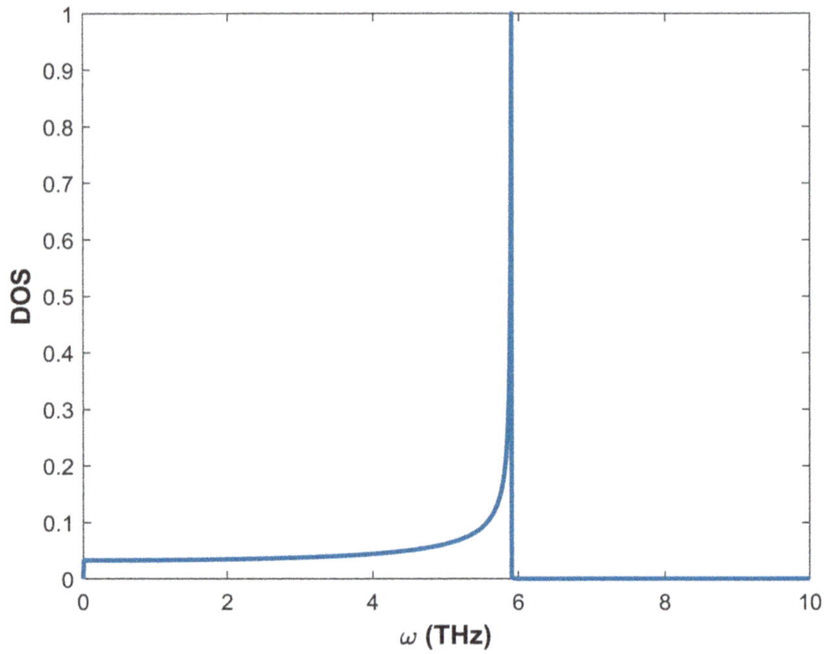

Figure 4.6. Density of states for a one-dimensional silicon chain.

4.3.1 From one dimension to three dimensions

Zhang *et al* [3, 4] first extended the traditional AGF method to thin films, which have an infinite size in two transverse directions. To avoid an infinite sized matrix, they wisely transformed the matrix from a site representation to a wavevector representation under an arbitrary two-dimensional transverse wavevector $\boldsymbol{k}_{\parallel}$ via Fourier transform with the assumption of ideal translation invariance in all transverse infinite directions. Only considering the interaction from the nearest neighbor, the intralayer and interlayer dynamical matrices are defined as

$$\tilde{D}_p(\boldsymbol{k}_{\parallel}) = \sum^{m} D_{im} \mathrm{e}^{-\mathrm{i}\boldsymbol{k}_{\parallel}\cdot \boldsymbol{R}_{im}}$$
$$\tilde{D}_{p,p+1}(\boldsymbol{k}_{\parallel}) = \sum^{n} D_{in} \mathrm{e}^{-\mathrm{i}\boldsymbol{k}_{\parallel}\cdot \boldsymbol{R}_{in}}.$$

Here, the index i is an arbitrary unit cell on layer p. For the intralayer dynamical matrix, index m includes the nearest-neighboring unit cell in the same layer p and the unit cell i itself, while for the interlayer dynamical matrix, index n includes the nearest-neighboring unit cell in the adjacent layer $p + 1$. Therefore, we convert the system with infinite size in the transverse direction back to the original model where the transverse direction is finite.

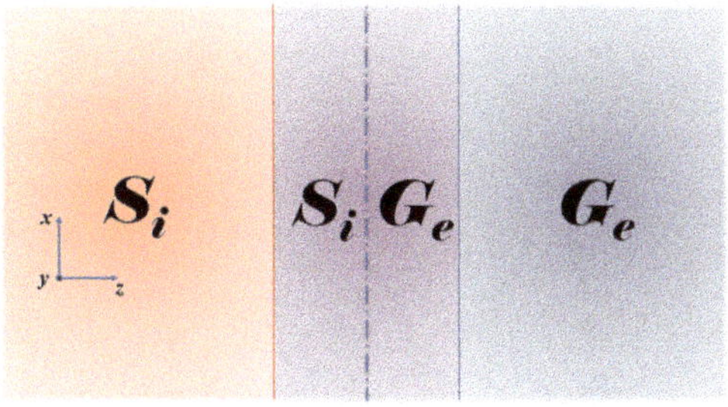

Figure 4.7. Three-dimensional infinite size extension model.

Then, we can use equation (4.10) to calculate the transmission $\Xi(\omega, \boldsymbol{k}_{\|})$ for every $\boldsymbol{k}_{\|}$ and the total transmission of the system should be

$$\Xi(\omega) = \int_{\boldsymbol{k}_{\|}} \Xi(\omega, \boldsymbol{k}_{\|}) \frac{\mathrm{d}\boldsymbol{k}_{\|}}{(2\pi)^2}.$$

For an $N \times N$ uniform mesh, the expression would be simplified to

$$\Xi(\omega) = \frac{1}{N^2} \sum_{\boldsymbol{k}_{\|}} \Xi(\omega, \boldsymbol{k}_{\|}).$$

We illustrate this method with an example of a three-dimensional silicon/germanium interface. In general, we usually define such an interface thermal transport problem as shown in figure 4.7.

Which is similar to the one-dimensional example we studied in section 4.3. The system is still divided into three regions, while each region contains a certain number of identically sized principle layers. However, the dimension of each layer is entirely different. Therefore, the procedure in section 4.2.1 cannot be directly applied here. As a result, we need to use the 3D AGF method and compute the interface transmission (figure 4.8) [5]. The random rough interface transmission results are also included in figure 4.8.

4.3.2 Polarization-specific transmission coefficient

Huang *et al* [17] first extended the traditional AGF to compute a polarization-specific transmission coefficient. They rewrite the level-width function as

$$\Gamma_{l=\mathrm{L,\,R}} = \mathrm{i}D_{\mathrm{C}l}(G_l^0 - G_l^{0\dagger})D_{l\mathrm{C}} = D_{\mathrm{C}l}A_c D_{l\mathrm{C}},$$

where A_c is defined as

$$A_c = \mathrm{i}(G_l^0 - G_l^{0\dagger}). \tag{4.34}$$

Considering that A_c is a Hermitian matrix, a spectrum decomposition can be performed:

$$A_c = \sum_i \lambda_i \varphi_i \varphi_i^\dagger,$$

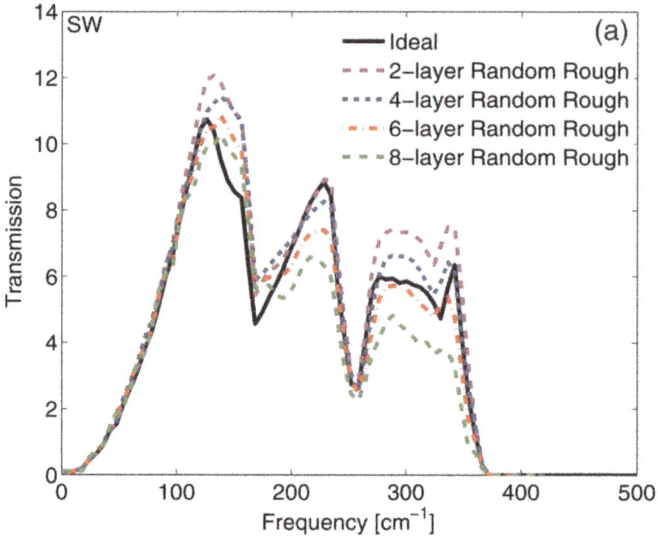

Figure 4.8. Total Si/Ge transmission with a rough interface. Reproduced with permission from [5]. Copyright 2012 the American Physical Society.

where λ_i and φ_i are the eigenvalue and corresponding eigenvector of matrix A_c. Therefore, we can define a polarization-based level-width function:

$$\gamma_{li} = D_{Cl}\lambda_i\varphi_i\varphi_i^{\dagger}D_{lC}.$$

Substituting the total level-width function Γ_L with γ_{Li} in equation (4.10), we can obtain the transmission function of one polarization mode from left to right:

$$\Xi_i(\omega) = \mathrm{Tr}(\gamma_{Li}G_C^{\dagger}\Gamma_R G_C). \tag{4.35}$$

Huang's method showed the possibility of calculating polarization-specific transmission. However, this method does not have a direct relation with bulk phonon modes or the surface Green's function of both leads [8, 9]. Despite this, this pioneering work highlights the profit of extension from the traditional AGF method.

Following this work, Ong *et al* [8] made a 'straightforward and efficient' extension to the traditional AGF method. The essence of their work is the idea of the 'Bloch matrix', which bridges the connection to the surface Green's functions of the leads. The relationship between the Bloch matrices and surface Green's functions is

$$F_L^{-1} = [D_L^{\text{hopping}}G_L]^{\dagger}$$
$$F_R = G_R D_R^{\text{hopping}\dagger},$$

where $D_{L,R}^{\text{hopping}}$ are the hopping matrices between two adjacent unit cells from one closer to the central region to the farther one inside the bulk leads, and $G_{L,R}$ is the surface Green's function of both leads.

After we obtain the expression of the Bloch matrices, we can obtain their eigenvalue matrix U and corresponding eigenstate matrix Λ:

$$F_{l=\mathrm{L,R}}U_l = U_l\Lambda_l.$$

Moreover, the velocity matrices are

$$V_{l=\mathrm{L,R}} = \frac{a_l}{2\omega}U_l^{\dagger}\Gamma_l U_l,$$

where a_l is the transverse length of lead l and Γ_l is the corresponding level-width function.

Finally, we can obtain the t matrix which presents the transmission between individual phonon modes:

$$t = \frac{2\mathrm{i}\omega}{\sqrt{a_\mathrm{L}a_\mathrm{R}}}V_\mathrm{R}^{\frac{1}{2}}U_\mathrm{R}^{-1}G_\mathrm{C}[U_\mathrm{R}^{\dagger}]^{-1}V_\mathrm{L}^{\frac{1}{2}}.$$

The square modulus element $|t_{m,n}|^2$ represents the transmission coefficient from the nth mode in the left lead to the mth mode in the right lead. A more detailed tutorial can be found in [18].

Ong *et al* have made significant progress in linking the bulk phonon modes of the leads with the polarization-based transmission. However, this method cannot give a equal transmission coefficient to degenerate the transverse acoustic (TA) modes [9]. Sadasivam *et al* [9] developed another kind of AGF extension which overcomes this kind of problem.

In their work, Sadasivam *et al* used the Lippmann–Schwinger equation and eigenvectors of dynamical matrices to calculate polarization-based interfacial transmissions. The relationship between the eigenvectors of the bulk leads and surface satisfies

$$\psi_\mathrm{L} = \phi_{\mathrm{L},n} - G_\mathrm{L}D_\mathrm{CR}\phi_{\mathrm{L},n+1}$$
$$\psi_\mathrm{R} = \phi_{\mathrm{R},n+1} - G_\mathrm{R}D_\mathrm{CL}\phi_{\mathrm{L},n},$$

where G_l is the total surface Green's function and $\phi_{l,n}$, $\phi_{l,n+1}$ ($l = \mathrm{L, R}$, n is an arbitrary integer) represents the eigenvectors of two adjacent unit cells in the bulk lead l which could be calculated by solving the generalized eigenvalue equation

$$\begin{pmatrix} \omega^2\mathbf{I} - D_{l,\mathrm{C}} & -D_{l,\mathrm{CR}} \\ \mathbf{I} & 0 \end{pmatrix}\begin{pmatrix} \phi_{l,n} \\ \phi_{l,n+1} \end{pmatrix} = \mathrm{e}^{-\mathrm{i}q_z a}\begin{pmatrix} D_{l,\mathrm{CL}} & 0 \\ 0 & \mathbf{I} \end{pmatrix}\begin{pmatrix} \phi_{l,n} \\ \phi_{l,n+1} \end{pmatrix}.$$

Here, q_z is the longitudinal wavevector and a denotes the periodicity in the propagating direction. $D_{l,c}$, $D_{l,\mathrm{CL}}$ and $D_{l,\mathrm{CR}}$ are the dynamical matrix of bulk lead l. After obtaining the eigenvectors, we can obtain the surface Green's function of individual modes

$$
\begin{aligned}
G_{\mathrm{L},\alpha}(\omega) &= -\frac{\mathrm{i}a}{2}\frac{\psi_{\mathrm{L},\alpha}\psi^{\dagger}_{\mathrm{L},\alpha}}{2\omega v_{g,z}(q_{z,\alpha})} \\
G_{\mathrm{R},\beta}(\omega) &= \frac{\mathrm{i}a}{2}\frac{\psi_{\mathrm{R},\beta}\psi^{\dagger}_{\mathrm{R},\beta}}{2\omega v_{g,z}(q_{z,\beta})},
\end{aligned}
\tag{4.36}
$$

where $v_{g,z}(q_z)$ is the group velocity of corresponding lead l:

$$
v_{g,z}(q_z) = \mathrm{i}a\frac{\phi^{\dagger}_{l,n}[-D_{l,\mathrm{CL}}\mathrm{e}^{-\mathrm{i}q_z a} + D_{l,\mathrm{CR}}\mathrm{e}^{\mathrm{i}q_z a}]\phi_{l,n}}{2\omega}.
$$

We can subsequently obtain the self-energy of individual modes $\Sigma_{l,\alpha}$ and the corresponding level-width function of individual modes $\Gamma_{l,\alpha}$. Finally, we can obtain the transmission from mode α of the left lead to mode β of the right lead:

$$
\Xi_{\alpha\beta}(\omega) = \mathrm{Tr}\left(\Gamma_{\mathrm{L},\alpha}G^{\dagger}_{\mathrm{C}}\Gamma_{\mathrm{R},\beta}G_{\mathrm{C}}\right),
$$

where G_{C} is the atomistic Green's function of the central region (figure 4.9).

4.3.3 Anharmonic Green's function

Recently, a many-body non-equilibrium AGF for phonons has been developed by Mingo. The thermal conduction through an anharmonic molecular junction between two solid surfaces is studied in [10]. This is a powerful method for studying non-equilibrium and interacting systems. It is rooted in quantum field theory and was originally designed for non-equilibrium electron problems. We refer the reader to Schwinger [19], Kadanoff and Baym [20], and most importantly Keldysh [21],

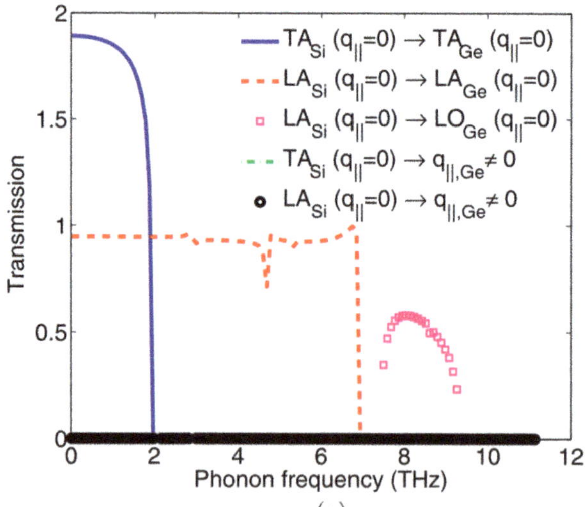

(a)

Figure 4.9. Mode-specific transmission of the Si/Ge system with a smooth interface. Reproduced with permission from [9]. Copyright 2017 the American Physical Society.

through which the phonon anharmonic Green's function scheme in this section is designed. The books by Datta [1] and Haug and Jauho [17] are recommended for detailed background information.

4.3.3.1 Hamiltonian

The quantum mechanical lattice properties of any atomically described system can be represented by the Hamiltonian

$$\hat{H} = \frac{1}{2}\sum_i M_i\hat{\dot{\varphi}}_i\hat{\dot{\varphi}}_i + \frac{1}{2}\sum_{ij} k_{ij}\hat{\varphi}_i\hat{\varphi}_j\hat{\varphi}_i + \sum_{ijk} \Lambda_{ijk}\hat{\varphi}_i\hat{\varphi}_j\hat{\varphi}_k + \dots. \tag{4.37}$$

The terms on the right are the kinetic, harmonic and anharmonic terms, respectively. $\hat{\varphi}_i$ is the Heisenberg displacement operator for the ith degree of freedom. k_{ij} and Λ_{ijk} are

$$k_{ij} \equiv \frac{\partial^2 E}{\partial u_i \partial u_j}, \ \Lambda_{ijk} \equiv \frac{\partial^3 E}{\partial u_i \partial u_j \partial u_k},$$

where E is the total energy and u_i is the atomic coordinates. M_i is the mass of the atom to which the ith degree of freedom belongs. For simplicity, we will assume the same mass M for every atom in the system. As can be seen, Λ_{ijk} is the third-order force constant included in this anharmonic problem.

4.3.3.2 Uncoupled atomistic Green's functions for free leads

Before we start to compute anharmonic phonon flows, it is necessary to introduce different types of non-equilibrium Green's function [22]:

$$\begin{aligned}
\mathrm{i}G^{\mathrm{R}}_{lm}(t_1 - t_2) &= \frac{1}{\hbar}\Theta(t_1 - t_2)\langle[\hat{\varphi}_l(t_1), \hat{\varphi}_m(t_2)]\rangle \\
\mathrm{i}G^{\mathrm{A}}_{lm}(t_1 - t_2) &= -\frac{1}{\hbar}\Theta(t_2 - t_1)\langle[\hat{\varphi}_l(t_1), \hat{\varphi}_m(t_2)]\rangle \\
\mathrm{i}G^{<}_{lm}(t_1 - t_2) &= \frac{1}{\hbar}\langle\hat{\varphi}_m(t_2)\hat{\varphi}_l(t_1)\rangle \\
\mathrm{i}G^{>}_{lm}(t_1 - t_2) &= \frac{1}{\hbar}\langle\hat{\varphi}_l(t_1)\hat{\varphi}_m(t_2)\rangle.
\end{aligned}$$

They are known as the retarded, advanced, greater and lesser Green's functions [1]. The reason we introduce them is that they are included in calculating the leads' self-energy matrices for the anharmonic problem. The following relation holds in both the frequency and time domains from the basic definitions [23]

$$\boldsymbol{G}^{\mathrm{R}} - \boldsymbol{G}^{\mathrm{A}} = \boldsymbol{G}^{>} - \boldsymbol{G}^{<}.$$

In addition, in systems with time translational invariance, $\boldsymbol{G}^{\mathrm{R}}$ and $\boldsymbol{G}^{\mathrm{A}}$ will be the Hermitian conjugates of each other:

$$\boldsymbol{G}^{\mathrm{R}}(\omega) = \boldsymbol{G}^{\mathrm{A}}(\omega)^{\dagger}.$$

Therefore, for general non-equilibrium steady-state cases, only two are independent. For the cases where the central region is decoupled from all the leads, there will be no current flow and the whole system will be in equilibrium. We denote the atomistic Green's function by $\boldsymbol{G}_0^<$, $\boldsymbol{G}_0^{\mathrm{R}}$ in such scenarios. There exist other relations between them for steady-state cases in the frequency domain:

$$\begin{aligned}
\boldsymbol{G}^<(\omega)^\dagger &= -\boldsymbol{G}^<(\omega) \\
\boldsymbol{G}^{\mathrm{R}}(-\omega) &= \boldsymbol{G}^{\mathrm{R}}(\omega)^* \\
\boldsymbol{G}^<(-\omega) = \boldsymbol{G}^>(\omega)^{\mathrm{T}} &= -\boldsymbol{G}^<(\omega)^* + \boldsymbol{G}^{\mathrm{R}}(\omega)^{\mathrm{T}} - \boldsymbol{G}^{\mathrm{R}}(\omega)^*.
\end{aligned}$$

In thermal equilibrium, there is an additional equation relating G_0^{R} and $G_0^<$:

$$\boldsymbol{G}_0^<(\omega) = f(\omega)(\boldsymbol{G}_0^{\mathrm{R}}(\omega) - \boldsymbol{G}_0^{\mathrm{A}}(\omega)),$$

where $f(\omega)$ is the Bose–Einstein distribution function.

With these relationships, one may notice that we only need to compute the retarded Green's function in the positive frequency domain:

$$\boldsymbol{G}_0^{\mathrm{R}}(\omega) = ((\omega + \mathrm{i}\eta)^2\boldsymbol{I} - \boldsymbol{D})^{-1}.$$

For details, we refer the reader to Mingo [10] and the book by Datta [1].

4.3.3.3 Coupled atomistic Green's functions

Once the leads are coupled to the central region, phonon flows from the hot lead to cold lead will be induced in this system. Also, the atomistic Green's functions in the coupled case satisfy the Dyson equations, similar to our deductions in previous sections. However, when the anharmonicity is included in the system, not only do the leads contribute to the self-energies, but the many-body interactions will also contribute to them.

The many-body self-energy is computed from Feynman diagrams, following the rules in [17], and enters the expression of the retarded Green's function of the central region:

$$\boldsymbol{G}^{\mathrm{R}}(\omega) = ((\omega + \mathrm{i}\eta)^2\boldsymbol{I} - \boldsymbol{D} - \Sigma_h^{\mathrm{R}} - \Sigma_c^{\mathrm{R}} - \Sigma_M^{\mathrm{R}})^{-1}, \tag{4.38}$$

where $\Sigma_{\mathrm{M}}^{\mathrm{R}}$ is the many-body self-energy due to the three-phonon process. The self-energy from the leads will be [1]

$$\begin{aligned}
\Sigma_{h(c)}^{\mathrm{R}} &= \boldsymbol{D}_h\boldsymbol{G}_{0,h(c)}^{\mathrm{R}}\boldsymbol{D}_h^{\mathrm{T}} \\
\Sigma_{h(c)}^< &= \boldsymbol{D}_h\boldsymbol{G}_{0,h(c)}^<\boldsymbol{D}_h^{\mathrm{T}}.
\end{aligned}$$

The three-phonon many-body self-energy can be computed in this way [1, 23]:

$$2\mathrm{Im}\Sigma_{\mathrm{M}}^{\mathrm{R}} = \Sigma_{\mathrm{M}}^> - \Sigma_{\mathrm{M}}^< \tag{4.39}$$

$$\mathrm{Re}\Sigma_{\mathrm{M}}^{\mathrm{R}} = \mathcal{H}(\mathrm{Im}\Sigma_{\mathrm{M}}^{\mathrm{R}}) \tag{4.40}$$

$$\Sigma_{\mathrm{M},in}^{<(>)}(\omega) = \mathrm{i}\hbar\sum\nolimits_{jklm}\int_{-\infty}^{\infty}\Lambda_{ijk}G_{jl}^{<(>)}(\omega')G_{km}^{<(>)}(\omega-\omega')\Lambda_{lmn}\mathrm{d}\omega' \tag{4.41}$$

$$G^{<(>)} = G^{\mathrm{R}}(\Sigma_h^{<(>)} + \Sigma_c^{<(>)} + \Sigma_M^{<(>)})G^{\mathrm{A}}. \tag{4.42}$$

The symbol $\mathcal{H}$ denotes the Hilbert transform. Here, a self-consistent calculation is required in this procedure. We can initially set the many-body self-energies to be zero and repeat the above calculation. The total error function is defined to be

$$\text{error} = \sum_{ij}\left[\Sigma_{\mathrm{M},ij}^{<,(n+1)} - \Sigma_{\mathrm{M},ij}^{<,(n)}\right]^2 \Big/ \sum_{ij}\left[\left(\Sigma_{\mathrm{M},ij}^{<,(n+1)}\right)^2 + \left(\Sigma_{\mathrm{M},ij}^{<,(n)}\right)^2\right], \tag{4.43}$$

where $\Sigma_{\mathrm{M},ij}^{<,(n)}$ is the lesser many-body self-energy we compute in the nth loop. The iteration process will stop until the total error is smaller than the threshold we set.

After it converges, we can compute the thermal conductance via the thermal current. Unlike the harmonic case, the thermal current is usually transformed into a current density matrix. As shown below, the current frequency distribution at the hot frontier and cold frontier will be [10]

$$J_{h(c)}(\omega) = +(-)\mathrm{Tr}\left(\Sigma_{h(c)}^{>} G^{<} - \Sigma_{h(c)}^{<} G^{>}\right)\hbar\omega. \tag{4.44}$$

Finally, we can obtain the expression for thermal conductance:

$$\sigma = \lim_{T_{\mathrm{L}} \to T_{\mathrm{R}}} \frac{1}{T_{\mathrm{L}} - T_{\mathrm{R}}} \frac{1}{2\pi} \int_0^{\infty} +(-)\mathrm{Tr}\left(\Sigma_{h(c)}^{<} G^{>} - \Sigma_{h(c)}^{>} G^{<}\right)\hbar\omega \mathrm{d}\omega. \tag{4.45}$$

4.3.3.4 Anharmonic behavior

To quantify the anharmonic effect, in Mingo's paper [10], a plot of the difference between the harmonic and anharmonic thermal conductance normalized by the harmonic thermal conductance as a function of temperature is given (figure 4.10).

As we can see, as the temperature increases, the anharmonic behavior becomes more important. The anharmonic deviation behaves linearly with respect to temperature, except in the low-temperature part.

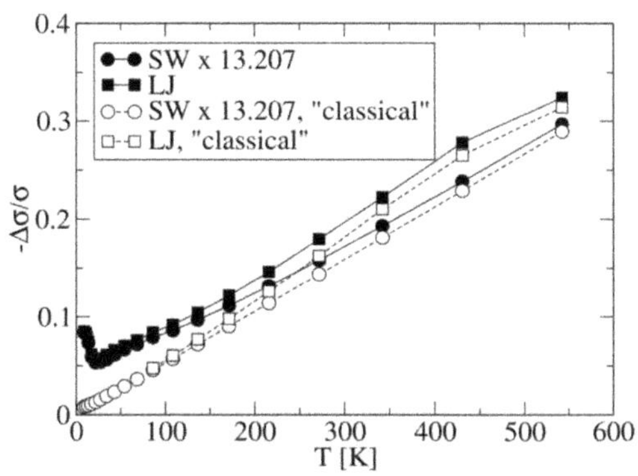

Figure 4.10. Relative thermal conductance deviation due to anharmonicity as a function of temperature. Reproduced with permission from [10]. Copyright 2006 American Physical Society.

In addition to the 1D atomic chain, a new AGF formalism named 3D anharmonic AGF has been developed by Dai and Tian [24]. This new method can apply to different types of three dimensional structures. The anharmonic behavior for three-dimensional systems has been discussed in detail in [24]. The application scope of the AGF is remarkably extended.

4.4 Summary

We have introduced the origin of AGF, applicable circumstances and a general model of interfacial transportation in section 4.1.

In section 4.2, the theory behind the AGF method is deduced, from which we obtain the expression of the atomistic Green's function of the central region, including the influence from the two leads through self-energies. Moreover, the relations between the atomistic Green's function of the central region and thermal conductance, and different methods to calculate the surface Green's function calculation are demonstrated. Also, a toy model example along with a more realistic example are presented to show the usage of the AGF method.

In section 4.3, we present the tremendous progress in generalizing the AGF method, including extension into the 3D model, mode-specific transmission calculation and correction to include the effect of anharmonicity. These extensions vitalized the AGF method and shed light on the deeper understanding of thermal transport.

Though AGF is a powerful tool, much work remains to achieve a practical model in thermal engineering. Limitations exist in involving higher orders of the phonon scattering process, accurate modeling of heat transfer at the heterogeneous interface and more confident predictions on unknown local geometries such as rough interfaces, dislocations, vacancies, etc. With the above understanding, in the long run we should develop more rigorous AGF models which match the corresponding experimental sample conductions and apply them to predict the thermal properties of different material systems efficiently. In summary, given the fast development of AGF along with the experimental status of thermal transport in nanostructures, with closer collaboration between experimental and computational modeling scientists, the two types of approaches will converge in the future.

Acknowledgments

This work was sponsored by the Department of the Navy, Office of Naval Research under ONR award number N00014-18-1-2724.

References

[1] Datta S 1997 *Electronic Transport in Mesoscopic Systems* (Cambridge: Cambridge University Press)

[2] Mingo N and Yang L 2003 Phonon transport in nanowires coated with an amorphous material: an atomistic Green's function approach *Phys. Rev.* B **68** 245406

[3] Zhang W, Fisher T S and Mingo N 2007 Simulation of interfacial phonon transport in Si–Ge heterostructures using an atomistic Green's function method *J. Heat Transfer* **129** 483

[4] Zhang W, Mingo N and Fisher T S 2007 Simulation of phonon transport across a non-polar nanowire junction using an atomistic Green's function method *Phys. Rev.* B **76** 195429
[5] Tian Z, Esfarjani K and Chen G 2012 Enhancing phonon transmission across a Si/Ge interface by atomic roughness: first-principles study with the Green's function method *Phys. Rev.* B **86** 235304
[6] Mingo N, Stewart D A, Broido D A and Srivastava D 2008 Phonon transmission through defects in carbon nanotubes from first principles *Phys. Rev.* B **77** 033418
[7] Esfarjani K and Stokes H T 2008 Method to extract anharmonic force constants from first principles calculations *Phys. Rev.* B **77** 144112
[8] Ong Z-Y and Zhang G 2015 Efficient approach for modeling phonon transmission probability in nanoscale interfacial thermal transport *Phys. Rev.* B **91** 174302
[9] Sadasivam S, Waghmare U V and Fisher T S 2017 Phonon-eigenspectrum-based formulation of the atomistic Green's function method *Phys. Rev.* B **96** 174302
[10] Mingo N 2006 Anharmonic phonon flow through molecular-sized junctions *Phys. Rev.* B **74** 125402
[11] Sadasivam S, Ye N, Feser J P, Charles J, Miao K, Kubis T and Fisher T S 2017 Thermal transport across metal silicide-silicon interfaces: first-principles calculations and Green's function transport simulations *Phys. Rev.* B **95** 085310
[12] Ryndyk D 2016 *Theory of Quantum Transport at Nanoscale* (Berlin: Springer), p 184
[13] Sadasivam S *et al* 2014 The atomistic Green's function method for interfacial phonon transport *Annu. Rev. Heat Transf.* **17** 89–145
[14] Teichert F, Zienert A, Schuster J and Schreiber M 2017 Improved recursive Green's function formalism for quasi one-dimensional systems with realistic defects *J. Comput. Phys.* **334** 607–19
[15] Sancho M L, Sancho J L and Rubio J 2001 Highly convergent scheme for the calculation of bulk and surface Green functions *J. Phys.* F **15** 851
[16] Velev J and Butler W 2004 On the equivalence of different techniques for evaluating the Green function for a semi-infinite system using a localized basis *J. Phys.: Condens. Matter.* **16** R637
[17] Huang Z, Murthy J Y and Fisher T S 2011 Modeling of polarization-specific phonon transmission through interfaces *J. Heat Transfer* **133** 114502
[18] Ong Z 2018 Tutorial: concepts and numerical techniques for modeling individual phonon transmission at interfaces *J. Appl. Phys.* **124** 151101
[19] Schwinger J 1961 Brownian motion of a quantum oscillator *J. Math. Phys.* **2** 407–32
[20] Kadanoff L P, Baym G and Trimmer J D 1963 Quantum statistical mechanics *Am. J. Phys.* **31** 309
[21] Keldysh L V 1965 Diagram technique for nonequilibrium processes *Sov. Phys. JETP* **20** 1018–26
[22] Economou E N 1983 *Green's Functions in Quantum Physics* (Berlin: Springer)
[23] Lifshitz E M and Pitaevskii L P 1981 *Course of Theoretical Physics: Physical Kinetics* (Oxford: Pergamon)
[24] Dai J and Tian Z 2019 Rigorous formalism of anharmonic atomistic Green's function for three-dimensional interfaces *Phys. Rev.* B **101** 041301

IOP Publishing

Nanoscale Energy Transport
Emerging phenomena, methods and applications
Bolin Liao

Chapter 5

Application of Bayesian optimization to thermal science

Jiang Guo, Shenghong Ju and Junichiro Shiomi

Thermal functional materials have been applied in various aspects in industry and society. Artificially designed thermal materials or structures can be tuned to match the needs of specific applications and have been shown to perform better than natural materials. For thermal energy applications, there are strong demands for materials with better heat transfer performance such as high/low thermal conductivity materials and wavelength-selective thermal emitters/absorbers. With the advances in nanotechnology in fabricating fine structures at the scale of heat carriers, there have been a number of works aiming to design nanostructures to further improve the performance. However, finding the optimal design of structure based on the classical searching method has been difficult due to the incapability to efficiently handle a large number of structure candidates. To this end, materials informatics (MI), which integrates the material simulation/experiment and the data-driven machine-learning method, can be extremely useful and has the potential to reform the current material design. In this chapter, we summarize the recent progress in MI for thermal science and engineering, including the control of thermal conductance and spectral thermal radiative properties.

5.1 Introduction

As the world's population continues to grow, with the fast-increasing demand for energy and food, the rapid depletion of fossil-based energy resources, and the concerns regarding CO_2 emission problems and global warming, it is urgent to find an efficient way to make the best use of energy resources [1–5]. For high efficiency energy harvesting and utilization, such as solar energy applications, maximizing the energy conversion efficiency and minimizing the unwanted thermal losses to the surroundings often inevitably involves engineering the thermal conductance or thermal emission properties of materials [6–12]. Thanks to the development of nanotechnology, we now are able to design and fabricate materials and structures at the nanoscale, or even at the atomic scale [13–16]. However, due to a lack of knowledge or large degrees of freedom in the underlying complex physics, we are still struggling to control thermal conduction and emission properties to be exactly as desired. The trial-and-error method is sometimes the only way to discover the best material or structure, but is time-consuming, costly and largely relies on luck. Today, inspired by the successful application of machine learning in the data science field [17–20], we are aware that it is not always necessary that we humans learn the underlying physics in advance to design a material or structure, instead, the machine-learning method can easily be used to classify or predict trends from seemingly disordered data. Contrasted with the use of the classical optimization algorithm in the past, the recently developed big-data-oriented algorithms can achieve self-learning from the data in hyper-dimensional space and also automated prediction with high efficiency.

The recently emerging field of MI, which combines material property calculations/measurements with informatics to accelerate material discovery/design, has attracted great attention. MI has been successfully applied in many fields, such as protein discovery in biology [21–24], drugs [25–27], polymer synthesis design in chemistry [28–30] and many others. Generally speaking, there are many searching algorithms that can be integrated into MI, including but not limited to the Bayesian algorithm, Monte Carlo tree searching, transfer learning and deep neural networks (NNs). Among these methods, Bayesian optimization methods have been extensively employed in MI design and have demonstrated high searching efficiency. Here, we will take Bayesian optimization as an example to illustrate the basic procedure of applying MI in solving thermal transport-related problems.

In this chapter, we summarized the recent progress in the application of Bayesian optimization in the heat transfer field. The chapter is organized as follows: in the first part, we introduce the fundamental theory of the Bayesian algorithm and explain how it was implemented as a close-loop and black-box tool to design materials automatically. In the second part, we summarize the recent successful applications of Bayesian optimization in designing thermal conductance and thermal radiation properties. We hope this review can provide new insights in dealing with designing or optimizing problems for researchers in the heat transfer field.

5.2 Bayesian optimization

5.2.1 Bayesian algorithm theory

Bayesian optimization is a statistics and probabilistics based algorithm which is particularly efficient when the objective function is not straightforward and the evaluation or experimental cost is expensive, but the observation of specific values is possible [31–33]. By taking advantage of historically observed data, Bayesian optimization is able to find the optimal results using only a limited number of observations compared to random search. The Bayesian algorithm relies on posterior probability theory which can be expressed simply as

$$P(A|B) \propto P(B|A)P(A), \tag{5.1}$$

which gives the information that the posterior probability $P(A|B)$ is proportional to the product of the likelihood $P(B|A)$ and the prior probability $P(A)$. Note that the marginal likelihood $P(B)$ is omitted here since it is assumed to be only a normalizing constant value in Bayes' theorem and does not have influence on the prediction. During the Bayesian optimization process, the prior probability is, however, unknown, but must be assumed either based on prior beliefs or based on a default universal Gaussian distribution. For example, if the prior belief is that the object function is smooth, the data with small deviation from the mean value will be more likely to be considered. An accurate prior belief can help to accelerate the search efficiently. For Bayesian optimization, the key point is to build an approximate surrogate model that can be updated with observed data and suggest the next observation points. Among various approximate probabilistic models, the Gaussian process (GP) is a powerful machine-learning model to predict the trend of unknown data based on collected data [34, 35]. The GP also allows a finite combination of random parameters governed by multiple Gaussian distributions, which means that multiple parameters can be optimized. Moreover, the prior estimation can be deduced in closed form if the GP is used, and thus the GP is often adopted in Bayesian optimization. To determine the next observation points, Bayesian optimization employs acquisition functions which efficiently balance the exploration (where the uncertainty area is likely to be high) and exploitation (where the object function value is likely to be high). The general Bayesian optimization process can be summarized as in algorithm 5.1.

Algorithm 5.1 Bayesian optimization

Input: Initial training data D_i.

1. Select x_t from the acquisition function, α such that:

$$x_t = \arg\max \alpha(x|D).$$

2. Sample the object function: $y_t = f(x_t) + \epsilon_t$.
3. Adding into the data: $D_t = \{D_{i:(t-1)} \cup (x_t, y_t)\}$.
4. Iterate from step 1 until an optimal result is targeted,

Here x_t is the observation point predicted by the acquisition function α, D_t is the data space, y_t is the object function to be optimized and ϵ_t is the noise factor.

The Bayesian optimization process can easily be understood from the demonstration in figure 5.1. First, an arbitrary target function $f(x)$ is defined here for optimizing. A set of random points is initialized to be observed, and their corresponding values are written in the surrogate model. The model here is approximated by the GP. Bayesian optimization will then provide the next possible points for evaluation according to the maximum acquisition function value position. The uncertain area clearly decreases with the updated observed data which means that a more accurate prediction can be given by the model. Here, a 95% confidence interval is used in the model. After several iterated steps, the next suggested

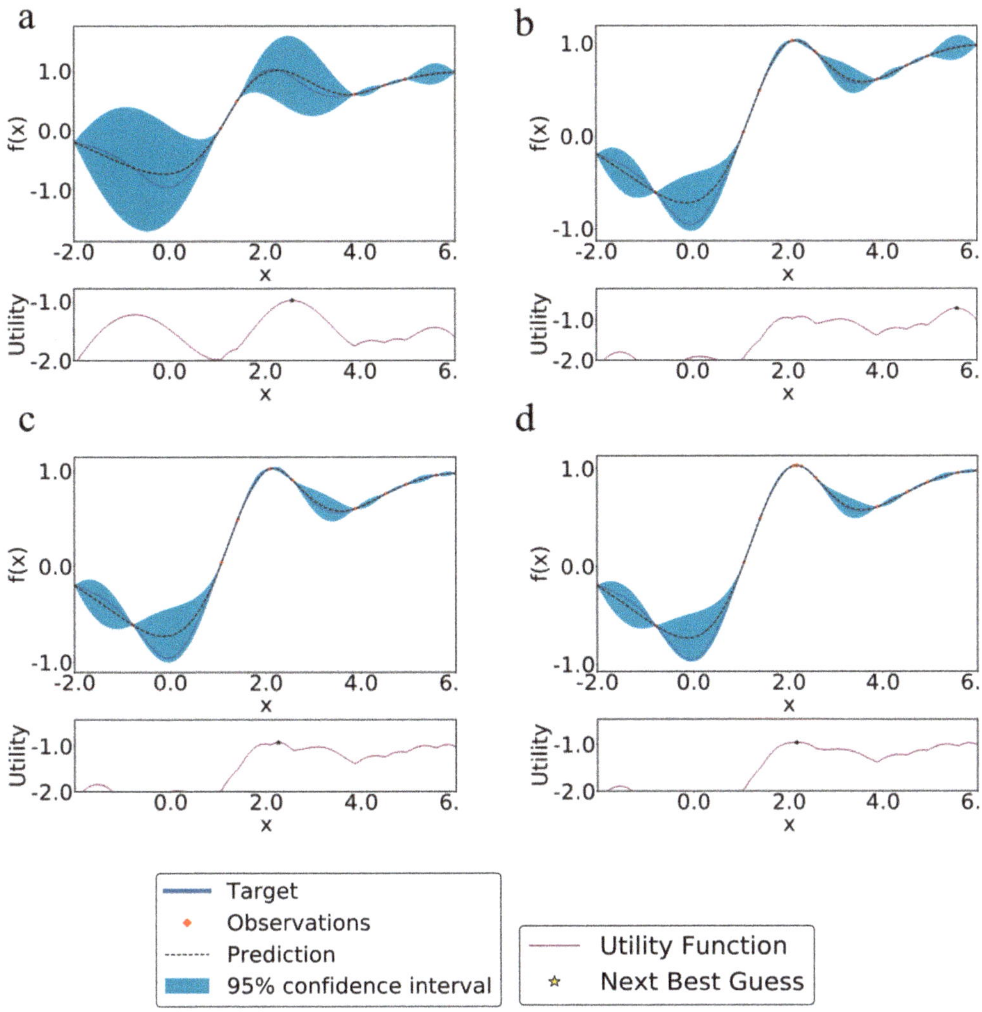

Figure 5.1. General Bayesian optimization process: (a) second, (b) fifth, (c) sixth and (d) tenth iterative optimization steps. The upper confidence bound (UCB) is used as an acquisition function (utility function) and five initial random observation points are applied.

observation points start to approach the peak position of the target function and the maximum value of the target function can be identified. Note that the number of initial randomly selected observation points has a great influence on the searching process, the more randomly distributed the points are, the higher the chance the global maximum or minimum values can be found by Bayesian optimization.

As for the acquisition function, it is used to give the suggested points or candidates for evaluation by balancing the exploration and exploitation. There are three typical acquisition functions used in Bayesian optimization: probability of improvement (PI), expected improvement and GP-lower confidence bound (GP-LCB). If the optimal object function is maximization targeted, the GP-LCB

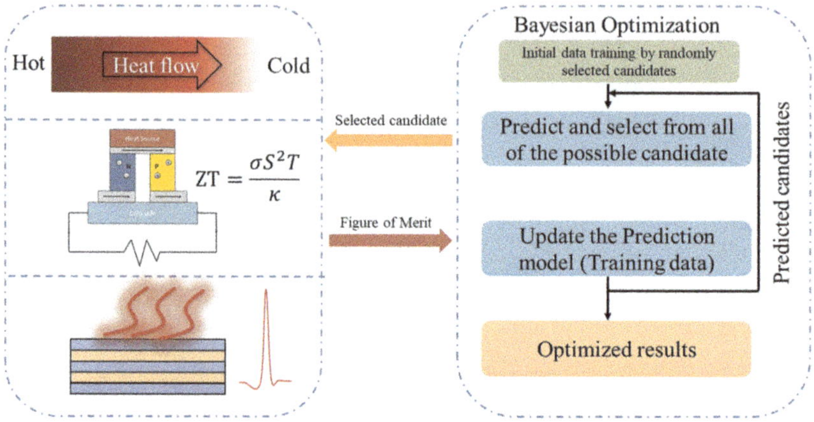

Figure 5.2. Bayesian optimization as a black-box tool for material design taking heat conduction, thermal electric material and thermal radiation material as examples.

is called the upper confidence bound (UCB). Further detailed information can be found in [36–38].

5.2.2 Bayesian optimization implemented as a black-box tool

There have been many developed open-source Bayesian optimization packages, such as scikit-learn [39], GPyOpt [40] and COMBO [41]. Here, we take COMBO as an example to explain the details of the implementation of Bayesian optimization. COMBO is developed as a black-box tool for material design with automated hyperparameter tuning for the kernel of the GP or acquisition function, and it has been successfully applied to various transport property optimizations [42–46]. Material scientists do not need professional knowledge in Bayesian optimization to solve their optimization problems. To use COMBO for the property optimization, three important elements should be well prepared: the descriptor, search space and figure of merit (FOM). The descriptors describe the parameters to be optimized, i.e. the structure one wants to design, the material components to be selected, etc. The design space includes all possible candidates and should be explicitly defined, which means one should list and label all of the possible candidates for running Bayesian optimization. The FOM defines the property to be optimized, and it should be a number which can be compared and acts as a feedback to update the approximate surrogate model. In COMBO, since UCB acquisition is used, the maximum positive FOM is always targeted. Figure 5.2 shows the flowchart of COMBO which works as a black-box tool for material design. First, COMBO will randomly pick M candidates (M can be defined by users according to the problem, generally M should not be too small a number) and retrieve the FOM from the material simulation/experiment part, then the candidates' descriptors and the corresponding values will be written into the GP model. After that, COMBO is initialized to run. During the optimization, COMBO will output N candidates (N is determined by the users to balance the total optimization time and iterated steps) which have the probability of high FOM at the current step. These N candidates' FOMs will then be added into the training model for better prediction. The learning rate can be controlled by setting the learning interval in the optimization process. After several iterations, the optimal candidates with the best FOM can be identified. Note that COMBO also provides a multi-probe choice that allows parallel optimization to accelerate the search in a relatively large candidate space.

5.3 Applications of Bayesian optimization in thermal science

With the development of nanotechnology in materials synthesis, it is possible to manipulate the energy transport at the characteristic length scale of heat carriers (phonons, electrons and photons). However, it is still challenging to identify the detailed optimal structure due to the coupled effects among multi-parameters as well as the unknown physical mechanisms. For instance, in phonon transport, the roughness [47, 48], vacancy defects [49, 50] and interfacial adhesion [51] have an important influence on heat transport. The electron–phonon coupling [52–56] and the constructive/deconstructive phonon interference [57–59] effects make the structure design and optimization rather more complicated. In photon transport, i.e. radiation heat transport, when the wavelength of the thermal radiation is comparable or even smaller than the length scale of the structure, the coupling of the evanescent wave or the tunnel phenomenon would occur, which is called the near-field effect [60–63]. In addition, tailoring the thermal radiation property also involves many complicated physical excitation phenomena, including the surface plasmon polariton [64, 65], surface phonon polariton [66–68], gap plasmon polariton [69–72], magnetic polariton [73–75], localized effect including localized surface plasmon polariton [76–78] and localized surface phonon polariton [79–82]. To solve the optimization and design problems involving the above-mentioned phenomena, an effective and efficient optimization method is indeed in demand. In this part, we summarize the recent successful design cases based on Bayesian optimization for manipulating the thermal conduction and radiation.

5.3.1 Thermal conductance modulation

Both high and low thermal conductance materials have widely ranging applications in our daily lives. For instance, Moore's law is largely constricted by the thermal dissipation speed for electronic devices [83]. In the recent NASA Parker Solar Probe project, scientists used light-weight carbon material as a heat shield, which helps to protect the majority of instruments keeping them at only 30 °C by withstanding a temperature of 1371 °C [84]. Although manipulating the heat carrier in energy transport is likely to be realized by current nanotechnology, again due to the complex physical mechanism and large degrees of freedom of the problem, a highly efficient optimization method is still needed to design the optimal structures by searching through a tremendous number of candidates.

Ju *et al* [43] first proposed designing an interfacial Si/Ge alloy structure to modulate the thermal conductance for Si–Si and Si–Ge interfaces at room temperature. Their interfacial region was composed of eight Si and eight Ge atoms.

According to combination theory, there were 12 870 possible candidates to be evaluated. To describe each candidate structure, the flag values '1' and '0' were used in the descriptors to represent the Ge and Si atoms, respectively. The atomistic Green's function [59, 85–89] was used to calculate the interfacial thermal conductance.

To check the robustness of the Bayesian optimization, ten rounds of optimization simulations were conducted using 20 different initially selected structures. As shown in figures 5.3(a) and (b), Bayesian optimization could obtain a global optimal structure quickly for all cases by calculating only 3.4% of the total candidate structures. From the optimal structures, only the maximum thermal conductance of Si–Si was intuitive since the continuous Si material in the interfacial region could provide a direct propagating path without much phonon scattering. The other optimal structures, however, gave new insights for thermal property modulation. The minimum conductance for both the Si–Si and Si–Ge interfaces was found to be in aperiodic multilayers, which have shown significantly low conductance compared to the periodic ones. Moreover, the optimal structure for maximum conductance at the Si–Ge interface could be viewed as a kind of rough interface which was consistent with the previous report that interfacial roughness can help to enhance the interfacial thermal conductance [47, 90].

To make the optimization more useful for realistic materials, Bayesian optimization was further conducted to design Si/Ge superlattices with a total thickness ranging from 8 to 16 unit layers (ULs) (1 UL = 5.43 Å). The finally obtained optimal structures were a kind of aperiodic superlattice, which showed significantly lower thermal conductance than even traditional well-known periodic superlattices, as compared in figure 5.4. By comparing the phonon transmission functions for both periodic and aperiodic superlattices shown in figure 5.4(b), it was clear to see that the designed aperiodic superlattice gives lower phonon transmission, which means lower thermal conductance than the other two types of periodic superlattices with different periodic layer thicknesses. Materials informatics can not only find an optimal structure with high efficiency but can also help us to explore new physical

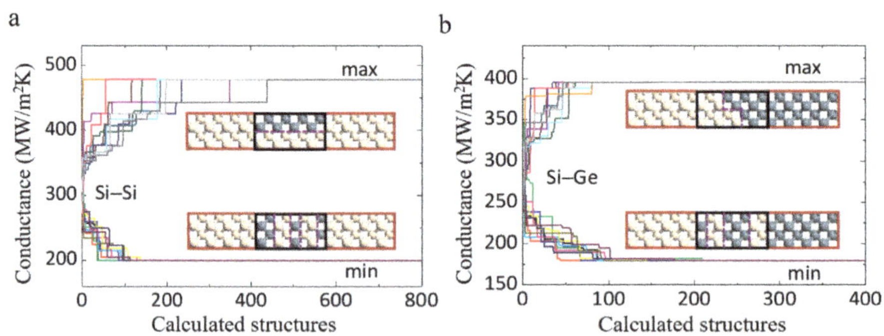

Figure 5.3. Bayesian optimization of interfacial Si/Ge alloy structures for maximum and minimum thermal conductances. Ten optimization runs with different initial choices of candidates for the (a) Si–Si and (b) Si–Ge cases, respectively. The insets show the corresponding optimal structures for maximum and minimum thermal conductance. Reproduced with permission from [43]. Creative Commons Attribution 4.0 International license.

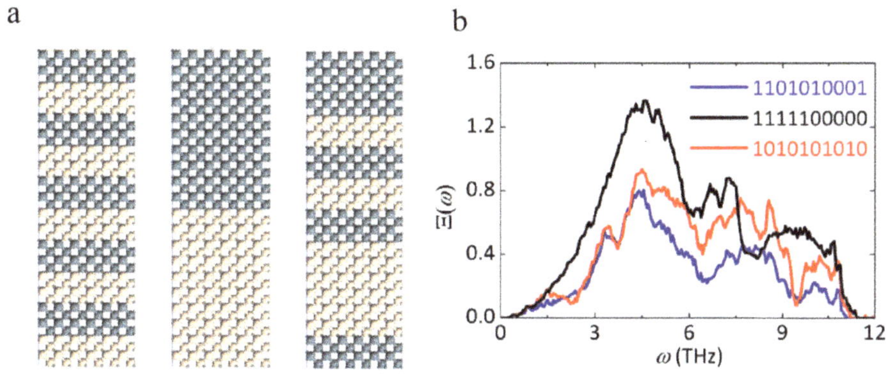

Figure 5.4. Designing a non-periodic Si/Ge superlattice with minimum thermal conductance. (a) Supperlattice structure with a different periodic thickness and designed aperiodic superlattice structure, and (b) calculated phonon transmission function of the periodic and designed aperiodic superlattice. The obtained thermal conductance is significantly lower than in the corresponding periodic superlattice. Reproduced with permission from [43]. Creative Commons Attribution 4.0 International license.

mechanisms behind the designed novel structures. It was found that with the increase of the layer thickness or the number of interfaces, the thermal conductance decreases and converges to a constant value. If the total thickness of the superlattice was fixed, balance was found between the layer thickness and the number of interfaces in order to obtain the minimum thermal conductance. The low thermal conductance of aperiodic superlattices is attributed to the balance of Fabry–Pérot wave interference [91, 92] and interfacial particle scattering [88, 93].

Bayesian optimization has also been successfully extended to thermoelectric applications. Yamawaki *et al* [44] optimized the defective graphene nanoribbons (GNRs) including periodic nanostructured GNR and antidot GNR to achieve an extremely high FOM for thermoelectric applications, as shown in figure 5.5. The periodic nanostructured GNR optimization for removing atoms m was performed by the Bayesian optimization, and the result was compared to a random search to check the optimization efficiency. In most cases, the top 0.5% could be identified by Bayesian optimization with only half the number of calculations required by the random search. The optimization for a larger system shows that the optimization efficiency was independent of the total number of candidate structures. The comparison of thermoelectric properties for the optimal, pristine and periodic GNR structures is shown in figure 5.5(d). The FOM for an aperiodic array of antidot structure increased to 11 times that of the pristine GNR. This demonstrates the versatile capability of Bayesian optimization in multi-transport property optimization with high efficiency and robustness.

5.3.2 Thermal radiation engineering

Tailoring thermal radiation, including spectral and direction control, can be beneficial for various energy harvesting applications, such as solar emitters/absorbers [94, 95], thermal functional textiles [96, 97], light incandescent sources [98], radiative cooling [99], etc. For example, the energy conversion efficiency of solar

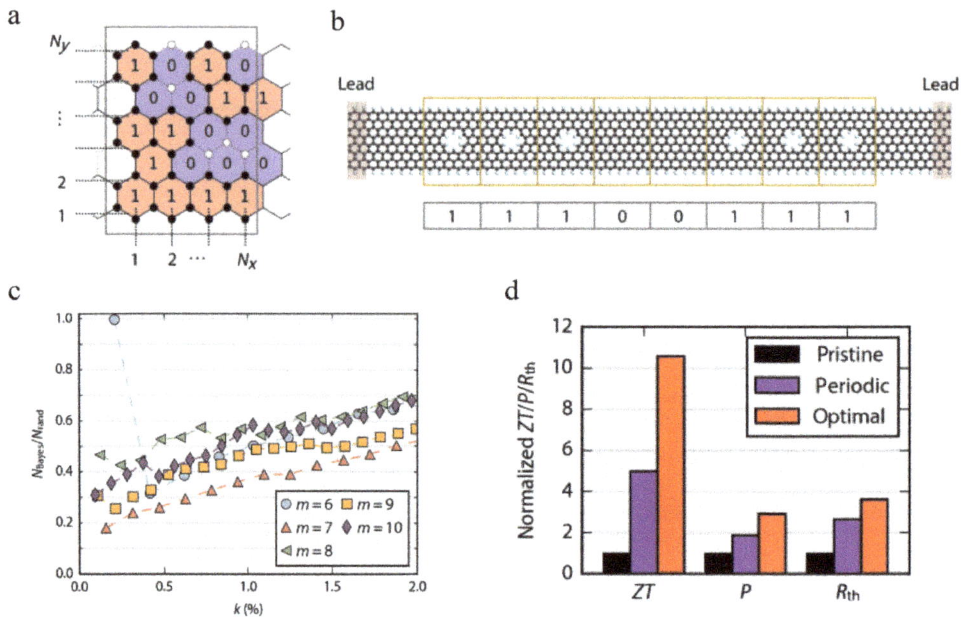

Figure 5.5. Designing a multifunctional structure of graphene for thermoelectric applications using Bayesian optimization. (a) Periodically nanostructured graphene nanoribbon model. (b) Antidote nanostructured graphene nanoribbon model. (c) Checking the Bayesian optimization efficiency by comparing to a random search. (d) Comparison of the normalized *ZT*, power factor (*P*) and thermal resistance (R_{th}) for the pristine, periodic and optimal structures. Reproduced from [44]. Creative Commons Attribution NonCommercial License 4.0 (CC BY-NC).

absorbers can be improved by designing wavelength-selective absorbers, which have high absorbance in 0.4–4 μm to capture most of the solar energy and low emittance beyond 4 μm to minimize their own thermal radiation losses. The micro/nano-structures can interact with a thermal radiative electromagnetic wave and allow tuning of the radiative properties if the structure length scale is comparable or smaller than the photon wavelength. Typical subwavelength structures used for tailoring radiative properties include a planar multilayer structure [75, 100], a nanoparticle embedded structure [101, 102], a periodic grating structure [63, 73], nanowires [103, 104] and two-dimensional materials [105–107]. Although we may understand the possible theoretical mechanisms to explain the tailored radiative properties, there is still a large gap between the theory and real applications. Recently, Wang and Zhang [108] proposed the so-called inductor and capacitor circuit model to predict and design thermal radiative properties, but the wavelength dependence of the dielectric function is ignored in this model, which reduces the practical applicability. Researchers have also tried to combine this approach with classical optimization algorithms such as the genetic algorithm, simulated annealing, or frog jump to design the radiative property, but the efficiency is still rather low due to the nature of the random search, and the searching becomes hopeless, in particular for the real three-dimensional photonic structure design given to the heavy calculation. The new development of MI gives us a new tool to surpass this barrier. Liu *et al* [109] recently proposed a forward model intergrated with a deep NN to deal with the fundamental property of non-unique electromagnetic scattering and applied it to optimize a tandem structure. Liu *et al* [110] adopted a generative adversarial network to design a metal–dielectric metamaterial and demonstrated the possibility to design a meta-surface in a systematic and efficient way. Ma *et al* [111, 112] applied a deep learning NN to optimize three-dimensional chiral metamaterials and deep generative modal with semi-supervised strategy to design two-dimensional metamaterial patterns, respectively. The proposed two bidirectional based neural networks, which are assembled by partial stacking strategy, can automatically design and optimize the chiral metamaterial with target chiroptical response. The effective inverse probabilistic representation based design strategy is reported to fast target the meta-surface patterns in the corresponding work. Although there are various informatics methods which can be incorporated with traditional material simulation, here we focus on Bayesian optimization and take it as an example.

In [45], Sakurai *et al* successfully realized a wavelength-selective thermal emitter with an ultra-narrow-band based on aperiodic multilayer metamaterials designed by Bayesian optimization. This was the first demonstration of the possibility of using a simple multilayer structure for an extremely high Q-factor. Initially, they targeted Ge, Si and SiO_2 as candidate materials, labeled as '1', '2' and '3' in the descriptors. After several pre-trial simulations, they fixed the total number of layers as 18 and the total thickness ranged from 3.6 to 4 μm, with an increment of 0.02 μm. For such a case, the total optimization candidate number was $3^{18} \times 21 = 8\ 135\ 830\ 269$, which was enormously large. Considering that all the possible candidates have to be listed in advance when using Bayesian optimization, this will inherently require a large memory to store these possible candidates. To reduce the memory costs, the total number of candidates was randomly divided into 42 000 subgroups and Bayesian optimization was conducted on each group. When all the optimization processes were finished, the structure with the highest FOM in each subgroup was collected to obtain the global highest FOM candidate structure. To evaluate the narrowness of the wavelength-selective property, the FOM was defined by the difference of the ratio of thermal energy emitted by the structure to the black-body falling in and out of the bandwidth. The general optimization process can easily be understood from figure 5.2. The simulation part provides the FOM of the selected descriptor as feedback to Bayesian optimization. For the Bayesian optimization part, after learning from the initial randomly chosen training data, a predictive FOM distribution can be estimated. Then, it outputs N candidate structures which have a high possibility to have a high FOM. These selected candidate structures were further evaluated by the transfer matrix method simulation and were added in training data for Bayesian optimization. After several iterations, the optimal structure can be identified. The optimal metamaterials with targeted wavelengths at 5, 6 and 7 μm are given in figures 5.6(a)–(c). It is interesting to see that although Si was included as a layer material option, the final optimized structure did not contain Si, and the obtained structure was aperiodic. This also gives new insights for the design of narrow-band emitters, since periodic structures have usually been adopted

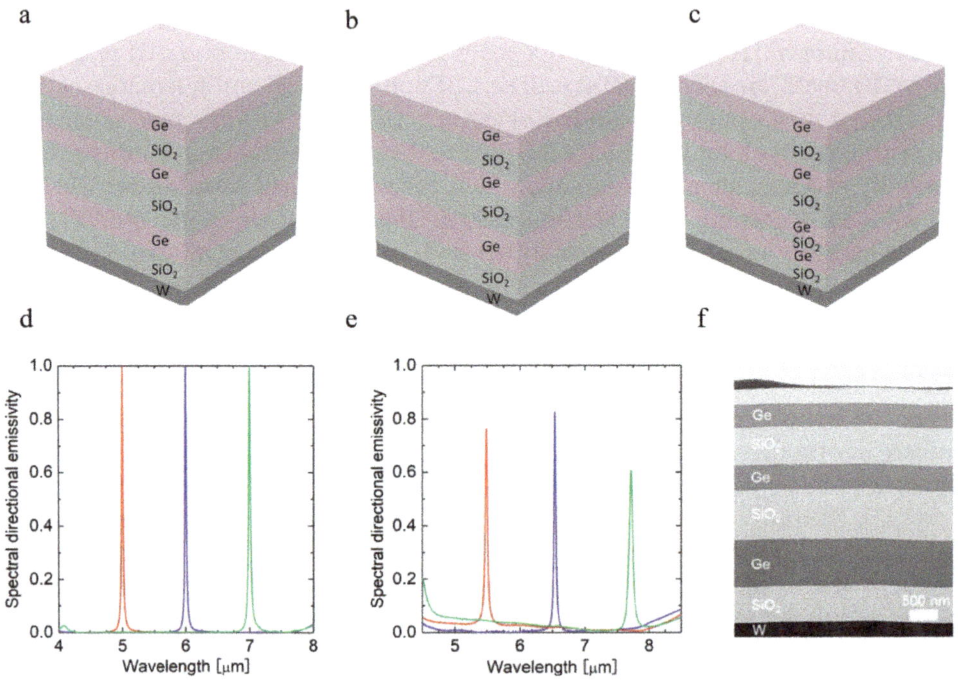

Figure 5.6. Optimized structures of the narrow-band thermal emitters for the target wavelengths of (a) 5.0 μm, (b) 6.0 μm and (c) 7.0 μm. (d) Calculated spectral directional emissivities of the optimized structures obtained with Bayesian optimization and (e) measured spectral directional emissivities of the fabricated structures' target wavelength aimed at 5.0 μm (red), 6.0 μm (blue) and 7.0 μm (green). (f) Cross-sectional TEM images of the fabricated sample for a target wavelength at 6.0 μm. Reproduced with permission from [45]. Copyright 2019 American Chemical Society.

in the past. In addition to obtaining the optimal structure by simulation, experimental validation was also conducted to demonstrate the effectiveness of the designed optimal structure via informatics. As shown in figures 5.6(d) and (e), the peaks for the designed optimal structures were targeted at 5, 6 and 7 μm with a high Q-factor of 217, 273 and 233. For experimental structures (figure 5.6(f)), a slight redshift for the targeted wavelength was observed and the amplitude for peak emissivity values were 0.76, 0.83 and 0.61, respectively, with a Q-factor of 132, 188 and 109 because the layer thicknesses of the actual fabricated samples somewhat deviated from the designed optimal case. Still, the obtained Q-factor of 188 is very high and is the best experimental value among the multilayered metamaterials reported so far.

5.4 Summary and perspectives

The successful application of Bayesian optimization on thermal science so far has given us much confidence in expanding this method to design better functionalized and cost-effective thermal materials, which may never have been picked out by the classical methods used in the past. The design of high/low thermal conductance or tailoring thermal radiative properties using Bayesian optimization has shown its superior high searching efficiency. Moreover, new physical mechanisms behind the optimization results can be explored and studied. However, just as coins always have two sides, the Bayesian algorithm also has its own drawbacks. For example, listing and saving all the possible candidates requires a large computational memory. With the increase of the searching space, the optimization speed will become slow. Most of the current MI algorithms are statistics-based and data-driven methods, which means that large-data training and accurate prediction models are necessary. However, it is sometimes very difficult and expensive to acquire such big-data input for the training. Researchers are now trying to develop more robust and accurate algorithms, such as transfer learning, Monte Carlo tree search and quantum annealing, which can learn from a small group of data and extend the search into a huge space. MI is still under development and we believe it will become much more powerful and may totally change the approach to materials design in the very near future. We hope more and more thermal science researchers can obtain insight from MI and apply MI to accelerate their research.

Acknowledgments

This work was supported by the 'Materials Research by Information Integration' Initiative (MI^2I) project of the Support Program for Starting Up Innovation Hub and CREST Grant No. JPMJCR16Q5 from the Japan Science and Technology Agency (JST), and KAKENHI (Grant Nos. 16H04274 and 19K14902) from the Japan Society for the Promotion of Science (JSPS).

References

[1] Chu S and Majumdar A 2012 Opportunities and challenges for a sustainable energy future *Nature* **488** 294–303

[2] Goldemberg J 2007 Ethanol for a sustainable energy future *Science* **315** 808–10

[3] Chu S, Cui Y and Liu N 2017 The path towards sustainable energy *Nat. Mater.* **16** 16–22

[4] Lund H 2007 Renewable energy strategies for sustainable development *Energy* **32** 912–9

[5] Baños R, Manzano-Agugliaro F, Montoya F G, Gil C, Alcayde A and Gómez J 2011 Optimization methods applied to renewable and sustainable energy: a review *Renew. Sustain. Energy Rev.* **15** 1753–66

[6] Sarı A and Karaipekli A 2007 Thermal conductivity and latent heat thermal energy storage characteristics of paraffin/expanded graphite composite as phase change material *Appl. Therm. Eng.* **27** 1271–7

[7] Mills A, Farid M, Selman J R and Al-Hallaj S 2006 Thermal conductivity enhancement of phase change materials using a graphite matrix *Appl. Therm. Eng.* **26** 1652–61

[8] Pei Q-X, Zhang Y-W, Sha Z-D and Shenoy V B 2013 Tuning the thermal conductivity of silicene with tensile strain and isotopic doping: a molecular dynamics study *J. Appl. Phys.* **114** 033526

[9] Guo Y, Cortes C L, Molesky S and Jacob Z 2012 Broadband super-Planckian thermal emission from hyperbolic metamaterials *Appl. Phys. Lett.* **101** 131106

[10] Marquier F, Joulain K, Mulet J P, Carminati R and Greffet J J 2004 Engineering infrared emission properties of silicon in the near field and the far field *Opt. Commun.* **237** 379–88

[11] Basu S, Zhang Z M and Fu C J 2009 Review of near-field thermal radiation and its application to energy conversion *Int. J. Energy Res.* **33** 1203–32
[12] Coppens Z J and Valentine J G 2017 Spatial and temporal modulation of thermal emission *Adv. Mater.* **29** 1701275
[13] Motayed A, Davydov A V, Vaudin M D, Levin I, Melngailis J and Mohammad S N 2006 Fabrication of GaN-based nanoscale device structures utilizing focused ion beam induced Pt deposition *J. Appl. Phys.* **100** 024306
[14] Xia Y, Rogers J A, Paul K E and Whitesides G M 1999 Unconventional methods for fabricating and patterning nanostructures *Chem. Rev.* **99** 1823–48
[15] Li X, Chang W-C, Chao Y J, Wang R and Chang M 2004 Nanoscale structural and mechanical characterization of a natural nanocomposite material: the shell of red abalone *Nano Lett.* **4** 613–7
[16] Schmitt J *et al* 1997 Metal nanoparticle/polymer superlattice films: fabrication and control of layer structure *Adv. Mater.* **9** 61–5
[17] Jordan M I and Mitchell T M 2015 Machine learning: trends, perspectives, and prospects *Science* **349** 255–60
[18] LeCun Y, Bengio Y and Hinton G 2015 Deep learning *Nature* **521** 436–44
[19] Ghahramani Z 2015 Probabilistic machine learning and artificial intelligence *Nature* **521** 452–9
[20] Mnih V *et al* 2015 Human-level control through deep reinforcement learning *Nature* **518** 529–33
[21] Lengauer T, Sander O, Sierra S, Thielen A and Kaiser R 2007 Bioinformatics prediction of HIV coreceptor usage *Nat. Biotechnol.* **25** 1407–10
[22] Schadt E E, Linderman M D, Sorenson J, Lee L and Nolan G P 2011 Cloud and heterogeneous computing solutions exist today for the emerging big data problems in biology *Nat. Rev. Genet.* **12** 224
[23] Murphy R F 2011 An active role for machine learning in drug development *Nat. Chem. Biol.* **7** 327–30
[24] Ross D T *et al* 2000 Systematic variation in gene expression patterns in human cancer cell lines *Nat. Genet.* **24** 227–35
[25] Sanchez-Lengeling B and Aspuru-Guzik A 2018 Inverse molecular design using machine learning: generative models for matter engineering *Science* **361** 360–5
[26] Chakradhar S 2017 Predictable response: finding optimal drugs and doses using artificial intelligence *Nat. Med.* **23** 1244–7
[27] Zhang L, Tan J, Han D and Zhu H 2017 From machine learning to deep learning: progress in machine intelligence for rational drug discovery *Drug Discovery Today* **22** 1680–5
[28] Raccuglia P *et al* 2016 Machine-learning-assisted materials discovery using failed experiments *Nature* **533** 73–6
[29] Ahneman D T, Estrada J G, Lin S, Dreher S D and Doyle A G 2018 Predicting reaction performance in C–N cross-coupling using machine learning *Science* **360** 186–90
[30] Langley P 1988 Machine learning as an experimental science *Mach. Learn.* **3** 5–8
[31] Berman H M *et al* 2000 The protein data bank *Nucleic Acids Res.* **28** 235–42
[32] Snoek J, Larochelle H and Adams R P 2012 Practical Bayesian optimization of machine learning algorithms *Advances in Neural Information Processing Systems* vol 25 ed F Pereira, C J C Burges, L Bottou and K Q Weinberger (New York: Curran Associates), pp 2951–9

[33] Riniker S and Landrum G A 2015 Better informed distance geometry: using what we know to improve conformation generation *J. Chem. Inf. Model.* **55** 2562–74
[34] Hernández-Lobato J M, Requeima J, Pyzer-Knapp E O and Aspuru-Guzik A 2017 Parallel and distributed Thompson sampling for large-scale accelerated exploration of chemical space *Int. Conf. on Machine Learning* pp 1470–9
[35] Srinivas N, Krause A, Kakade S and Seeger M 2010 Gaussian process optimization in the bandit setting: no regret and experimental design *Int. Conf. on Machine Learning* pp 1015–22
[36] Shahriari B, Swersky K, Wang Z, Adams R P and de Freitas N 2016 Taking the human out of the loop: a review of Bayesian optimization *Proc. IEEE* **104** 148–75
[37] Brochu E, Cora V M and de Freitas N 2010 A tutorial on Bayesian optimization of expensive cost functions, with application to active user modeling and hierarchical reinforcement learning (arXiv:1012.2599)
[38] Swersky K, Snoek J and Adams R P 2013 Multi-task Bayesian optimization *Advances in Neural Information Processing Systems* vol 26 ed C J C Burges, L Bottou, M Welling, Z Ghahramani and K Q Weinberger (New York: Curran Associates), pp 2004–12
[39] Pedregosa F *et al* 2011 Scikit-learn: machine learning in Python *J. Mach. Learn. Res.* **12** 2825–30
[40] GPy 2012 GPy: A Gaussian process framework in python http://github.com/SheffieldML/GPy
[41] Ueno T, Rhone T D, Hou Z, Mizoguchi T and Tsuda K 2016 COMBO: an efficient Bayesian optimization library for materials science *Mater. Discovery* **4** 18–21
[42] Yamashita T, Sato N, Kino H, Miyake T, Tsuda K and Oguchi T 2018 Crystal structure prediction accelerated by Bayesian optimization *Phys. Rev. Mater.* **2** 013803
[43] Ju S, Shiga T, Feng L, Hou Z, Tsuda K and Shiomi J 2017 Designing nanostructures for phonon transport via Bayesian optimization *Phys. Rev.* X **7** 021024
[44] Yamawaki M, Ohnishi M, Ju S and Shiomi J 2018 Multifunctional structural design of graphene thermoelectrics by Bayesian optimization *Sci. Adv.* **4** eaar4192
[45] Sakurai A *et al* 2019 Ultranarrow-band wavelength-selective thermal emission with aperiodic multilayered metamaterials designed by Bayesian optimization *ACS Cent. Sci.* **5** 319–26
[46] Seko A, Togo A, Hayashi H, Tsuda K, Chaput L and Tanaka I 2015 Prediction of low-thermal-conductivity compounds with first-principles anharmonic lattice-dynamics calculations and Bayesian optimization *Phys. Rev. Lett.* **115** 205901
[47] Tian Z, Esfarjani K and Chen G 2012 Enhancing phonon transmission across a Si/Ge interface by atomic roughness: first-principles study with the Green's function method *Phys. Rev.* B **86** 235304
[48] Merabia S and Termentzidis K 2014 Thermal boundary conductance across rough interfaces probed by molecular dynamics *Phys. Rev.* B **89** 054309
[49] Liu Y, Hu C, Huang J, Sumpter B G and Qiao R 2015 Tuning interfacial thermal conductance of graphene embedded in soft materials by vacancy defects *J. Chem. Phys.* **142** 244703
[50] Arora A, Hori T, Shiga T and Shiomi J 2017 Thermal rectification in restructured graphene with locally modulated temperature dependence of thermal conductivity *Phys. Rev.* B **96** 165419

[51] Sakata M, Oyake T, Maire J, Nomura M, Higurashi E and Shiomi J 2015 Thermal conductance of silicon interfaces directly bonded by room-temperature surface activation *Appl. Phys. Lett.* **106** 081603

[52] Lanzara A *et al* 2001 Evidence for ubiquitous strong electron–phonon coupling in high-temperature superconductors *Nature* **412** 510–4

[53] Yan J, Zhang Y, Kim P and Pinczuk A 2007 Electric field effect tuning of electron–phonon coupling in graphene *Phys. Rev. Lett.* **98** 166802

[54] Lin Z, Zhigilei L V and Celli V 2008 Electron–phonon coupling and electron heat capacity of metals under conditions of strong electron–phonon nonequilibrium *Phys. Rev.* B **77** 075133

[55] Brorson S D *et al* 1990 Femtosecond room-temperature measurement of the electron–phonon coupling constant γ in metallic superconductors *Phys. Rev. Lett.* **64** 2172–5

[56] Hameau S *et al* 1999 Strong electron–phonon coupling regime in quantum dots: evidence for everlasting resonant polarons *Phys. Rev. Lett.* **83** 4152–5

[57] Sood A K, Chandrabhas N, Muthu D V S and Jayaraman A 1995 Phonon interference in $BaTiO_3$: high-pressure Raman study *Phys. Rev.* B **51** 8892–6

[58] Han H, Potyomina L G, Darinskii A A, Volz S and Kosevich Y A 2014 Phonon interference and thermal conductance reduction in atomic-scale metamaterials *Phys. Rev.* B **89** 180301

[59] Feng L, Shiga T, Han H, Ju S, Kosevich Y A and Shiomi J 2017 Phonon-interference resonance effects by nanoparticles embedded in a matrix *Phys. Rev.* B **96** 220301

[60] Lim M, Lee S S and Lee B J 2015 Near-field thermal radiation between doped silicon plates at nanoscale gaps *Phys. Rev.* B **91** 195136

[61] Tang G and Wang J-S 2018 Heat transfer statistics in extreme-near-field radiation *Phys. Rev.* B **98** 125401

[62] Ilic O, Jablan M, Joannopoulos J D, Celanovic I, Buljan H and Soljačić M 2012 Near-field thermal radiation transfer controlled by plasmons in graphene *Phys. Rev.* B **85** 155422

[63] Messina R, Jin W and Rodriguez A W 2016 Strongly coupled near-field radiative and conductive heat transfer between planar bodies *Phys. Rev.* B **94** 121410

[64] Worthing P T and Barnes W L 2001 Efficient coupling of surface plasmon polaritons to radiation using a bi-grating *Appl. Phys. Lett.* **79** 3035–7

[65] Zhong R, Yu C, Hu M and Liu S 2018 Surface plasmon polaritons light radiation source with asymmetrical structure *AIP Adv.* **8** 015327

[66] Le Gall J, Olivier M and Greffet J-J 1997 Experimental and theoretical study of reflection and coherent thermal emissionby a SiC grating supporting a surface-phonon polariton *Phys. Rev.* B **55** 10105–14

[67] Lee B J and Zhang Z M 2008 Lateral shifts in near-field thermal radiation with surface phonon polaritons *Nanoscale Microscale Thermophys. Eng.* **12** 238–50

[68] Mulet J-P, Joulain K, Carminati R and Greffet J-J 2002 Enhanced radiative heat transfer at nanometric distances *Microscale Thermophys. Eng.* **6** 209–22

[69] Gramotnev D K and Bozhevolnyi S I 2014 Nanofocusing of electromagnetic radiation *Nat. Photonics* **8** 13–22

[70] Nielsen M G, Pors A, Albrektsen O and Bozhevolnyi S I 2012 Efficient absorption of visible radiation by gap plasmon resonators *Opt. Express* **20** 13311–9

[71] Ginzburg P, Arbel D and Orenstein M 2006 Gap plasmon polariton structure for very efficient microscale-to-nanoscale interfacing *Opt. Lett.* **31** 3288–90

[72] Bharadwaj P, Bouhelier A and Novotny L 2011 Electrical excitation of surface plasmons *Phys. Rev. Lett.* **106** 226802
[73] Yang Y and Wang L 2016 Spectrally enhancing near-field radiative transfer between metallic gratings by exciting magnetic polaritons in nanometric vacuum gaps *Phys. Rev. Lett.* **117** 044301
[74] Wang L P and Zhang Z M 2009 Resonance transmission or absorption in deep gratings explained by magnetic polaritons *Appl. Phys. Lett.* **95** 111904
[75] Lee B J, Wang L P and Zhang Z M 2008 Coherent thermal emission by excitation of magnetic polaritons between periodic strips and a metallic film *Opt. Express* **16** 11328–36
[76] Pendry J B, Martín-Moreno L and Garcia-Vidal F J 2004 Mimicking surface plasmons with structured surfaces *Science* **305** 847–8
[77] Zayats A V and Smolyaninov I I 2003 Near-field photonics: surface plasmon polaritons and localized surface plasmons *J. Opt. A: Pure Appl. Opt.* **5** S16–50
[78] Maier S A, Andrews S R, Martín-Moreno L and García-Vidal F J 2006 Terahertz surface plasmon–polariton propagation and focusing on periodically corrugated metal wires *Phys. Rev. Lett.* **97** 176805
[79] Caldwell J D *et al* 2015 Low-loss, infrared and terahertz nanophotonics using surface phonon polaritons *Nanophotonics* **4** 44–68
[80] Feng K *et al* 2015 Localized surface phonon polariton resonances in polar gallium nitride *Appl. Phys. Lett.* **107** 081108
[81] Huber A J, Ocelic N and Hillenbrand R 2008 Local excitation and interference of surface phonon polaritons studied by near-field infrared microscopy *J. Microsc.* **229** 389–95
[82] Hillenbrand R, Taubner T and Keilmann F 2002 Phonon-enhanced light–matter interaction at the nanometre scale *Nature* **418** 159–62
[83] Meijer G I 2010 Cooling energy-hungry data centers *Science* **328** 318–9
[84] Garner R 2018 Cutting-edge heat shield installed on NASA's Parker Solar Probe, NASA, Available from: http://nasa.gov/feature/goddard/2018/cutting-edge-heat-shield-installed-on-nasa-s-parker-solar-probe (Accessed: 28 February 2019)
[85] Zhang W, Fisher T S and Mingo N 2006 Simulation of interfacial phonon transport in Si–Ge heterostructures using an atomistic Green's function method *J. Heat Transfer* **129** 483–91
[86] Wang J-S, Wang J and Lü J T 2008 Quantum thermal transport in nanostructures *Eur. Phys. J.* B **62** 381–404
[87] Oyake T, Feng L, Shiga T, Isogawa M, Nakamura Y and Shiomi J 2018 Ultimate confinement of phonon propagation in silicon nanocrystalline structure *Phys. Rev. Lett.* **120** 045901
[88] Ju S, Shiga T, Feng L and Shiomi J 2018 Revisiting PbTe to identify how thermal conductivity is really limited *Phys. Rev.* B **97** 184305
[89] Gaskins J T *et al* 2018 Thermal boundary conductance across heteroepitaxial ZnO/GaN interfaces: assessment of the phonon gas model *Nano Lett.* **18** 7469–77
[90] Jia L, Ju S, Liang X and Zhang X 2016 Tuning phonon transmission and thermal conductance by roughness at rectangular and triangular Si/Ge interface *Mater. Res. Express* **3** 095024
[91] Hopkins P E, Norris P M, Tsegaye M S and Ghosh A W 2009 Extracting phonon thermal conductance across atomic junctions: nonequilibrium Green's function approach compared to semiclassical methods *J. Appl. Phys.* **106** 063503

[92] Hyldgaard P 2004 Resonant thermal transport in semiconductor barrier structures *Phys. Rev.* B **69** 193305

[93] Wang J, Li L and Wang J-S 2011 Tuning thermal transport in nanotubes with topological defects *Appl. Phys. Lett.* **99** 091905

[94] Wäckelgård E and Hultmark G 1998 Industrially sputtered solar absorber surface *Sol. Energy Mater. Sol. Cells* **54** 165–70

[95] Rephaeli E and Fan S 2008 Tungsten black absorber for solar light with wide angular operation range *Appl. Phys. Lett.* **92** 211107

[96] Hsu P-C *et al* 2016 Radiative human body cooling by nanoporous polyethylene textile *Science* **353** 1019–23

[97] Hsu P-C *et al* 2017 A dual-mode textile for human body radiative heating and cooling *Sci. Adv.* **3** e1700895

[98] Ilic O, Bermel P, Chen G, Joannopoulos J D, Celanovic I and Soljačić M 2016 Tailoring high-temperature radiation and the resurrection of the incandescent source *Nat. Nanotechnol.* **11** 320–4

[99] Raman A P, Anoma M A, Zhu L, Rephaeli E and Fan S 2014 Passive radiative cooling below ambient air temperature under direct sunlight *Nature* **515** 540–4

[100] Lee B J and Zhang Z M 2006 Design and fabrication of planar multilayer structures with coherent thermal emission characteristics *J. Appl. Phys.* **100** 063529

[101] Huang Z and Ruan X 2017 Nanoparticle embedded double-layer coating for daytime radiative cooling *Int. J. Heat Mass Transfer* **104** 890–96

[102] Atiganyanun S *et al* 2018 Effective radiative cooling by paint-format microsphere-based photonic random media *ACS Photonics* **5** 1181–7

[103] Basu S and Wang L 2013 Near-field radiative heat transfer between doped silicon nanowire arrays *Appl. Phys. Lett.* **102** 053101

[104] Chang J-Y, Wang H and Wang L 2017 Tungsten nanowire metamaterials as selective solar thermal absorbers by excitation of magnetic polaritons *J. Heat Transfer* **139** 052401

[105] Xia F, Wang H, Xiao D, Dubey M and Ramasubramaniam A 2014 Two-dimensional material nanophotonics *Nat. Photonics* **8** 899–907

[106] Zhao B and Zhang Z M 2015 Strong plasmonic coupling between graphene ribbon array and metal gratings *ACS Photonics* **2** 1611–8

[107] Dai S *et al* 2014 Tunable phonon polaritons in atomically thin van der Waals crystals of boron nitride *Science* **343** 1125–9

[108] Wang L P and Zhang Z M 2011 Phonon-mediated magnetic polaritons in the infrared region *Opt. Express* **19** A126–35

[109] Liu D, Tan Y, Khoram E and Yu Z 2018 Training deep neural networks for the inverse design of nanophotonic structures *ACS Photonics* **5** 1365–9

[110] Liu Z, Zhu D, Rodrigues S P, Lee K-T and Cai W 2018 Generative model for the inverse design of metasurfaces *Nano. Lett.* **18** 6570–6

[111] Ma W, Cheng F and Liu Y 2018 Deep-learning-enabled on-demand design of chiral metamaterials *ACS Nano.* **12** 6326–34

[112] Ma W, Cheng F, Xu Y, Wen Q and Liu Y 2019 Probabilistic Representation and inverse design of metamaterials based on a deep generative model with semi-supervised learning strategy *Adv. Mater.* **31** 1901111

IOP Publishing

Nanoscale Energy Transport
Emerging phenomena, methods and applications
Bolin Liao

Chapter 6

Phonon mean free path spectroscopy: theory and experiments

Chengyun Hua

Phonon mean free path (MFP) spectroscopy is an experimental technique to reveal microscopic information on phonons, such as MFP distribution, spectral transmissivity across an interface and specularity parameters at a surface. This technique is based on observations of nonlocal thermal transport behavior, which occurs if a temperature gradient exists over a length scale comparable to the phonon MFPs. This chapter presents an overview of the theoretical basis of this technique based on the Peierls–Boltzmann transport equation. We discuss how macroscopic observables reveal microscopic information on phonons and how various mathematical frameworks are applied to extract mode-dependent phonon information from these observables. Furthermore, this chapter provides an overview of thermal characterization techniques that have reported observations of nonlocal thermal transport.

6.1 Introduction

Phonons, the normal modes of lattice vibrations, are the dominant heat carriers in nonmagnetic dielectrics, semiconductors and insulators. The ability to control and manipulate the properties of phonons is of critical importance for a variety of technologies, including thermoelectrics for waste heat recovery and solid-state refrigeration, effective thermal management in microelectronic devices, and solid-state batteries. Of particular relevance here is the characterization of phonon MFPs, knowledge of which is crucial for engineering thermal conductivity in nanostructured materials such as nanowires, superlattices and nanocomposites with strongly reduced thermal conductivities due to phonon scattering at interfaces and surfaces. Recent advances in computational power and numerical methods have enabled us to predict the MFPs of phonons in single crystals from first-principles calculations. However, few experimental techniques exist to measure them. Inelastic neutron [1–4]

doi:10.1088/978-0-7503-1738-2ch6

and x-ray scattering [5, 6] have been used to measure the vibrational properties of solids, such as the phonon density of states, dispersion and lifetimes in single crystal samples. Raman spectroscopy has been employed to measure the linewidths (inverse of phonon lifetimes) of Raman modes in semiconductor crystals [7] and later in more exotic systems [8–13]. However, Raman spectroscopy is limited to zone center optic modes which do not carry heat, and most of these techniques are not suitable for thin films or materials with nanostructures such as polycrystalline materials or superlattices. Although the thermal conductivity of two-dimensional materials has been measured by Raman thermometry [14–18], it is not suitable for extracting the spectral information of phonons. Thus, our knowledge of phonon MFPs remains incomplete, and a complementary technique that allows us to measure phonon MFPs in a wide range of material classes is desirable.

Recently, phonon MFP spectroscopy has been developed to measure phonon MFP distributions over a wide range of materials. This technique, first developed by Minnich [19], is based on the observation of size effects in thermal transport, which occurs at scales comparable to phonon wavelengths and MFPs. These observations have been used to reconstruct microscopic phonon information including phonon MFP distributions in various materials [19–24], spectral phonon transmission coefficients across a metal–semiconductor interface [25], mode-dependent specularity parameters of phonons at a rough surface of silicon membranes [26] and phonon relaxation times in a silicon thin film [27].

This chapter aims to provide a basic foundation for the physical origin, a theoretical formalism, and a working principle of phonon MFP spectroscopy and the various mathematical frameworks used to extract microscopic information on phonons from measured observables. Furthermore, this chapter will provide an overview of a few key applications of MFP spectroscopy and their challenges.

6.2 Principles of MFP spectroscopy

The principle of MFP spectroscopy is based on the observation of size effects of heat conduction or nonlocal heat conduction, which occur at length scales comparable to phonon wavelengths and MFPs. There are two types of size effects. The first type is caused by the macroscopic dimensions of a bulk sample and was first considered by Casimir [28]. Here, we define a key dimensionless measure of this, the Knudsen number, $\mathrm{Kn} \equiv \Lambda/L$, which compares phonon MFP, Λ, with a characteristic length, L. When $\mathrm{Kn} \sim 1$, the characteristic dimension of the structure approaches the phonon MFPs, and the thermal conductivity can be substantially smaller than the bulk value due to scattering from boundaries [29, 30]. Significant thermal conductivity reductions have been observed in a number of nanoscale systems, including nanowires [31–33], nanotubes [34], nanoribbons [18], superlattices [35] and thin membranes [20]. Information about MFPs is obtained by measuring the thermal conductivity over variable lengths of nanostructures. Such an approach has been used to reconstruct phonon MFPs in different nanoscale silicon membranes [20] and graphene ribbons, for example [36].

The second type of size effect is caused by a temperature gradient that exists over length scales comparable to the phonon MFPs. In this type of nonlocal heat conduction, the heat flux and temperature deviate from those predicted by heat diffusion theory based on Fourier's law. These discrepancies were first observed at a localized hotspot created by a doped resistor thermometer in a suspended silicon membrane [37] and more recently in optical pump–probe experiments including soft x-ray diffraction from nanoline arrays [38, 39], transient thermal grating (TTG) [26, 40, 41] and thermoreflectance methods [21, 24, 25, 42–45]. These observations of nonlocal thermal conduction have been used to reconstruct the MFP distribution of phonons in various materials using a mathematical framework first introduced by Minnich [19]. In addition to phonon MFP spectra, experimental observations of nonlocal transport behavior also reveal insights on the microscopic processes of phonon transport across grain boundaries and interfaces such as mode-dependent phonon transmission coefficients and specularity parameters when combined with *ab initio* modeling.

The basic principle of the second type of size effect is illustrated in figure 6.1, where a Gaussian hotspot with a diameter D imposes a temperature gradient on the surface of a material. When D is much larger than the phonon MFPs, i.e. $\mathrm{Kn} \ll 1$, heat conduction follows a diffusion description governed by Fourier's law. When D is much smaller than the phonon MFPs, i.e. $\mathrm{Kn} \gg 1$, heat conduction becomes ballistic, which is governed by thermal radiation. Between these two regimes is where nonlocal heat conduction occurs ($\mathrm{Kn} \sim 1$). In experiments, nonlocal heat conduction is observed as the discrepancy in the measured heat flux or temperature from the Fourier's law prediction. For example, the measured surface temperature in figure 6.1 would be higher than the surface temperature predicted by Fourier's law.

MFP spectroscopy can be based on the observation of either type of size effect. The technique consists of observing the discrepancies in the experimental observables, Z_i, as compared to the bulk material properties or Fourier's law predictions, as a thermal length, D_i, is systematically varied from $\mathrm{Kn} \ll 1$ to $\mathrm{Kn} > 1$. The mathematical formulation of this principle, first proposed by Minnich, can be expressed as [19]

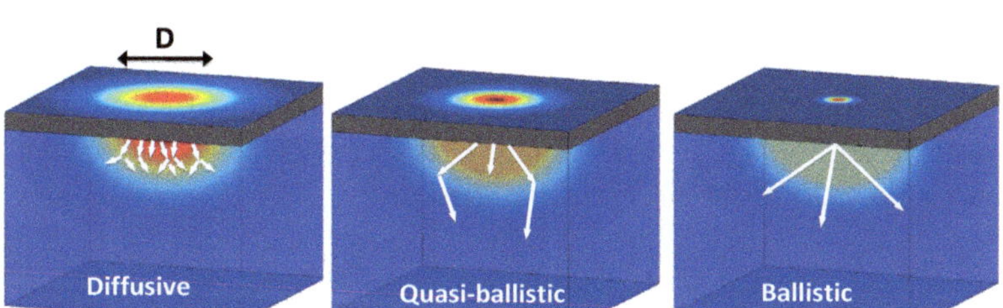

Figure 6.1. Size effect caused by nonlocal heat conduction. When the characteristic thermal length scale, such as the Gaussian heating diameter D in this illustration, becomes comparable to or smaller than the phonon MFPs, nonlocal heat conduction occurs. Experimental observables such as the measured heat flux and temperature deviate from those predicted by Fourier's law. Image credit: A J Minnich.

$$Z_i = \sum_{\mu} \mathcal{H}_\mu(\mathcal{T}_\mu, D_i), \tag{6.1}$$

where $\mathcal{H}_\mu$ is called the microscopic transfer function and $\mathcal{T}_\mu$ is the desired microscopic property for phonon state μ, such as a phonon MFP spectrum [19, 20], the spectral phonon transmission profile across an interface [25], or mode-dependent specularity parameters at a surface [26]. Given Z_i and the microscopic transfer function, $\mathcal{H}_\mu$, equation (6.1) can be solved as an inverse problem to obtain T_μ.

There are three necessary components of an MFP spectroscopic experiment. First, it requires an experimental technique that allows the users to obtain a set of experimental observables, Z_i, that deviate from macroscopic measurements, by systematically changing either the dimension of the sample's structure or the length scales of the temperature gradient imposed on the sample. Examples of the former case include thickness-dependent measurements of the thermal conductivity of nanoscale silicon membranes [20] and length-dependent measurements of the thermal conductivity of graphene ribbons [36]. In the latter case, a systematic variation of the length scale of the temperature gradient can be achieved in optical pump–probe experiments, such as thermoreflectance techniques, x-ray diffraction and thermal transient grating techniques.

The second requirement is a known microscopic transfer function, which relates the experimental observables to the microscopic phonon properties. Finally, an appropriate mathematical framework to map the experimental observables back to the microscopic phonon properties, usually an ill-posed inverse problem, is equally critical to successfully determine the microscopic phonon properties using MFP spectroscopy.

Here, a hypothetical scenario for determining the phonon MFP spectrum through a systematic variation of grain size in a polycrystalline material is presented. The physics of thermal conductivity is commonly interpreted using kinetic theory:

$$k^j = \sum_{\mu} C_\mu v_{\mu j} \Lambda_{\mu j}, \tag{6.2}$$

where μ indicates a phonon state and j is a directional index, i.e. x, y or z. C_μ is the mode-specific heat, $v_{\mu j}$ is the group velocity along direction j, and $\Lambda_{\mu j}$ is the effective MFP in the same direction. $\Lambda_{\mu j}$ includes all possible scattering mechanisms that limit the phonon MFPs in a bulk sample, $\Lambda_{\mu j,\mathrm{bulk}}$, as well as additional scattering due to grain boundaries, $\Lambda_{\mu j,\mathrm{bdy}}$, which are combined using Matthiessen's rule: $\Lambda_{\mu j}^{-1} = \Lambda_{\mu j,\mathrm{bulk}}^{-1} + \Lambda_{\mu j,\mathrm{bdy}}^{-1}$. The simplest model to describe grain boundary scattering is a gray model given by $\Lambda_{\mu j,\mathrm{bdy}} = D$, where D is the average size of the grains. Using Matthiessen's rule, equation (6.2) can be written as

$$k^j(D) = \sum_{\mu} C_\mu v_{\mu j} \Lambda_{\mu j,\mathrm{bulk}} \frac{1}{1 + \frac{\Lambda_{\mu j,\mathrm{bulk}}}{D}}, \tag{6.3}$$

where the thermal conductivity is now a function of the grain size. Here, $\mathcal{T}_\mu = C_\mu v_{\mu j} \Lambda^j_{\mu,\mathrm{bulk}}$ (the mode-specific thermal conductivity in a bulk sample) is the desired distribution, and the microscopic transfer function $\mathcal{H}_\mu$ is given by $\mathcal{T}_\mu(1 + \mathrm{Kn}^j_{\mu,\mathrm{bulk}})^{-1}$ with $\mathrm{Kn}^j_{\mu,\mathrm{bulk}} = \Lambda^j_{\mu,\mathrm{bulk}}/D$. Given the values of k^j at various D, $\mathcal{T}_\mu$ is then the solution to the ill-posed problem of the linear system described by equation (6.3).

In reality, this hypothetical experiment is difficult to achieve due to a lack of precise control of grain size during the fabrication of polycrystalline materials and more accurate description of grain boundary scattering is also more complicated than the above description. In the rest of this chapter, we will focus on the transport theory that allows derivation of the necessary microscopic transfer function $\mathcal{H}_\mu$ for different geometries and the applications of this in various existing experiments.

6.3 Theory

This section builds a theoretical underpinning of phonon MFP spectroscopy through a thorough overview of phonon transport based on the Peierls–Boltzmann transport equation (PBE) and various mathematical frameworks that allow for the extraction of microscopic information from macroscopic observables.

6.3.1 Nonlocal theory of heat conduction

To understand the theory behind phonon MFP spectroscopy, it is necessary to first discuss the physical origin of nonlocal heat conduction. To obtain a microscopic view of nonlocal phonon transport, we start with the PBE, which can describe heat conduction at length scales comparable to phonon MFPs.

The PBE, first derived by Peierls [46], is an integro-differential equation of time, real space and phase space. Due to its high dimensionality, solving the PBE for a general space–time-dependent problem remains a challenging task. Guyer and Krumhansl [47] first performed a linear response analysis of the PBE and applied their solution to develop a phenomenological coupling between phonons and elastic dilatational fields caused by lattice anharmonicity. The variational principle was also used to solve the PBE with Umklapp scattering incorporated [48, 49]. Levinson developed a nonlocal diffusion theory of thermal conductivity from a solution of the PBE with three-phonon scattering in the low-frequency limit [50]. Lattice thermal conductivity has been computed from the steady-state PBE by imposing a constant temperature gradient and using an iterative method [51–55] or a variational approach [56]. Chaput [57] presented a direct solution to the transient linearized PBE imposed with a linear temperature profile. Cepellotti and Marzari [58] used an eigendecomposition method, first used by Hardy in his study of second sound [59, 60], to diagonalize the collision matrix and project the PBE into the eigenspace of the collision matrix. They applied this treatment to solve steady-state problems in two-dimensional systems with a constant temperature gradient imposed in one direction [61].

Solving the PBE with the full collision operator, even in its linearized form, is difficult for complicated geometries. Therefore, various theoretical frameworks based on a simplified PBE have been developed to describe nonlocal thermal transport for general problems. Non-diffusive responses observed in experiments [24, 40, 42, 43, 62, 63] have been explained separately using a phonon-hydrodynamic heat equation [64], a truncated Levy formalism [65], a two-channel model in which low- and high-frequency phonons are described by the PBE and a heat equation [66], and a Mckelvey–Shockley flux method [67]. Methods based on solving the PBE under the relaxation time approximation (RTA), where each phonon mode relaxes towards thermal equilibrium at a characteristic relaxation rate independent of the other phonon populations, have been developed to investigate nonlocal transport in an infinite domain system [68–71], a finite one-dimensional slab [72, 73], and experimental configurations including transient grating [71, 74] and thermoreflectance experiments [75–77]. An efficient Monte Carlo scheme has also been used to solve the PBE under the RTA in complicated geometries involving multiple boundaries [78–80].

Here, we begin by briefly reviewing the derivation of the transport solution to the space–time-dependent PBE under the RTA, which is given by

$$\frac{\partial g_\mu(\mathbf{x}, t)}{\partial t} + \mathbf{v}_\mu \cdot \nabla g_\mu(\mathbf{x}, t) = -\frac{g_\mu(\mathbf{x}, t) - g_0(T(\mathbf{x}, t))}{\tau_\mu} + \dot{Q}_\mu(\mathbf{x}, t), \tag{6.4}$$

where $g_\mu(\mathbf{x}, t) = \hbar\omega_\mu(f_\mu(\mathbf{x}, t) - f_0(T_0))$ is the deviational energy distribution function at position $\mathbf{x}$ and time t for phonon states μ ($\mu \equiv (\mathbf{k}, s)$, where $\mathbf{k}$ is the wavevector and s is the phonon branch index). f_0 is the equilibrium Bose–Einstein distribution, and $g_0(T(\mathbf{x}, t)) = \hbar\omega_\mu(f_0(T(\mathbf{x}, t)) - f_0(T_0)) \approx C_\mu \Delta T(\mathbf{x}, t)$, where $T(\mathbf{x}, t)$ is the local temperature, T_0 is the global equilibrium temperature, $\Delta T(\mathbf{x}, t) = T(\mathbf{x}, t) - T_0$ is the local temperature deviation from equilibrium and $C_\mu = \hbar\omega_\mu \frac{\partial f_0}{\partial T}|_{T_0}$ is the mode-dependent specific heat. Here, we assume that $\Delta T(\mathbf{x}, t)$ is small such that $g_0(T(\mathbf{x}, t))$ is approximated to be the first term of its Taylor expansion around T_0. Finally, $\dot{Q}_\mu(\mathbf{x}, t)$ is the heat input rate per mode, $\mathbf{v}_\mu = (v_{\mu x}, v_{\mu y}, v_{\mu z})$ is the phonon group velocity vector and τ_μ is the phonon relaxation time.

To close the problem, energy conservation is used to relate $g_\mu(\mathbf{x}, t)$ to $\Delta T(\mathbf{x}, t)$ as

$$\frac{\partial E(\mathbf{x}, t)}{\partial t} + \nabla \cdot \mathbf{q}(\mathbf{x}, t) = \dot{Q}(\mathbf{x}, t), \tag{6.5}$$

where $E(\mathbf{x}, t) = V^{-1}\sum_\mu g_\mu(\mathbf{x}, t)$ is the total volumetric energy, $\mathbf{q}(\mathbf{x}, t) = V^{-1}\sum_\mu g_\mu(\mathbf{x}, t)\mathbf{v}_\mu$ is the directional heat flux and $\dot{Q}(\mathbf{x}, t) = V^{-1}\sum_\mu \dot{Q}_\mu(\mathbf{x}, t)$ is the volumetric mode-specific heat input rate. Here, the sum over μ denotes a sum over all phonon modes in the Brillouin zone, and V is the volume of the crystal. The solution of equation (6.4) yields a distribution function, from which heat flux fields can be obtained. The temperature is obtained by simultaneously solving equations (6.4) and (6.5). For a general multi-dimensional geometry, equation (6.4) is typically solved using numerical methods such as finite difference [81] or Monte Carlo schemes [78, 79, 82]. In [83], a generalized constitutive relation was derived that links the temperature gradient to the modal heat flux for a general geometry using a coordinate transformation from which the temperature field can be solved numerically. Here, we will derive an analytical expression of the heat flux in infinite transverse geometries based on the work in [83] by recognizing that, for most of the experimental configurations where nonlocal thermal transport has been observed, the transverse directions (y and z) can be regarded as infinite.

We begin by performing a Fourier transform in time t and two transverse directions, y and z, on equation (6.4), which gives

$$\Lambda_{\mu x}\frac{\partial \tilde{g}_{\mu}}{\partial x} + \alpha_{\mu}\tilde{g}_{\mu} = C_{\mu}\Delta\tilde{T} + \tilde{Q}_{\mu}\tau_{\mu}, \tag{6.6}$$

where $\alpha_{\mu} = 1 + \mathrm{i}\xi_y\Lambda_{\mu y} + \mathrm{i}\xi_z\Lambda_{\mu z} + \mathrm{i}\eta\tau_{\mu}$. In equation (6.6), η is the Fourier temporal variable, ξ_y and ξ_z are the Fourier spatial variables in the y and z directions, respectively, and $\Lambda_{\mu x}$, $\Lambda_{\mu y}$ and $\Lambda_{\mu z}$ are the phonon MFPs along the x, y and z directions, respectively. Γ defines the boundary of the problem in the x direction. Without losing generality, the domain is set to be $x \in [0, L]$. To facilitate the derivation below, we define a dimensionless variable $\chi \equiv x/L$ and directional Knudsen numbers as $\mathrm{Kn}_{\mu x} \equiv \Lambda_{\mu x}/L$, $\mathrm{Kn}_{\mu y} \equiv \xi_y\Lambda_{\mu y}$ and $\mathrm{Kn}_{\mu z} \equiv \xi_z\Lambda_{\mu z}$, which compare phonon directional MFP with a characteristic length in its corresponding direction, i.e. L in the x direction, ξ_y^{-1} in the y direction and ξ_z^{-1} in the z direction.

Using the dimensionless terms defined above, equation (6.6) has the following solution:

$$\tilde{g}_{\mu}(\chi, \xi_y, \xi_z, \eta) = P_{\mu}^{+}\mathrm{e}^{-\alpha_{\mu}\frac{\chi}{\mathrm{Kn}_{\mu x}}} + \int_0^{\chi}\frac{C_{\mu}\Delta\tilde{T}+\tilde{Q}_{\mu}\tau_{\mu}}{\mathrm{Kn}_{\mu x}}\mathrm{e}^{-\alpha_{\mu}\frac{\chi-\chi'}{\mathrm{Kn}_{\mu x}}}\mathrm{d}\chi' \quad \text{for } v_{\mu x} > 0, \tag{6.7a}$$

$$\tilde{g}_{\mu}(\chi, \rho, \zeta, \eta) = P_{\mu}^{-}\mathrm{e}^{\alpha_{\mu}\frac{1-\chi}{\mathrm{Kn}_{\mu x}}} - \int_{\chi}^{1}\frac{C_{\mu}\Delta\tilde{T}+\tilde{Q}_{\mu}\tau_{\mu}}{\mathrm{Kn}_{\mu x}}\mathrm{e}^{-\alpha_{\mu}\frac{\chi-\chi'}{\mathrm{Kn}_{\mu x}}}\mathrm{d}\chi' \quad \text{for } v_{\mu x} < 0. \tag{6.7b}$$

P_{μ}^{+} and P_{μ}^{-} are functions of ξ_y, ξ_z and η and are determined by the boundary conditions at $\chi = 0$ and $\chi = 1$, respectively. Using the symmetry of $v_{\mu x}$ about the center of the Brillouin zone, i.e. $v_{\mu x} = -v_{-\mu x}$, equations (6.7a) and (6.7b) can be combined into the following form:

$$\tilde{g}_{\mu}(\chi, \xi_y, \xi_z, \eta) = P_{\mu}\mathrm{e}^{-\alpha_{\mu}\frac{\chi}{\mathrm{Kn}_{\mu x}}} + \int_{\Gamma}\frac{C_{\mu}\Delta\tilde{T}+\tilde{Q}_{\mu}\tau_{\mu}}{|\mathrm{Kn}_{\mu x}|}\mathrm{e}^{-\alpha_{\mu}\left|\frac{\chi'-\chi}{\mathrm{Kn}_{\mu x}}\right|}\mathrm{d}\chi', \tag{6.8}$$

where

$$P_{\mu} = \begin{cases} P_{\mu}^{+} & \text{if } v_{\mu x} > 0 \\ P_{\mu}^{-}\mathrm{e}^{\frac{\alpha_{\mu}}{\mathrm{Kn}_{\mu x}}} & \text{if } v_{\mu x} < 0 \end{cases} \tag{6.9}$$

and

$$\Gamma \in \begin{cases} [0, \chi) & \text{if } v_{\mu x} > 0 \\ (\chi, 1] & \text{if } v_{\mu x} < 0 \end{cases}. \tag{6.10}$$

A governing equation for the temperature field can be obtained by inserting equation (6.8) into equation (6.5), from which it can be numerically determined.

More physical insights can be obtained by examining the heat flux in non-diffusive regimes. The heat flux along the x direction can be expressed as

$$\tilde{q}_x = \frac{1}{V}\sum_{\mu} g_\mu v_{\mu x} = \frac{1}{V}\sum_{\mu}\left[P_\mu v_{\mu x} e^{-\alpha_\mu \frac{\chi}{\mathrm{Kn}_{\mu x}}} + \int_\Gamma \frac{C_\mu v_{\mu x}\Delta\tilde{T} + \tilde{Q}_\mu v_{\mu x}\tau_\mu}{|\mathrm{Kn}_{\mu x}|} e^{-\alpha_\mu\left|\frac{\chi'-\chi}{\Lambda_{\mu x}}\right|} d\chi'\right]. \quad (6.11)$$

Applying integration by parts to the second term in equation (6.11) and using the symmetry of $v_{\mu x}$ about the zone center gives

$$\tilde{q}_x = -\int_0^1 \kappa_x(\chi - \chi', \xi_y, \xi_z, \eta)\frac{1}{L}\frac{\partial\Delta\tilde{T}(\chi', \xi_y, \xi_z, \eta)}{\partial\chi'} d\chi' + B_x(\chi, \xi_y, \xi_z, \eta), \quad (6.12)$$

where

$$\begin{aligned} B_x(\chi, \xi_y, \xi_z, \eta) = {} & \frac{1}{V}\sum_{\mu}\left[P_\mu v_{\mu x} e^{-\alpha_\mu \frac{\chi}{\mathrm{Kn}_{\mu x}}} + \mathrm{sgn}(v_{\mu x})\int_\Gamma \tilde{Q}_\mu(\chi', \xi_y, \xi_z, \eta) e^{-\alpha_\mu\left|\frac{\chi-\chi'}{\mathrm{Kn}_{\mu x}}\right|} d\chi'\right] \\ & + \frac{1}{V}\sum_{\mu_x>0,\mu_y,\mu_z} \frac{C_\mu v_{\mu x}}{\alpha_\mu}\left[\Delta\tilde{T}(1) e^{-\alpha_\mu \frac{1-\chi}{\mathrm{Kn}_{\mu x}}} - \Delta\tilde{T}(0) e^{-\alpha_\mu \frac{\chi}{\mathrm{Kn}_{\mu x}}}\right] \end{aligned} \quad (6.13)$$

is solely determined by the boundary condition and the volumetric heat input rate. $\kappa_{\mu x}(x, \xi_y, \xi_z, \eta)$ is the thermal conductivity along the x direction given by

$$\kappa_{\mu x}(\chi, \xi_y, \xi_z, \eta) = \frac{1}{V}\sum_{\mu_x>0,\mu_y,\mu_z} C_\mu v_{\mu x}\mathrm{Kn}_{\mu x}\frac{e^{-\alpha_\mu\frac{|\chi|}{\mathrm{Kn}_{\mu x}}}}{\alpha_\mu \mathrm{Kn}_{\mu x}}. \quad (6.14)$$

Similarly, heat flux along the y direction can be expressed as

$$\begin{aligned} \tilde{q}_y = \frac{1}{V}\sum_{\mu} g_\mu v_{\mu y} = {} & \frac{1}{V}\sum_{\mu}\left[P_\mu v_{\mu y} e^{-\alpha_\mu \frac{\chi}{\mathrm{Kn}_{\mu x}}} + \int_\Gamma \frac{\tilde{Q}_\mu \Lambda_{\mu y}}{|\mathrm{Kn}_{\mu x}|} e^{-\alpha_\mu\left|\frac{\chi'-\chi}{\mathrm{Kn}_{\mu x}}\right|} d\chi'\right] \\ & + \frac{1}{V}\sum_{\mu_x>0,\mu_y,\mu_z}\int_0^1 \frac{C_\mu v_{\mu y}}{\mathrm{Kn}_{\mu x}} e^{-\alpha_\mu\frac{|\chi'-\chi|}{\mathrm{Kn}_{\mu x}}}\Delta\tilde{T}(\chi', \xi_y, \xi_z, \eta) d\chi'. \end{aligned} \quad (6.15)$$

Using the symmetry of $v_{\mu y}$ and $v_{\mu z}$ about the zone center, equation (6.15) becomes

$$\begin{aligned} \tilde{q}_y = {} & \frac{1}{V}\sum_{\mu}\left[P_\mu v_{\mu y} e^{-\alpha_\mu \frac{\chi}{\mathrm{Kn}_{\mu x}}} + \int_\Gamma \frac{\tilde{Q}_\mu \Lambda_{\mu y}}{|\mathrm{Kn}_{\mu x}|} e^{-\alpha_\mu\left|\frac{\chi'-\chi}{\mathrm{Kn}_{\mu x}}\right|} d\chi'\right] \\ & - \frac{1}{V}\sum_{\mu_{x,y,z}>0}\int_0^1 C_\mu v_{\mu y}\frac{4e^{-\gamma_\mu\frac{|\chi'-\chi|}{\mathrm{Kn}_{\mu x}}}}{\mathrm{Kn}_{\mu x}}\cosh\left(i\mathrm{Kn}_{\mu z}\frac{|\chi-\chi'|}{\mathrm{Kn}_{\mu x}}\right) \\ & \times \sinh\left(i\mathrm{Kn}_{\mu y}\frac{|\chi-\chi'|}{\mathrm{Kn}_{\mu x}}\right)\Delta\tilde{T}(\chi', \xi_y, \xi_z, \eta) d\chi', \end{aligned} \quad (6.16)$$

where $\gamma_\mu = 1 + i\eta\tau_\mu$. Using the identities $\cosh(ix) = \cos(x)$ and $\sinh(ix) = \mathrm{i}\sin(x)$, we can rearrange the above equation into the following form:

$$\tilde{q}_y = -\int_0^1 \mathrm{i}\xi_y \Delta\widetilde{T}(\chi', \xi_y, \xi_z, \eta)\kappa_y(\chi - \chi', \xi_y, \xi_z, \eta)\mathrm{d}\chi' + B_y(\chi, \xi_y, \xi_z, \eta), \quad (6.17)$$

where

$$B_y(\chi, \xi_y, \xi_z, \eta) = \frac{1}{V}\sum_\mu \left[P_\mu v_{\mu y} \mathrm{e}^{-\alpha_\mu \frac{\chi}{\mathrm{Kn}_{\mu x}}} + \int_\Gamma \frac{\tilde{Q}_\mu \Lambda_{\mu y}}{|\mathrm{Kn}_{\mu x}|} \mathrm{e}^{-\alpha_\mu \left|\frac{\chi' - \chi}{\mathrm{Kn}_{\mu x}}\right|} \mathrm{d}\chi' \right] \quad (6.18)$$

is again solely determined by the boundary condition and the volumetric heat input rate. $\kappa_y(\chi, \xi_y, \xi_z, \eta)$ is the thermal conductivity along the y direction given by

$$\kappa_y(\chi, \xi_y, \xi_z, \eta) = \frac{1}{V}\sum_{\mu_{x,y,z}>0} C_\mu v_{\mu y} \Lambda_{\mu y} \frac{4\mathrm{e}^{-\gamma_\mu \frac{|\chi|}{\mathrm{Kn}_{\mu x}}}}{\mathrm{Kn}_{\mu x}} \cos\left(\mathrm{Kn}_{\mu z} \frac{|\chi|}{\mathrm{Kn}_{\mu x}}\right) \frac{\sin\left(\mathrm{Kn}_{\mu y} \frac{|\chi|}{\mathrm{Kn}_{\mu x}}\right)}{\mathrm{Kn}_{\mu y}}. \quad (6.19)$$

The inverse Fourier transform of $\mathrm{i}\xi_y \Delta\tilde{T}(\chi, \xi_y, \xi_z, \eta)$ in ξ_y gives a temperature gradient along the y direction, $\partial\Delta T/\partial y$, in real space. Following a similar derivation, the heat flux along the z direction can be written as

$$\tilde{q}_z = -\int_0^1 \mathrm{i}\xi_z \Delta\tilde{T}(\chi', \xi_y, \xi_z, \eta)\kappa_z(\chi - \chi', \xi_y, \xi_z, \eta)\mathrm{d}\chi' + B_z(\chi, \xi_y, \xi_z, \eta), \quad (6.20)$$

where

$$B_z(\chi, \xi_y, \xi_z, \eta) = \frac{1}{V}\sum_\mu \left[P_\mu v_{\mu z} \mathrm{e}^{-\alpha_\mu \frac{\chi}{\mathrm{Kn}_{\mu x}}} + \int_\Gamma \frac{\tilde{Q}_\mu \Lambda_{\mu z}}{|\mathrm{Kn}_{\mu x}|} \mathrm{e}^{-\alpha_\mu \left|\frac{\chi' - \chi}{\mathrm{Kn}_{\mu x}}\right|} \mathrm{d}\chi' \right] \quad (6.21)$$

and

$$\kappa_z(\chi, \xi_y, \xi_z, \eta) = \frac{1}{V}\sum_{\mu_{x,y,z}>0} C_\mu v_{\mu z} \Lambda_{\mu z} \frac{4\mathrm{e}^{-\gamma_\mu \frac{|\chi|}{\mathrm{Kn}_{\mu x}}}}{\mathrm{Kn}_{\mu x}} \cos\left(\xi_y \Lambda_{\mu y} \frac{|\chi|}{\mathrm{Kn}_{\mu x}}\right) \frac{\sin\left(\xi_z \Lambda_{\mu z} \frac{|\chi|}{\mathrm{Kn}_{\mu x}}\right)}{\mathrm{Kn}_{\mu z}}. \quad (6.22)$$

To obtain the heat flux back to the real space, one simply needs to perform an inverse Fourier transform of equations (6.12), (6.17) and (6.20). Therefore, the heat flux vector in real space can be summarized as

$$\boldsymbol{q} = -\frac{1}{V_D}\iiint_D \underline{\boldsymbol{\kappa}}(x - x', y - y', z - z', \eta)\nabla T(x', y', z', \eta)\mathrm{d}x'\mathrm{d}y'\mathrm{d}z' + \boldsymbol{B}, \quad (6.23)$$

where $D = \{0 < x < L, -\infty < y, z < \infty\}$ and V_D represents the volume of the integrated domain. The thermal conductivity matrix, $\underline{\boldsymbol{\kappa}}(x, y, z, \eta)$, is diagonal

$$\begin{bmatrix} \mathcal{F}^{-1}(\kappa_x) & 0 & 0 \\ 0 & \mathcal{F}^{-1}(\kappa_y) & 0 \\ 0 & 0 & \mathcal{F}^{-1}(\kappa_z) \end{bmatrix}$$

and vector $\boldsymbol{B}$ is given by $[\mathcal{F}^{-1}(B_x), \mathcal{F}^{-1}(B_y), \mathcal{F}^{-1}(B_z)]^{-1}$, where symbol $\mathcal{F}^{-1}$ indicates the inverse Fourier transform in ξ_y and ξ_z.

Equation (6.23) is the constitutive relation of nonlocal heat conduction that links the temperature gradient to the heat flux in infinite transverse geometries, i.e. $y, z \to \pm\infty$. There are two parts in equation (6.23). The first part represents a convolution between the temperature gradient along a certain direction and its corresponding directional thermal conductivity that depends on both time and space. In contrast to prior works that have regarded $\underline{\boldsymbol{\kappa}}(x, y, z)$ as specific for different experimental geometries, the function of this space- and time-dependent thermal conductivity is independent of specific boundary conditions and source terms. The dependence of heat flux on boundary conditions and source terms is accounted for by the second part of equation (6.23), which has been neglected in most previous works but is critical to correctly interpret observations of nonlocal heat conduction.

In the diffusive regime, the spatial and temporal dependences of thermal conductivity disappear and asymptotically approach a constant, and the inhomogeneous term approaches zero. To demonstrate this limit, we first identify key dimensionless parameters in equation (6.23). First are the directional Knudsen numbers, $\mathrm{Kn}_{\mu j}$, where $j = x, y, z$. The second is the transient number, $\Xi_\mu = \eta\tau_\mu$, which compares the phonon relaxation times with a characteristic time, η^{-1} in this case. In the diffusive limit, both Ξ and $\mathrm{Kn}_{\mu x,y,z}$ are much less than unity. In this limit, we can perform the following simplifications:

$$\lim_{\Xi,\, \mathrm{Kn}_{\mu x,y,z} \to 0} \alpha_\mu \approx 1, \tag{6.24}$$

$$\lim_{\Xi,\, \mathrm{Kn}_{\mu x,y,z} \to 0} \mathrm{e}^{-\alpha_\mu \frac{\chi}{|\mathrm{Kn}_{\mu x}|}} \approx 0, \tag{6.25}$$

$$\lim_{\Xi,\, \mathrm{Kn}_{\mu x,y,z} \to 0} \frac{\mathrm{e}^{-\alpha_\mu \frac{|\chi - \chi'|}{\mathrm{Kn}_{\mu x}}}}{2\mathrm{Kn}_{\mu x} L} \approx \delta(\chi - \chi'), \tag{6.26}$$

$$\lim_{\Xi,\, \mathrm{Kn}_{\mu x} \to 0} \frac{|\chi - \chi'| \mathrm{e}^{-\gamma_\mu \frac{|\chi - \chi'|}{\mathrm{Kn}_{\mu x}}}}{2\mathrm{Kn}_{\mu x}^2 L^2} \approx \delta(\chi - \chi'), \tag{6.27}$$

$$\lim_{\Xi,\, \mathrm{Kn}_{\mu x,j} \to 0} \cos\left(\mathrm{Kn}_{\mu j} \frac{|\chi - \chi'|}{\mathrm{Kn}_{\mu x}}\right) \approx 1 \text{ with } j = y \text{ or } z, \tag{6.28}$$

$$\lim_{\Xi,\,\mathrm{Kn}_{\mu x,j}\to 0}\frac{\sin\left(\mathrm{Kn}_{\mu j}\dfrac{|\chi-\chi'|}{\mathrm{Kn}_{\mu x}}\right)}{\mathrm{Kn}_{\mu j}}\approx\frac{|\chi-\chi'|}{\mathrm{Kn}_{\mu x}}\ \text{with } j=y \text{ or } z. \tag{6.29}$$

Therefore, $B_j \to 0$ and the diagonal elements of the thermal conductivity matrix in equation (6.23) become $\kappa^j = \sum_\mu C_\mu v_{\mu j}\Lambda_{\mu j}\delta(x - x')$ with $j = x, y, z$. Equation (6.23) in the diffusive limit becomes $\boldsymbol{q} = -\underline{\boldsymbol{\kappa}}\nabla T$, recovering Fourier's law.

Given the mathematical expression of the external heat source and the appropriate boundary conditions for a particular problem, the relation between heat flux and temperature, the two measurable quantities, is now well defined by equation (6.23). This constitutive relation provides a formal definition of nonlocal thermal conductivity in the non-diffusive regime and a theoretical basis to study nonlocal thermal transport using existing experimental methods. All the intrinsic properties of phonons in a bulk material, i.e. phonon relaxation times and the MFP spectrum, are contained in $\underline{\boldsymbol{\kappa}}(x, y, z, \eta)$, while $\boldsymbol{B}$ contains the microscopic information of phonon interactions with interfaces and surfaces, i.e. phonon transmission coefficients and specularity parameters at an interface. To close the problem and obtain the necessary microscopic transfer function $\sum_\mu \mathcal{H}_\mu$, the equation of energy conservation, equation (6.5), is required to solve for either the heat flux or temperature field of a given problem. After obtaining the appropriate microscopic transfer function, it is a matter of choosing the right mathematical method to extract the desired phonon information from the experimental observables.

6.3.2 Solving the inverse problem

The appropriate method to solve the inverse problem posed by equation (6.1) depends on the complexity of the microscopic transfer function that links the desired phonon information to the observables. This section aims to provide a summary of the methods that have been used to extract the microscopic phonon information in the existing literature.

6.3.2.1 Convex optimization

Perhaps the simplest case in MFP spectroscopy is to directly measure an effective thermal conductivity, which is linked to the desired MFP spectrum through

$$\kappa(L_i) = \sum_\mu f(\Lambda_{\mu,\mathrm{bulk}})S\left(\frac{\Lambda_{\mu,\mathrm{bulk}}}{L_i}\right), \tag{6.30}$$

where $f(\Lambda_{\mu,\mathrm{bulk}}) = C_\mu v_\mu \Lambda_{\mu,\mathrm{bulk}}$ is the differential MFP distribution and S is a suppression function derived from equation (6.23). Examples of thermal conductivity measurements using this include a length/thickness-dependent experiment [84] and a transient grating experiment [20]. Both are discussed in detail in section 6.4. The matrix given by $S(\Lambda_{\mu,\mathrm{bulk}}/L_i)$ is singular and therefore cannot be solved by inversion. Moreover, equation (6.30) cannot be reformulated as an inverse problem because there is little knowledge of the form of the differential MFP distribution. To overcome this challenge, Minnich [19] first proposed transforming the above equation to its cumulative form using summation by parts given by

$$\kappa(L) = \frac{1}{L}\sum_{\mu} F(\Lambda_\mu)K(x), \tag{6.31}$$

where $x = \Lambda_\mu/L$, $F(\Lambda_\mu) = \lim_{\Delta\Lambda\to 0} \sum_{\Lambda_1}^{\Lambda_\mu} f(\Lambda')\Delta\Lambda$ is the accumulative MFP distribution and $K(x) = -\lim_{\Delta x\to 0} \Delta S(x)/\Delta x = -\mathrm{d}S/\mathrm{d}x$ is the kernel. Here, we use the fact that $F(0) = 0$ and $S(\infty) = 0$ to eliminate the constant introduced in the summation by parts. The cumulative distribution function, $F(\Lambda_\mu)$, is non-negative, smooth and monotonically increasing with limits of $F(\Lambda_{\min}) = 0$ and $F(\Lambda_{\max}) = 1$. These requirements set the constraints to the inverse problem which now can be solved using convex optimization. To allow equation (6.31) to be numerically solved, we first write it into its matrix form:

$$\sum_{n=1}^{N} A_{i,n}F_n = \kappa_i, \tag{6.32}$$

where N is the number of phonon modes, n is the index of the phonon MFP array given by $\{\Lambda_n\} = [\Lambda_1 = \Lambda_{\min}, \ldots, \Lambda_N = \Lambda_{\max}]$, $F_n = F(\Lambda_n)$ is the desired distribution, and $A_{i,n} = K(\Lambda_n/L_i)/L_i$ is the kernel matrix with dimensions $N \times M$. Now, we seek to minimize an objective function $\mathcal{P}$ defined as

$$\mathcal{P} = ||\underline{\boldsymbol{A}}\boldsymbol{F} - \boldsymbol{\kappa}||_2^2 + \epsilon||\Delta^2\boldsymbol{F}||_2^2 \tag{6.33}$$

subject to the following constraints:

$$F_1 = 0 \tag{6.34}$$

$$F_N = 1 \tag{6.35}$$

$$F_{n+1} \geqslant F_n \geqslant 0 \text{ with } n = 1, \ldots, n-1. \tag{6.36}$$

In equation (6.33), $||\cdot||_2$ is the ℓ^2-norm and the second difference operator $\Delta^2\boldsymbol{F}$ for a non-uniformly spaced array $\{\Lambda_n\}$ is given by

$$\Delta^2\boldsymbol{F} = \frac{2\left[F_{n+1} + \frac{h_n}{h_{n-1}}F_{n-1} - \left(1 + \frac{h_n}{h_{n-1}}\right)F_n\right]}{h_n h_{n-1}\left(1 + \frac{h_n}{h_{n-1}}\right)}, \tag{6.37}$$

where $h_n = \Lambda_{n+1} - \Lambda_n$ and $h_{n-1} = \Lambda_n - \Lambda_{n-1}$.

The first term on the right-hand side of equation (6.33) enforces the solution to satisfy equation (6.31), while the second term enforces smoothness in the solution. ϵ controls the relative weight of the smoothness penalty. If ϵ is too small, the solution will have unphysical jumps, while if ϵ is too large, the solution will not satisfy equation (6.31). Therefore, the choice of ϵ must be justified for each problem. The numerical implementation of the optimization can be carried out using CVX, a package for specifying and solving convex problems [85, 86].

6.3.2.2 Global optimization

The advantage of convex optimization is its relatively straightforward implementation, but the major drawback is that it does not address the ill-posedness of the inverse problems nor does it quantify any uncertainty associated with this optimization process. To quantify uncertainty, a few non-derivative based methods have been proposed.

A combination of particle swarm optimization (PSO) and Gibbs sampling has been implemented to search for the probability density plot of the optimal profile of the target microscopic phonon spectrum. The goal of this approach is to iteratively minimize an objective function $\mathcal{P}$, similar to equation (6.33), that is defined as

$$\mathcal{P} = ||\sum_{\mu} \mathcal{H}_{\mu}(\mathcal{T}_{\mu}, D_i) - Z_i(D_i)||_2^2 + \epsilon||\Delta^2\mathcal{T}||_2^2. \tag{6.38}$$

The first part of the equation evaluates the ℓ^2-norm of the difference between experimentally measured Z_i and PBE-simulated signals $\sum_{\mu}\mathcal{H}_{\mu}$ given a microscopic phonon spectrum $\mathcal{T}_{\mu}$. The second part of the equation, serving a smoothness penalty, evaluates the L^2-norm of the second derivative of the phonon spectrum given by

$$||\Delta^2\mathcal{T}||_2^2 = \int_0^{\infty} \frac{\partial^2\mathcal{T}}{\partial\Omega^2}\mathrm{d}\Omega, \tag{6.39}$$

where Ω is the relevant independent variable associated with the desired phonon spectrum, i.e. MFP Λ_{μ} or phonon frequency ω_{μ}. The smoothness condition is imposed by physical constraints that properties of phonons should not change abruptly compared to those of their neighboring modes. The smoothing parameter ϵ determines the relative importance of the second part of equation (6.38) to the first part. If $\epsilon = 0$, then no smoothness constraint is imposed.

To search for the optimal profile that minimizes the objective function, the PSO algorithm randomly initializes a collection of microscopic phonon spectra and evolves them in steps throughout the phase space that contains all possible spectra. At each step and for each spectrum, the algorithm evaluates the objective function defined above. After this evaluation, the algorithm decides how each spectrum should evolve according to the current best spectrum. The spectrum evolves, then the algorithm re-evaluates. The algorithm stops when the objective function reaches the desired value. The phonon spectrum that achieves the minimum value of the objective function is the optimal spectrum that explains the data.

However, since the inverse problem is ill-posed, a unique solution does not exist. Gibbs sampling can be used to explore adjacent regions of the optimal spectra obtained by PSO and generate a probability density plot for the target spectrum. To implement Gibbs sampling, a large collection of spectra are first generated by perturbing the optimal spectra. The objective function is then evaluated at all the perturbed profiles and values are recorded. At each iteration, a spectrum, a, from the stored population is randomly drawn and the value of its corresponding objective function, $\mathcal{P}_n$, is compared to the one from the previous step, $\mathcal{P}_{n-1}$, evaluated at spectrum b. If $\mathcal{P}_n$ is less than $\mathcal{P}_{n-1}$, a is accepted and $\mathcal{P}_n$ is kept. If not, a random number r is drawn and compared to $u = p/(1 + p)$, where

$$p = \exp\left(\frac{\mathcal{P}_n - \mathcal{P}_{n-1}}{T_0}\right). \tag{6.40}$$

If r is smaller than u, then a is accepted and $\mathcal{P}_n$ is kept. If not, a is rejected and $\mathcal{P}_n$ is updated to be $\mathcal{P}_{n-1}$. The system temperature, T_0, is chosen such that the stationary distribution is gradually changing. Here, T_0 is set to the mean value of the objective functions of all the perturbed samples. By tracking how many times each spectrum was chosen at each iteration, a histogram of the occurrence frequency of each profile is generated. The sampling process is stopped when the histogram becomes stationary. This occurrence frequency is also called the likelihood of the target spectrum. The higher the value of a spectrum's likelihood is, the better is the fit with the experimentally measured signals. Thus, by combining PSO with a Gibbs sampling algorithm, the most likely microscopic phonon spectrum can be determined.

6.3.2.3 Bayesian inference

Bayesian inference is another commonly used method to quantify the uncertainty in the process of optimization. Bayesian inference is based on Bayes' theorem which derives the posterior probability as a consequence of a prior probability and a likelihood function derived from a statistical model for the observed data. Mathematically, it can be expressed as

$$P(H|E) \propto P(E|H)P(H), \tag{6.41}$$

where E represents some measurable data (called the evidence below) and H represents the target parameters to be optimized (called the hypothesis below). $P(H)$, the prior probability, is the estimate of the probability of hypothesis H before the current evidence E is observed. $P(E|H)$ is called the likelihood, which is the probability of observing E given H. $P(H|E)$, the posterior probability, is the probability of a hypothesis H given the observed evidence E, which is the quantity to be determined here.

Applying Bayes' theorem to the inverse problem posed by equation (6.1), H corresponds to $\mathcal{T}_\mu$, the unknown microscopic phonon spectrum and E corresponds to the experimental observables, Z. Equation (6.41) is re-written into

$$P(\mathcal{T}_\mu|Z) \propto P(Z|\mathcal{T}_\mu)P(\mathcal{T}_\mu). \tag{6.42}$$

The likelihood of an independent measurement, $P_i(Z_i|\mathcal{T}_\mu)$, is defined as a normal distribution centered at the measured Z_i at a given thermal length, D_i, with the standard deviations, σ_i, given by the errors associated with that measurement. Mathematically, it is expressed as

$$P_i(Z_i|\mathcal{T}_\mu) = \frac{1}{\sqrt{2\pi\sigma_i}} \exp\left(-\frac{\left(Z_i - \sum_\mu \mathcal{H}_\mu(\mathcal{T}_\mu, D_i)\right)^2}{2\sigma_i^2} \right), \tag{6.43}$$

where $\sum_\mu \mathcal{H}_\mu(\mathcal{T}_\mu, D_i)$ represents the PBE-simulated measurable for a given trial phonon spectrum. The total likelihood function accounting for all the measurements is then given by

$$P(Z|\mathcal{T}_\mu) \propto \exp\left(-\sum_i^N \frac{\left(Z_i - \sum_\mu \mathcal{H}_\mu(\mathcal{T}_\mu, D_i)\right)^2}{2\sigma_i^2} \right). \tag{6.44}$$

The prior probability, $P(\mathcal{T}_\mu)$, is built on the constraints imposed on the desired phonon spectrum. Similar to the previous two optimization methods, smoothness of the target spectrum is required in most of the applications and the second derivative of the target spectrum should be minimized. Therefore, the prior probability is defined as a probability distribution for the second derivative of the target spectrum. A normal distribution centered at 0 is used here (corresponding to perfect smoothness). To mathematically represent this normal distribution, we first consider the second derivative of one segment of the spectrum at a given state μ:

$$\Delta^2\mathcal{T}_\mu = \left.\frac{\partial^2\mathcal{T}_\mu}{\partial\Omega^2}\right|_\mu \tag{6.45}$$

$$P_\mu(\mathcal{T}_\mu) \propto \exp\left(-\frac{\Delta^2\mathcal{T}_\mu\Delta\Omega}{2\gamma^2} \right), \tag{6.46}$$

where Ω is the relevant independent variable associated with the desired phonon spectrum and γ is a parameter representing how far from 0 the second derivatives are likely to deviate. $\Delta\Omega$ quantifies how much of the chain segment is being considered and scales the penalty with the amount of the curve that has a high curvature. To build the total prior probability distribution of the entire spectrum, all $P_\mu(\mathcal{T}_\mu)$ terms are multiplied together. Mathematically, this can be represented as a sum over the arguments of the exponentials:

$$P(\mathcal{T}_\mu) \propto \exp\left(-\frac{1}{2\gamma^2} \sum_\mu^M \Delta^2\mathcal{T}_\mu\Delta\Omega \right). \tag{6.47}$$

In the limit of $\Delta\Omega \to 0$, the above probability distribution can be written as

$$P(\mathcal{T}_\mu) \propto \exp\left(-\frac{1}{2\gamma^2} \int_0^\infty \Delta^2\mathcal{T}_\mu \mathrm{d}\Omega \right). \tag{6.48}$$

Therefore, the posterior probability is given by

$$P(\mathcal{T}_\mu|Z) \propto \exp\left(-\sum_i^N \frac{\left(Z_i - \sum_\mu \mathcal{H}_\mu(\mathcal{T}_\mu, D_i)\right)^2}{2\sigma_i^2}\right)\exp\left(-\frac{1}{2\gamma^2}\int_0^\infty \Delta^2\mathcal{T}_\mu \mathrm{d}\Omega\right). \quad (6.49)$$

It is worth noting that taking the negative natural logarithm of equation (6.49) gives back an objective function defined by equation (6.38). The outcome of Bayesian inference is equivalent to that of PSO since minimizing the negative logarithm of likelihood is equivalent to maximizing likelihood.

To perform Bayesian inference and estimate the posterior distribution, a Metropolis–Hastings (MH) Markov chain Monte Carlo (MCMC) algorithm [87] has been used in several works [26, 41] for reconstructing microscopic phonon distribution. The idea behind an MH-MCMC algorithm is to implement a random walk in parameter space, sampling various proposed parameters ($\mathcal{T}_\mu$). Subsequent steps are chosen according to a proposed distribution, $K(\mathcal{T}|\mathcal{T}_{\mathrm{prev}})$, a distribution of possible next steps ($\mathcal{T}$) from the walker's current position ($\mathcal{T}_{\mathrm{prev}}$). Each step is then independently accepted or rejected according to the relative posterior probability at the new point. This allows the walker to spend more time in regions of high probability density. The posterior probability distribution is then constructed as the algorithm explores parameter space.

In addition to convex optimization techniques and global optimization methods, recently the Nelder–Mead (NM) algorithm [88], a simplex-based search method that is free from gradient computation, has been implemented to minimize equation (6.38) and reconstruct the phonon relaxation times from TTG experiments [27, 89, 90]. It is necessary to point out that the methods described above are not the only methods to be used in MFP spectroscopy. Regardless of a specific method, an appropriate optimization method for MFP spectroscopy should be able to minimize the objective function defined by equation (6.38) as well as to quantify the uncertainty of the obtained microscopic phonon spectrum.

6.4 Experiments

This section aims to provide an overview of a few key experiments that have been commonly used as the platform to perform MFP spectroscopy.

6.4.1 Size-dependent thermal conductivity measurements

Here, the example of using observations of the size effect caused by system size variation to determine the phonon MFP spectra in graphene ribbons is given by [36, 84]. Thermal conductivity measurements over variable lengths on suspended single-layer graphene ribbons were performed at room temperature [84]. In the experiment, a set of thermal conductivities $k_i(L)$ as a function of ribbon length

L_i was obtained. Zhang and co-workers [36] derived a heat flux suppression function $S(\Lambda/L_i)$ based on the PBE that links the MFP spectrum $f(\Lambda)$ in a bulk crystal to the length-dependent thermal conductivity for a steady heat conduction problem where a two-dimensional crystal with thickness L is imposed by cold and hot black-body walls. The mathematical expression used in this work (a specific case of equation (6.1)) is given by

$$k_i(L_i) = \int_0^\infty S(\Lambda/L_i)f(\Lambda)\mathrm{d}\Lambda. \tag{6.50}$$

To obtain the MFP spectrum, equation (6.50) was posed as an inverse problem solved by convex optimization as discussed in section 6.3.2.1 [19], from which the accumulative MFP distribution $F(\Lambda) \equiv \int_0^\Lambda f(\Lambda)\mathrm{d}\Lambda$ was obtained (shown in figure 6.2). The reconstructed accumulative MFP distribution reveals that phonons with MFPs longer than 1 μm carry the majority of the heat in suspended single-layer graphene.

6.4.2 TTG spectroscopy

The second key experimental platform for MFP spectroscopy is TTG spectroscopy, a pump–probe technique that uses a pair of pump pulses to generate a spatially periodic temperature profile in a sample and monitors the decay of this temperature grating through thermal transport via diffraction of a probe laser beam [40]. The major advantage of TTG as a platform for MFP spectroscopy is the simplicity of its experimental configuration. The periodic temperature profile is induced by the interference of two pump beams in a sample and its period is given by

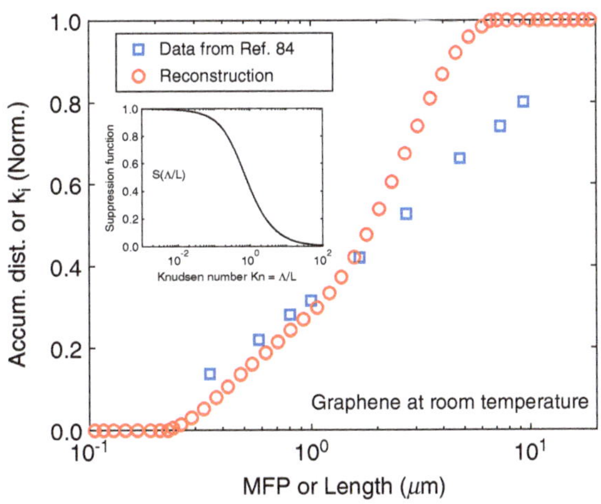

Figure 6.2. Experimentally measured length-dependent thermal conductivity (blue open squares; [84]) and the corresponding reconstructed accumulative thermal conductivity as a function of phonon MFP (red open circles) in suspended graphene samples. All these thermal conductivities are normalized to thermal conductivities of 'bulk' graphene flakes, which are calculated using the extrapolating method in [91]. Phonons with MFPs longer than 1 μm carry the majority of the heat in suspended graphene. Inset: heat flux suppression function in a two-dimensional crystal.

$L = \lambda/2\sin(\theta/2)$, defined by the pump wavelength λ and the angle θ between the beams. If the grating period is much smaller than the absorption depth of the excitation light, thermal transport is nearly one-dimensional and no interface is involved with a measurement for a single crystal sample. Moreover, the single spatial frequency greatly facilitates theoretical analysis [66, 92].

In this experiment, the heat generation rate has a spatial profile of $e^{i\beta x}$ in an infinite domain, where $\beta \equiv 2\pi/L$. The boundary term vanishes, i.e. $x \in (-\infty, \infty)$, and both the distribution function and the temperature field exhibit the same spatial dependence. Since it is a one-dimensional problem, equation (6.23) simplifies to

$$\tilde{q}_x(x, \eta) = i\beta\tilde{T}(\eta)e^{i\beta x}\sum_{\mu x>0}\frac{\kappa_{\mu x}}{\alpha_\mu^2 + \Lambda_{\mu x}^2\beta^2} + \sum_{\mu x>0}\frac{Q_\mu}{\delta}\frac{e^{i\beta x}\alpha_\mu\Lambda_{\mu x}}{\alpha_\mu^2 + \Lambda_{\mu x}^2\beta^2}, \tag{6.51}$$

where the total volumetric energy deposited on a sample is given by $\sum_\mu Q_\mu$ and the duration of the energy deposition is δ.

The time-scale of a typical TTG experiment is of the order of a few hundred nanoseconds while relaxation times of phonons are typically less than a nanosecond for many semiconductors at room temperature. Therefore, we assume that $\Xi \ll 1$, and equation (6.51) is simplified to

$$\tilde{q}_x(x, \eta) = i\beta\tilde{T}(\eta)e^{i\beta x}\sum_{\mu_x>0}\frac{\kappa_{\mu x}}{1 + \Lambda_{\mu x}^2\beta^2} + \sum_{\mu_x>0}\frac{Q_\mu}{\delta}\frac{e^{i\beta x}\Lambda_{\mu x}}{1 + \Lambda_{\mu x}^2\beta^2}, \tag{6.52}$$

which is consistent with what has been derived in previous works [19, 66, 92]. The first part of equation (6.52) represents the conventional understanding of nonlocal thermal transport, a Fourier-type relation with a reduced thermal conductivity given by

$$\kappa_x = \sum_{\mu_x>0}\frac{\kappa_{\mu x}}{1 + \Lambda_{\mu x}^2\beta^2}, \tag{6.53}$$

while the second part of the equation represents the contribution from the heat source to the total heat flux, which increases as the Knudsen number $\Lambda_{\mu x}/L$ increases. In a TTG experiment, the presence of a single spatial frequency simplifies the convolutions in equation (6.23) into products, and the only time dependence of the heat flux comes from the temperature. Therefore, the decay rate of the measured transient temperature profile is directly proportional to the reduced thermal conductivity, which justifies the approach that a Fourier-type relation with a modified thermal conductivity is used to interpret the non-diffusive thermal transport observed in a TTG experiment.

Johnson and co-workers first demonstrated that room temperature thermal transport in a silicon membrane significantly deviates from the diffusion model already at microscale distances using TTG [40]. By changing the period of the thermal grating from 10 to 1 μm on a 370 nm thick silicon membrane sample, they observed a reduction in the measured thermal conductivity (red open squares in

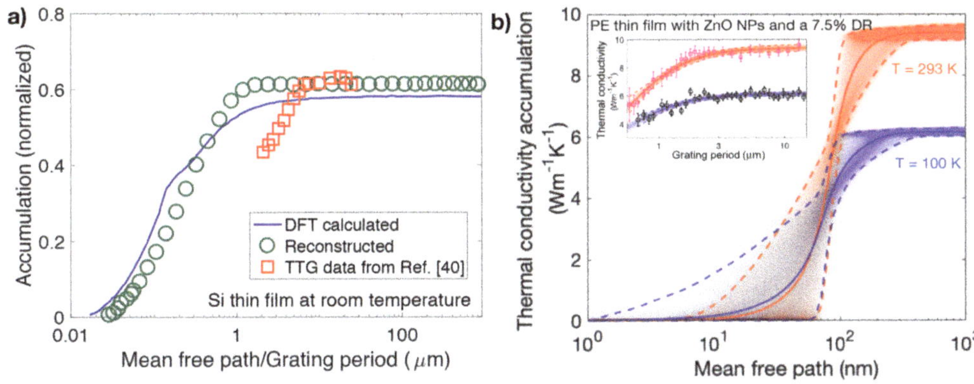

Figure 6.3. (a) Experimentally measured grating-period-dependent thermal conductivity (red squares; [40]) and the corresponding reconstructed accumulative thermal conductivity as a function of phonon MFP (green circles) in a 390 nm thick silicon membrane. These thermal conductivities are normalized to thermal conductivities of bulk silicon. The reconstructed distribution is in good agreement with an analytic calculation (solid line), which was computed from a first-principles calculation of the MFP distribution in Si and accounting for diffuse scattering from the membrane boundaries. The figure is regenerated from [19]. (b) Posterior probability density for reconstructed MFP spectra for a polyethylene (PE) thin film with ZnO nanoparticle (NP) fillers and a 7.5% draw ratio (DR) at 295 K (red shaded region) and 100 K (blue shaded region). For each MFP, the darker regions have a higher probability density. The solid lines correspond to the mean spectrum, and the dotted lines enclosed the 95% credible interval. Inset: the corresponding grating-dependent thermal conductivities overlaid with the calculated thermal conductivities from the spectra. The error bars refer to 68% confidence intervals. Reproduced with permission from [41]. Copyright 2019 the National Academy of Sciences.

figure 6.3(a)). Minnich then solved equation (6.53) as an inverse problem using convex optimization (discussed in section 6.3.2.1) to reconstruct the MFP spectrum of the silicon membrane (green circles in figure 6.3(a)) [19]. An analytic calculation, which was computed from a first-principles calculation of the MFP distribution in Si and accounting for diffuse scattering from the membrane boundaries using a Fuchs–Sondheimer relation [93, 94], agrees with the reconstructed distribution within approximately 15%. Their work demonstrated for the first time that a phonon MFP distribution can be accurately reconstructed from the observations of nonlocal thermal transport without assumptions regarding the phonon scattering mechanisms.

Recently, Robbins *et al* conducted TTG experiments to measure the thermal conductivity of a series of aligned semi-crystalline PE thin films with different DRs and NP fillers at various temperatures [41]. They observed a 50% reduction in the thermal conductivity of the PE film with ZnO NP fillers and a 7.5% DR as the grating period varied from 15.7 μm to 577 nm as shown in figure 6.3(b). The posterior probability distribution for the reconstructed MFP distribution was calculated from the measured thermal conductivity using Bayesian inference via an MH-MCMC algorithm discussed in section 6.3.2.3. The reconstructed MFP distribution shows the existence of long MFP phonons in these aligned molecular solids, many of which substantially exceed the crystalline domain sizes. Their work indicates that phonons propagate ballistically within and across the nanocrystalline domains and nearly a third of the thermal conductivity is attributed to those long cross-domain MFP phonons, providing insights into the microscopic origins of high thermal conductivity in semi-crystalline polymer materials.

In addition to phonon MFP spectra, the experimental observation of nonlocal thermal transport in membranes using TTG also reveals insights of the microscopic processes of phonon transport at the boundaries when combined with *ab initio* modeling. Juffe *et al* conducted thermal conductivity measurements of suspended Si membranes ranging from 15 to 1500 nm thickness using TTG at a fixed grating period [20]. Assuming a Fuchs–Sondheimer suppression relation [93, 94] between the membrane thickness and phonon MFPs, they claimed that the MFP distribution in bulk Si can be accurately constructed using convex optimization. However, the Fuchs–Sondheimer relation assumes that all phonons are reflected diffusely at the boundary. The validation of such an assumption is under question because phonons of terahertz frequencies can be reflected specularly from a surface and therefore preserve their phase. However, the specularity parameter, p_λ, the probability of specular reflection of a phonon mode with wavelength λ from a surface, is not readily available.

Navaneetha *et al* demonstrated that phonon wavelength-dependent specularity parameters were extracted from macroscopic thermal conductivity measurements in a TTG experiment by the interpretation of the measured observables with an *ab initio* description of phonon transport [26]. In their work, equation (6.53) is written as

$$\kappa_x(\beta) = \sum_{\mu_x>0} \kappa_{\mu x} S(\beta \Lambda_{\mu_x}, \Lambda_{\mu_x}/d, p_\lambda), \tag{6.54}$$

where d is the thickness of the Si membrane, $\kappa_{\mu x}$ is the bulk mode-dependent thermal conductivity, and $S(\beta \Lambda_{\mu_x}, \Lambda_{\mu_x}/d, p_\lambda)$ is the suppression function that includes the size effects caused by both membrane thickness and a large temperature gradient [95]. The specularity parameter p_λ is treated as an unknown distribution, while $\kappa_{\mu x}$ is obtained from *ab initio* calculations. Bayesian inference via MH-MCMC was implemented to calculate the posterior distribution probability for the reconstructed specularity distribution from the grating-period-dependent thermal conductivity of a Si membrane. Figure 6.4 gives the specularity parameter as a function of phonon wavelength, showing that phonons with wavelengths less than 20 Å are nearly entirely diffusely reflected, while phonons with wavelengths longer than 60 Å are reflected nearly completely specularly.

6.4.3 Thermoreflectance and diffraction techniques

Many recent experiments have investigated nonlocal thermal transport using optical pump–probe methods to heat a metal transducer film or lithographically patterned nanoscale heaters. Koh and Cahill were the first to report the modulation-frequency-dependent thermal conductivity of semiconductor alloys as measured by time-domain thermoreflectance (TDTR) [96]. TDTR is an optical pump–probe technique

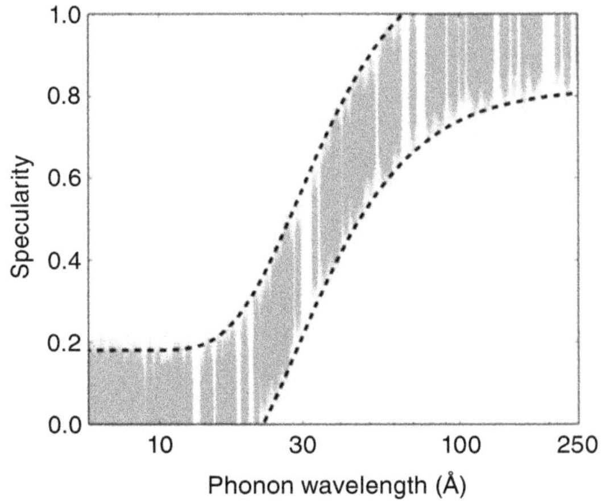

Figure 6.4. Specularity parameter versus phonon wavelength obtained from Bayesian inference. The gray region is an intensity plot of the posterior probability distribution for the specularity parameter with dashed lines indicating a 95% credible interval [26]. Reproduced with permission from [26]. Copyright 2018 the American Physical Society.

that is used to characterize thermal properties. In this experiment, a sample consists of a metal transducer film on a substrate. A pulsed laser beam from an ultrafast oscillator is split into a pump and a probe beam. The pump pulse train is modulated at a frequency from 1 to 15 MHz to enable lock-in detection, and is then used to impulsively heat the metal film coated on the sample. The transient temperature decay, $Z(t)$, at the surface is detected as a change in optical reflectance by the probe beam [99]. This transient signal is related to the desired thermal properties by a macroscopic transfer function based on a multilayer heat diffusion model [100]. This function maps thermal properties such as substrate thermal conductivity and metal–substrate interface conductance to the TDTR signal, and thus these properties are obtained by varying these parameters until the simulated results match the measured datasets.

The modulation frequency, f, of the pump pulse train defines a thermal penetration depth into the substrate given by $d = \sqrt{k/\pi Cf}$, where k is the thermal conductivity of the substrate and C is the volumetric heat capacity. As illustrated in the inset of figure 6.5(a), if this thermal penetration depth is much longer than the phonon MFPs in the substrate, thermal transport is diffusive. As the modulation frequency increases, the penetration depth decreases. If some MFPs become comparable to the thermal penetration depth, nonlocal thermal transport occurs. When a heat diffusion model is applied to interpret the measured signals, then a reduction in the fitted thermal conductivity compared to its bulk value is observed, as shown in figure 6.5(a). Similar observations of frequency-dependent thermal conductivity of silicon have also been reported using the frequency-domain thermoreflectance technique (FDTR) [24], a steady-state version of TDTR.

Reducing the size of the laser beam is another way to observe nonlocal thermal transport in TDTR. The beam-size-dependent thermal conductivity of silicon was

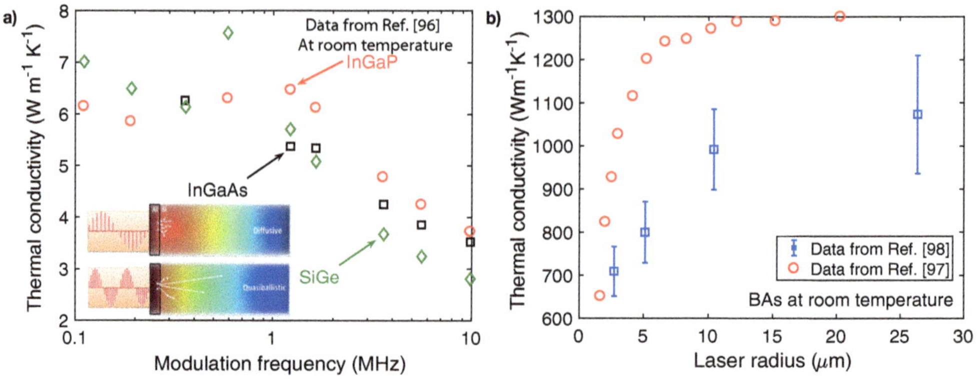

Figure 6.5. (a) Room temperature thermal conductivity of 2010 nm thick InGaP, 3330 nm thick $In_{0.49}Ga_{0.53}As$ and 6000 nm thick $Si_{0.4}Ge_{0.6}$ are shown as open circles, filled circles and open triangles, respectively. The figure is regenerated from [96]. Inset: a schematic of the principle underlying the measurement of frequency-dependent thermal conductivity [25]. The modulation frequency, f, of the pump pulse train defines a thermal penetration depth into the substrate given by $d = \sqrt{k/\pi C f}$, where k is the thermal conductivity of the substrate and C is the volumetric heat capacity. If this thermal penetration depth is much longer than the phonon MFPs in the substrate, thermal transport is diffusive. If some MFPs become comparable to the thermal penetration depth, nonlocal thermal transport occurs. (c) Room temperature thermal conductivity of BAs as a function of laser beam diameter. The data are from [97] (circles) and [98] (squares).

first reported by Minnich *et al* [44] and later confirmed by Ding *et al* [101]. Recent thermal measurements in single crystal BAs using TDTR not only confirmed the ultrahigh thermal conductivity of BAs [97, 98, 102] but also revealed the transport length scales of phonons in BAs through a series of observations of nonlocal thermal transport [97, 98]. As shown in figure 6.5(b), a reduction of as much as 50% in the thermal conductivity compared to the bulk BAs thermal conductivity was observed when the $1/e^2$ radius of the pump beam was systematically reduced from 21 to 1.6 μm [97]. This observation confirmed the prediction from density functional theory [103–105] that a large portion of phonons in BAs have long MFPs.

The typical smallest thermal length scale that can be achieved by either increasing modulation frequency or decreasing the laser beam size is greater than 1 μm. Deterioration in the signal-to-noise ratio occurs as the modulation frequency increases and the diffraction limit of the laser source limits access to smaller thermal length scales. However, phonons in a large portion of solids have MFPs much smaller than 1 μm. To gain access to a smaller thermal length scale, Siemens *et al* were the first to use a lithographically patterned metallic layer to heat the substrate underneath [39]. In their work, an infrared laser beam at 800 nm heated periodic nickel lines with linewidths varying from 65 to 2000 nm that were fabricated on a sapphire substrate and a fused silica substrate using electron beam lithography and a lift-off process. By interferometrically monitoring displacement in the nickel lines using diffraction of soft x-ray light, the cooling dynamics of the heat source into its bulk surroundings was detected. A multi-dimensional and multi-layered heat diffusion model was used to fit to the measured surface dynamics where interface resistivity was treated as a fitting parameter.

As shown in figure 6.6(a), the best fit to the data demonstrates that the measured interface resistivity of the sapphire substrate increases as the heated region becomes smaller while that of the fused silica substrate remained constant. The strong ballistic effects in sapphire are attributed to the long MFP phonons with an average value of 120 nm as estimated by the authors, while the average MFP of fused silica is around 2 nm. Subsequently, lithographically patterned metallic heaters were adapted for the TDTR experiments. Hu *et al* [21] and Zeng *et al* [22] reported large and monotonic increases in thermal resistance as the dimensions of individual heaters in patterned arrays become smaller than MFPs as shown in figure 6.6(b).

In general, these experiments report that the thermal transport from heaters of characteristic lengths smaller than phonon MFPs is increasingly impeded compared to the Fourier's law prediction as the heater size decreases. However, recent experiments have shown that the situation may be more complicated than the conventional viewpoint [25]. Hoogeboom-Pot *et al* probed thermal transport in sapphire and silicon using arrays of nickel nanowires as heat sources and showed that when the separation between nanoscale line heaters was small compared to the dominant phonon MFPs, the measured thermal boundary resistance recovered to the diffusive limit [23]. Another recent work used aluminum nanoline arrays as heating sources in TDTR experiments and showed that the measured thermal conductivity of sapphire reached a constant even as the linewidth decreased [106]. These experiments suggest that the actual dependence on the heater dimensions and geometry cannot simply be attributed to a single characteristic dimension of the heater [107].

In fact, the nonlocal thermal transport in a thermoreflectance or diffraction experiment described by equation (6.23) is more complicated than that in a TTG experiment due to the presence of the interface. After applying the interface

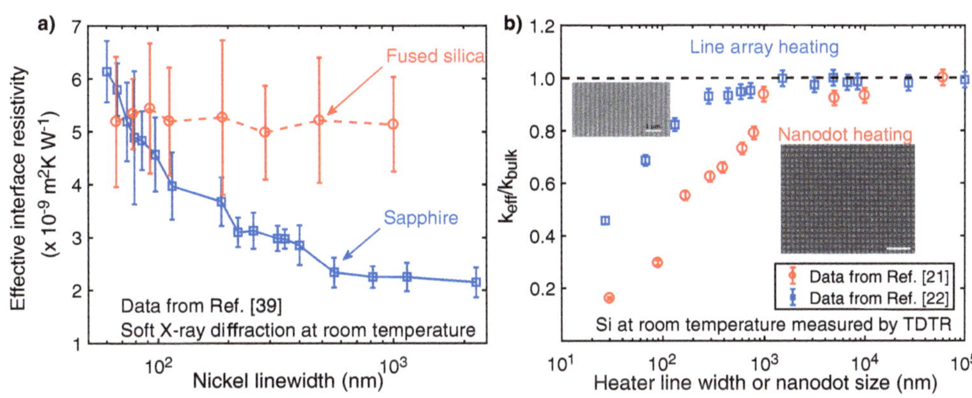

Figure 6.6. (a) Measured effective thermal resistivity versus linewidth L at room temperature for nickel nanostructures deposited on fused silica (red circles) and sapphire (blue square) substrates using soft x-ray diffraction. The figure is regenerated from [39]. (b) Normalized effective thermal conductivity of silicon at room temperature versus heater linewidth (blue squares; [22]) or nanodot size (red circles; [21]) using TDTR.

conditions to equation (6.23), the constitutive relation along the cross-plane direction in the substrate is given by

$$\widetilde{q}_x(x, \xi_y, \xi_z, \eta) = -\int_{L_1}^{L_2} \kappa_x(x - x', \xi_y, \xi_z, \eta)\frac{\partial T}{\partial x'}\mathrm{d}x' + \sum_{\mu}\widetilde{B}_\mu(x, \xi_y, \xi_z, \eta), \tag{6.55}$$

where thermal conductivity κ_x is given by

$$\kappa_x(x, \xi_y, \xi_z, \eta) = \sum_{\mu_x>0,\mu_y,\mu_z} \kappa_{\mu x}\frac{e^{-\frac{1+\mathrm{i}\Xi_\mu+\mathrm{i}\xi_y\Lambda_{\mu y}+\mathrm{i}\xi_z\Lambda_{\mu z}}{|\Lambda_{\mu x}|}x}}{(1 + \mathrm{i}\Xi_\mu + \mathrm{i}\xi_y\Lambda_{\mu y} + \mathrm{i}\xi_z\Lambda_{\mu z})|\Lambda_{\mu x}|}. \tag{6.56}$$

The detailed discussion of the interface conditions and the exact expression of $B_\mu(x, \xi_y, \xi_z, \eta)$ can be found in [83]. In this case, both the temperature field and the inhomogeneous term (the second part of equation (6.55)) have spatial and temporal dependences. Therefore, a Fourier-type relation with a reduced thermal conductivity is not appropriate in this particular experimental setting. Moreover, the dependence on the boundary conditions and the heat source should be accounted for when extracting the intrinsic thermal conductivity from the observables such as total heat flux or an average temperature.

Combining the constitutive relation given by equation (6.55) with the *ab initio* properties of aluminum and silicon, Hua *et al* treated the spectral transmission coefficient profile of phonons across the Al/Si interface as an unknown distribution in equation (6.38). They applied a PSO algorithm with a Gibbs sampling technique as discussed in section 6.3.2.2 to calculate the probability density plots of the transmission coefficient profiles for three Al/Si samples with distinct interface features. As shown in figure (6.7), the measured spectral transmission coefficient

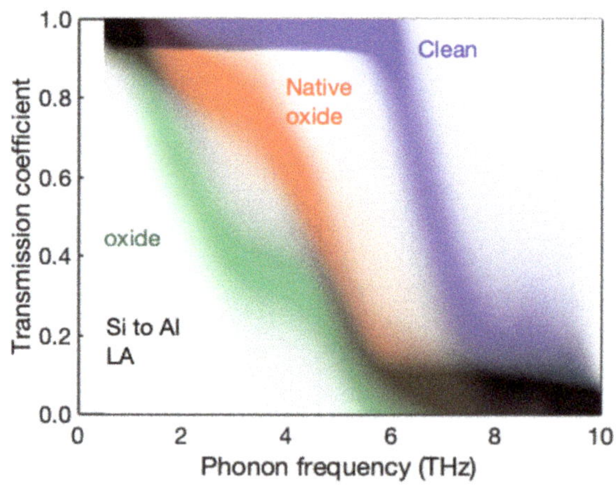

Figure 6.7. Phonon transmission coefficients versus phonon frequency for longitudinal acoustic phonons at Si/Al interfaces: clean (blue region), with an oxide layer (green region) and with a native oxide layer (red region). Long MFP phonons may be less affected by interfaces or grain boundaries, and disorder at the interfaces will decrease transmission coefficients at all phonon frequencies. The figure is regenerated from [25].

profile indicates that low-frequency, long MFP phonons are transmitted to the maximum extent allowed by the principle of detailed balance, while high-frequency, short MFP phonons are nearly completely reflected at the interface. As the disorder at an interface increases, such as adding an oxidized layer, transmission coefficients at all phonon frequencies decrease and more phonons, in particular those with long wavelength, are reflected at the interface.

6.5 Summary

In summary, this chapter has presented details of the physical origins and theoretical formalism of phonon MFP spectroscopy in the context of the PBE as well as various mathematical frameworks to extract the desired microscopic phonon information from experimental observations of nonlocal thermal transport. Several experimental platforms to carry out MFP spectroscopy were also presented. The key message from this chapter is that microscopic information on phonons and potentially other energy carriers can be obtained through the interwoven application of experimental measurements in the nonlocal heat conduction regime and *ab initio* transport modeling based on the PBE. This knowledge of the spectral content of heat will permit the rational understanding and control of heat transport at the length scales of the energy carriers, a capability that could impact numerous applications.

References

[1] Phelan D, Millican J N, Thomas E L, Leão J B, Qiu Y and Paul R 2009 Neutron scattering measurements of the phonon density of states of $FeSe_{1-x}$ superconductors *Phys. Rev.* B **79** 014519

[2] Piovano A 2015 Inelastic neutron scattering applied to materials for energy *EPJ Web Conf.* **104** 01006

[3] Mukhopadhyay S, Bansal D, Delaire O, Perrodin D, Bourret-Courchesne E, Singh D J and Lindsay L 2017 The curious case of cuprous chloride: giant thermal resistance and anharmonic quasiparticle spectra driven by dispersion nesting *Phys. Rev.* B **96** 100301

[4] Li C W *et al* 2014 Phonon self-energy and origin of anomalous neutron scattering spectra in SnTe and PbTe thermoelectrics *Phys. Rev. Lett.* **112** 175501

[5] Tian Z, Li M, Ren Z, Ma H, Alatas A, Wilson S D and Li J 2015 Inelastic x-ray scattering measurements of phonon dispersion and lifetimes in $PbTe_{1-x}$ Se_x alloys *J. Phys. Condens. Matter.* **27** 375403

[6] Ma H, Li C, Tang S, Yan J, Alatas A, Lindsay L, Sales B C and Tian Z 2016 Boron arsenide phonon dispersion from inelastic x-ray scattering: potential for ultrahigh thermal conductivity *Phys. Rev.* B **94** 220303

[7] Menendez J and Cardona M 1984 Temperature dependence of the first-order Raman scattering by phonons in Si, Ge, and α-Sn: anharmonic effects *Phys. Rev.* B **29** 2051–9

[8] Cusca R, Gil B, Cassabois G and Artus L 2016 Temperature dependence of Raman-active phonons and anharmonic interactions in layered hexagonal BN *Phys. Rev.* B **94** 155435

[9] Ponosov Y S and Streltsov S V 2017 Raman-active E_{2g} phonon in MgB_2: electron–phonon interaction and anharmonicity *Phys. Rev.* B **96** 214503

[10] Ziegler F, Gibhardt H, Leist J and Eckold G 2015 High-resolution polarised Raman scattering study on spin-phonon coupling in multiferroic $MnWO_4$ *Mater. Res. Exp.* **2** 096103

[11] Blumberg G, Klein M V, Börjesson L, Liang R and Hardy W N 1994 Investigation of the temperature dependence of electron and phonon Raman scattering in single crystal $YBa_2Cu_3O_{6.952}$ *J. Superconduct* **7** 445–8

[12] Perez-Osorio M A, Lin Q, Phillips R T, Milot R L, Herz L M, Johnston M B and Giustino F 2018 Raman spectrum of the organic–inorganic halide perovskite CH_3NH_3 PbI_3 from first principles and high-resolution low-temperature Raman measurements *J. Phys. Chem.* C **122** 21703–17

[13] Mukhopadhyay S, Parker D S, Sales B C, Puretzky A A, McGuire M A and Lindsay L 2018 Two-channel model for ultralow thermal conductivity of crystalline Tl_3VSe_4 *Science* **360** 1455–8

[14] Yan R, Simpson J R, Bertolazzi S, Brivio J, Watson M, Wu X, Kis A, Luo T, Hight Walker A R and Xing H G 2014 Thermal conductivity of monolayer molybdenum disulfide obtained from temperature-dependent Raman spectroscopy *ACS Nano* **8** 986–93

[15] Calizo I, Balandin A A, Bao W, Miao F and Lau Temperature C N 2007 Dependence of the Raman spectra of graphene and graphene multilayers *Nano Lett.* **7** 2645–9

[16] Neumann C *et al* 2015 Raman spectroscopy as probe of nanometre-scale strain variations in graphene *Nat. Commun.* **6** 8429

[17] Lee J-U, Yoon D, Kim H, Lee S W and Cheong Thermal H 2011 Conductivity of suspended pristine graphene measured by Raman spectroscopy *Phys. Rev.* B **83** 081419

[18] Chen S *et al* 2011 Raman measurements of thermal transport in suspended monolayer graphene of variable sizes in vacuum and gaseous environments *ACS Nano* **5** 321–8

[19] Minnich A J 2012 Determining phonon mean free paths from observations of quasiballistic thermal transport *Phys. Rev. Lett.* **109** 205901

[20] Cuffe J *et al* 2015 Reconstructing phonon mean-free-path contributions to thermal conductivity using nanoscale membranes *Phys. Rev.* B **91** 245423

[21] Hu Y, Zeng L, Minnich A J, Dresselhaus M S and Chen G 2015 Spectral mapping of thermal conductivity through nanoscale ballistic transport *Nat. Nanotechnol.* **10** 701–6

[22] Zeng L *et al* Measuring phonon mean free path distributions by probing quasiballistic phonon transport in grating nanostructures *Sci. Rep.* **5** 17131

[23] Hoogeboom-Pot K M *et al* A new regime of nanoscale thermal transport: collective diffusion increases dissipation efficiency**112** 4846–51

[24] Regner K, Sellan D, Su Z, Amon C, McGaughey A and Malen J 2012 Broadband phonon mean free path contributions to thermal conductivity measured using frequency-domain thermoreflectance *Nat. Commun.* **4** 1640

[25] Hua C, Chen X, Ravichandran N K and Minnich A J 2017 Experimental metrology to obtain thermal phonon transmission coefficients at solid interfaces *Phys. Rev.* B **95** 205423

[26] Ravichandran N K, Zhang H and Minnich A J 2018 Spectrally resolved specular reflections of thermal phonons from atomically rough surfaces *Phys. Rev.* X **8** 041004

[27] Forghani M and Hadjiconstantinou N G 2019 Phonon relaxation time reconstruction from transient thermal grating experiments and comparison with density functional theory predictions *Appl. Phys. Lett.* **114** 023106

[28] Casimir H B G 1938 Note on the conduction of heat in crystals *Physica* **5** 495–500

[29] Lan Y, Jerome Minnich A, Chen G and Ren Z 2010 Enhancement of thermoelectric figure-of-merit by a bulk nanostructuring approach *Adv. Funct. Mater.* **20** 357–76
[30] Biswas K, He J, Blum I V, Wu C I, Hogan T P, Seidman D N, Dravid V P and Kanatzidis M G 2012 High-performance bulk thermoelectrics with all-scale hierarchical architectures *Nature* **489** 414–8
[31] Hsiao T-K, Chang H-K, Liou S-C, Chu M-W, Lee S-C and Chang C-W 2013 Observation of room-temperature ballistic thermal conduction persisting over 8.3 μm in SiGe nanowires *Nat. Nanotechnol* **8** 534–8
[32] Li D, Wu Y, Kim P, Shi L, Yang P and Majumdar A 2003 Thermal conductivity of individual silicon nanowires *Appl. Phys. Lett.* **83** 2934–6
[33] Boukai A I, Bunimovich Y, Tahir-Kheli J, Yu J-K, Goddard W A III and Heath J R 2008 Silicon nanowires as efficient thermoelectric materials *Nature* **451** 168–71
[34] Chang C W, Okawa D, Garcia H, Majumdar A and Zettl A 2008 Breakdown of Fourier's law in nanotube thermal conductors *Phys. Rev. Lett.* **101** 075903
[35] Mei S and Knezevic I 2015 Thermal conductivity of III–V semiconductor superlattices *J. Appl. Phys.* **118** 175101
[36] Zhang H, Hua C, Ding D and Minnich A J 2015 Length dependent thermal conductivity measurements yield phonon mean free path spectra in nanostructures *Sci. Rep.* **5** 9121
[37] Sverdrup P G, Sinha S, Asheghi M, Uma S and Goodson K E 2001 Measurement of ballistic phonon conduction near hotspots in silicon *Appl. Phys. Lett.* **78** 3331–3
[38] Highland M, Gundrum B C, Koh Y K, Averback R S, Cahill D G, Elarde V C, Coleman J J, Walko D A and Landahl E C 2007 Ballistic-phonon heat conduction at the nanoscale as revealed by time-resolved x-ray diffraction and time-domain thermoreflectance *Phys. Rev.* B **76** 075337
[39] Siemens M E, Li Q, Yang R, Nelson K A, Anderson E H, Margaret M and Kapteyn H C 2010 Quasi-ballistic thermal transport from nanoscale interfaces observed using ultrafast coherent soft x-ray beams *Nat. Mater.* **9** 29–30
[40] Johnson J A, Maznev A A, Cuffe J, Eliason J K, Minnich A J, Kehoe T, Sotomayor Torres C M, Chen G and Nelson K A 2015 Direct measurement of room-temperature nondiffusive thermal transport over micron distances in a silicon membrane *Phys. Rev. Lett.* **110** 025901
[41] Robbins A B, Drakopoulos S X, Ignacio M-F, Ronca S and Minnich A J 2019 Ballistic thermal phonons traversing nanocrystalline domains in oriented polyethylene *Proc. Natl Acad. Sci.* **116** 17163–8
[42] Cahill D G *et al* 2014 Nanoscale thermal transport II: 2003–2012 *Appl. Phys. Rev.* **1** 011305
[43] Koh Y K and Cahill D G 2007 Frequency dependence of the thermal conductivity of semiconductor alloys *Phys. Rev.* B **76** 075207
[44] Minnich A J, Johnson J A, Schmidt A J, Esfarjani K, Dresselhaus M S, Nelson K A and Chen G 2015 Thermal conductivity spectroscopy technique to measure phonon mean free paths *Phys. Rev. Lett.* **107** 095901
[45] English T S, Phinney L M, Hopkins P E and Serrano J R 2013 Mean free path effects on the experimentally measured thermal conductivity of single-crystal silicon microbridges *J. Heat Transfer* **135** 091103
[46] Peierls R 1929 On the kinetic theory of thermal conduction in crystals *Ann. Phys.* **3** 1055
[47] Guyer R A and Krumhansl J A 1966 Solution of the linearized phonon Boltzmann equation *Phys. Rev.* **148** 766–78

[48] Hamilton R A H and Parrott J E 1969 Variational calculation of the thermal conductivity of germanium *Phys. Rev.* **178** 1284–92

[49] Srivastava G P 1976 Derivation and calculation of complementary variational principles for the lattice thermal conductivity *J. Phys. C: Solid State Phys.* **9** 3037

[50] Levinson I B 1980 Nonlocal phonon heat conductivity *J. Exp. Theor. Phys.* **52** 704

[51] Ward A, Broido D A, Stewart D A and Deinzer G 2009 *Ab initio* theory of the lattice thermal conductivity in diamond *Phys. Rev.* B **80** 125203

[52] Broido D A, Ward A and Mingo N 2005 Lattice thermal conductivity of silicon from empirical interatomic potentials *Phys. Rev.* B **72** 014308

[53] Li W, Carrete J, Katcho N A and Mingo N 2014 ShengBTE: a solver of the Boltzmann transport equation for phonons *Comput. Phys. Commun.* **185** 1747–58

[54] Carrete J, Vermeersch B, Katre A, van Roekeghem A, Wang T, Madsen G K H and Mingo N 2017 almaBTE: a solver of the space-time dependent Boltzmann transport equation for phonons in structured materials *Comput. Phys. Commun.* **220** 351–62

[55] Omini M and Sparavigna A 1995 An iterative approach to the phonon Boltzmann equation in the theory of thermal conductivity *Physica* B **212** 101–12

[56] Fugallo G, Lazzeri M, Paulatto L and Mauri F 2013 *Ab initio* variational approach for evaluating lattice thermal conductivity *Phys. Rev.* B **88** 045430

[57] Chaput L 2013 Direct solution to the linearized phonon Boltzmann equation *Phys. Rev. Lett.* **110** 265506

[58] Cepellotti A and Marzari N 2016 Thermal transport in crystals as a kinetic theory of relaxons *Phys. Rev.* X **6** 041013

[59] Hardy R J 1970 Phonon Boltzmann equation and second sound in solids *Phys. Rev.* B **2** 1193–207

[60] Hardy R J 1965 Lowest-order contribution to the lattice thermal conductivity *J. Math. Phys.* **6** 1749–61

[61] Cepellotti A and Marzari N 2017 Boltzmann transport in nanostructures as a friction effect *Nano Lett.* **17** 4675–82

[62] Wilson R B, Feser J P, Hohensee G T and Cahill D G 2013 Two-channel model for nonequilibrium thermal transport in pump–probe experiments *Phys. Rev.* B **88** 144305

[63] Yang F and Dames C 2015 Heating-frequency-dependent thermal conductivity: an analytical solution from diffusive to ballistic regime and its relevance to phonon scattering measurements *Phys. Rev.* B **91** 165311

[64] Torres P, Ziabari A, Torelló A, Bafaluy J, Camacho J, Cartoixà X, Shakouri A and Alvarez F X 2018 Emergence of hydrodynamic heat transport in semiconductors at the nanoscale *Phys. Rev. Mater* **2** 076001

[65] Vermeersch B, Carrete J, Mingo N and Shakouri A 2015 Superdiffusive heat conduction in semiconductor alloys. I. Theoretical foundations *Phys. Rev.* B **91** 085202

[66] Maznev A A, Johnson J A and Nelson K A 2011 Onset of nondiffusive phonon transport in transient thermal grating decay *Phys. Rev.* B **84** 195206

[67] Maassen J and Lundstrom M 2015 Steady-state heat transport: ballistic-to-diffusive with Fourier's law *J. Appl. Phys.* **117** 035104

[68] Mahan G D and Claro F 1988 Nonlocal theory of thermal conductivity *Phys. Rev.* B **38** 1963–9

[69] Hua C and Minnich A J 2014 Analytical Green's function of the multidimensional frequency-dependent phonon Boltzmann equation *Phys. Rev.* B **90** 214306

[70] Allen P B and Perebeinos V 2018 Temperature in a Peierls–Boltzmann treatment of nonlocal phonon heat transport *Phys. Rev.* B **98** 085427
[71] Collins K C, Maznev A A, Tian Z, Esfarjani K, Nelson K A and Chen G 2013 Non-diffusive relaxation of a transient thermal grating analyzed with the Boltzmann transport equation *J. Appl. Phys.* **114** 104302
[72] Hua C and Minnich A J 2015 Semi-analytical solution to the frequency-dependent Boltzmann transport equation for cross-plane heat conduction in thin films *J. Appl. Phys.* **117** 175306
[73] Kan Koh Y, Cahill D G and Sun B 2014 Nonlocal theory for heat transport at high frequencies *Phys. Rev.* B **90** 205412
[74] Ramu A T and Ma Y 2014 An enhanced Fourier law derivable from the Boltzmann transport equation and a sample application in determining the mean-free path of nondiffusive phonon modes *J. Appl. Phys.* **116** 093501
[75] Regner K T, McGaughey A J H and Malen J A 2014 Analytical interpretation of nondiffusive phonon transport in thermoreflectance thermal conductivity measurements *Phys. Rev.* B **90** 064302
[76] Zeng L and Chen G 2014 Disparate quasiballistic heat conduction regimes from periodic heat sources on a substrate *J. Appl. Phys.* **116** 064307
[77] Vermeersch B, Carrete J, Mingo N and Shakouri A 2015 Superdiffusive heat conduction in semiconductor alloys. I. Theoretical foundations *Phys. Rev.* B **91** 085202
[78] Peraud J-P M and Hadjiconstantinou N G 2011 Efficient simulation of multidimensional phonon transport using energy-based variance-reduced Monte Carlo formulations *Phys. Rev.* B **84** 205331
[79] Peraud J-P M and Hadjiconstantinou N G 2012 An alternative approach to efficient simulation of micro/nanoscale phonon transport *Appl. Phys. Lett.* **101** 15311
[80] Hua C and Minnich A J 2014 Importance of frequency-dependent grain boundary scattering in nanocrystalline silicon and silicongermanium thermoelectrics *Semicond. Sci. Technol.* **29** 124004
[81] Minnich A J, Chen G, Mansoor S and Yilbas B S 2011 Quasiballistic heat transfer studied using the frequency-dependent Boltzmann transport equation *Phys. Rev.* B **84** 235207
[82] Péraud J-P M and Hadjiconstantinou N G 2015 Adjoint-based deviational Monte Carlo methods for phonon transport calculations *Phys. Rev.* B **91** 235321
[83] Hua C, Lindsay L, Chen X and Minnich A A generalized Fourier's law for non-diffusive thermal transport: theory and experiment *Phys. Rev.* B **100** 085203
[84] Xu X *et al* 2014 Length-dependent thermal conductivity in suspended single-layer graphene *Nat. Commun.* **5** 3689
[85] Grant M and Boyd S 2014 CVX: Matlab software for disciplined convex programming, version 2.1. Available from: http://cvxr.com/cvx March 2014
[86] Grant M and Boyd S 2008 Graph implementations for nonsmooth convex programs *Recent Advances in Learning and Control, Lecture Notes* ed V Blondel, S Boyd and H Kimura (Berlin: Springer), pp 95–110 http://stanford.edu/boyd/graph_dcp.html
[87] Hastings W K 1970 Monte Carlo sampling methods using Markov chains and their applications *Biometrika* **57** 97–109
[88] Nelder J A and Mead R 1965 A simplex method for function minimization *Comput. J.* **7** 308–13

[89] Forghani M, Hadjiconstantinou N G and Péraud J-P M 2016 Reconstruction of the phonon relaxation times using solutions of the Boltzmann transport equation *Phys. Rev.* B **94** 155439

[90] Forghani M and Hadjiconstantinou N G 2018 Reconstruction of phonon relaxation times from systems featuring interfaces with unknown properties *Phys. Rev.* B **97** 195440

[91] Wei Z, Yang J, Chen W, Bi K, Li D and Chen Y 2014 Phonon mean free path of graphite along the *c*-axis *Appl. Phys. Lett.* **104** 081903

[92] Hua C and Minnich A J 2014 Transport regimes in quasiballistic heat conduction *Phys. Rev.* B **89** 094302

[93] Sondheimer E H 1952 *The Mean Free Path of Electrons in Metals* (Oxford: Taylor and Francis)

[94] Fuchs K 1938 The conductivity of thin metallic films according to the electron theory of metals *Math. Proc. Camb. Phil. Soc.* **34** 100–8

[95] Ravichandran N K and Minnich A J 2016 Role of thermalizing and nonthermalizing walls in phonon heat conduction along thin films *Phys. Rev.* B **93** 035314

[96] Koh Y K and Cahill D G 2007 Frequency dependence of the thermal conductivity of semiconductor alloys *Phys. Rev.* B **76** 075207

[97] Kang J S, Li M, Wu H, Nguyen H and Hu Y 2018 Experimental observation of high thermal conductivity in boron arsenide *Science* **361** 575–8

[98] Li S, Zheng Q, Lv Y, Liu X, Wang X, Huang P Y, Cahill D G and Lv B 2018 High thermal conductivity in cubic boron arsenide crystals *Science* **361** 579–81

[99] Capinski W S and Maris H J 1996 Improved apparatus for picosecond pump and probe optical measurements *Rev. Sci. Instrum.* **67** 2720–6

[100] Schmidt A J, Chen X and Chen G 2008 Pulse accumulation, radial heat conduction, and anisotropic thermal conductivity in pump–probe transient thermoreflectance *Rev. Sci. Instrum.* **79** 114902

[101] Ding D, Chen X and Minnich A J 2014 Radial quasiballistic transport in time-domain thermoreflectance studied using Monte Carlo simulations *Appl. Phys. Lett.* **104** 143104

[102] Tian F *et al* 2018 Unusual high thermal conductivity in boron arsenide bulk crystals *Science* **361** 582–5

[103] Lindsay L, Broido D A and Reinecke T L 2013 First-principles determination of ultrahigh thermal conductivity of boron arsenide: a competitor for diamond? *Phys. Rev. Lett.* **111** 025901

[104] Broido D A, Lindsay L and Reinecke T L 2013 *Ab initio* study of the unusual thermal transport properties of boron arsenide and related materials *Phys. Rev.* B **88** 214303

[105] Feng T, Lindsay L and Ruan X 2017 Four-phonon scattering significantly reduces intrinsic thermal conductivity of solids *Phys. Rev.* B **96** 161201

[106] Chen X, Hua C, Zhang H, Ravichandran N K and Minnich A J 2018 Quasiballistic thermal transport from nanoscale heaters and the role of the spatial frequency *Phys. Rev. Appl.* **10** 054068

[107] Hua C and Minnich A J 2018 Heat dissipation in the quasiballistic regime studied using the Boltzmann equation in the spatial frequency domain *Phys. Rev.* B **97** 014307

IOP Publishing

Nanoscale Energy Transport
Emerging phenomena, methods and applications
Bolin Liao

Chapter 7

Thermodynamics of anharmonic lattices from first principles

Keivan Esfarjani and Yuan Liang

Self-consistent phonon (SCP) theory and its application in computing thermodynamic properties of materials are reviewed from a historical perspective. Various more recent implementations based on first-principles electronic structure methods using the density functional theory have been discussed. The SCP equations can be derived either from a diagrammatic perturbation theory or a variational approach based on free-energy minimization. These methods can also be used to predict phase change due to phonon softening and can be extended to study the coupling of phonons to other degrees of freedom in the system.

7.1 Introduction

7.1.1 Motivation

Atomic vibrations play an important role in describing the thermodynamics of materials. In crystalline solids, we refer to these excitations as phonons. A quantum description based on phonons allows the calculation of thermodynamic properties such as free energy, heat capacity and entropy, as well as Raman intensities, all of which can be compared to experimental data on neutron scattering and Raman spectra in order to better understand the underlying physics and properties of a material. The most commonly employed modern methods for phonon spectra calculation include harmonic and quasi-harmonic approximations (QHAs), in which the potential energy is expanded up to second order in powers of atomic displacements (assumed to be small) about their equilibrium positions. The inclusion of only harmonic forces in the system implies that the phonons have an infinite lifetime. This limits the possibility to obtain properties such as thermal conductivity. Furthermore, at elevated temperatures (compared to the Debye temperature, for instance), where atomic displacements become larger, anharmonic contributions to the forces are not negligible anymore and one needs to go beyond the harmonic approximation.

doi:10.1088/978-0-7503-1738-2ch7

Additionally, phonon frequencies vary with temperature and collisions between them limits their lifetime. These effects can be accounted for by treating anharmonic effects as a perturbation. The need for such theories is also present in systems near a phase transition where atomic displacements are large and therefore anharmonicity is more important. Another case of interest which originally motivated such theories was the study of rare-gas crystals such as neon or argon (helium was exceptionally challenging) where the quantum zero-point motion of atoms is large and atoms probe anharmonic regions of the potential. The advantage of rare-gas atom systems is that analytical pair interatomic potentials (typically of Lennard-Jones type) can accurately describe their properties, and therefore different theories can easily be tested.

In this review, after going over the general lattice dynamics formalism, we will start by explaining the basic idea behind the self-consistent phonon (SCP) in section 7.1.2. This will be followed by a simple example to illustrate the method and its power in section 7.1.3. The historical development of SCP theory is detailed in section 7.2 by briefly going into more detail over the work of famous contributors, including Born, Hooton, Horton, Cowley, Choquard, Koehler, Werthammer, Gillis and many others. Next, modern implementations using density functional theory (DFT) calculations will be described in section 7.3. This section will first deal with the model parameter extraction from DFT calculations in real or reciprocal spaces, and then its actual implementations will be discussed. We will review the merits and shortcomings of the two approaches. Finally, in section 7.4, we propose an extension which includes internal and cell relaxations and coupling to other order parameters (OPs). This method provides a theory of phase transitions incorporating the coupling of OPs with vibrational and strain degrees of freedom and also accounts for possible internal atomic relaxations and structural change at a low computational cost.

7.1.2 Lattice dynamics theory and the self-consistent phonon idea

In the Born–von-Kármán approximation, atomic nuclei possess a mass and are connected by massless electron 'springs'. The fundamental assumption from Born–von-Kármán's theory is that the oscillation of atomic nuclei is confined to a small region relative to the nuclear–nuclear separation. So to leading order, a set of harmonic equations of motion can be constructed to describe the system using the quadratic coefficients called the force constants. It is the basis of what is now known as the theory of lattice dynamics, whose extensions will be discussed in this chapter:

$$H \approx H_{\text{Harmonic}} = \sum_{i,\alpha} \frac{P_{i\alpha}^2}{2m_i} + V_0 + \frac{1}{2} \sum_{i,j} \Phi_{i\alpha,j\beta} U_{i\alpha} U_{j\beta}. \tag{7.1}$$

In this equation, $U_{i\alpha}$ is the displacement of atom i about its equilibrium position in the direction α, $P_{i\alpha}$ is its conjugate momentum, and $\Phi_{i\alpha,j\beta}$ is called the harmonic force constant between atoms i and j. Greek letters represent the Cartesian components x, y, z and they are summed over if repeated. After decoupling the dynamical displacement variables by the appropriate unitary transformation $\epsilon_{i\alpha,\lambda}$, the Hamiltonian becomes 'diagonal':

$$H_{\text{Harmonic}} = \sum_{\lambda} \frac{P_\lambda^2}{2} + V_0 + \frac{1}{2}\sum_{\lambda} \omega_\lambda^2 U_\lambda^2. \tag{7.2}$$

The transformation is the one that has diagonalized the rescaled, symmetrized force constant matrix: $\tilde{\Phi}_{i\alpha,j\beta} = \Phi_{i\alpha,j\beta}/\sqrt{m_i m_j}$. In other words,

$$\sum_{j,\beta} \tilde{\Phi}_{i\alpha,j\beta}\, \epsilon_{j\beta,\lambda} = \omega_\lambda^2\, \epsilon_{i\alpha,\lambda} \tag{7.3}$$

with $\sqrt{m_i}\, U_{i\alpha} = \sum_\lambda \epsilon_{i\alpha,\lambda} U_\lambda$ and the orthogonality and completeness relations among the eigenvectors coming, respectively, from $\epsilon^\dagger\epsilon = \mathbb{1}$ and $\epsilon\epsilon^\dagger = \mathbb{1}$.[1]

This is the standard harmonic lattice dynamics theory, from which one can extract eigenvalues ω_λ^2 (phonon frequencies squared) and eigenvectors (phonon polarization vectors). Normal coordinates play an important role in simplifying the mathematical formulation of thermodynamics of the harmonic model, since it reduces an interacting atomic Hamiltonian to N independent or non-interacting harmonic oscillators, each having their own frequencies and polarization vectors. In other words, phonon modes can be thought of as independent harmonic oscillators. The translational invariance of the lattice structure leads to wave-like solutions and a dispersion relation between frequencies and wave-vectors (see appendix B). This harmonic model is accurate at low-enough temperatures and for heavy-enough atomic masses since the resulting atomic displacements remain small. If quantized, this model will also correctly predict the heat capacity at low temperatures. Anharmonic effects, however, have to be taken into consideration at higher temperatures. Also, for cases like rare-gas/quantum crystals, where the zero-point motion is large, this approximation is not valid [1]. This can also be because for some interatomic potentials, the second-order term in the Taylor expansion of potential energy may not be positive definite. Conventional harmonic lattice dynamics does not work in this scenario as it leads to imaginary phonon frequencies. To make the structure stable, one needs to incorporate higher-order terms in the Taylor expansion of the potential energy. The standard perturbation theory approach [2], where cubic and higher-order terms in the Taylor expansion are treated as perturbation, allows the calculation of renormalized phonon frequencies and their lifetime and temperature dependence. The effectiveness of this approach depends on two major assumptions: (i) that the harmonic force constants form a positive-definite matrix so that the resulting harmonic frequencies are real and positive numbers, and (ii) that anharmonic effects are comparably smaller than the harmonic terms. In other words, the starting reference atomic positions are a local minimum of the potential energy and the dynamical displacements of nuclei should be strictly limited within the vicinity of equilibrium position for the perturbative method to work. However, the small displacement assumption will not necessarily remain valid at high temperatures, in particular at or near a phase transition, nor for any other

[1] For a real symmetric matrix Φ the eigenvectors ϵ form an orthogonal matrix, while for a Hermitian matrix, they are unitary: $\sum_{i\alpha} \epsilon^*_{i\alpha,\lambda}\epsilon_{i\alpha,\mu} = \delta_{\lambda,\mu}$; $\sum_\lambda \epsilon^*_{i\alpha,\lambda}\epsilon_{j\beta,\lambda} = \delta_{i,j}\delta_{\alpha,\beta}$.

forms of instability in a lattice system. Even at low temperatures if the zero-point motion is large enough, anharmonic terms can become important.

The physics of the above problem can be explained as follows. For an atom in a crystal with low vibrational energy, within its oscillation scope around a local minimum, the potential curve can be nicely fitted with a positive curvature parabola. There are cases, however, in which as the temperature increases, the atom may possess an energy level higher than the nearest potential hump, thus at that point the harmonic approximation will give a parabola pointing downward, which leads to imaginary phonon frequencies, as shown in figure 7.1, although on average, the atom oscillates around the high-symmetry but unstable position. In order to overcome this difficulty, a new renormalized perturbation expansion, later to be recognized as the SCP theory, was first formally constructed by D J Hooton 1955, in his journal article co-authored with his advisor, Max Born, 'Statistische Dynamik mehrfach periodischer Systeme' [3].

The basic concept of self-consistent harmonic approximation (SCHA, also called SC1 by some authors) is to fit the actual potential energy curve by an effective positive-definite parabola, i.e. to take the thermally weighted average of the second derivative of the potential energy instead of taking the second derivative at the equilibrium position, which could be negative:

$$\Phi_{\text{actual}} \to \Phi_{\text{SCHA}} \equiv \langle \Phi_{\text{actual}} \rangle = \langle \Phi(R_{ij} + U_i - U_j) \rangle \tag{7.4}$$

$$= \Phi(R_{ij}) + \frac{1}{2}\langle UU \rangle \frac{\partial^2 \Phi}{\partial R^2} + \frac{3}{4!}\langle UU \rangle \langle UU \rangle \frac{\partial^4 \Phi}{\partial R^4} + \cdots. \tag{7.5}$$

$\langle \Phi_{\text{actual}} \rangle$ means a thermal average over force constants with bond lengths fluctuating about their thermal equilibrium value. As a shorthand, U symbolically represents $U_i - U_j$. We will explicitly show the derivation of this equation as done originally by Boccara and Sarma [5], in appendix C. By Taylor expanding the force constant in powers of atomic displacements U_i^α we see that we can calculate this average if all even-order derivatives are known, because displacements have a Gaussian distribution, and averages of the type $\langle U_i U_j \ldots U_l \rangle$ can be exactly evaluated by using Wick's theorem [6]. More formally, keeping in mind that the averaging is only over the dynamical parameters U, which has a Gaussian distribution, and replacing for the sake of brevity $U_i - U_j$ by U_{ij}, this can be written as follows:

$$\langle \Phi(R_{ij} + U_{ij}) \rangle = \langle e^{U_{ij} \cdot \nabla_R} \rangle \Phi(R_{ij}) = e^{\frac{1}{2}\langle U_{ij} U_{ij} \rangle : \nabla_R^2} \Phi(R_{ij}).$$

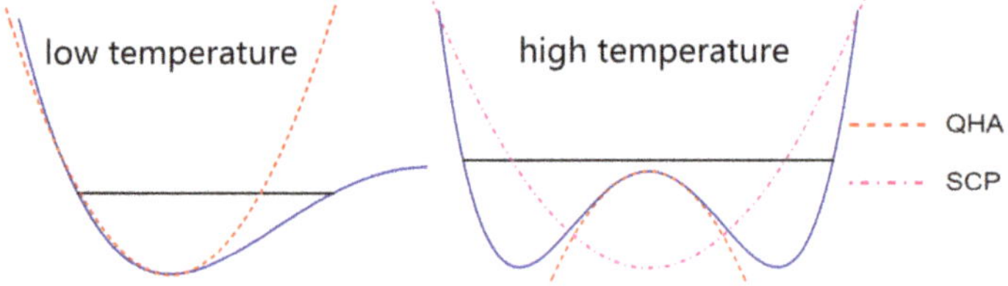

Figure 7.1. Intuitive illustration of Born's insight [4]. While the harmonic potential has a negative second derivative, the SCP potential is positive definite and leads to 'stable' phonons.

The ∇_R operator, which is used in the exponent to express the translation, acts on the function $\Phi(R)$. This equation is formally equivalent to equation (7.5). The self-consistency comes in the evaluation of the $\langle UU \rangle$ term which is expressed in terms of eigenvalues and eigenvectors of Φ_{SCHA} itself. In terms of Feynman diagrams, this approximation can be described by the sum of the diagrams in figure 7.2, where the four-, six- and eight-point vertices correspond, respectively, to the fourth, sixth and eighth derivatives of the potential energy with respect to atomic displacements and the dashed and solid lines correspond to the bare harmonic and self-consistent harmonic propagators (or Green's functions), respectively.

Hoerner has shown that this self-consistent perturbation approach is equivalent to the variational approach [7] originally proposed by Boccara and Sarma [5], which suggests the question: what are the 'best' trial harmonic force constants that minimize the system's variational free energy? In a mathematical sense, we are using a quadratic 'trial' potential V_{trial} to best fit the 'actual' potential according to the actual displacements taking place at a given temperature. The equivalence is also shown by Gillis in 1968 [8], who used the variational formulation and showed that the best trial force constants can in fact be expressed in terms of the sum of even-order force constants, as shown in equation (7.5). We will provide a derivation of these equations using the variational approach later in the text.

7.1.3 Implementation example of the variational approach

7.1.3.1 Variational formulation

We will first illustrate the variational formulation with a simple example where the free energy of a one-dimensional anharmonic oscillator is minimized in accordance with the Bogoliubov inequality (BI). BI is a strong tool that provides the mean-field equations treating a system at finite temperature. The Bogoliubov inequality upon which the variational formulation is based is as follows:

$$F_{\text{actual}} \leqslant F_0 + \langle H_{\text{actual}} - H_{\text{trial}} \rangle_0. \tag{7.6}$$

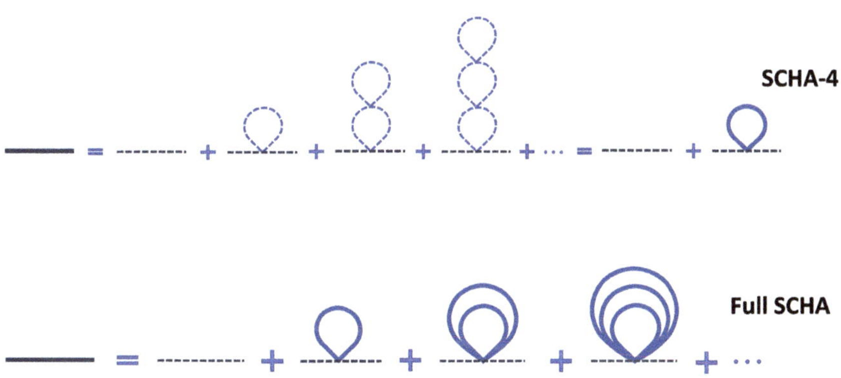

Figure 7.2. Feynman diagrams for the phonon propagator corresponding to the SCHA. Dashed lines represent the bare harmonic propagator, while solid lines represent the self-consistent ones. The top is what we call SCHA-4 where only the quartic term is kept, and the bottom corresponds to keeping all (even) derivatives in the expansion in equation (7.5).

Here, F is the actual (exact) free energy and H is the actual Hamiltonian, H_{trial} is any trial solvable Hamiltonian, meaning for which the density matrix is known, and the free energy F_0 is readily calculated. The sharp brackets $\langle\ \rangle_0$ mean the thermal average taken with a density matrix defined with H_{trial}:

$$\langle A\rangle_0 = \text{Tr}\,\rho_{\text{trial}}A \tag{7.7}$$

$$\rho_{\text{trial}} = \frac{e^{-H_{\text{trial}}/k_B T}}{Z_0} \tag{7.8}$$

$$Z_0 = \text{Tr}\,e^{-H_{\text{trial}}/k_B T}; \quad F_0 = -k_B T \ln Z_0. \tag{7.9}$$

Thus, the effect of temperature is introduced through the averaging procedure with respect to a known density matrix or distribution function.

The best choice of trial Hamiltonian is the one that leads to the lowest value of the right-hand side of equation (7.6), which is therefore the closest possible to the true free energy. So in practice one chooses a solvable trial Hamiltonian with some parameters which will be determined by minimizing the right-hand side of equation (7.6). This is the meaning of the variational approach.

In SCP and Feynman's example introduced below, the trial Hamiltonian $H_0 = H_{\text{trial}}$ always takes the form of a harmonic oscillator which is exactly solvable. After applying the variational formulation, the trial mass can be shown to be the same as the real mass of the particle. The $\langle H_{\text{actual}} - H_{\text{trial}}\rangle_0$ can therefore be replaced by the potential energy part $\langle V_{\text{actual}} - V_{\text{trial}}\rangle_0$. In his book [9], Feynman gave an elegant proof of equation (7.6) using the Baker–Campbell–Hausdorff expansion relating the product of two exponentials to the exponential of the sum and correction terms involving commutators. It can also be derived from a simple matrix and algebra, and perturbation theory using the Cauchy–Schwarz inequality [10].

7.1.3.2 The single-oscillator case

Consider a single particle in a general bounded potential $V_{\text{actual}}(U)$, with U being the dynamical displacement from the equilibrium position. We adopt a trial Hamiltonian with a displaced equilibrium position u_0 (in order to incorporate thermal expansion) and a displaced oscillation frequency ω_0, both of which need to be calculated by minimizing the right-hand side of equation (7.6) with respect to these two parameters. Feynman showed that the minimization results in two concise equations [9]:

$$\partial F_{\text{trial}}/\partial u_0 = 0 \Rightarrow \langle V'_{\text{actual}}(U)\rangle_0 = 0 \tag{7.10}$$

$$\partial F_{\text{trial}}/\partial \omega_0 = 0 \Rightarrow \langle U\ V'_{\text{actual}}(U)\rangle_0 = \langle P^2/m\rangle_0. \tag{7.11}$$

Equation (7.10) is the statement that at equilibrium, the net average force on the particle should be zero. It is derived from the minimization with respect to the displaced equilibrium position u_0. Equation (7.11) is more interesting since it is a statement of the virial theorem. It is derived from the differential of the trial free energy with respect to the effective oscillator frequency. More generally, the virial theorem can be derived from a consideration of the change in the total energy under volume or more generally under a length scale change. Expressing that this change, which is essentially the pressure, is zero at equilibrium, one reaches equation (7.11).

Below, we will use a toy model to showcase how those formulas work. For the sake of simplicity and to reduce heavy notation, we choose a particle of unit mass in an anharmonic potential with cubic and quartic terms in the potential energy V_{actual}:

$$V_{\text{actual}} = \frac{1}{2}\omega^2\, U^2 + \frac{1}{6}a\, U^3 + \frac{1}{24}b\, U^4 \tag{7.12}$$

$$V_{\text{trial}} = \frac{1}{2}\omega_0^2\, (U - u_0)^2. \tag{7.13}$$

V_{actual} is the toy model potential with $U(t)$ being the atomic dynamical displacement from its equilibrium position ($U = 0$). Both a and b, which are known parameters, can be thought of as anharmonic spring constants. V_{trial} is the trial potential with variational parameters ω_0 and u_0 which we want to determine from free-energy minimization. The former can be seen as the 'renormalized' phonon frequency and the latter is related to the thermal expansion. Since the trial system is purely harmonic, one can find the exact expression for F_{trial} and $\langle V_{\text{trial}} \rangle$ as shown in equations (A.3) and (B.7), respectively. We can substitute and rewrite the Bogoliubov inequality as

$$F_{\text{actual}} \leqslant \frac{1}{\beta}\ln\left[2\sinh\left(\frac{\beta\hbar\omega_0}{2}\right)\right] + \langle V_{\text{actual}} \rangle - \frac{\hbar\omega_0}{4}\coth\left(\frac{\beta\hbar\omega_0}{2}\right), \tag{7.14}$$

where $\beta = 1/k_B T$ and k_B is the Boltzmann constant. The right-hand side of this equation, the variational free energy, represents the upper bound of actual free energy. One can directly take its first derivatives with respect to the two variables ω_0 and u_0 to set up the following two coupled equations:

$$\frac{\partial \langle V_{\text{actual}} \rangle_0}{\partial u_0} = 0 \tag{7.15}$$

$$\frac{1}{\hbar}\frac{\partial \langle V_{\text{actual}} \rangle_0}{\partial \omega_0} + \frac{\beta\hbar\omega_0}{8}\frac{1}{\sinh^2(\beta\hbar\omega_0/2)} = 0. \tag{7.16}$$

Note that the thermal average is with respect to the distribution function of H_{trial}. The potential energy term $\langle V_{\text{actual}} \rangle_0$ involves powers of U which can readily be calculated. To simplify the calculation of the above derivatives, one can use the following results which hold for the variable $U - u_0$ having a Gaussian distribution:

$$\begin{aligned} \langle (U - u_0)^{2n+1} \rangle_0 &= 0 \\ \langle (U - u_0)^{2n+2} \rangle_0 &= (2n + 1)\langle (U - u_0)^{2n} \rangle_0 \langle (U - u_0)^2 \rangle_0 \\ \langle (U - u_0)^2 \rangle_0 &= \frac{\hbar}{2\omega_0}\,\text{Coth}\left(\frac{\beta\hbar\omega_0}{2}\right). \end{aligned} \tag{7.17}$$

The two variational parameters (ω_0, u_0) can thus be found by either numerically solving the above two coupled equations, or analytically with some approximations and in some specific limits. If we were to use the small oscillation limit $u_0 \ll 1$ and high-temperature limit $\omega_0 \ll k_B T/\hbar$, consistent with the former constraint, the transcendental equations can be simplified and solved analytically:

$$\begin{aligned} u_0 &\approx -\frac{aT}{2\omega^4} + O\left(\frac{1}{T}\right) \\ \omega_0^2 &\approx \omega^2 - \frac{bT}{2a\omega^2} + O(\omega^{-4}). \end{aligned} \tag{7.18}$$

The classical thermal expansion for a one-dimensional harmonic chain [11] is thus recovered. With more accurate calculation of u_0 and ω_0, the Helmholtz free energy can thus be computed and plotted as a function of temperature. For more complicated systems such as ferroelectrics, one could investigate possible structural phase transitions by looking at the contour plots of the free energy in the (T, u_0) plane. Above is just the simplest implementation of the SCHA (or SC1). In a real lattice system, various computational techniques based on DFT or molecular dynamics are practised to approximate high-order FCs. The 'Gaussian smeared' $\langle V_{\text{actual}} \rangle_0$ is a notable feature for self-consistent theories since the trial Hamiltonian is quadratic and displacements about the equilibrium sites have a Gaussian distribution.

7.2 Overview: historical development

Now that the main ideas behind the SCP have been clarified with an example, we proceed to give an overview of the method building on the harmonic theory.

Attempts such as *ab initio* molecular dynamics (AIMD) [12], path-integral molecular dynamics (PIMD) [13] and imaginary-time path-integral Monte Carlo (PIMC) to treat quantum anharmonic systems are either computationally intensive (PIMC and PIMD), or do not properly include quantum effects (AIMD) or are limited to equilibrium property calculations, which severely limit their scope of applicability. Thus, many approaches to overcome these limitations have been developed by different groups based on the renormalized perturbation expansion method, later to be known as the SCP theory or SCHA first introduced by Born [3] and then his student, Hooton [4, 14], who in his second paper introduced a variational formulation of the problem. In the context of phonon softening and phase transitions, the variational approach was also adopted by Boccara and Sarma in 1965 [5], where they defined a harmonic Hamiltonian with displaced coordinates and effective force constants, to be determined by minimizing the free energy. Later, in 1966, Koehler introduced a similar approach [15], and applied the self-consistent harmonic phonon approach to solid Ne at zero temperature, where, due to large zero-point motions, atoms probed anharmonic regions of the potential energy. His results produced a lower value of the ground-state energy [15] compared to those of Bernades [16], and Nosanow and Shaw [17] who used an uncorrelated Einstein model to describe localized excitations in solid Ne. In that publication, his method was basically identical to the quasi-harmonic phonon approximation which he referred to as a 'self-consistent Hamiltonian', although the energy optimization is based on a set of uncorrelated Gaussian wave functions. He also claimed that additional features of self-consistency were presented despite being inherently similar to Nosanow and Werthamer's paper [18] in 1965 on sound velocity calculation.

SCP provides an alternative to the earlier perturbation theory approaches, in that it is a mean-field theory consisting essentially in replacing the harmonic force constants (or dynamical matrix) by their average over atomic displacements taking place at a given temperature. Naturally, this average is defined by anharmonic terms in the expansion of the forces in powers of atomic displacements. The 'self-consistency' condition comes from expressing that the average over atomic displacements has to be performed with respect to the renormalized phonon modes and not the original harmonic ones. This theory can be derived in two different ways. One is from the diagrammatic perturbation method, where a series of diagrams are added to infinite order, and the other is from a variational approach wherein the free energy of the crystal is minimized with respect to some effective force constants, and other variational parameters if needed, in order to obtain the self-consistent or mean-field equations to be solved.

In 1965, Ranninger [19] published a work based on diagrammatic perturbation theory, which he attributed to Choquard. Choquard's book [20] on the subject appeared two years later in 1967. He had generalized the method to second-order, which was a more complete theory since it included phonon damping. Later, by implementing a selective re-summation of diagrammatic perturbation theory, Horner showed that the two formulations were identical [7]. The free-energy minimization is, however, more elegant and on solid ground, and more amenable to future extensions (e.g. to describe phase transitions) if need be.

The first complete SCP theory was, however, first worked out as a mean-field theory by Gillis *et al* in 1968 [8]. It offered a starting point for quantum crystal studies. Techniques such as Green's function methods, variational approaches or diagram summation all lead to the full SCHA scheme. Gillis and Werthammer's SCHA (or SC1) theory was established based on several authors' previous work as well as Koehler's computational implementations.

Gillis, Werthamer and Koehler's SCHA theory (see figure 7.2) was limited due to the omission of odd derivatives of the potential. An improved self-consistent phonon approximation was introduced by Goldman, Horton and Klein in 1968 [21]. The cubic term has been directly added as a second-order correction term to the Helmholtz free energy. This is done after the free-energy minimization, which has produced effective phonon frequencies. This was shown to correspond to the leading correction term in Choquard's first-order self-consistent equations. With this addition, the free energy is not variational anymore but may well exceed the first-order theory in accuracy. This method had been carefully tested on a selection of rare gases, with the adoption of a phenomenological nearest-neighbor central-force potential. The results show that quantities such as the high-temperature heat capacity C_p agree best with the experiment, compared to the SCHA and simple perturbation theory of Feldman and Horton [22]. They also agree well with this perturbation theory at roughly 1/3 of the melting temperature, thus the validity of this method holds at low temperature for rare gases.

One cannot avoid Werthamer's contribution while discussing early SCP development. After Horner's paper in February of 1967, Werthamer published a paper on the phonon frequencies and thermodynamic properties of crystalline bcc He calculations, co-authored with Wette and Nosanow, in May 1967 [1]. Regardless of the more detailed mathematical formalism, their method was similar to Koehler's —using the time-dependent Hartree approximation together with the results of variational calculations of the ground-state energy using correlated trial wave functions. The advantage of the Hartree approximation is that it yields a self-consistent solution in which the particles are indeed localized. This is a convenient way to introduce the structure of the crystal into the calculation. Both short- (atomic repulsion) and long-range (phonon) correlations are considered. They assumed that with external imposed disturbances, the system wave function can still be factorized for all times into a product of single-particle functions. The phonon frequencies are defined as poles of the response function, just like the collective excitations in the electron gas. For short-range correlations, they adopt the results of variational calculation of the ground-state energy using cluster-expansion techniques. Half a year later, Werthamer along with Gillis and Koehler [8] constructed the complete self-consistent formalism, later to be known as SC1 theory. Nevertheless, their formulas for the SCPs are equivalent to those of Ranninger, Choquard and Horner. In Werthamer's 1970 paper [23], he organized the previous authors' work in a single functional variational formalism and extended it to not only give the leading cumulants of the free energy, but also the first-order thermodynamic derivatives. Thus, a phonon dynamical model with adjustable parameters to fit the lower part of the excitation spectrum of the model system as closely as possible was constructed. Werthamer also took damping of phonons into account, by considering a trial action with force constants non-local in time.

Once actual free energy is specified by this SCP approximation, an equation can then be given to calculate the pressure for a particular choice of lattice constant. Evaluations of isothermal elastic constants, specific heats and thermal strain were also given in this paper. Werthamer then moved toward the second-order term, which gives the leading contribution to the phonon damping rate and also produces a frequency shift connected with the damping through the Kramers–Kronig relations. With the inclusion of the second-order term, now odd numbers of derivatives of the potential enter into SCP theory. A substantially simplified version for the second-order approximation approach was followed by Goldman [21] and Koehler [24] in their numerical calculations, with a slightly improved agreement with experiment over the first-order phonon spectrum.

Later practices in full second-order theory led to divergence difficulties, with the physical reason explained by Horton in his book with Maradudin, *Dynamical Properties of Solids* [25]. He showed that the Gaussian pair distribution function that occurs in the thermal averaging penetrates too far into the repulsive core of the potential. The fully established SCP theory offers a practical approach for computing the structural and dynamical properties of a general quantum or classical many-body system that also incorporates anharmonic effects and has advantages over standard harmonic approximation, although convincing demonstration of its accuracy was limited to rare gases in its early history.

In 1973, Samathiyakanit and Glyde offered an alternative path-integral formulation of the anharmonic lattice problem [26]. Their derivation uses a trial harmonic action. The exact partition function is then expanded in cumulants about the trial harmonic partition function. Successive orders of the self-consistent theory are then obtained by keeping successive cumulants and requiring that their contribution to the crystal dynamics vanishes. The method provides a description of both the vibrational dynamics and thermodynamics. This procedure is similar to Choquard's who also used a cumulant expansion method. Their theory also agrees with Werthamer's [23] up to first order, but not to second order where it was claimed the latter theory has omitted some extra terms.

7.3 Modern interpretations and implementations

Theory development for anharmonic lattices reached its limit in the 1970s and it was implemented for rare-gas atoms where the interparticle potential is of two-body type and relatively well-known. With the advent of DFT methods, which enabled accurate calculation of many-body forces and force constants in real materials, these theories have been revived over the past decade. The major differences among different developments and implementations have been in two directions: (i) the selection and extraction of force constants, and (ii) the sampling of phase space in order to calculate the thermal averages.

7.3.1 Selection and extraction of force constants

There are essentially two different approaches to obtain the force constants: (a) the reciprocal space method, which is based on the DFPT and the $2n + 1$ theorem [27–29], and (b) the real-space method which is based on finite displacements of atoms in a supercell [30–37].

7.3.1.1 Force constant from DFPT (reciprocal-space formulation)

A predictive theoretical scheme that originates from the first-principles approach, with its emphasis on anharmonic decay of phonons was developed by Debernadi *et al* in 1995 [29]. This method is based on DFPT [28, 38], which worked well for semiconductors, and the use of the $2n + 1$ theorem, which, in particular, states that cubic force constants, which are third derivatives of the potential energy, can be expressed in terms of only first derivatives of the electronic wavefunctions. The method was then implemented for the calculation of phonon lifetimes in diamond, Si and Ge. The importance of this method was that it offered a reliable third-order force constant calculation scheme that could demonstrate the anharmonic decay of phonons in semi-conductors. A comparison between computational and experimental data was provided to support the methodology. The advantage of using DFPT over standard lattice-dynamical calculations in the harmonic approximation is that the former offers an accessible way, in terms of computational cost, to describe anharmonic phonon lifetimes, in particular considering it may not easily be evaluated in experiments.

A decade later, in 2007 [39], Broido came up with a novel method, also employing DFPT, to compute intrinsic lattice thermal conductivity. Due to anharmonic phonon–phonon scattering and difficulties in accurately describing interatomic forces, a theoretical approach to predict lattice thermal conductivity has been difficult since Peierls first introduced the phonon Boltzmann equation. Broido's treatment involved no adjustable parameters with the only input required being anharmonic FCs, which are determined from first principles using DFPT. In the three-phonon process, the scattering matrix element is as follows:

$$V_{q_1\lambda_1,q_2\lambda_2,q_3\lambda_3} = \frac{1}{N}\left(\frac{\hbar}{2}\right)^{3/2} \sum\nolimits_{1,2,3} \frac{\Psi_{R_1\tau_1,R_2\tau_2,R_3\tau_3}}{\sqrt{m_{\tau_1}m_{\tau_2}m_{\tau_3}}} e^{i(q_1\cdot R_1+q_2\cdot R_2+q_3\cdot R_3)} \times\frac{\varepsilon_{R_1\tau_1,q_1\lambda_1}\,\varepsilon_{R_2\tau_2,q_2\lambda_2}\,\varepsilon_{R_3\tau_3,q_3\lambda_3}}{\sqrt{\omega_{q_1\lambda_1}\omega_{q_2\lambda_2}\omega_{q_3\lambda_3}}} \tag{7.19}$$

where R labels a unit cell and τ labels an atom in the unit cell, of mass m_τ, while $\Psi_{1,2,3}$ is the third-order anharmonic interatomic force constant for the indicated triplets of atoms, and ε are polarization (or eigen-) vectors of the dynamical matrix as defined in equation (7.3). They form a unitary transformation from the atomic and Cartesian $R\tau$ basis to the normal mode $q\lambda$ basis, and can be written as the product $e^{iq.R}\epsilon_{\tau,\lambda}(q)$ where the latter are the so-called polarization vectors (on atom τ) of mode λ at wavevector q. This scattering matrix element term is used to determine the three-phonon scattering rate using Fermi's golden rule. From the Peierls–Boltzmann equation, nonequilibrium distribution functions can be solved via an iterative approach, and the lattice thermal conductivity tensor can be obtained. In this theoretical calculation scheme, the only input required is force constants: harmonic $\Phi_{1,2}$ and anharmonic $\Psi_{1,2,3}$. In that sense, it is crucial to have accurate FCs. In the DFPT approach, their Fourier transform is directly calculated in the reciprocal space with the help of the $2n + 1$ theorem. The $2n + 1$ theorem [28] states that the $2n + 1$th derivatives of the total energy can be obtained from knowledge of the nth derivatives of the wave functions. Broido worked with a method first developed by Deinzer in 2003 [40] which calculates Fourier coefficients V of the cubic force constants Ψ at arbitrary wavenumbers from a Fourier interpolation. The method was first tested on silicon and germanium and provided an outstanding agreement with experimental measurements in terms of lattice thermal conductivities. However, this framework is not able to compute fourth-order terms, which would require the second derivatives of the wavefunctions. The alternative to higher-order DFPT is to use a finite difference on low-order DFPT force constants. This was first implemented by Rousseau and Bergara in 2010 [41] who used the frozen-phonon approach to extract the cubic FCs from a finite difference of dynamical matrices. Their purpose was to see the effect of cubic anharmonicity on the strength of the electron–phonon coupling in aluminum hydride superconductors. The DFPT approach could be applied to various materials within a considerable scope of temperature, without introducing any extra parameters, if thermal expansion is negligible. However, since the FCs are calculated in reciprocal space, this approach becomes impractical for bulk materials with very large primitive cells as we will clarify later.

7.3.1.2 Extraction of force constants in real space

The frozen-phonon method. The first calculations of phonon spectra from DFT originated in the 1980s, and the method to extract harmonic force constants was the so-called frozen-phonon method. This is a finite difference method in which atoms in a supercell, which we label here with $R\tau$, are moved according to a frozen-phonon mode of wavenumber k commensurate with the superlattice, and forces on atoms are calculated. In essence, the displacement $\boldsymbol{U}_{R\tau}(k) = \epsilon_k \cos \mathrm{k}(\mathrm{R} + \tau)$ is imposed and the force $\boldsymbol{F}_{R'\tau'}(k)$ is calculated on all atoms in the supercell. The ratio between the two gives directly the dynamical matrix elements at that wavenumber k as we must have $\boldsymbol{F}_{R'\tau'}(k) = -\sum_{R\tau} \Phi_{R'\tau',R\tau}(k) \boldsymbol{U}_{R\tau}(k)$. Since there is only one term in this sum, the force constant Φ is obtained by taking the ratio of the force to displacement. This formula does not have a finite difference if the atoms are originally relaxed so that the force for $\boldsymbol{U}_\tau = 0$ is zero. This must, however, be done for many commensurate k vectors and a Fourier transformation will give the force constants in the real space from which the dynamical matrix at any arbitrary k vector may be computed.

Small displacement method. This method is implemented using the codes PHONON [32], PHON [42], ShengBTE [34, 43] and PHONOPY [35]. In 1997, Parlinski [32] presented a direct approach for phonon dispersion calculation, using an *ab initio* force constant method. The method uses the Hellmann–Feynman theorem to calculate forces on displaced atoms in a supercell and extracts all the harmonic force constants definable in that supercell. The displacements are not along a phonon mode polarization and can be reduced if the symmetry of the crystal is used. Effectively, this method assumes the cutoff range of interactions equals the supercell size. As a result, it reproduces the dynamical matrix at q points commensurate with the supercell. For instance, if the supercell is of size $2 \times 2 \times 2$, the dynamical matrix is exact at the zone center and zone boundaries which correspond to $G_1/2$, $G_2/2$ and $G_3/2$. Then, the dynamical matrix and phonon dispersion are obtained at an arbitrary q-point by using the Fourier interpolation method. In this method, the dynamical matrices are calculated on a coarse q-mesh ($2 \times 2 \times 2$ in this example, corresponding to the second neighbor range in real space). They are then Fourier transformed and their Fourier coefficients are in effect the force constants, which will then be used to compute the dynamical matrix at any arbitrary q-point. The force constants are determined by considering symmetry-inequivalent displacements of atoms in the supercell. This scheme was applied to calculate phonons in the cubic phase of ZrO_2 (zirconia), the Hellmann–Feynman force constants were acquired using the CASTEP program, adopting the local-density-functional approximation (LDA) on an fcc supercell of 96 atoms. The amplitude of atomic displacements had been limited to $u_0 = \pm 0.010 a_0$, with a_0 being the lattice parameter. In this method, the range of force constants is determined by the supercell size. Cubic terms are not considered in this approach either. In the case of ionic materials, the Born charges Z^* and the dielectric constant ϵ_∞ need to be calculated (from a separate DFPT calculation) and their contribution added as a non-analytical correction to the short-range part of the dynamical matrix. This correctly leads to the splitting of TO from LO modes [44–46]. In such calculations, imaginary phonon frequencies indicate a lattice instability implying the structure is not at an energy minimum and there are displacements that can further lower the energy of the structure.

In 2018, Parlinski's new paper [47] formulated a method to model anharmonicity without actually computing the cubic force constants. In this method, larger-amplitude symmetry-adapted atomic displacements in a supercell are used to compute various dispersions corresponding to various large-amplitude displacements. The standard deviation in the obtained phonon dispersions will give an indication of the inverse lifetimes.

Inclusion of anharmonic FCs within a cutoff. This method is also implemented in ALAMODE [48], TDEP [49] and the lesser known, but equally performant, `hiPhive` package developed in Erhart's group [50].

In 2008, Esfarjani and Stokes [33] were the first to introduce a method for extraction of higher-order force constants of, in principle, any rank and up to any neighbor shell from first-principles calculations in one or more supercells. This is in contrast to DFPT which can calculate up to cubic terms. If higher-rank terms were needed, major coding would be required to compute second derivatives of wave-functions and the resulting fourth-/fifth-order force constants. Another alternative within DFPT would be to use a finite difference method to obtain the quartic terms from finite difference of cubic ones.

In this method, the force on each atom in the supercell is expanded in powers of atomic displacements:

$$\begin{aligned} \boldsymbol{F}_{R\tau} = &- \Pi_\tau - \sum_{R_1\tau_1} \Phi_{R\tau, R_1\tau_1} \boldsymbol{U}_{R_1\tau_1} \\ &- \sum_{R_1\tau_1, R_2\tau_2} \Psi_{R\tau, R_1\tau_1, R_2\tau_2} \boldsymbol{U}_{R_1\tau_1} \boldsymbol{U}_{R_2\tau_2} + \cdots, \end{aligned} \tag{7.20}$$

where, as before, R refers to a primitive cell within the supercell and τ refers to an atom in the primitive cell, so that the pair $R\tau$ refers to an arbitrary atom within the supercell. The force constants Φ and Ψ are the unknown parameters of the model.

The methodology takes the following steps. (i) Symmetry operations of the crystal are calculated and, accordingly, the irreducible unknown force constants are identified. The remaining ones can be deduced from irreducible ones using symmetry operations $\Phi^{\alpha\beta}_{S(\tau),S(\tau')} = \sum_{\alpha'\beta'} S_{\alpha\alpha'} S_{\beta\beta'} \Phi^{\alpha'\beta'}_{\tau,\tau'}$ which simply say: to obtain the FC tensor of a 'rotated' bond, one simply needs to rotate the 3×3 tensor of the original bond. (ii) A cutoff distance is chosen for each rank of force constants, and thus the number of unknown force constants is fixed. (iii) For appropriately chosen atomic displacements, the force on all atoms $\boldsymbol{F}_{R\tau}$ in the supercell is calculated using any first-principles method. Several sets of such displacements are chosen in order to generate enough 'equations' from which the unknown FCs can be deduced. (iv) There are additional constraints of translational and rotational invariance which are imposed as additional linear equations the FCs Φ, Ψ, ... must exactly satisfy. These relations express the fact that if the crystal is translated or rotated by an *arbitrary* amount, the total energy should remain unchanged. Given that the force–displacement relation itself (equation (7.20)) is also linear in the FCs, one ends up with an over-determined set of *linear* equations on FCs. These equations are solved using a singular-value decomposition algorithm, which in essence means that the difference between DFT forces and anharmonic model forces of equation (7.20) above is minimized with respect to the unknown selected FCs. The difference between this method and what we called the 'small displacements method' is that the harmonic FCs are also chosen to be non-zero within a cutoff neighbor shell, whereas the previously mentioned methods include all harmonic force constants that exist within the supercell, and in fact they can only recover the combination $\tilde{\Phi}_{0\tau,R'\tau'} = \sum_L \Phi_{0\tau+L,R'\tau'}$ where L is a translation vector of the supercell. The latter method can, however, in principle, extract all force constants independently if force–displacement data on several supercells of different size and shape are provided. The major merits of this real-space method are: (i) the methodology can employ force–displacement data in one or many supercells of various sizes and shapes in order to handle longer-range FCs, and (ii) the method, as implemented, includes cubic and quartic anharmonic terms which play a crucial role in phonon lifetime calculation, crystal stability and second-order SCP theory. This, however, can be extended to higher-rank FCs as well.

The choice of atomic displacements is important if not crucial in a correct and reliable determination of the force constants. They can be obtained either by moving atoms of the primitive cell one by one along the Cartesian directions, from an MD simulation, from random displacements of the given magnitude or, if one wants to obtain effective force constants at a given temperature, one can sample the canonical ensemble at that temperature. The choice ultimately depends on the purpose of the simulations. For a pure harmonic FC extraction in order to obtain the phonon spectra, one can generate a minimal set of small displacements, typically 1% or 2% of the bond length. Atoms can either be moved one at a time in each 'snapshot' or even collectively. Symmetry considerations can reduce the number of needed displacements. For instance, in Si where both atoms are identical and the three Cartesian directions are equivalent, it is enough to move only one Si atom along the x direction, in order to obtain all the harmonic force constants. Collective displacements may also be made according to the irreducible representations of the space group of the crystal.

Convergence checks always need to be performed with respect to the chosen cutoff range and also the supercell size. In the reciprocal space (DFPT) approach, this is done by increasing the size of the q-mesh used for the DFPT calculations. This is reflective of the equivalent supercell size. For instance, a q-mesh of $4 \times 4 \times 4$ is equivalent to a supercell which is four times longer in each direction and, accordingly, the range of the FCs goes to the fourth neighbor cell.

The real-space method provided the accurate prediction of the force constants of a purely analytical Lennard-Jones potential [33], providing therefore a validation of the approach. This implies that first-principles forces need to be well-converged in order to achieve an accurate fit. However, there is no systematic approach for the choice of the supercell size and the cutoff range of the FCs. The recent work by Marianetti's group [51] has, however, considered this important question and has a comprehensive discussion of this topic, and we refer the reader to this work for more details on how to adopt a systematic approach.

This scheme is automatically well-suited for lattice distortions, because for small distortions one can use the higher-rank force constants to predict the lower-rank force constant of the distorted structure using the Taylor expansion: $\Phi_{\text{new}} = \Phi + \Psi \cdot \eta + \cdots$, where ϵ is the strain field and $\eta = \epsilon \cdot R$ represents the corresponding atomic distortion.

At the end, we should note that recently a real-space formalism for linear response calculations (DFPT) is presented and applied to harmonic phonon calculations by Carbogno and Scheffler [52].

7.3.1.3 Discussion of the differences between FC extraction techniques

Before moving on to more recent SCP calculation methods, we give a brief comparison of the real-space and reciprocal space methods. The major two differences are (i) computational load and (ii) symmetry and invariance conservation.

Some of the major contributors to QUANTUM ESPRESSO, Baroni *et al*, in their review article 'Phonons and related crystal properties from DFPT' [53], give a detailed comparison between the computational loads of DFPT and snapshot supercell DFT calculations. Suppose the range of FCs is $\mathcal{R}_{\text{FC}}$, thus the supercell should be of that size and the number of atoms $N_{\text{at}}^{\text{sc}}$ in such a supercell should be proportional to its cube: $N_{\text{at}}^{\text{sc}} \propto \mathcal{R}_{\text{FC}}^3$. In a single snapshot real-space supercell DFT calculation, the computational cost is proportional to the cube of the number of plane waves N_{pw}. This number is decided by the supercell length L and the minimum wavelength $\lambda_{\min}$: $N_{\text{pw}} = (L^3/\lambda_{\min}^3)$. Notice that the numerator L^3 as the volume of this supercell is proportional to the number of atoms in the supercell $N_{\text{at}}^{\text{sc}}$. Thus, the single snapshot computation time is on the order of $\mathcal{R}_{\text{FC}}^9$. A complete FC extraction using the real-space method requires at least $3N_{\text{at}}$ snapshots, where N_{at} is the number of atoms in the primitive cell, so the total computational load for real-space methods is on the order of $N_{\text{at}} \times \mathcal{R}_{\text{FC}}^9$. On the other hand, the reciprocal space method based on DFPT requires the evaluation of the dynamical matrix on a q-mesh, where the interval between neighboring q points is $\Delta q \approx 2\pi/\mathcal{R}_{\text{FC}}$, so that the total number of $\boldsymbol{q}$ points N_q is roughly $N_q \propto \mathcal{R}_{\text{FC}}^3$. As for the calculation of the dynamical matrix $\mathcal{D}(\vec{q})$ for each $\vec{q}$, the cost is on the order of N_{at}^4. Such a calculation for $\mathcal{D}(\vec{q})$ is performed over the irreducible first Brillouin zone, so a DFPT calculation will require a CPU time on the order of $N_{\text{at}}^4 \times \mathcal{R}_{\text{FC}}^3$. The ratio of these two CPU times is therefore:

$$\frac{\text{CPUTime}_{\text{DFPT}}}{\text{CPUTime}_{\text{SC}}} = O\left(\frac{N_{\text{at}}^4 \times \mathcal{R}_{\text{FC}}^3}{N_{\text{at}} \times \mathcal{R}_{\text{FC}}^9}\right) = O\left(\frac{N_{\text{at}}^3}{N_q^2}\right). \tag{7.21}$$

From equation (7.21), for systems with small unit-cells that require a relative large q-mesh size, a DFPT implementation will work better. On the other hand, for systems with a large primitive cell, the real-space supercell approach will be more efficient. The size of the supercell or the range of FCs is for most materials on the same order of a nanometer or less, whether they have a large primitive cell or a small one does not matter. Another point to keep in mind is that in the DFPT method, which provides an exact equation for the FCs, translational and rotational invariances should be, in principle, satisfied. In practice, however, due to numerical inaccuracies due to convergence and finite basis sets, these invariance relations can be violated. We have not seen any systematic study of this issue. In routine DFPT calculations, however, the ASR (or the acoustic sum rule coming from translational invariance) is enforced by other means, implying that such violations exist and are not necessarily small.

7.3.2 Sampling of the configuration space for effective theories at finite temperature

The above approaches would in principle provide force constants at zero temperature or the 'bare' force constants. DFPT is a linear response theory, so it assumes atomic displacements are infinitesimally small, although atoms are not even moved in this calculation as this is a linear response. In real-space methods, displacements are finite but very small so that the contribution of higher-rank terms becomes almost negligible compared to the harmonic ones and the Taylor series are convergent. In practical applications, however, one needs the value of these parameters at finite and even high temperatures. This can be achieved using several sampling methods. The different existing codes and approaches they use differ in their sampling approach. At first sight, one may think that this problem is not so well-defined because when temperature and therefore atomic displacements become large, anharmonic terms start to contribute more, and it is not clear that the Taylor series as we wrote in equation (7.5) would even remain convergent. One would really need a different expansion if displacements and the corresponding anharmonic terms are large. It turns out that the correct 'Taylor expansion' at high temperatures corresponds to the SCP theory. Assuming there is no atomic diffusion and atoms vibrate around their equilibrium site at all times, we need to use a mean-field theory to describe the vibration of atoms. Terms in the Taylor expansion need to be rearranged so that the series converge. If the harmonic term is replaced by an effective one, one then requires that the remaining part of the potential energy be zero on average: $\langle V_{\text{total}}(U)\rangle = 1/2\Phi_{\text{eff}}\langle UU\rangle$. Here, the average is with respect to the equilibrium state defined by the *effective* normal modes; the corresponding density matrix is $\rho_{\text{eff}} = \mathrm{e}^{-\beta\Phi_{\text{eff}}UU/2}/Z_{\text{eff}}$. We can readily see that the effective harmonic FC is a sum of only even-order terms and averages of even powers of displacements. This is illustrated in figure 7.2. In this case, the remaining higher-order anharmonic terms have a smaller contribution since their mean is zero by construction. To illustrate this better, let us consider symbolically a Taylor expansion up to only a fourth-order term and regroup some terms:

$$V = V_0 + \frac{1}{2}\sum \Phi U^2 + \frac{1}{4!}\sum \chi U^4. \tag{7.22}$$

Requiring this to have an average equal to $1/2\Phi_{\text{eff}}\langle U^2\rangle$ implies

$$\frac{1}{2}\sum \Phi_{\text{eff}}\langle U^2\rangle = \frac{1}{2}\sum \Phi\langle U^2\rangle + \frac{1}{4!}\sum \chi\langle U^4\rangle$$
$$= \frac{1}{2}\sum \Phi\langle U^2\rangle + \frac{1}{8}\sum \chi\langle U^2\rangle\langle U^2\rangle, \tag{7.23}$$

where we used properties of variable U having a Gaussian distribution. This equation uniquely defines the effective harmonic force constants, which are obtained by taking the derivative of the above equation with respect to $\langle U^2\rangle$:

$$\Phi_{\text{eff}} = \Phi + \frac{1}{2}\sum \chi\langle U^2\rangle. \tag{7.24}$$

We now see that even for large values of displacements U, the contribution of the second term involving only χ in the regrouped version is smaller because the potential energy now has the form

$$V = V_0 + \frac{1}{2}\sum \Phi_{\text{eff}} U^2 + \frac{1}{4!}\sum \chi U^2(U^2 - 6\langle U^2\rangle), \tag{7.25}$$

where the second term has zero average by construction. So, although the first version may be divergent, the second formulation seems to have better convergence properties. This is the essence of the SCP theory and the reason behind its success. So the FCs at finite temperatures acquire a new meaning which is different from their zero temperature version as derivatives of the potential energy evaluated at the equilibrium position. We are in essence changing the theoretical model and need a different way to obtain its parameters.

At a fixed temperature, the system is sampling the canonical ensemble, and atoms will have displacements of amplitude commensurate with that temperature. The methods we outline below are different ways of performing this sampling.

7.3.2.1 SCAILD

The first implementation of SCP using first-principles forces was reported by Souvatzis *et al* [54, 55] and was labeled as SCAILD, which stands for self-consistent *ab initio* lattice dynamics. This method was introduced to explain the vibrational entropy stabilization of group IV materials such as Ti and Zr at high temperatures in their bcc phase. In this method, atoms are displaced in a supercell and forces on them are computed in order to obtain an initial phonon dispersion, albeit with imaginary modes. From the eigenmodes at wavevectors commensurate with the supercell, a set of displacements are derived. For these displacements, the forces on all atoms are then computed and Fourier transformed. New phonon frequencies are then derived from the relation between the DFT forces and polarization vectors:

$$\begin{aligned} \omega_{k\lambda} &= \sqrt{-\boldsymbol{F}_k^\tau \cdot \boldsymbol{\epsilon}_{k\lambda}^\tau / m A_{k\lambda}^\tau} \\ \text{since } \boldsymbol{F}_k^\tau &= -\sum_\lambda m\omega_{k\lambda}^2 \boldsymbol{\epsilon}_{k\lambda}^\tau A_{k\lambda}^\tau \\ \text{with } A_{k\lambda}^\tau &= \pm\sqrt{\hbar(2n_{k\lambda}(T) + 1)/2m_\tau \omega_{k\lambda}}. \end{aligned} \tag{7.26}$$

As can be seen, $A^{\tau}_{k\lambda}$ is the displacement amplitude of mode $k\lambda$ on atom τ at temperature T, which is included through the Bose–Einstein distribution function $n_{k\lambda}(T)$. Every time a new set of frequencies is obtained, they are properly symmetrized according to the symmetry of the k-points:

$$\Omega^2_{k\lambda} = \frac{1}{p_k} \sum_{S \in S(k)} \omega^2_{S^{-1}k\lambda}$$
$$\omega^2_{k\lambda} = \frac{1}{N} \sum_{i=1}^{N} \Omega^2_{k\lambda}(i), \tag{7.27}$$

where $S(k)$ is the symmetry group of the wave vector k, p_k is the number of elements of this group and $\Omega_{k\lambda}(i)$ are the symmetry restored frequencies at iteration i. The average of this distribution due to several iterations is taken as the definition of the phonon frequency according to equation (7.27). In this sum, one may want to only include the last few iterations. Once a new set of frequencies are obtained in this way, atoms in the supercell are moved again with the updated amplitudes $A_{k\lambda}$ according to

$$\boldsymbol{U}_{R\tau} = \frac{1}{\sqrt{N}} \sum_{k\lambda} A^{\tau}_{k\lambda}\, \boldsymbol{\epsilon}^{\tau}_{k\lambda}\, \mathrm{e}^{\mathrm{i}k.R}. \tag{7.28}$$

This process is repeated until the frequencies converge. Also notable is the fact that the eigenvectors are not updated during the iterations. This method was later applied by Zhang *et al* to treat the stability of the bcc and fcc phases of tungsten at high temperatures [56].

7.3.2.2 Temperature-dependent effective potential method

The temperature-dependent effective potential method (TDEP) was proposed by Hellman *et al* in 2011 [49]. This method used the atomic configurations from a molecular dynamics run and fitted the DFT forces to an effective harmonic model. They used the symmetry properties of the crystal to reduce the number of FCs and performed a least squares fit to extract the inequivalent FCs, in a similar spirit to the work of Esfarjani and Stokes [33]. Rotational invariance constraints did not seem to have been imposed. This might be fine for a purely harmonic model but has consequences if anharmonic FCs also need to be included, as rotational invariances relate the two sets. The main difference with the latter method is that the effective FCs are temperature dependent as the MD snapshots were from a constant temperature run and, as such, the TDEP method, similar to SCAILD, gives the best harmonic FCs as effective couplings at a given temperature and can provide effective phonon dispersions at that temperature. The method was validated on bcc phases of Li and Zr crystals. Two years later, in 2013, Hellman *et al* extended their method to include free-energy calculations [37] by including anharmonic corrections to the free energy through thermodynamic integration. A later work that year included temperature-dependent cubic anharmonic FC extraction [57] which was applied to Si and FeSi. In this work, the effective harmonic and cubic force constants were extracted simultaneously from the minimization of the difference between the DFT forces and the anharmonic model forces, similar to the original work. Due to the high computational cost on initialization of canonical ensemble at temperature T, and the length of an AIMD simulation, in a later work in 2018, Hellman *et al* employed a formulation of atomic displacements sampling the canonical ensemble, proposed originally by Estreicher [58] and called it 'stochastically initialized temperature-dependent effective potential' (s-TDEP) [59]. The following equation was used to generate random displacements modeling the canonical ensemble at temperature T. This requires knowledge of the phonon frequencies $\omega_{k\lambda}$ and eigenmodes $\boldsymbol{\epsilon}_{k\lambda}^{\tau}$, which are initially deduced from minimal displacements in the same supercell:

$$\boldsymbol{U}_\tau = \sum_{k\lambda} \sqrt{\frac{\hbar(2n_{k\lambda}(T) + 1)}{2Nm_\tau\, \omega_{k\lambda}}} \sqrt{-2\ln \zeta_1}\, \sin(2\pi\zeta_2)\, \boldsymbol{\epsilon}_{k\lambda}^{\tau}, \tag{7.29}$$

where N is the number of k-points, τ refers to an atom *in the supercell,* unlike in previous notations, and ζ_1, ζ_2 are random numbers chosen uniformly in [0, 1] so that the product $\sqrt{-2\ln \zeta_1}\, \sin(2\pi\zeta_2)$ has a normal distribution of mean 0 and standard deviation 1. This feature saves major CPU time as the AIMD simulations to reach thermal equilibrium and several uncorrelated snapshots were the most time-consuming part of the work. Another disadvantage of the MD method is that it treats ions classically, and therefore the atomic snapshots generated from MD are a valid description of the canonical ensemble only at temperatures above the Debye temperature, whereas equation (7.29) includes the effect of zero-point vibrations at low temperatures. Assuming the snapshots used to extract FCs are correctly generated, this scheme does not require self-consistency as in the SCP theory, and includes implicitly through the DFT force all other interactions with phonons, namely that of electrons (assuming the adiabatic approximation).

7.3.2.3 SCHA-4

The SCP theory to lowest order will only contain the quartic term in the potential energy, similar to the example we used in equations (7.22). For this reason, we will refer to it as SCHA-4. In terms of Feynman diagrams, it only includes the first two diagrams in figure 7.2. In this theory, the effective harmonic force constants, as we showed in equation (7.24), can be shown to be $\Phi_{\text{eff}} = \Phi + \chi\langle UU\rangle/2$, where the average is with respect to the self-consistent ground state defined by Φ_{eff}. This approach combined with first-principles was first applied by Vanderbilt *et al* [30, 60] to the case of carbon, silicon and germanium, where FCs were obtained using the frozen-phonon approximation. More recently, Rousseau *et al* applied this theory in 2010 to study the effect of cubic anharmonicity on the strength of the electron–phonon coupling in aluminum hydride superconductors [41]. The most recent implementation of this theory was performed by Ravichandran and Broido in 2018 who referred to the method as the 'phonon renormalization approach' [61]. The motivation for his work was to include temperature effects beyond the quasi-harmonic approximation in the description of phonons and be able to describe thermal expansion and include this effect in the calculation of other properties such as lattice thermal conductivity. Although previous approaches which used the bare harmonic and anharmonic force constants were successful for many materials, they did not produce good results for simple anharmonic lattices which have large thermal expansion, such as NaCl, even at moderate temperatures.

It can be shown that the relationship between the effective force constants Θ and the zero temperature ones denoted by Φ is given by the equations defining the SCHA-4 approximation displayed in figure 7.2 and defined in equation (7.24). To be more precise, after substitution of the mean square displacements by their expression in terms of *renormalized phonon eigenvectors* ϵ, the relationship, with all the indices explicitly included, becomes

$$\Theta_{1,2}^{\alpha\beta} = \Phi_{1,2}^{\alpha\beta} + \frac{\hbar}{4N_q}\sum_{\vec{q}\lambda}\sum_{3,4} X_{1,2,3,4}^{\alpha\beta\gamma\delta}\frac{\epsilon_{\vec{q}\lambda}^{\tau_3\gamma}\epsilon_{-\vec{q}\lambda}^{\tau_4\delta}}{\Omega_{\vec{q}\lambda}\sqrt{m_{\tau_3}m_{\tau_4}}}e^{i\vec{q}\cdot(\vec{R}_3-\vec{R}_4)}(2n_{\vec{q}\lambda}+1), \tag{7.30}$$

where the numerical indices 1, 2, 3, 4 denote atom sites within the supercell, $\vec{R}_i$ are the primitive lattice translation vectors, $\vec{\tau}_i$ corresponds to atoms within the primitive cell, N_q is the number of q-points in the Brillouin zone (equivalent to number of unit cells in the supercell), $\Omega_{\vec{q}\,\lambda}$ is the 'renormalized' phonon frequency for mode λ at $\vec{q}$, $\epsilon_{\vec{q}\lambda}$ is the corresponding eigenvector and $n_{\vec{q}\lambda}$ is the Bose–Einstein occupation number. Since the renormalized phonon frequencies and eigenvectors depend in turn on renormalized FCs Θ, this equation needs to be solved self-consistently.

In their approach, similar to the s-TDEP, atomic configurations taken from the canonical ensemble are generated in a supercell according to equation (7.29). Then, forces are calculated using DFT and fitted to a quartic Hamiltonian to produce Φ and X. Starting with the harmonic eigenvalues and eigenvectors used in equation (7.30), a first set of renormalized harmonic FCs Θ are generated. In the next iteration, Φ and X are kept fixed and and Θ is updated with the new set of eigenmodes. This procedure is continued until convergent. New snapshots according to equation (7.29) but with the renormalized eigenmodes and new fitting to DFT forces did not change the results significantly. Once the renormalized harmonic FCs are fixed, they proceed to generate a few more snapshots using the eigenmodes of the latter in equation (7.29), and fit the DFT forces to a nonlinear force model of the form

$$F_\tau = -\sum_{\tau'}\Theta_{\tau\tau'}U_{\tau'} - \sum_{\tau'1}\Psi'_{\tau\tau'1}U_{\tau'}U_1 - \sum_{\tau'12}\chi'_{\tau\tau'12}U_{\tau'}U_1U_2 \tag{7.31}$$

with fixed Θ. This defines the new renormalized cubic and quartic FCs Ψ' and χ', different from their zero-temperature counterparts. This Taylor expansion of the force is now a well-behaved form as the deviations from the harmonic part of the forces are, by construction, the smallest possible (zero on the average). This was not necessarily the case for the zero-temperature force constants.

This renormalization framework was applied to NaCl at temperatures of 80 and 300 K, where traditional QHA shows an obvious softening effect and large lifetimes. The renormalized phonon dispersion calculated in this way matched well with inelastic neutron scattering results. It also reproduced features such as relative temperature independence of the optical phonons. Furthermore, compared to QHA, thermal expansion, phonon–phonon scattering and lattice thermal conductivity studied using renormalized phonons produced better agreement with experimental datasets.

SCP theory once again shows its robustness for describing finite temperature phonons and corresponding thermal properties. Similar tests were also performed using diamond to further solidify the soundness of this renormalization approach.

7.3.2.4 Stochastic SCHA

In 2013, Mauri's group came up with a stochastic implementation of the SSCHA [62]. This method has several noticeable features compared to previous ones. First, the SSCHA equations are derived from a free-energy minimization using the Bogoliubov inequality. The Hamiltonian is written as the sum of a reference effective harmonic Hamiltonian and the anharmonic corrections to it. The variational parameters include two groups of parameters: the equilibrium position of ions $\mathbf{R}_{eq}$ and the trial harmonic FCs $\Phi^{\alpha\beta}_{R\tau}$. The minimization is performed using the conjugate-gradients method. The potential energy and forces are not expanded in Taylor series but kept from DFT calculations. This has the advantage that there is no truncation and large anharmonicities can be handled exactly, but at the cost of more DFT calculations in the supercell. To alleviate the number of DFT calculations, thermal averages are computed using stochastic methods, similar to the Monte Carlo importance sampling of the canonical ensemble, as we will describe below. To remind the reader, if the variational parameters are the equilibrium lattice positions S_τ and the trial force constants Φ, and the free energy to be minimized is (see also equation (7.6)) $F_0[\Phi] + \langle V[S] - V_{\text{trial}}[\Phi]\rangle$. First, a set of equilibrium atomic positions and a trial harmonic force constant (S^0, Φ^0) are guessed, and its normal modes calculated. Then, many snapshots are generated from the canonical ensemble at temperature T for instance from MD, or according to equation (7.29) given the normal modes of the trial FCs. These snapshots are used to compute the thermal averages of different terms needed, namely the first derivatives of the free energy which need to be set to zero in order to find the variational parameters (S, Φ). These two minimization equations to be solved are

$$0 = \frac{\partial \langle V[S]\rangle_\Phi}{\partial S} \tag{7.32}$$

$$0 = \frac{\partial F_0}{\partial \Phi} + \frac{\partial \langle V[S]\rangle_\Phi}{\partial \Phi} - \frac{1}{2}\langle yy\rangle_\Phi, \tag{7.33}$$

where atomic displacements were written as $U(t) = S + y(t)$. In the second equation the derivative of the potential energy with respect to the trial FC is not zero as the averaging is performed with respect to the trial density matrix $\rho_0 = \mathrm{e}^{-\beta H_0[\Phi]}/Z_0$ involving Φ itself: $\langle V[S]\rangle = \mathrm{Tr}\,\rho_0[\Phi]\mathrm{V[S]}$. The trace is an integral over the dynamical variables y and their conjugate momenta. Such integrals over all possible y weighted by ρ_0 are replaced by a sum over a few atomic snapshots where the y take different values. After the averages are calculated and the equations solved, one finds a new pair of variational parameters (S^1, Φ^1) as a solution, for which the above process is repeated. This is because with the new density matrix defined by the latter, averages change and the gradients are not zero anymore. In the second and later iterations, however, new snapshots will not be generated. The same snapshots will be used, but there will be a reweighting of the probabilities according to $\rho_0[\Phi^1]/\rho_0[\Phi^0] = e^{-\beta(V_0[\Phi^1]-V_0[\Phi^0])}Z_0[\Phi^0]/Z_0[\Phi^1]$. The same reweighting will hold after iteration l, where Φ^1 only needs to be replaced by Φ^l, the FC after iteration number l. In these equations, subscript 0 refers to the trial harmonic quantities.

To check for the consistency of the reweighting procedure, the ratio $\frac{1}{N_s}\sum_{j=1}^{N_s}\rho_0[\Phi^l](y_j)/\rho_0[\Phi^0](y_j)$ is checked to stay near 1. If it is very different from 1, then at that iteration, new DFT snapshots need to be generated according to the latest density matrix and trial FC, since this indicates that the original sampling was not appropriate. The question of how many snapshots are needed in order to obtain a decent sampling of the canonical ensemble is a subtle one and has to be found by checking the convergence of the results.

This method was later proven to be robust through various test examples such as PdH and PtH. Later the SSCHA was combined with $2n + 1$ theorem [63]. The scheme was capable of giving a leading-order approximation of phonon line broadening as well as possessing the ability to accurately describe phonon spectra for the accurate description of phonon spectra.

The advantage of this approach is that, similarly to SCAILD and TDEP, the thermal average of the 'exact' potential energy (or force) is evaluated and does not require a Taylor expansion. The evaluation of the thermal average is, however, stochastic, as instead of an integral one is using a summation over a finite set of randomly generated snapshots.

7.4 A recent extension to SCHA-4

In this section, we will present, using the variational approach, an extension which will enable us to predict phase changes in anharmonic systems not only as a function of pressure but also the more challenging parameter of temperature. This approach is based on the SCP theory, and therefore will use a parametrization of the potential energy as a truncated Taylor expansion. The disadvantage compared to the stochastic method is that the evaluation of the potential energy will be approximate. But the advantage is that the phase transition can be predicted with a minimal anharmonic model, and calculations are lighter. The extension is in the inclusion of strain and internal atomic relaxations as the temperature and unit cell shape and volume are changing. In this case, traditionally dropped cubic terms become important and are explicitly present in the formulation. Coupling to other OPs such as orbital ordering or magnetization can also be incorporated in the model.

7.4.1 Formulation

We will assume an anharmonic Hamiltonian in which the potential energy is a polynomial in atomic displacements about a reference equilibrium position. The polynomial can be of degree 4 or more if needed. The thermal average of any power of displacements can readily be calculated. The variational approach consists in adopting a trial but solvable Hamiltonian, which we take to be harmonic but allow displaced equilibrium positions and deformed unit cell, the optimal values of which at any given temperature can be found by minimizing the variational free energy at that temperature.

The absolute displacement from the equilibrium site is denoted by $\vec{U}_{R\tau}(t)$. Since we want to describe phase transitions, the shape of the unit cell may change with T and as a result the internal atoms denoted by τ may find a new equilibrium position in the strained cell. For this reason, we will write the general displacement of atoms as a sum of static strain represented by the strain tensor $\bar{\bar{\eta}}$, an internal relaxation term $\boldsymbol{u}_\tau$, and a dynamical displacement $\vec{y}_{R\tau}(t)$ as follows:

$$\vec{U}_{R\tau}(t) = \bar{\bar{\eta}} \cdot (\vec{R} + \vec{\tau}) + \vec{u}_\tau + \vec{y}_{R\tau}(t). \tag{7.34}$$

Here, the dynamical variable $\vec{y}(t)$ has a Gaussian distribution. R corresponds to the reference lattice vector, R and τ uniquely define an atom in the system. With this in mind, the typical Taylor expansion of the potential energy (equation (7.35)) for the crystal structure can be written as

$$\begin{aligned} V(\vec{U}) = V_0 &+ \sum_{1} \Pi_1^\alpha U_1^\alpha + \frac{1}{2!}\sum_{1,2} \Phi_{1,2}^{\alpha\beta}\, U_1^\alpha U_2^\beta \\ &+ \frac{1}{3!}\sum_{1,2,3} \Psi_{1,2,3}^{\alpha\beta\gamma}\, U_1^\alpha U_2^\beta U_3^\gamma + \frac{1}{4!}\sum_{1,2,3,4} \chi_{1,2,3,4}^{\alpha\beta\gamma\delta}\, U_1^\alpha U_2^\beta U_3^\gamma U_4^\delta + \dots, \end{aligned} \tag{7.35}$$

where Φ, Ψ, X are conventional Greek letter notations for force constants of rank 2, 3 and 4, respectively. The integers 1, 2, 3, 4, which we use for brevity, refer to atoms $R_1\tau_1$, $R_2\tau_2$, … in the system. The first-order FCs Π should vanish if the expansion is done around equilibrium positions. In order to implement Bogoliubov (equation (7.6)) inequality,

$$V_{\text{trial}}(\vec{y};\, [K, \bar{\bar{\eta}}, u]) = V_0 + \frac{1}{2!}\sum_{1,2} K_{1,2}^{\alpha\beta} y_1^\alpha y_2^\beta. \tag{7.36}$$

a harmonic trial potential model is defined with $(K, \bar{\bar{\eta}}, u)$ as variational parameters. Here K is referred to as a 'trial' force constant and dictates the distribution of $\vec{y}$. The variational parameters at a given temperature are found by minimizing the trial free with respect to them:

$$\frac{\partial F_{\text{trial}}}{\partial K_{R\tau,R'\tau'}^{\alpha\beta}} = 0 \tag{7.37}$$

$$\frac{\partial F_{\text{trial}}}{\partial \bar{\bar{\eta}}} = 0 \tag{7.38}$$

$$\frac{\partial F_{\text{trial}}}{\partial u_\tau} = 0, \tag{7.39}$$

where the trial free energy is defined as $F_{\mathrm{trial}} = F_0 + \langle V - V_{\mathrm{trial}} \rangle$ and thermal averages were defined in equation (7.9). Note that in the standard SCP approach, all odd-order terms vanished because displacements had a Gaussian distribution and only their even powers yielded non-zero thermal averages. In our formulation with strain, however, all powers of U have a non-zero thermal average, and therefore cubic terms also explicitly appear in the results. Furthermore, the derivatives with respect to the static variables $\bar{\bar{\eta}}$ and u_τ only appear in the potential energy V but not the trial potential nor the harmonic free energy F_0.

As is shown in previous variational treatments [5, 26, 62, 64], the derivatives with respect to K are tedious and require the use of the chain rule. However, the trial free energy has a more explicit dependence on $Y = \langle yy \rangle$, which has the same number of components as K. If we consider

$$\frac{\partial \langle F_{\mathrm{trial}} \rangle}{\partial K} = \sum \frac{\partial \langle F_{\mathrm{trial}} \rangle}{\partial \langle yy \rangle} \frac{\partial \langle yy \rangle}{\partial K} \tag{7.40}$$

we realize that $\frac{\partial \langle F_{\mathrm{trial}} \rangle}{\partial K} = 0$ is equivalent to $\frac{\partial \langle F_{\mathrm{trial}} \rangle}{\partial \langle yy \rangle} = 0$ as the Jacobian of the transformation $(\frac{\partial \langle yy \rangle}{\partial K})$ is not singular because physically, any change in K results in a change in $Y = \langle yy \rangle$ and vice versa. So instead of K we will minimize the trial free energy with respect to Y. In this case, the Harmonic free energy F_0 does not depend on Y, so that the minimization equation (7.37) is replaced by the following equation:

$$\frac{\partial \langle V \rangle}{\partial Y} - \frac{\partial \langle V_{\mathrm{trial}} \rangle}{\partial Y} = \frac{\partial \langle V \rangle}{\partial Y} - \frac{1}{2} K = 0, \tag{7.41}$$

which gives an explicit equation for the effective harmonic force constant in terms of derivatives of the potential energy which is a polynomial in y and can be readily calculated. Boccara and Sarma have shown that in general for any pair potentials the derivative $\frac{\partial \langle V \rangle}{\partial Y}$ is equal to $\frac{1}{2} \langle \frac{\partial^2 V}{\partial y \partial y} \rangle$, providing the interpretation that the optimal trial or effective harmonic FC is the thermal average of the second derivative of the potential energy with respect to atomic displacements. Their other simplifying contribution was to consider Y as additional variational parameters which produced precisely equation (7.41) avoiding all the chain rules. This is in spirit similar to a Legendre transform changing the variables K to $Y = \langle yy \rangle$.

7.4.2 Minimization equations with strain included

If we denote the static part of displacement U by S, so that $U(t) = S + y(t)$, we can easily derive the expression for the effective FC K. To make the equations simpler, we drop the indices and first write the thermal average of the potential energy $\langle V - V_0 \rangle$ in terms of S and $Y = \langle yy \rangle$ symbolically as

$$\frac{1}{2}\Phi(Y + S^2) + \frac{1}{6}\psi(3SY + S^3) + \frac{1}{24}\chi(S^4 + 6S^2 Y + 3YY) \tag{7.42}$$

so that the minimization equation leading to K becomes

$$K = \Phi + \psi S + \frac{1}{2}\chi(S^2 + Y). \tag{7.43}$$

Putting back the indices, and keeping in mind that $S_1^\alpha = \sum_\beta \bar{\bar{\eta}}_{\alpha\beta}(R_1 + \tau_1)^\beta + u_{\tau_1}^\alpha$, we obtain the expression of the effective (or renormalized) harmonic FCs as well as the second minimization equation leading to the thermal average of the potential energy derivative, i.e. force, to be zero:

$$K_{1,2}^{\alpha\beta} = \Phi_{1,2}^{\alpha\beta} + \sum_{3,\gamma} \psi_{1,2,3}^{\alpha\beta\gamma} S_3^\gamma + \frac{1}{2}\sum_{34,\gamma\delta} \chi_{1,2,3,4}^{\alpha\beta\gamma\delta}\left(S_3^\gamma S_4^\delta + \left\langle y_3^\gamma y_4^\delta \right\rangle\right) \tag{7.44}$$

$$\begin{aligned} 0 = \frac{\partial\langle V\rangle}{\partial S_1^\alpha} &= \Pi_1^\alpha + \sum_2 \Phi_{1,2}^{\alpha\beta} S_2^\beta + \frac{1}{2!}\sum_{2,3} \Psi_{1,2,3}^{\alpha\beta\gamma}\left(S_2^\beta S_3^\gamma + \left\langle y_2^\beta y_3^\gamma \right\rangle\right) \\ &+ \frac{1}{3!}\sum_{2,3,4} \chi_{1,2,3,4}^{\alpha\beta\gamma\delta} S_2^\beta\left(S_3^\gamma S_4^\delta + 3\left\langle y_3^\gamma y_4^\delta \right\rangle\right) + \cdots. \end{aligned} \tag{7.45}$$

Note that once the derivatives with respect to S are known, those with respect to u are identical, and for $\partial/\partial\bar{\bar{\eta}}$ one can use the following chain rule equation:

$$\frac{\partial}{\partial\bar{\bar{\eta}}_{\alpha\beta}} = \frac{1}{2}\left(\frac{\partial}{\partial S^\alpha}(R + \tau)^\beta + \frac{\partial}{\partial S^\beta}(R + \tau)^\alpha\right). \tag{7.46}$$

There are some advantages to this approach which we note below. Since $\langle y_i y_j \rangle$ has direct correspondence with IFCs, the same point-group symmetry could be applied to significantly reduce the total number of terms. The updated trial FCs $\boldsymbol{K}$ are directly given by equation (7.44) similar to SCHA-4 and Ravichandran's [61] 'renormalized' phonons, but also include cubic terms coupled with other OPs such as strain and internal relaxations. The real-space calculations are short-ranged and fast, and only the calculation of thermal averages of $\langle yy \rangle$ are performed independently in reciprocal-space using the normal modes. We use Broyden's method to solve the gradient equations in an iterative scheme since the updated K will have new normal modes which define the $\langle yy \rangle$ on the right-hand sides of equations (7.44) and (7.45).

The only input required in addition to harmonic and anharmonic FCs at finite temperature are the structure and symmetry information. A brief flowchart for this computational protocol is presented in figure 7.3.

7.4.3 Application to a simple model

We applied our protocol first on a fabricated double-well potential in a paper studied by Morris and Gooding [65]. Consider a one-dimensional atomic chain with the following potential energy V:

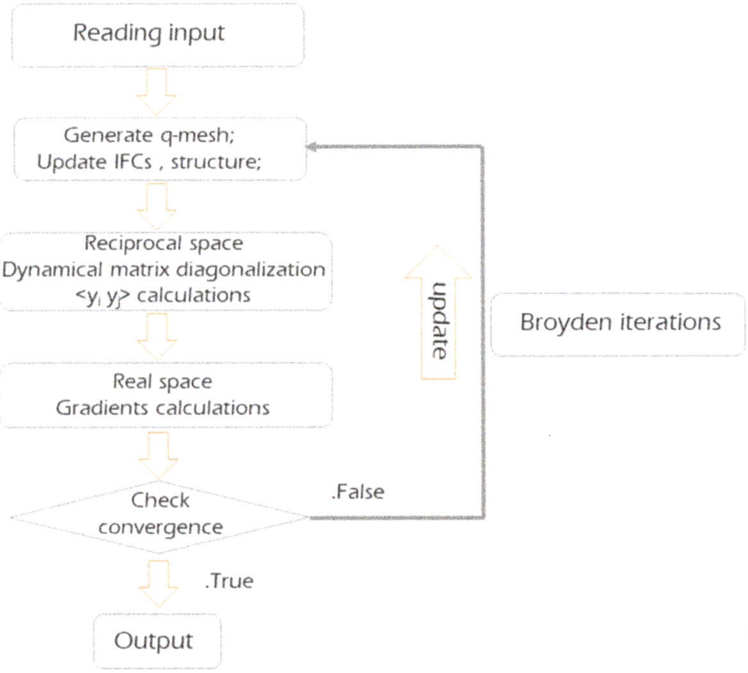

Figure 7.3. Computational protocol. The roots of the gradient equations are in effect found by Broyden's method, which updates the Hamiltonian parameters S, Φ in order to make the gradients zero.

$$
\begin{aligned}
V &= V_{\text{site}} + V_{\text{pair}} \\
V_{\text{site}} &= \frac{1}{2}AU_i^2 - \frac{1}{4}BU_i^4 + \frac{1}{6}CU_i^6 \\
V_{\text{pair}} &= \frac{k}{2}(U_i - U_{i+1})^2 + \frac{k'}{4}\left(U_i^2 + U_{i+1}^2\right)(U_i - U_{i+1})^2 \\
U_i(t) &= y_i(t) + u_0,
\end{aligned}
\tag{7.47}
$$

where V_{site} is the on-site potential with A, B and C all being positive numbers chosen to have a metastable minimum at $u_0 = 0$ and a doubly degenerate stable minimum at $u_0 = \pm 1$. For pair potential V_{pair}, the only non-zero intersite couplings are nearest-neighbor sites for the sake of simplicity. The force constant factors k, k' are set to make V_{pair} be much larger than the on-site potential's localization energy, to simulate the displacive limit where only weakly anharmonic oscillations happen. Notice that U_i is the absolute displacement in our formalism that contains two parts: (i) u_0 stands for internal atomic shift $\vec{u}_\tau$ (since it is mono-atomic, it is the same parameter for every i), and $y_i(t)$ is the harmonic vibration displacement. Now we can construct the trial potential energy as V_{trial} which is in the standard form:

$$
V_{\text{trial}} = \frac{1}{2}\sum_{i,j} K_{ij}\, y_i y_j \tag{7.48}
$$

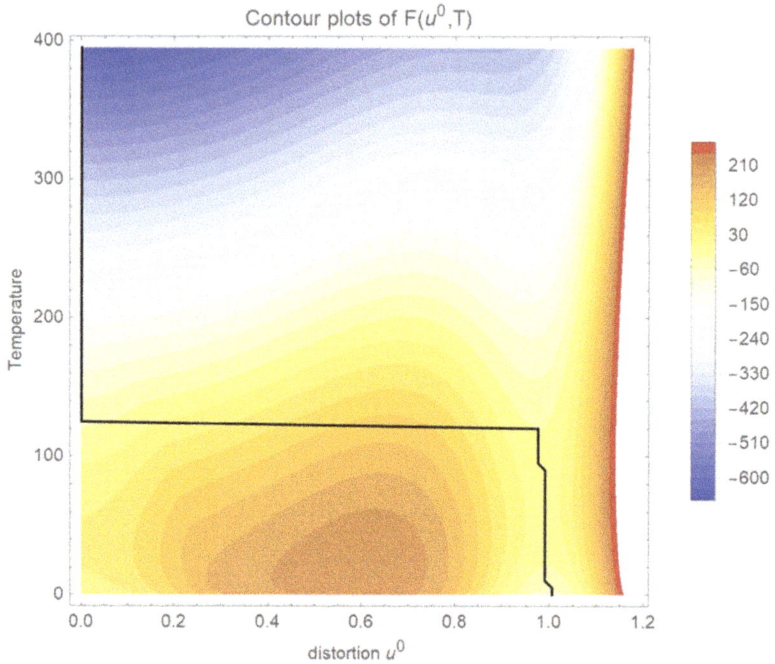

Figure 7.4. Contour plot of free energy F with respect to atomic shift u^0 and temperature T. One can notice the phase transition from non-zero ferroic order parameter u^0 at low T, to $u^0 = 0$ at $T > 140$.

since the intersite coupling is limited to nearest neighbors and in this case it is a one-dimensional mono-atomic model. There are only two independent trial force constants K_{00}, K_{01} and so are the correlation terms $\langle y_i y_i \rangle$, $\langle y_i y_{i+1} \rangle$:

$$\langle V_{\text{trial}} \rangle/\text{atom} = \frac{1}{2} K_{00} \langle y_i y_i \rangle + \frac{1}{2} K_{01} \langle y_i y_{i+1} \rangle. \tag{7.49}$$

After taking the thermal average of V, we have the complete expression for the effective free energy F_{eff}. The goal here is to optimize it with respect to three variational parameters, K_{00}, K_{01} and u_0, at every temperature. Once the convergence threshold in self-consistent iterations have been met, we use those parameters to obtain an approximation of real free energy values at different temperatures and by comparison, we plot the phase transition diagram in a contour plot, as shown in figure 7.4). The analytically calculated phase transition temperature that is around 140 K in the papers by Morris [66, 67] has been successfully reproduced using this variational approach.

7.5 Conclusions

With the advent of accurate force and total energy calculations using DFT methods, it is now possible to implement the SCP theory to compute lattice-dynamical properties of real materials at high temperatures. Several methods were described which showed the great success of this approach in predicting the thermodynamic properties of a few materials at high temperatures. Few systems have so far been studied using this approach, and a lot more are awaiting the application and testing of SCP to see how accurately it can predict the properties of complex materials. This approach is still in its infancy and further extensions will be implemented to predict phase transitions and thermal properties of strongly anharmonic systems.

Acknowledgement

KE would like to acknowledge useful discussions with N Ravichandran.

Appendix A Thermodynamic properties of harmonic oscillators

The partition function for a single harmonic oscillator with frequency $\omega_\lambda = \epsilon_\lambda/\hbar$ is

$$Z_\lambda = \sum_{n=0}^{\infty} \mathrm{e}^{-\beta(n+1/2)\epsilon_\lambda} = \frac{\mathrm{e}^{-\beta\epsilon_\lambda/2}}{1 - \mathrm{e}^{-\beta\epsilon_\lambda}}. \tag{A.1}$$

For each atom in the lattice, it has three vibrational degrees of freedom, thus for an N-atom unit cell, the partition function describes $3N$ independent oscillators:

$$Z_N = \prod_{\lambda=1}^{3N} \frac{\mathrm{e}^{-\beta\epsilon_\lambda/2}}{1 - \mathrm{e}^{-\beta\epsilon_\lambda}}. \tag{A.2}$$

From this, we can acquire both the free energy $F = -k_\mathrm{B}T \ln Z$ and phonon vibrational entropy $S = -\frac{\partial F}{\partial T}$:

$$F_0 = k_\mathrm{B}T \sum_{\lambda}^{\infty} \ln\left[2 \sinh\left(\frac{\beta\epsilon_\lambda}{2}\right)\right] \tag{A.3}$$

$$S = k_\mathrm{B} \sum_{\lambda}^{3N}\left(\frac{\beta\epsilon_\lambda}{2} \coth\left(\frac{\beta\epsilon_\lambda}{2}\right) - \ln\left[2 \sinh\left(\frac{\beta\epsilon_\lambda}{2}\right)\right]\right). \tag{A.4}$$

Appendix B Normal modes and Gaussian averages

In the formulation of SCP theory, a trial Hamiltonian that has only harmonic terms is used. In SCHA theories, averages of the type $\langle yy \rangle = \mathrm{Tr}(\rho_0\, yy)/Z_0$ need to be computed, where the density matrix itself is an exponential of a quadratic function of the variable y, and the trace represents a sum over all degree of freedom, which in the classical case is the integral $\int_{-\infty}^{\infty} \mathrm{d}y$ (see below for the quantum version). Such Gaussian integrals are readily computed after writing the real-space variables y in terms of their Fourier components

$$y_{\boldsymbol{R}\tau}(t) = \frac{1}{\sqrt{N_q m_\tau}} \sum_{\boldsymbol{q}} y_{\boldsymbol{q}\tau}(t)\mathrm{e}^{\mathrm{i}\boldsymbol{q}\cdot\boldsymbol{R}} \tag{B.1}$$

$$= \frac{1}{\sqrt{N_q m_\tau}} \sum_q \mathrm{e}^{\mathrm{i}\boldsymbol{q}.\boldsymbol{R}} \sum_\lambda y_{q\lambda}(t)\epsilon_{q\lambda}^{\tau}. \tag{B.2}$$

In this way, the harmonic potential energy is first decoupled and the density matrix can be written as a product over independent modes

$$\begin{aligned} V &= \frac{1}{2} \sum_{RR';\tau\tau'} K_{R\tau,R'\tau'} : \boldsymbol{y}_{R\tau}\boldsymbol{y}_{R'\tau'} \\ &= \frac{1}{2} \sum_{\boldsymbol{q};\tau\tau'} D_{\tau\tau'}(\boldsymbol{q}) : y_{q\tau} y_{-q\tau'}, \end{aligned} \tag{B.3}$$

where D is the dynamical matrix defined by

$$D_{\tau\tau'}(\boldsymbol{q}) = \sum_{\boldsymbol{R}} K_{0\tau,R\tau'}\, \mathrm{e}^{\mathrm{i}\boldsymbol{q}\cdot\boldsymbol{R}}.$$

After diagonalization, its eigenvalues and eigenvectors are defined by

$$\sum_{\tau'} D_{\tau\tau'}(\boldsymbol{q}) \cdot \epsilon_{q\lambda}^{\tau'} = \omega_{q\lambda}^2\, \epsilon_{q\lambda}^{\tau} \qquad \forall \lambda. \tag{B.4}$$

This allows us, after quantization of positions and momenta, to write

$$y_{q\tau} = \sum_\lambda \left(\frac{\hbar}{2\omega_{q\lambda}}\right)^{1/2} \left(a_{-q\lambda}^{\dagger} + a_{q\lambda}\right) \epsilon_{q\lambda}^{\tau},$$

where $a^\dagger$ and a are phonon creation and annihilation operators satisfying $\langle a^\dagger a\rangle = n$; $\langle aa^\dagger\rangle = n + 1$. Finally, from the properties of the harmonic oscillator, the orthogonality of the eigenvectors and the above results, we obtain the (quantum) thermal average of the displacements squared:

$$\langle y_{q\tau} y_{-q\tau'}\rangle = \sum_\lambda \left(\frac{\hbar}{2\omega_{q\lambda}}\right) (2n_{q\lambda} + 1)\, \epsilon_{q\lambda}^{\tau} \epsilon_{-q\lambda}^{\tau'} \tag{B.5}$$

$$\langle \boldsymbol{y}_{R\tau} \boldsymbol{y}_{R'\tau'}\rangle = \sum_{q\lambda} \frac{\hbar(2n_{q\lambda} + 1)}{2N_q\omega_{q\lambda}} \frac{\epsilon_{q\lambda}^{\tau}\epsilon_{-q\lambda}^{\tau'}}{\sqrt{m_\tau m_{\tau'}}}\, e^{i\boldsymbol{q}\cdot(\boldsymbol{R}-\boldsymbol{R}')}. \tag{B.6}$$

From the above, it can be deduced that the displacement amplitude $A_{q\lambda}$ of mode $\boldsymbol{q}\lambda$ satisfies $A_{q\lambda}^2 = \frac{\hbar(2n_{q\lambda} + 1)}{2\omega_{q\lambda}}$. One can also recover the equipartition theorem:

$$\left\langle \frac{p_{q\lambda}^2}{2} \right\rangle = \frac{1}{2}\omega_{q\lambda}^2 A_{q\lambda}^2 = \frac{\hbar\omega_{q\lambda}}{4} \coth\left(\frac{\beta\hbar\omega_{q\lambda}}{2}\right). \tag{B.7}$$

Appendix C Formal SCHA equations

If the Hamiltonian of the crystal is

$$\mathcal{H} = \sum_i -\frac{1}{2m_i}\nabla_i^2 + \sum_{i,j} V(\boldsymbol{U}_i - \boldsymbol{U}_j), \tag{C.1}$$

where V is a pair potential, we use a harmonic trial Hamiltonian of the form

$$\mathcal{H}_{\text{trial}} = \sum_i -\frac{1}{2m_i}\nabla_i^2 + \frac{1}{2}\sum_{ij} \boldsymbol{y}_i \cdot \boldsymbol{K}_{ij} \cdot \boldsymbol{y}_j, \tag{C.2}$$

where i, j are atomic labels in the lattice and $\boldsymbol{K}_{ij}$ is the trial force constants to be determined variationally. From the trial Hamiltonian, one can write the density matrix ρ_{trial} as

$$\rho_{\text{trial}} = \frac{\mathrm{e}^{-\beta\mathcal{H}_{\text{trial}}}}{\mathrm{Tr}\,\mathrm{e}^{-\beta\mathcal{H}_{\text{trial}}}} \tag{C.3}$$

and the actual free energy F can be approximated as

$$F \approx \mathrm{Tr}[\rho_{\text{trial}}(\mathcal{H} + \beta^{-1}\ln\rho_{\text{trial}})] \equiv \langle \mathcal{H} + \beta^{-1}\ln\rho_{\text{trial}}\rangle \tag{C.4}$$

$$= F_{\text{trial}} + \langle \mathcal{H} - \mathcal{H}_{\text{trial}}\rangle. \tag{C.5}$$

Below, we define the correlation function $\boldsymbol{Y}_{ij}$ for dynamical displacement pairs:

$$\boldsymbol{Y}_{ij} = \langle \boldsymbol{y}_i\, \boldsymbol{y}_j\rangle. \tag{C.6}$$

The exact expression for this term was presented in the previous section (equation (B.6)). Now, we can apply Taylor expansion around the static equilibrium positions $\boldsymbol{S}_{ij} = \boldsymbol{R}_i + \boldsymbol{\tau}_i - \boldsymbol{R}_j - \boldsymbol{\tau}_j$ to calculate the thermal averaged term in actual free energy:

$$\begin{aligned}\langle \phi(\boldsymbol{U}_i - \boldsymbol{U}_j)\rangle &= \langle \phi(\boldsymbol{S}_{ij} + \boldsymbol{y}_i - \boldsymbol{y}_j)\rangle \\ &= \langle (\mathrm{e}^{(\boldsymbol{y}_i - \boldsymbol{y}_j)\cdot\nabla})\phi(\boldsymbol{S}_{ij})\rangle \\ &= \mathrm{e}^{\frac{1}{2}\boldsymbol{Y}_{ij}:\nabla\nabla}\phi(\boldsymbol{S}_{ij}).\end{aligned} \tag{C.7}$$

Following Boccara and Sarma [5], the free energy F can be written into a functional of both trial force constants $\boldsymbol{K}_{ij}$ and correlation functions $\boldsymbol{Y}_{ij}$, while the trial free energy F_{trial} is related only to $\boldsymbol{K}_{ij}$, or rather its eigenvalues. $\boldsymbol{K}_{ij}$ and $\boldsymbol{Y}_{ij}$ can be varied independently to give two variational equations minimizing F. Keeping in mind that the thermal averages depend only on $\boldsymbol{K}$, and $\langle \mathcal{H}_{\text{trial}}\rangle = \sum_{ij}\boldsymbol{K}_{ij} : \boldsymbol{Y}_{ij}/2$, and using equation (C.7), we have

$$\frac{\partial F}{\partial \boldsymbol{K}_{ij}} = \frac{\partial F_{\text{trial}}}{\partial \boldsymbol{K}_{ij}} - \frac{1}{2}\boldsymbol{Y}_{ij} = 0 \tag{C.8}$$

$$\frac{\partial F}{\partial \boldsymbol{Y}_{ij}} = \frac{1}{2}\langle \nabla\nabla\phi(\boldsymbol{U}_i - \boldsymbol{U}_j)\rangle - \frac{1}{2}\boldsymbol{K}_{ij} = 0. \tag{C.9}$$

The last equation can then be given an interpretation for the effective or trial force constants $\boldsymbol{K}$—they are the thermal average of the actual FCs over thermally displaced atoms, while equation (C.8) is a general property of harmonic systems. Using the definition of normal modes as in equation (B.1), one can express the correlation function $\boldsymbol{Y}$ in terms of normal modes. This is exactly equation (B.6). The set of equations to be solved, which constitute the SCHA approximation are then:

- The assumption of normal modes (starting with the harmonic ones).
- The computation of the correlation functions $\boldsymbol{Y}$ from equation (B.6).
- Obtaining the new trial FCs $\boldsymbol{K}$ from equation (C.9) in which use must be made of the Taylor expansion in equation (C.7) and the calculated expression of $\boldsymbol{Y}$.
- The calculation of normal modes of this new $\boldsymbol{K}$.
- Reiterating if the normal modes did not converge.

References

[1] De Wette F W, Nosanow L H and Werthamer N R 1967 Calculation of phonon frequencies and thermodynamic properties of crystalline bcc helium *Phys. Rev.* **162** 824

[2] Maradudin A A and Fein A E 1962 Scattering of neutrons by an anharmonic crystal *Phys. Rev.* **128** 2589

[3] Born M and Hooton D J 1955 Statistische Dynamik mehrfach periodischer Systeme *Z. Phys.* **142** 201

[4] Hooton D 1955 LI. a new treatment of anharmonicity in lattice thermodynamics: I *Lond. Edinb. Dubl. Philos. Mag. J. Sci.* **46** 422

[5] Boccara N and Sarma G 1965 Theorie microscopique des transitions s'accompagnant d' une modification de la structure cristalline *Phys. Phys. Fiz.* **1** 219

[6] Wick G C 1950 The evaluation of the collision matrix *Phys. Rev.* **80** 268

[7] Horner H 1967 Lattice dynamics of quantum crystals *Z. Phys.* **205** 72

[8] Gillis N S, Werthamer N R and Koehler T R 1968 Properties of crystalline argon and neon in the self-consistent phonon approximation *Phys. Rev.* **165** 951

[9] Feynman R P 1972 *Statistical Mechanics, a Set of Lectures* (Reading, MA: Benjamin/Cummings)

[10] Isihara A 1968 The Gibbs–Bogoliubov inequality *J. Phys. A: Gen. Phys.* **1** 539

[11] Bauer E and Wu T Y 1956 Thermal expansion of a linear chain *Phys. Rev.* **104** 914

[12] Car R and Parrinello M 1985 Unified approach for molecular dynamics and density-functional theory *Phys. Rev. Lett.* **55** 2471

[13] Ceperley D M 1995 Path integrals in the theory of condensed helium *Rev. Mod. Phys.* **67** 279

[14] Hooton D J 1958 The use of a model in anharmonic lattice dynamics *Philos. Mag. J. Theor. Exp. Appl. Phys.* **3** 49

[15] Koehler T R 1966 Theory of the self-consistent harmonic approximation with application to solid neon *Phys. Rev. Lett.* **17** 89

[16] Bernardes N 1958 Theory of solid Ne, A, Kr, and Xe at 0 K *Phys. Rev.* **112** 1534

[17] Nosanow L H and Shaw G L 1962 Hartree calculations for the ground state of solid He and other noble gas crystals *Phys. Rev.* **128** 546

[18] Nosanow L H and Werthamer N R 1965 Calculations of sound velocities in crystalline helium at zero temperature *Phys. Rev. Lett.* **15** 618

[19] Ranninger J 1965 Lattice thermal conductivity *Phys. Rev.* **140** A2031

[20] Choquard P 1967 *The Anharmonic Crystal* (New York: W A Benjamin)

[21] Goldman V V, Horton G K and Klein M L 1968 An improved self-consistent phonon approximation *Phys. Rev. Lett.* **21** 1527

[22] Feldman J L and Horton G K 1967 Anharmonic contributions to the Helmholtz free energy of a simple crystal model: high-temperature limit and zero-point energy *Proc. Phys. Soc.* **92** 227

[23] Werthamer N R 1970 Self-consistent phonon formulation of anharmonic lattice dynamics *Phys. Rev.* B **1** 572

[24] Koehler T R 1969 Theoretical temperature-dependent phonon spectra of solid neon *Phys. Rev. Lett.* **22** 777

[25] Horton G K and Maradudin A A 1974 *Dynamical Properties of Solids* vol 1 (Amsterdam: North Holland) p 433

[26] Samathiyakanit V and Glyde H R 1973 Path integral theory of anharmonic crystals *J. Phys. C: Solid State Phys.* **6** 1180

[27] Baroni S, Giannozzi P and Testa A 1987 Green's-function approach to linear response in solids *Phys. Rev. Lett.* **58** 1861

[28] Gonze X and Vigneron J P 1989 Density-functional approach to nonlinear-response coefficients of solids *Phys. Rev.* B **39** 13120

[29] Debernardi A, Baroni S and Molinari E 1995 Anharmonic phonon lifetimes in semiconductors from density-functional perturbation theory *Phys. Rev. Lett.* **75** 1819

[30] Vanderbilt D, Louie S G and Cohen M L 1984 Calculation of phonon–phonon interactions and the absence of two-phonon bound states in diamond *Phys. Rev. Lett.* **53** 1477

[31] Narasimhan S and Vanderbilt D 1991 Anharmonic self-energies of phonons in silicon *Phys. Rev.* B **43** 4541

[32] Parlinski K, Li Z Q and Kawazoe Y 1997 First-principles determination of the soft mode in cubic (ZrO_2) *Phys. Rev. Lett.* **78** 4063

[33] Esfarjani K and Stokes H T 2008 Method to extract anharmonic force constants from first principles calculations *Phys. Rev.* B **77** 144112

[34] Li W, Carrete J, Katcho N A and Mingo N 2014 ShengBTE: a solver of the Boltzmann transport equation for phonons *Comp. Phys. Commun.* **185** 1747–58

[35] Togo A and Tanaka I 2015 First principles phonon calculations in materials science *Scr. Mater.* **108** 1

[36] Tadano T and Tsuneyuki S 2018 First-principles lattice dynamics method for strongly anharmonic crystals *J. Phys. Soc. Jpn.* **87** 041015

[37] Hellman O, Steneteg P, Abrikosov I A and Simak S I 2013 Temperature dependent effective potential method for accurate free energy calculations of solids *Phys. Rev.* B **87** 104111

[38] Giannozzi P, de Gironcoli S, Pavone P and Baroni S 1991 *Ab initio* calculation of phonon dispersions in semiconductors *Phys. Rev.* B **43** 7231

[39] Broido D A, Malorny M, Birner G, Mingo N and Stewart D A 2007 Intrinsic lattice thermal conductivity of semiconductors from first principles *Appl. Phys. Lett.* **91** 231922

[40] Deinzer G, Birner G and Strauch D 2003 *Ab initio* calculation of the linewidth of various phonon modes in germanium and silicon *Phys. Rev.* B **67** 144304

[41] Rousseau B and Bergara A 2010 Giant anharmonicity suppresses superconductivity in AlH_3 under pressure *Phys. Rev.* B **82** 104504

[42] Alfè D 2009 PHON: a program to calculate phonons using the small displacement method *Comput. Phys. Commun.* **180** 2622

[43] Li W, Mingo N, Lindsay L, Broido D A, Stewart D A and Katcho N A 2012 Thermal conductivity of diamond nanowires from first principles *Phys. Rev.* B **85** 195436

[44] Detraux F, Ghosez P H and Gonze X 1998 Long-range Coulomb interaction in ZrO_2 *Phys. Rev. Lett.* **81** 3297

[45] Parlinski K, Li Z Q and Kawazoe Y 1998 Parlinski, Li, and Kawazoe reply *Phys. Rev. Lett.* **81** 3298

[46] Wang Y, Wang J J, Wang W Y, Mei Z G, Shang S L, Chen L Q and Liu Z K 2010 A mixed-space approach to first-principles calculations of phonon frequencies for polar materials *J. Phys. Condens. Matter* **22** 202201

[47] Parlinski K 2018 *Ab initio* determination of anharmonic phonon peaks *Phys. Rev.* B **98** 054305

[48] Tadano T, Gohda Y and Tsuneyuki S 2014 Anharmonic force constants extracted from first-principles molecular dynamics: applications to heat transfer simulations *J. Phys. Condens. Matter* **26** 225402

[49] Hellman O, Abrikosov I A and Simak S I 2011 Lattice dynamics of anharmonic solids from first principles *Phys. Rev.* B **84** 180301

[50] Eriksson F, Fransson E and Erhart P 2019 The Hiphive package for the extraction of high-order force constants by machine learning *Adv. Theory Simul.* **2** 1800184

[51] Fu L, Kornbluth M, Cheng Z and Marianetti C A 2019 Group theoretical approach to computing phonons and their interactions *Phys. Rev.* B **100** 014303

[52] Shang H, Carbogno C, Rinke P and Scheffler M 2017 Lattice dynamics calculations based on density-functional perturbation theory in real space *Comput. Phys. Commun.* **215** 26

[53] Baroni S, de Gironcoli S, Dal Corso A and Giannozzi P 2001 Phonons and related crystal properties from density-functional perturbation theory *Rev. Mod. Phys.* **73** 515

[54] Souvatzis P, Eriksson O, Katsnelson M I and Rudin S P 2008 Entropy driven stabilization of energetically unstable crystal structures explained from first principles theory *Phys. Rev. Lett.* **100** 095901

[55] Souvatzis P, Eriksson O, Katsnelson M and Rudin S 2009 The self-consistent *ab initio* lattice dynamical method *Comput. Mater. Sci.* **44** 888

[56] Zhang H Y, Niu Z W, Cai L C, Chen X R and Xi F 2018 *Ab initio* dynamical stability of tungsten at high pressures and high temperatures *Comput. Mater. Sci.* **144** 32

[57] Hellman O and Abrikosov I A 2013 Temperature-dependent effective third-order interatomic force constants from first principles *Phys. Rev.* B **88** 144301

[58] West D and Estreicher S K 2006 First-principles calculations of vibrational lifetimes and decay channels: hydrogen-related modes in Si *Phys. Rev. Lett.* **96** 115504

[59] Yang F C, Hellman O, Lucas M S, Smith H L, Saunders C N, Xiao Y, Chow P and Fultz B 2018 Temperature dependence of phonons in (Pd_3Fe) through the Curie temperature *Phys. Rev.* B **98** 024301

[60] Vanderbilt D, Louie S G and Cohen M L 1986 Calculation of anharmonic phonon couplings in C, Si, and Ge *Phys. Rev.* B **33** 8740

[61] Ravichandran N K and Broido D 2018 Unified first-principles theory of thermal properties of insulators *Phys. Rev.* B **98** 085205

[62] Paulatto L, Mauri F and Lazzeri M 2013 Anharmonic properties from a generalized third-order *ab initio* approach: theory and applications to graphite and graphene *Phys. Rev.* B **87** 214303

[63] Paulatto L, Errea I, Calandra M and Mauri F 2015 First-principles calculations of phonon frequencies, lifetimes, and spectral functions from weak to strong anharmonicity: the example of palladium hydrides *Phys. Rev.* B **91** 054304
[64] Glyde H R and Goldman V V 1976 Dynamics of quantum crystals *J. Low Temp. Phys.* **25** 601
[65] Morris J R and Gooding R J 1990 Exactly solvable heterophase fluctuations at a vibrational-entropy-driven first-order phase transition *Phys. Rev. Lett.* **65** 1769
[66] Morris J R and Gooding R J 1991 Vibrational entropy effects at a diffusionless first-order solid-to-solid transition *Phys. Rev.* B **43** 6057
[67] Morris J R and Ho K M 1995 Calculating accurate free energies of solids directly from simulations *Phys. Rev. Lett.* **74** 940

Part II

Measurements and applications

Chapter 8

Experimental approaches for probing heat transfer and energy conversion at the atomic and molecular scales

Longji Cui, Edgar Meyhofer and Pramod Reddy

Recent work has achieved tremendous progress toward experimentally probing energy transport at the nanoscale. In examining the limits of the applicability of classical laws, atomic- and molecular-scale structures have emerged as paradigmatic systems and have revealed exotic transport phenomena that arise in the quantum regime. Moreover, understanding how thermal energy is transported, converted and dissipated at the most fundamental level is of great importance for achieving high-performance energy conversion technologies and rationally designing thermally robust circuits when approaching the limit of electronic miniaturization. In this chapter, we present a brief review of the basic concepts and experimental progress made in atomic- and molecular-scale thermal science. In particular, we focus on the recent development of high-resolution scanning thermal microscopy probes and their applications in understanding heat conduction in atomic and single-molecule junctions as well as thermoelectric energy conversion and heat dissipation in molecular junctions.

8.1 Introduction

Atomic-sized structures such as atomic junctions, gaps and molecular chains represent the ultimate limit for the miniaturization of physical devices. Over the past half century, the modern electronics industry has been significantly advanced by the growing demand for faster, larger, and more affordable storage and computing hardware. The physical size of functional units such as transistors has continuously shrunk over the last few decades, approaching the physically allowed size limit—the atomic scale. Whereas classical laws have proven to be useful in describing energy transport phenomena in macroscopic devices and materials, the laws governing transport at the atomic scale remain largely unexplored. The study of energy

doi:10.1088/978-0-7503-1738-2ch8

transport at the atomic scale provides unique opportunities to test widely accepted rules and frameworks and holds great promise to strongly affect a wide range of contemporary technologies ranging from electronics and photonics to chemical catalysis and energy conversion. To date, a large amount of theoretical and experimental work has been directed toward understanding the electrical transport properties and mechanisms in atomic-sized structures and devices [1, 2]. Although essential for the functionalities and performance of materials, the thermal transport properties of materials and devices ranging in size from a few nanometers (comparable to the characteristic size of the smallest commercially available transistors) down to the single-atom limit have remained largely unexplored. This is due to the fact that, at this length scale, thermophysical properties such as heat flow rates, temperature and energy conversion are quite challenging to measure. Developing novel experimental techniques and approaches to enable quantitative calibration of these critical characteristics remains a key challenge in the field of atomic-scale thermal science.

In this chapter, we focus on reviewing the current state-of-the-art in thermal energy transfer and conversion at the atomic scale. Specifically, we first introduce in section 8.2 some basic concepts of energy transfer and conversion in nano- and mesoscopic devices. From sections 8.3 to 8.6, we review the recent development of scanning thermal microscopy probes and the experimental work that has revealed quantum thermal transport in single-atom junctions, heat dissipation in single-molecule junctions, the observation of the Peltier effect in molecular junctions, and the recent experimental advance enabling the first measurement of thermal conductance of single-molecule junctions. Finally, we conclude this chapter by highlighting some open challenges in this field.

8.2 Theoretical concepts

8.2.1 Energy transport in atomic-scale junctions

Atomic junctions refer to functional devices with one or a few atoms bridging two electrodes (figure 8.1(a)). Similarly, individual molecules can form molecular junctions by connecting two electrodes through chemical and physical bonds (figure 8.1(b)). For these atomic-sized devices, energy transport via basic energy carriers such as electrons and phonons is often described using Landauer's quantum transport framework [3–5]. Specifically, when applying a voltage bias (V) across atomic-scale junctions, the resultant electrical current (I) is given by [6]

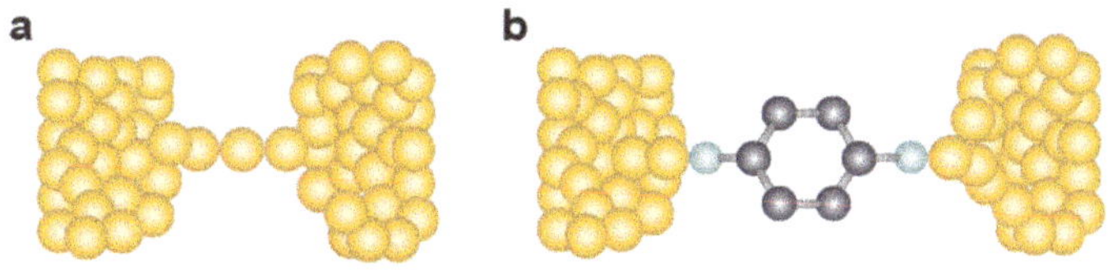

Figure 8.1. Schematics of the atomic-sized structures discussed in this chapter. (a) An atomic junction is formed when a one-atom or few-atom wide wire is sandwiched between two macroscopic electrodes. (b) A single-molecule junction.

$$I = \frac{2e}{h}\int_{-\infty}^{+\infty} (f_{\rm L} - f_{\rm R})\tau(E)\mathrm{d}E, \tag{8.1}$$

where e and h are the electron charge and Planck's constant, $f_{\rm L}$ and $f_{\rm R}$ are the Fermi–Dirac distributions of the left and right electrodes, respectively, and $\tau(E)$ is the transmission function, which describes the energy-dependent transmission probability of electrons through the junction. For the specific atomic structures, $\tau(E)$ can be estimated by employing density functional theory in conjunction with non-equilibrium Green's function techniques [7].

In the approximation of a small bias voltage and low temperatures, equation (8.1) can be further reduced to

$$G_{\rm e} = \frac{I}{V} = \frac{2e^2}{h}\tau_{E=E_{\rm F}}, \tag{8.2}$$

where $G_{\rm e}$ is the electrical conductance of the junction, $E_{\rm F}$ is the Fermi energy of the electrodes and $\tau_{E=E_{\rm F}}$ is the transmission probability of all contributing electronic channels (modes) in the junction. Under the circumstance that $\tau_{E=E_{\rm F}} = 1$, which indicates the existence of one fully transparent ballistic channel for electron transport, the electrical conductance is quantized and is given by $G_{\rm e} = G_0 = 2e^2/h = 1/(12.9\ \mathrm{k\Omega})$.

Thermal energy transport at the atomic scale can also be readily described using the Landauer formalism. Basically, when a temperature difference (ΔT) is applied across the atomic-sized junction, a corresponding heat current flows through the junction, driven by both electrons and phonons. The electronic heat current ($J_{\rm el}$) and the phononic heat current ($J_{\rm ph}$) are given by [4, 8]

$$J_{\rm el} = \frac{2}{h}\int_{-\infty}^{\infty} (E - E_{\rm F})\tau(E)(f_{\rm L} - f_{\rm R})\mathrm{d}E \tag{8.3}$$

and

$$J_{\rm ph} = \int_{0}^{\infty} (h\omega)\tau_{\rm ph}(\omega)(b_{\rm L} - b_{\rm R})\mathrm{d}\omega, \tag{8.4}$$

where ω is the phonon frequency, $\tau_{\rm ph}(\omega)$ is the energy-dependent transmission function for phonons, and $b_{\rm L}$ and $b_{\rm R}$ are the Bose–Einstein distributions for the left and the right thermal reservoirs, respectively.

Interestingly, a thermal analog of the quantized electrical conductance can also be derived from equations (8.3) and (8.4) for atomic-scale heat conduction. For ideal thermal coupling between the ballistic atomic conductor and the thermal reservoirs, which results in a transmission probability equaling unity for a single channel, both electronic and phononic contributions to thermal transport are given by

$$G_{\rm th}^{\rm el} = G_{\rm th}^{\rm ph} = g_0 = \frac{\pi^2 k_{\rm B}^2 T}{3h}, \tag{8.5}$$

where $G_{\rm th}^{\rm el} = J_{\rm el}/\Delta T$ and $G_{\rm th}^{\rm ph} = J_{\rm ph}/\Delta T$ are the electronic and phononic thermal conductances of the atomic junctions, respectively, $k_{\rm B}$ is the Boltzmann constant and T is the temperature of the thermal reservoirs. Here, $g_0 = (9.456 \times 10^{-13} \times T) W/K$ is the quantized thermal conductance that has an expression independent of any material parameters and particle statistics (universal for bosons, fermions and anions) [9–13]. For fermions specifically, the ratio of the quantized thermal and electrical conductance obeys the Wiedemann–Franz law, which relates the electrical transport and the electronic contribution to thermal transport in a material [14] via

$$\frac{G_{\rm th}^{\rm el}}{G_{\rm e}} = LT, \tag{8.6}$$

where

$$L = \frac{\pi^2 k_{\rm B}^2}{3e^2}. \tag{8.7}$$

Probing the electron contribution to heat transfer provides a window into the nature of electron transport in a material. While in most conventional metallic materials the Wiedemann–Franz law is found to be valid and provides quantitatively good estimates of electronic contributions, there have been recent observations of strong violations in several quasi-one-dimensional and nanomaterials [15, 16], which if understood may provide additional means for controlling energy conversion and transport in future devices.

8.2.2 Heat dissipation and thermoelectric energy conversion in molecular junctions

When an electrical current (I) flows across a molecular junction, the electrical energy is converted to other energy forms and is eventually dissipated into heat. To better understand this question in nanoscale devices, one may ask a series of questions: How does the energy conversion happen? What is the efficiency of a specific form of energy conversion? Where does the heat dissipate? In Landauer's framework which accounts for elastically transported electrons, i.e. when no energy exchange happens inside the molecular region, the heat dissipation in the left and right electrodes $Q_{\rm L}$ and $Q_{\rm R}$, respectively, is given by [17, 18]

$$Q_{\rm L} = \frac{2}{h}\int_{-\infty}^{\infty} (\mu_{\rm L} - E)\tau(E)(f_{\rm L} - f_{\rm R}){\rm d}E \tag{8.8}$$

and

$$Q_{\rm R} = \frac{2}{h}\int_{-\infty}^{\infty} (E - \mu_{\rm R})\tau(E)(f_{\rm L} - f_{\rm R}){\rm d}E, \tag{8.9}$$

where $\mu_{\rm L/R}$ represent the chemical potentials of the left/right electrodes, respectively.

The thermoelectric energy conversion properties of molecular junctions can also be described within the Landauer transport framework. The open circuit voltage (ΔV) developed across a molecular or atomic junction in the presence of a temperature difference (ΔT) enables the estimation of the Seebeck coefficient (S) of the junction, which is defined by $S = -\Delta V/\Delta T$, and can be related to the electronic transmission characteristics of the molecular junction by [19, 20]

$$S = -\frac{\pi^2 k_B^2 T}{3|e|} \frac{\partial \ln(\tau(E))}{\partial E}\bigg|_{E=E_F}. \tag{8.10}$$

The sign of the Seebeck coefficient indicates the identity of the dominant charge carrier. Specifically, electron-dominated transport results in a negative Seebeck coefficient, whereas hole-dominated transport results in a positive Seebeck coefficient. Further, it can be seen from equation (8.10) that the magnitude of the Seebeck coefficient is determined by the slope of the transmission function at the Fermi level of the junction.

According to the Thompson relation, the Peltier coefficient (Π) of the molecular junction can also be related to the Seebeck coefficient by

$$\Pi = TS, \tag{8.11}$$

where T is the absolute temperature of the junction. The Peltier effect describes the reversible heating and cooling effects that arise when an electric current flows across the isothermal molecular junction.

The thermoelectric energy conversion efficiency of a material is usually described by a quantity called the figure of merit (ZT). It is defined as $ZT = S^2\sigma T/\kappa$, where σ is the electrical conductivity and κ is the thermal conductivity. The corresponding ZT of a molecular junction can be defined similarly as

$$ZT = \frac{S^2 G_e T}{G_{th}}, \tag{8.12}$$

where G_e and G_{th} represent the electrical and thermal conductances, respectively. Given the temperatures of the left and right electrodes, T_L and T_R (assuming $T_L < T_R$), the heat-to-electricity conversion efficiency of molecular junction (η) is given by

$$\eta = \eta_C \frac{\sqrt{1 + ZT} - 1}{\sqrt{1 + ZT} + T_L/T_R}, \tag{8.13}$$

where $\eta_C = 1 - T_L/T_R$ is the Carnot efficiency, which gives the highest energy conversion efficiency any heat engine can reach.

8.3 Heat transfer and energy conversion at the atomic scale: experiments

Performing thermal transport measurements at the atomic scale is challenging for a number of reasons. First, atomic-scale structures, due to their nanoscale dimensions, cannot be visually accessed using optical microscopy tools. This frustrates efforts to conduct *in situ* observations and manipulations of the studied objects while simultaneously performing transport measurements, for example, using well-established thermal characterization tools based on optical approaches [21, 22]. Second, the magnitude of heat flow in such structures is exceedingly small, requiring high-resolution thermal sensing techniques to reliably record the signals of interest. In order to overcome these challenges, researchers have developed scanning thermal microscopy (SThM)-based approaches [23–26] where custom-fabricated high-resolution thermal sensors are integrated into scanning probes to achieve a nanowatt or even picowatt (pW) scale heat current resolution. Further, these tools enable reliable creation and manipulation of atomic-sized objects. Below, we present a brief review of the experimental progress made in applying SThM-based techniques to address open questions in atomic-scale thermal science. Specifically, we will focus on recent work on the measurement of the quantum of heat transport in single-atom junctions, heat dissipation in single-molecule junctions, the Peltier cooling in molecular junctions, as well as the measurement of thermal conductance of single alkanedithiol molecule junctions. For a detailed review of the theoretical and experimental progress in these topics, readers are referred to recent literature [27, 28].

8.3.1 Quantum heat transport in single-atom junctions

Key to probing heat conduction in atomic junctions is the development of probes that can both enable the creation of atomic junctions and resolve heat flows in such junctions. Recently, calorimetric scanning thermal microscopy (C-SThM) [29] probes, which are indeed capable of achieving the above-described goals, were developed. As shown in figure 8.2, the micro-fabricated probes feature a suspended island that is isolated by long 'T' shaped beams that feature a large thermal resistance (R_P ~ 1.3 × 10^6 K/W). Further, within the suspended island a high-resolution Pt thermometer with a temperature resolution of <1 mK (in a 10 Hz bandwidth) is embedded. These combined characteristics make it possible to measure a thermal conductance variation of ~25 pW K^{-1}, when a temperature difference of ~20 K is applied across atomic-scale junctions. Moreover, the two long, T-shaped SiN_x beams, which are critical for achieving a large thermal resistance, also feature very high stiffness values (>10^4 N m^{-1} in the normal direction) which greatly reduces thermal fluctuation [26] and enhances the stability of atomic junctions.

In order to probe heat transport, atomic junctions were created by employing approaches similar to those established in past pioneering works [30, 31]. Specifically, an STM break junction technique [32] is employed to stably create and maintain the metallic atomic junctions. Briefly, the Au-coated tip of the C-SThM probe is displaced toward a planar Au substrate by piezoelectric actuation. When Au on the tip makes contact with Au in the substrate the electrical conductance of the junction reduces to a few hundred ohms (i.e. electrical conductance values that are significantly larger than G_0). The tip is subsequently slowly withdrawn from the contact (typically at a rate smaller than 1 nm s^{-1}). During this process, the nanoscale contact region thins continuously by mechanical stretching until the formation of a single-atom junction as established in several past works [32]. A voltage bias and a temperature differential are applied across the tip

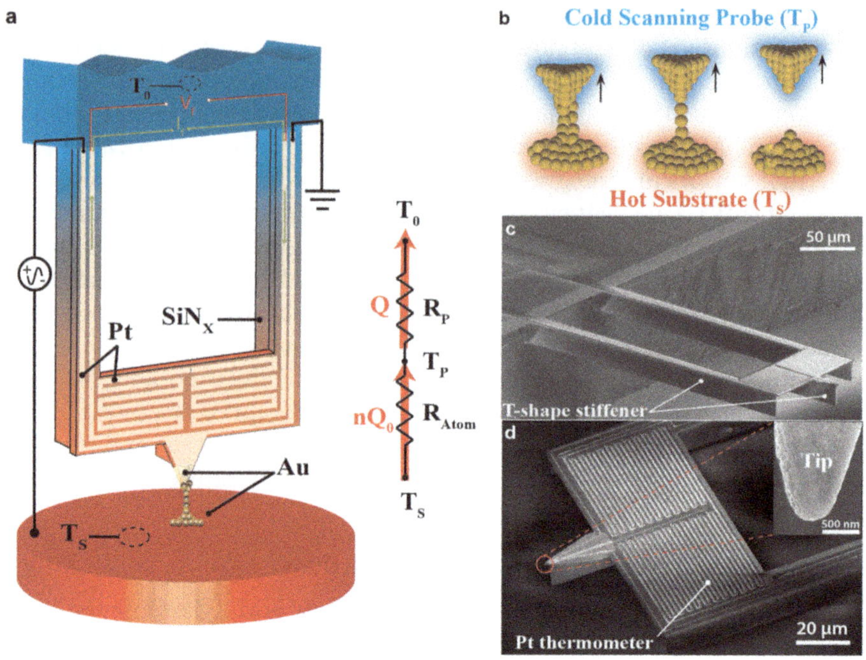

Figure 8.2. Experimental set-up and strategy for measuring heat transport in atomic junctions. (a) Schematic of a micro-fabricated C-SThM probe. Simultaneous thermal and electrical transport measurements are performed by monitoring the temperature change of the probe and the tunneling current flowing across the atomic junction, respectively. A resistance network is shown to illustrate the energy flow (Q) and relevant thermal resistance in the system, where R_P and R_{Atom} are the thermal resistance of the C-SThM probe and the atomic junction, and T_0, T_P and T_S correspond to the temperature of the thermal reservoir, the probe and the substrate, respectively. (b) Schematics showing the forming, thinning and breaking of an atomic junction during the withdrawal process of the C-SThM probe from the heated substrate. (c) SEM image (side view) of a C-SThM probe with two long and stiff T-shaped SiN_x beams. (d) SEM image (top view) of the probe featuring an Au-coated tip (inset) and an integrated Pt thermometer on the suspended island. Reproduced with permission from [29]. Copyright 2017 American Association for the Advancement of Science.

(at temperature T_P) and the substrate (temperature T_S) to quantify the electrical conductance and thermal conductance simultaneously. Specifically, the electrical current is measured using a current amplifier connected in series with the junction, while the heat current flowing in or out of the C-SThM probe is quantified by measuring the small temperature change (~1 mK or smaller) of the suspended island (ΔT_P), which is monitored via the integrated Pt thermometer. The thermal conductance of the atomic junctions (G_{Th}) can be related to the temperature change by $G_{Th} = \Delta T_P/[R_P(T_S - T_P - \Delta T_P)]$, where R_P is the thermal resistance of the probe, and T_S and T_P are the temperatures of the substrate and the probe, respectively.

Representative results from electrical and thermal transport measurements on Au atomic junctions are summarized in figure 8.3. It can be seen that, as the separation between the tip and substrate is increased, the electrical and thermal conductances of the junction decrease in steps. It can be seen that the steps are found to favor values

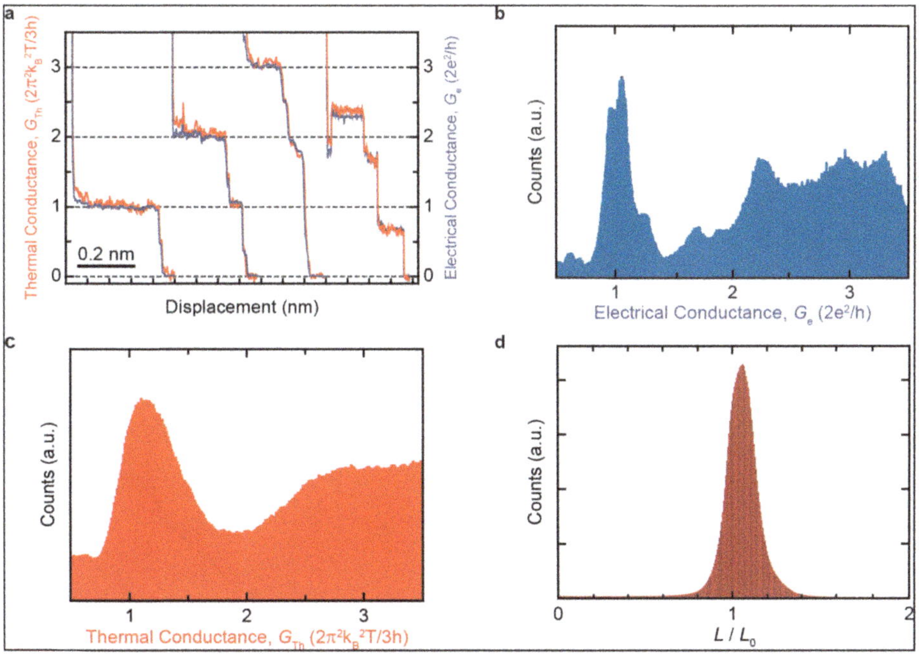

Figure 8.3. Experimentally measured quantized thermal transport in Au single-atom junctions. (a) Typical measured traces of thermal (red) and electrical (blue) conductances, in the units of $2\pi^2k_B{}^2T/3h$ ($2g_0$, twice the thermal conductance quantum) and $2e^2/h$ (G_0, the electrical conductance quantum), respectively. (b) and (c) Histograms of electrical and thermal conductances, obtained from 2000 consecutively measured conductance traces. (d) Analysis of the measured histograms of thermal and electrical conductances shows that the Wiedemann–Franz law is valid at the atomic scale to predict the thermal conductance of Au atomic junctions (the peak is at 1.06). Figure reproduced with permission from [29]. Copyright 2017 American Association for the Advancement of Science.

corresponding to the multiple integers of the quantized conductances ($G_0 = 2e^2/h$ for electrical transport, and $G_{0,\mathrm{Th}} = 2g_0 = 2\pi^2k_B^2T/3h$ for thermal transport, where a factor of 2 comes from the spin degeneracy in electron transport). By combining ~2000 consecutively obtained traces into the histogram of electrical and thermal conductance, clear peaks are observed at $1G_0$ and $1G_{0,\mathrm{Th}}$, which provide a definite proof of thermal transport quantization in single-atom junctions.

In order to robustly estimate the electrical conductance of the junctions, an unbiased statistical analysis was performed where histograms were constructed from several hundred traces. Such an analysis is motivated by the fact that the transport properties of atomic junctions are sensitive to the atomic-scale details of the geometries. Further motivation comes from the fact that previous studies [33–35] have shown that the presence of plateaus in conductance measurements is insufficient evidence of transport quantization.

It is to be noted that strong quantum confinement effects drive the energy level spacing in metallic atomic junctions enabling energy level spacing of a few electron

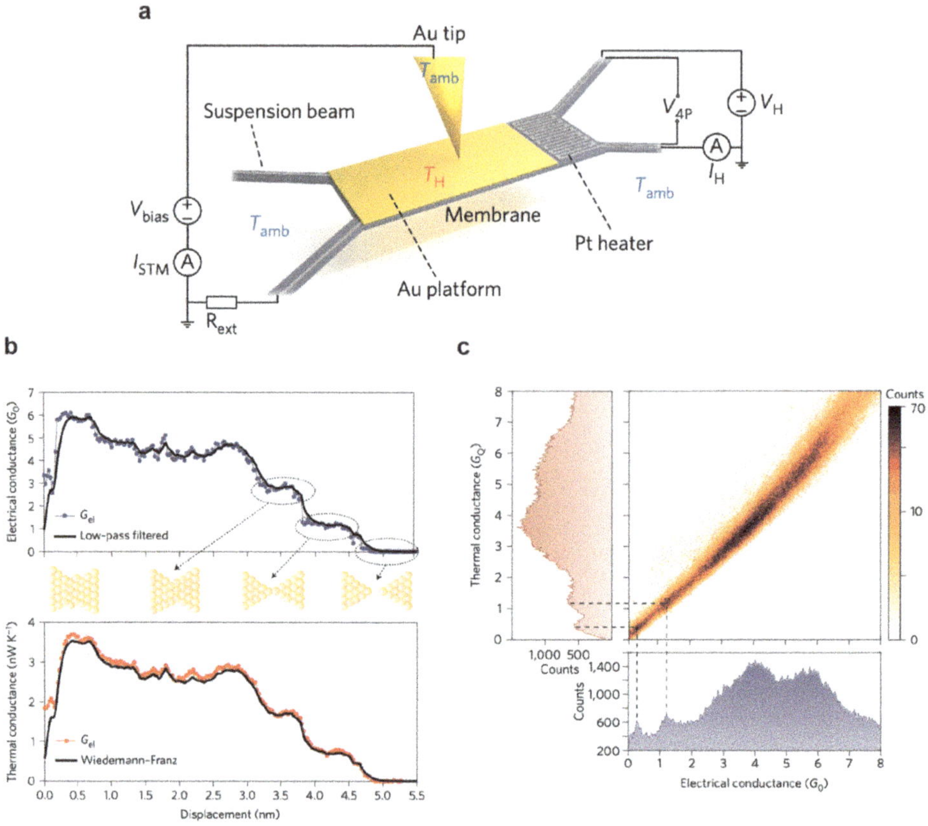

Figure 8.4. Experimental set-up for measuring heat transport in Au atomic junctions. (a) Schematics of the experimental set-up. (b) Measured electrical (upper panel) and thermal (lower panel) conductance trace, showing the stepwise behavior and the comparison with the prediction from the Wiedeman–Franz law. (c) Histograms of thermal and electrical conductances built from 2000 out of 5000 measured traces. Reproduced with permission from [36]. Copyright 2017 Macmillan Publishers.

volts, i.e. much larger than the thermal energy, making possible the above-described observations of quantized thermal transport at room temperature.

The developed C-SThM technique also makes it possible to investigate the validity of the Wiedemann–Franz law at the single-atom limit. The histogram in figure 8.4(d) illustrates the ratio of the measured thermal to electrical conductance, normalized by temperature and L_0, $G_{\mathrm{Th}}/G_{\mathrm{e}}TL_0$. The Wiedemann–Franz law predicts this ratio to be equal to one and indeed it can be seen that the most probably value is very close to 1 (~1.06). According to the *ab initio* calculations performed for the Au single-atom junctions, the observed deviation (~5%–10%) from the predicted value can be attributed to the phononic thermal conduction, which is small due to the mismatch of the vibrational energy spectrum between the atomic wire and the macroscopic electrodes.

In addition to the experiments described above, Mosso *et al* [36] have also studied heat transport across Au atomic-scale junctions using a different experimental technique that is shown in figure 8.4. Briefly, they employed a suspended device with an integrated thermal sensor to measure the temperature change upon the rupture of an atomic junction, formed between an Au STM tip and an Au-coated region of the suspended device. Similar to the data shown in figure 8.3(a), they showed a strong correlation between the measured thermal and electrical conductance traces, which established a stepwise behavior during the breaking process of the atomic junctions (figure 8.4(b)). The Wiedemann–Franz law was also found to be valid at the atomic scale by applying a histogram analysis on both the thermal and electrical signals. Moreover, pronounced conductance peaks were observed at conductance values of $0.4G_{0,Th}$ and $1.3G_{0,Th}$, rather than $1G_{0,Th}$, which is expected from the thermal conductance quantization in such atomic system [29]. This discrepancy was attributed to the potential existence of contaminant molecules (CO or H_2O) absorbed on the Au membranes.

8.4 Heat dissipation in atomic- and molecular-scale junctions

To understand how heat is dissipated (the Joule effect) in atomic-sized junctions, Lee *et al* [37] developed an experimental approach that employed nanoscale thermocouple integrated scanning thermal microscope probes. In contrast to the C-SThM probes described above, these probes feature a sensitive nanoscale thermal sensor made of the Au–Cr junction (figure 8.5) which is located in close proximity to the apex of the tip. Multiple layers of dielectric and metallic thin films are deposited to form a sharp tip with a diameter of less than 300 nm.

In order to probe heat dissipation, an atomic or single-molecule junction is formed between the tip (coated with Au) and a flat Au substrate. A voltage bias (V) applied across the junction leads to a temperature rise of the nanoscale thermocouple (ΔT) due to the energy dissipation. The temperature rise (ΔT) is related

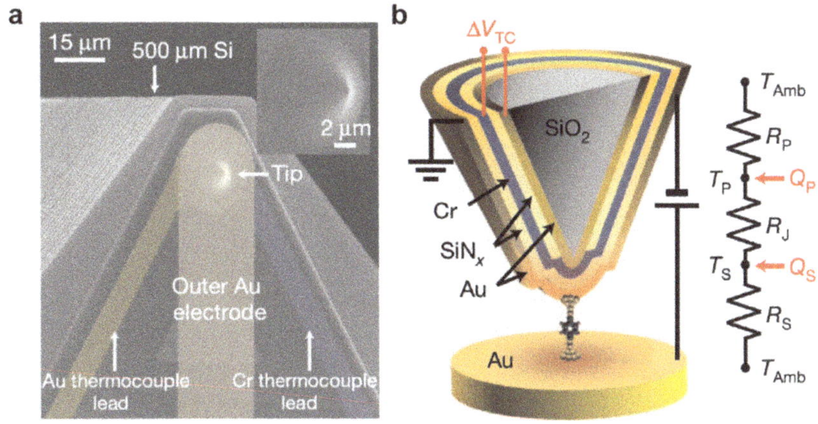

Figure 8.5. Experimental set-up for measuring heat dissipation in atomic-sized junctions. (a) SEM image of a nanoscale thermocouple integrated scanning thermal microscope probe. (b) Schematics of a single-molecule junction formed between the SThM tip (cross-sectional view) and a flat Au substrate. A thermal resistance network is shown on the right that represents the dominant thermal resistance to the dissipated heat flow. Figure reproduced with permission from [37]. Copyright 2013 Macmillan Publishers.

to the measured thermoelectric voltage (ΔV_{TC}) across the thermocouple by $\Delta T = \Delta V_{TC}/S_{Au-Cr}$, where S_{Au-Cr} is the Seebeck coefficient of the Au–Cr thermocouple. The heat dissipation (Q) in the Au electrode of SThM probe can subsequently be quantified using $Q = \Delta T/R_{probe}$, where R_{probe} is the thermal resistance of the SThM probe.

This work revealed an intimate relationship between the heat dissipation and the electronic transmission characteristics of atomic-sized junctions. As summarized in figure 8.6, for single-molecule junctions that have transmission characteristics that are strongly energy-dependent, asymmetric heat dissipation was observed in the two electrodes between which the molecular junction is formed. Moreover, the heat dissipation was found to depend strongly on both the bias polarity and the identity of the majority charge carriers (electron- or hole-dominated transport in molecular junctions). In contrast, for Au atomic junctions, which have weak energy-dependent electronic transmission characteristics, the measured heat dissipation demonstrated no appreciable asymmetry. These observations, in conjunction with the first-principles calculations, provided the first experimental evidence that validates the use of the Landauer transport framework to understand heat dissipation in systems where elastic electron transport is dominant.

8.5 Peltier cooling in molecular-scale junctions

Current flow across molecular junctions results in both heat dissipation via the Joule effect and cooling via the Peltier effect. Net refrigeration can only be observed when the Peltier cooling power is larger in magnitude than Joule heating. To probe the

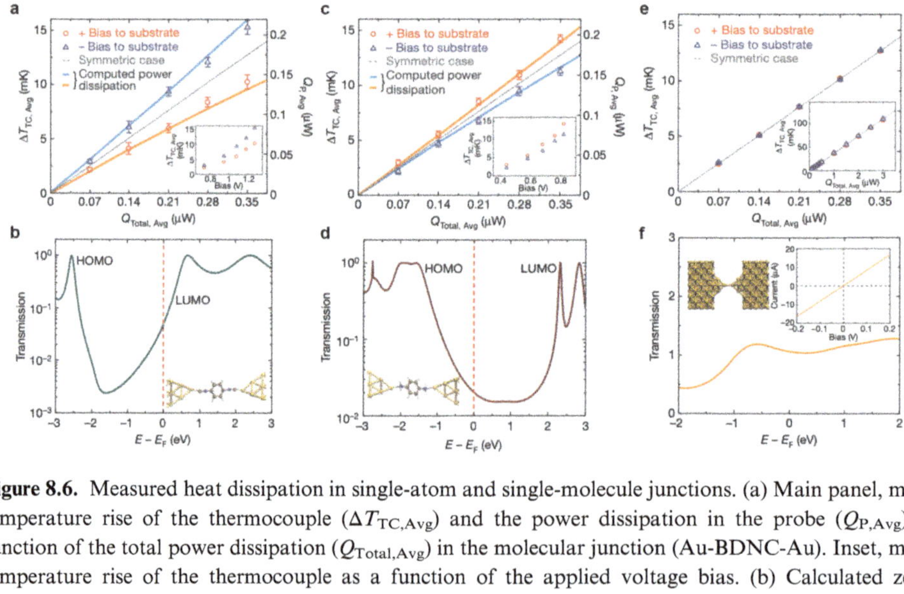

Figure 8.6. Measured heat dissipation in single-atom and single-molecule junctions. (a) Main panel, measured temperature rise of the thermocouple ($\Delta T_{TC,Avg}$) and the power dissipation in the probe ($Q_{P,Avg}$) as the function of the total power dissipation ($Q_{Total,Avg}$) in the molecular junction (Au-BDNC-Au). Inset, measured temperature rise of the thermocouple as a function of the applied voltage bias. (b) Calculated zero-bias electronic transmission function of the BDNC junction. HOMO: highest occupied molecular orbital; LUMO: lowest unoccupied molecular orbital. Reproduced with permission from [37]. Copyright 2013 Macmillan Publishers.

cooling effect at the molecular scale, we developed an experimental platform, as illustrated in figure 8.7. Specifically, we custom-fabricated calorimetric microdevices with embedded high-resolution Pt thermometers that can resolve a temperature change as small as 0.1 mK [38]. The microdevices feature four long and doubly clamped SiN_x beams to achieve excellent thermal isolation. These characteristics enabled the measurement of heating and cooling power with ~30 pW resolution.

During the measurement, a sharp Au-coated contact-mode AFM tip (with a radius of ~100 nm) is used to make a soft contact with a monolayer of organic

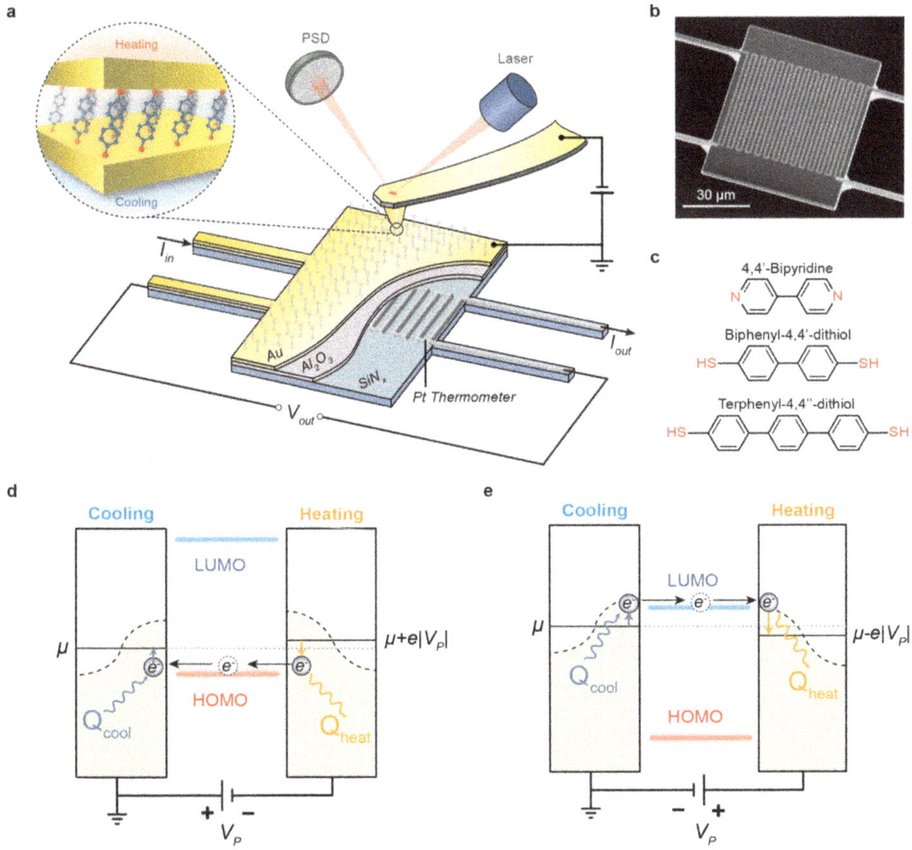

Figure 8.7. Experimental set-up of probing Peltier cooling in molecular junctions. (a) Schematic of the set-up. Molecular junctions are created by contacting an Au-coated atomic force microscope (AFM) tip with a monolayer of molecules self-assembled on an Au-coated calorimetric microdevice. The *I–V* characteristic of molecular junctions is measured by supplying a small voltage bias across the junctions and monitoring the resultant electric current. The heating or cooling effect in the current-flowing molecular junctions is measured by recording the temperature change of the microdevices using the embedded Pt thermometer. (b) SEM image of the microdevice. (c) Chemical structures of the molecules studied including BP, BPDT and TPDT. (d)–(e) Description of the Peltier effect in a molecular junction in which transport is HOMO and LUMO-dominated, respectively. A Lorentzian shaped transmission function is depicted around the HOMO and LUMO levels of the molecular junctions. μ_{cal}, Q_{heat} and Q_{cool} denote chemical potential, heating and cooling power, respectively. Reproduced with permission from [38]. Copyright 2018 Macmillan Publishers.

molecules that are self-assembled on the top Au layer of the microdevice. It can be estimated that approximately 100 molecules are contained in the nanoscopic volume between the AFM tip and the substrate. A small voltage bias (<20 mV) is applied across the molecular junctions to generate heating and cooling in the electrodes, which is captured on the side of the microdevice by the embedded Pt thermometer. The electrical conductance of the molecular junction can be quantified by measuring the resultant electrical current driven by the applied voltage bias, whereas the Seebeck coefficient of the molecular junctions could also be measured by applying suitable temperature differentials and probing the resulting thermoelectric voltage.

The observation of Peltier cooling in prototypical molecular junctions (Au–biphenyl-4,4′-dithiol (BPDT)–Au, Au–terphenyl-4,4′′(TPDT)-dithiol–Au, and Au–4,4′-bipyridine (BP)–Au) is shown in figure 8.8. It can be seen that net refrigeration occurs within a narrow voltage interval at small biases. The maximum cooling power, observed in Au–BPDT–Au molecular junctions in this work, is around 300 pW for approximately 100 individual molecules. Depending on the identity of the majority charge carrier in the molecular junction, the voltage bias at which net refrigeration occurs can be either positive (for LUMO-dominated molecular junctions such as Au–BP–Au junctions) or negative (for HOMO-dominated molecular junctions such as Au–BPDT–Au and Au–TPDT–Au

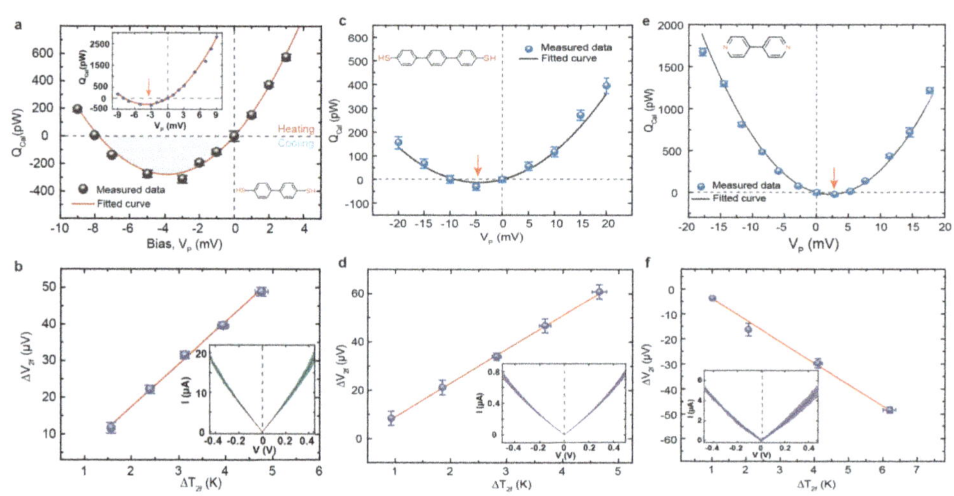

Figure 8.8. Measurement of the Peltier cooling, Seebeck effect and *I*–*V* characteristic of molecular junctions. (a) Main panel: measured thermal power as the function of the applied voltage bias across the BPDT molecular junctions. The solid red line indicates the fitted curve using equation $Q = Q_{\text{Pel}} + Q_{\text{Joule}} = \text{GTSV} + \text{GV}^2/2$ and the measured Seebeck coefficient and electrical conductance. The shaded blue region indicates the voltage region where net refrigeration is measured. Inset: the measured data and the fitted curve for voltage bias from –9 to 9 mV. The red arrow indicates the voltage at which the maximum cooling is observed. (b) Measured Seebeck coefficient (*S*) of the BPDT junctions. The red solid line indicates the best linear fit, with the slope showing the Seebeck coefficient. Inset: the *I*–*V* characteristics of the junctions. (c) and (e) The same as (a), but for TPDT and BP molecular junctions, respectively. (d) and (f) The same as (b), but for TPDT and BP molecular junctions, respectively. Reproduced with permission from [38]. Copyright 2018 Macmillan Publishers.

junctions). The parabolic shape of the heating or cooling power versus voltage bias curve can be understood readily in the Landauer transport framework (section 8.2.2), that is, in the small bias limit the total heating or cooling power of the molecular junction is the summation of the Peltier effect (Q_{Pel} = GTSV) and the Joule effect (Q_{Joule} = $GV^2/2$), where G is the electrical conductance, S is the Seebeck coefficient, V is the applied bias and T is the absolute temperature. As described above, our developed platform is capable of characterizing multiple transport parameters including electrical conductance and the Seebeck coefficient. Further experimental investigations have demonstrated the remarkable consistency between the measured heating and cooling power and the predictions of the Landauer framework.

8.6 Measurement of thermal conductance of single-molecule junctions

The study of heat transport in single-molecule junctions offers unique opportunities to reveal the fundamental mechanisms of vibrational (phonon) energy transport. Unlike heat transport in macroscopic and most mesoscopic materials where incoherent phonons govern the energy processes, thermal phonons in short molecule junctions transport coherently, even at room temperature. Peculiar phenomena are expected to arise, such as quantum interference of thermal transport [39]. Moreover, as natural one-dimensional physical systems, individual molecules may help provide deep insight into understanding the Fermi–Pasta–Ulam problem [40], an over half-century-old research topic regarding the possibility of anomalous infinitely large thermal conductivity in one-dimensional systems.

As a first step in the understanding of heat transfer at the single-molecule scale, our recent experimental study [41] has measured the thermal conductance of a series of single alkanedithiol molecule junctions and revealed the dependence of thermal transport on the number of molecular units. Specifically, we leveraged the scanning thermal microscopy SThM set-up (figure 8.9(a)) developed to measure the quantized thermal transport in single metallic-atom junctions. Once a molecule junction is trapped between the SThM tip and the substrate, the pre-applied temperature and voltage bias differential will allow both the heat current and electrical current flowing through the molecule junction. Figure 8.9(b) shows the measured electrical conductance and thermal conductance traces of a single hexanedithiol (C6) junction right before and after the spontaneous junction rupture. It can be seen that while the electrical signal exhibited a sudden drop by ~6×10^{-4} G_0 indicating the existence of a single-molecule junction, a corresponding conductance change is not discernible in the simultaneously recorded thermal signal. To improve the signal-to-noise ratio of thermal measurement, we implemented a time-averaging scheme by first performing hundreds of electrical and thermal measurements and then using the electrical conductance versus time traces to identify the time point when the junction rupture occurs. Using the electrical conductance signals as references, we then aligned and averaged the measured thermal signal to obtain a time-averaged thermal conductance curve. In figure 8.9(c), the results of thermal conductance after averages over

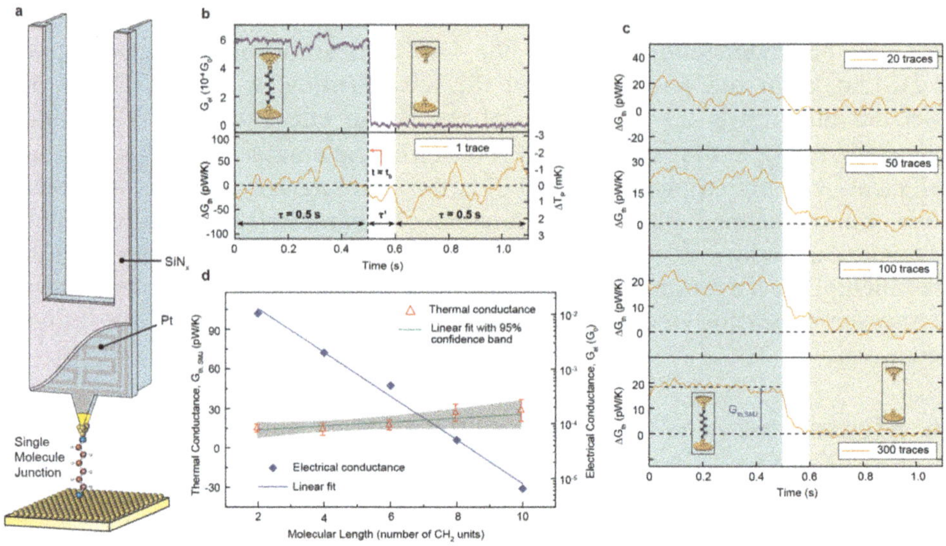

Figure 8.9. Measurements of the thermal conductance of single-molecule junctions. (a) Schematics of a single molecule trapped between an Au-coated tip of the heated SThM probe and an Au substrate at room temperature. The thermal conductance of single-molecule junctions is quantified by recording the temperature change of the Pt heater-thermometer when a single-molecule junction is broken. (b) Experimental protocol for measuring the thermal conductance of a single hexanedithiol (C6) junction. Upper panel: measured single electrical conductance versus time trace showing junction rupture at $t = t_b$ accompanied by a sudden drop of the conductance value. Lower panel: the simultaneously measured thermal conductance change (ΔG_{th}) and temperature change (ΔT_P) showing a large noise background in which the expected thermal conductance change is buried. (c) Improvement of the signal-to-noise ratio is clearly seen from the time-averaging of 20, 50, 100 and 300 traces. $G_{th,SMJ}$ in the last panel indicates the measured thermal conductance of Au–C6–Au single-molecule junctions. (d) Length-dependent electrical (blue diamonds) and thermal conductance (red triangles) of single alkanedithiol molecule junction. Reproduced with permission from [41]. Copyright 2019 Macmillan Publishers.

20, 50, 100 and 300 traces are illustrated, and reveal a clear thermal conductance jump at the time point when the junction ruptures.

Following the same experimental protocol, we have studied a series of alkanedithiol molecule junctions with different molecular units (–CH_2, from 2 to 10). As shown in figure 8.9(d), the measured electrical conductance exhibits an exponentially decaying behavior, as expected from the tunneling electron transport-dominated mechanism in such junctions. However, the thermal conductance of alkanedithiol molecules with different molecular lengths is observed to be nearly constant at a value of approximately 20 pW K^{-1}, strongly suggesting that phonon transport in these junctions is ballistic. We have also performed first-principles calculations which provide strong support for the experimental observations and detailed mechanisms in terms of the coherent vibrational energy transport in the single-molecule junctions.

8.7 Concluding remarks and outlook

Recently developed high-resolution scanning thermal microscopy techniques have made significant progress in probing energy transport in atomic-sized structures and testing the limitations of existing theoretical frameworks. These techniques and approaches set the stage for exploration of thermal and electrical properties of a broad range of atomic- and nano-structures and materials. Below, we briefly highlight some of the open questions in the field of atomic- and molecular-scale thermal science and, in light of the recent experimental advancements, we believe the first sets of measurements to address these open challenges are possible in the near future and would provide important insight into our current understanding of energy processes at the fundamental limit.

A central challenge in the field of thermoelectrics is to search for materials with a high thermoelectric figure of merit ZT. Bismuth telluride, the mostly applied thermoelectric material, possesses a ZT of ~1, corresponding to <20% of Carnot efficiency in energy conversion applications. This is significantly lower than the thermodynamic efficiency of energy conversion technologies such as internal combustion engines. A $ZT > 3$ is considered favorable to compete with traditional technologies, in particular at or slightly above room temperature. One fundamental question that remains to be answered in molecular systems is how high a value of ZT can be achieved in principle. Recent theoretical studies [42–47] have predicted high ZT values in specifically designed or synthesized molecular junctions due to quantum effects that have no classical analogues. One may ask, can we design proper experiments to validate these theoretical proposals? How can we rationally design the chemical structure of molecules to selectively enhance favorable transport properties and simultaneously suppress others for thermoelectric applications?

Another open question is to deepen our understanding of thermal (specifically phonons) transport in one-dimensional molecular systems such as individual polymer chains. Although previous experimental efforts [48–57] were made to characterize thermal transport in ensembles of molecules, such as self-assembled monolayers and polymer nanofibers, the recent study of single-molecule junctions allows one to resolve the issue of ensemble-based molecular measurements, such as uncertainties due to the heterogeneities of the molecules as well as potential influences of intermolecular interactions [58]. We expect that systematic experimental studies of single-molecule heat transport will enable comparisons among competing theories and help set the design principles for organic molecules for optimal thermal transport in soft materials.

References

[1] Agrait N, Yeyati A L and van Ruitenbeek J M 2003 *Phys. Rep.* **377** 81–279

[2] Aradhya S V and Venkataraman L 2013 *Nat. Nanotechnol.* **8** 399–410

[3] Imry Y 1997 *Introduction to Mesoscopic Physics* (New York: Oxford University Press), p xiii

[4] Landauer R 1957 *IBM J. Res. Dev.* **1** 223–31

[5] Landauer R 1989 *J. Phys. Condens. Mater.* **1** 8099–110

[6] Datta S 2005 *Quantum Transport: Atom to Transistor* (Cambridge: Cambridge University Press) p xiv

[7] Cuevas J C and Scheer E 2017 *Molecular Electronics: An Introduction to Theory and Experiment* (Singapore: World Scientific) p xix

[8] Esfarjani K, Zebarjadi M and Kawazoe Y 2006 *Phys. Rev.* B **73** 085406

[9] Chiatti O, Nicholls J T, Proskuryakov Y Y, Lumpkin N, Farrer I and Ritchie D A 2006 *Phys. Rev. Lett.* **97** 056601

[10] Jezouin S, Parmentier F D, Anthore A, Gennser U, Cavanna A, Jin Y and Pierre F 2013 *Science* **342** 601–4

[11] Meschke M, Guichard W and Pekola J P 2006 *Nature* **444** 187–90

[12] Partanen M, Tan K Y, Govenius J, Lake R E, Makela M K, Tanttu T and Mottonen M 2016 *Nat. Phys.* **12** 460–4

[13] Schwab K, Henriksen E A, Worlock J M and Roukes M L 2000 *Nature* **404** 974–7

[14] Ashcroft N W and Mermin N D 1976 *Solid State Physics* (New York: Holt) p xxi

[15] Crossno J *et al* 2016 *Science* **351** 1058–61

[16] Wakeham N, Bangura A F, Xu X, Mercure J F, Greenblatt M and Hussey N E 2011 *Nat. Commun.* **2** 396

[17] Sivan U and Imry Y 1986 *Phys. Rev.* B **33** 551–8

[18] Lake R and Datta S 1992 *Phys. Rev.* B **46** 4757–63

[19] Butcher P N 1990 *J. Phys. Condens. Mat.* **2** 4869–78

[20] Paulsson M and Datta S 2003 *Phys. Rev.* B **67** 241403

[21] Cahill D G 2004 *Rev. Sci. Instrum.* **75** 5119–22

[22] Balandin A A 2011 *Nat. Mater.* **10** 569–81

[23] Majumdar A 1999 *Annu. Rev. Mater. Sci.* **29** 505–85

[24] Kim K, Jeong W H, Lee W C and Reddy P 2012 *ACS Nano* **6** 4248–57

[25] Menges F, Mensch P, Schmid H, Riel H, Stemmer A and Gotsmann B 2016 *Nat. Commun.* **7** 10874

[26] Cui L, Jeong W, Fernandez-Hurtado V, Feist J, Garcia-Vidal F J, Cuevas J C, Meyhofer E and Reddy P 2018 *Nat. Commun.* **9** 14479

[27] Jeong W, Hur S, Meyhofer E and Reddy P 2015 *Nanosc. Microsc. Therm.* **19** 279–302

[28] Cui L, Miao R, Jiang C, Meyhofer E and Reddy P 2017 *J. Chem. Phys.* **146** 092201

[29] Cui L, Jeong W, Hur S, Matt M, Klockner J C, Pauly F, Nielaba P, Cuevas J C, Meyhofer E and Reddy P 2017 *Science* **355** 1192–5

[30] Yanson A I, Bollinger G R, van den Brom H E, Agrait N and van Ruitenbeek J M 1998 *Nature* **395** 783–5

[31] Scheer E, Agrait N, Cuevas J C, Yeyati A L, Ludoph B, Martin-Rodero A, Bollinger G R, van Ruitenbeek J M and Urbina C 1998 *Nature* **394** 154–7

[32] Xu B Q and Tao N J J 2003 *Science* **301** 1221–3

[33] Olesen L, Laegsgaard E, Stensgaard I, Besenbacher F, Schiotz J, Stoltze P, Jacobsen K W and Norskov J K 1994 *Phys. Rev. Lett.* **72** 2251–4

[34] Krans J M, Muller C J, Vanderpost N, Postma F R, Sutton A P, Todorov T N and Vanruitenbeek J M 1995 *Phys. Rev. Lett.* **74** 2146

[35] Olesen L, Laegsgaard E, Stensgaard I, Besenbacher F, Schiotz J, Stoltze P, Jacobsen K W and Norskov J K 1995 *Phys. Rev. Lett.* **74** 2147

[36] Mosso N, Drechsler U, Menges F, Nirmalraj P, Karg S, Riel H and Gotsmann B 2017 *Nat. Nanotechnol.* **12** 430–3

[37] Lee W, Kim K, Jeong W, Zotti L A, Pauly F, Cuevas J C and Reddy P 2013 *Nature* **498** 209–13
[38] Cui L, Miao R, Wang K, Thompson D, Zotti L A, Cuevas J C, Meyhofer E and Reddy P 2018 *Nat. Nanotechnol.* **13** 122–7
[39] Markussen T 2013 *J. Chem. Phys.* **139** 244101
[40] Fermi E, Pasta J and Ulam S 1955 Los Alamos Report, LA-1940 978
[41] Cui L, Hur S, Akbar Z A, Klockner J C, Jeong W, Pauly F, Jang S Y, Reddy P and Meyhofer E 2019 *Nature* **572** 628–33
[42] Murphy P, Mukerjee S and Moore J 2008 *Phys. Rev.* B **78** 161406
[43] Sadeghi H, Sangtarash S and Lambert C J 2015 *Nano Lett.* **15** 7467–72
[44] Karlstrom O, Linke H, Karlstrom G and Wacker A 2011 *Phys. Rev.* B **84** 113415
[45] Bergfield J P, Solis M A and Stafford C A 2010 *ACS Nano* **4** 5314–20
[46] Finch C M, Garcia-Suarez V M and Lambert C J 2009 *Phys. Rev.* B **79** 033405
[47] Hou S J, Wu Q Q, Sadeghi H and Lambert C J 2019 *Nanoscale* **11** 3567–73
[48] Xu Y F, Wang X X, Zhou J W, Song B, Jiang Z, Lee E M Y, Huberman S, Gleason K K and Chen G 2018 *Sci. Adv.* **4** eaar3031
[49] Wang X J, Ho V, Segalman R A and Cahill D G 2013 *Macromolecules* **46** 4937–43
[50] Meier T, Menges F, Nirmalraj P, Holscher H, Riel H and Gotsmann B 2014 *Phys. Rev. Lett.* **113** 060801
[51] Shen S, Henry A, Tong J, Zheng R T and Chen G 2010 *Nat. Nanotechnol.* **5** 251–5
[52] Wang R Y, Segalman R A and Majumdar A 2006 *Appl. Phys. Lett.* **89** 173113
[53] Wang Z H, Carter J A, Lagutchev A, Koh Y K, Seong N H, Cahill D G and Dlott D D 2007 *Science* **317** 787–90
[54] Losego M D, Grady M E, Sottos N R, Cahill D G and Braun P V 2012 *Nat. Mater.* **11** 502–6
[55] Majumdar S, Sierra-Suarez J A, Schiffres S N, Ong W L, Higgs C F, McGaughey A J H and Malen J A 2015 *Nano Lett.* **15** 2985–91
[56] Ge Z B, Cahill D G and Braun P V 2006 *Phys. Rev. Lett.* **96** 186101
[57] O'Brien P J, Shenogin S, Liu J X, Chow P K, Laurencin D, Mutin P H, Yamaguchi M, Keblinski P and Ramanath G 2013 *Nat. Mater.* **12** 118–22
[58] Majumdar S, Malen J A and McGaughey A J H 2017 *Nano Lett.* **17** 220–7

IOP Publishing

Nanoscale Energy Transport
Emerging phenomena, methods and applications
Bolin Liao

Chapter 9

Ultrafast thermal and magnetic characterization of materials enabled by the time-resolved magneto-optical Kerr effect

Dustin M Lattery, Jie Zhu, Dingbin Huang and Xiaojia Wang

As traditional complementary metal oxide semiconductors struggle to extend previous industrial trends, new technologies must be researched and delivered. One of the most important aspects that must be considered is the transport of heat within the material. By advancing the design of materials and interfaces, heat transfer within electronic devices can be improved. At the same time, novel technologies that rely on the magnetism of thin films also need to have their transient magnetic behavior optimized. By measuring the magnetic response of the materials, engineers can select the best-matched materials to design and fabricate devices with lower power consumption and higher processing speed and thus improved performance. Such material transport studies require new methods and metrology development that can provide highly sensitive and accurate characterization of the materials. The time-resolved magneto-optical Kerr effect (TR-MOKE) technique is capable of probing both the thermophysical and magnetic properties of a variety of materials, and it offers superb spatial (micrometer) and temporal (sub-picosecond) resolutions. In this chapter, we provide information about this technique through examples of its application in the study of thermal transport and magnetization dynamics. We then highlight several novel research directions that are potentially enabled by this technique, thus expanding the applications of TR-MOKE to form a more comprehensive picture of energy transport.

doi:10.1088/978-0-7503-1738-2ch9

9.1 Introduction

9.1.1 Background and motivation

Transport properties (e.g. electrical conductivity, thermal conductivity and the transfer of magnetic moment of materials) are of critical importance in a broad range of engineering applications. In this chapter, we highlight the state-of-the-art time-resolved magneto-optical Kerr effect (TR-MOKE) methodology, based on the ultrafast pump–probe technique, for characterizing the thermal and magnetic transport properties of several representative materials. These materials are of technological importance, serving as building blocks for the next generation of electronic, spintronic and data storage devices. For decades, these device components have been manufactured following Moore's law which states that the number of transistors per chip should double every two years [1]. Semiconductor industries have pushed to maintain this trend, but they are finally being limited by the power density of device operation, or more simply, heat extraction [2]. By moving electrons through more closely spaced transistors at faster switching speeds, these devices are producing progressively more dense heat loads, imposing a continually growing need for thermal management (the capability of redistributing and removing heat). The solutions proposed by researchers have followed two main paths: (i) developing new technologies that require less power and (ii) engineering new materials and better interfaces that can be scaled down without increasing heat generation or impeding heat transfer.

Following the first path, the field of spintronics (spin-electronics) has proven a promising direction since the discovery of giant magnetoresistance [3, 4]. On the fundamental level, spintronics focuses on advancing materials by manipulating the magnetization (or spin) in magnetic materials to achieve so-called 'beyond CMOS' (complementary metal oxide semiconductor) technologies. Theoretically, spintronic devices have the benefits of minuscule amounts of power being required for switching, fast switching speeds and non-volatility (i.e. they do not require power to retain information), making them ideal for both processing and memory. Spintronics have already been adopted in widespread applications. The most common application can be found in magnetic random-access memory (MRAM), which has rapidly gone from utilizing the magnetic field to switch memory [5], to spin-transfer torque-MRAM [6, 7] and to spin-orbit torque-MRAM [8, 9], making use of cutting-edge physics along the way. The unique advantages of these memory technologies have further enabled advanced applications in all-spin logic (using only spin transport for computation) [10, 11], probabilistic computing [12], spin torque oscillators [13–16] and heat-assisted magnetic recording (HAMR) [17–19], among others. For these technologies, it is crucial to understand the magnetic properties (such as the Gilbert damping α) of materials at short time scales (e.g. sub-nanosecond) to guide further research and development.

The characterization of magnetic material properties has often exploited the technique of ferromagnetic resonance (FMR) [20]. Generally, FMR uses a microwave signal to excite a magnetic sample. The change in magnetic susceptibility of the sample is measured as the sample goes through its resonance condition, during which the external field and frequency agree with the Kittel dispersion for resonance [21]. The resulting microwave absorption, as a function of frequency or field, can be fitted to a Lorentzian or anti-Lorentzian function, where the width of the Lorentzian (the so-called linewidth) is dependent on the damping parameter α [22]. While this highly versatile technique has adapted advancements (such as stripline FMR and others [23]), it has difficulty characterizing new technologically relevant materials with large perpendicular magnetic anisotropy (PMA). The large anisotropy requires a high-power input to excite the magnetization, and the relatively large damping of metallic materials intrinsically leads to large linewidths. The search for alternative measurement techniques has been motivated by these challenges faced by FMR.

Along the second path of material engineering solutions, electronics research has continued in the categories of materials innovation and heat dissipation (particularly in high-flux and high-voltage power electronics). Recent developments in materials and in device miniaturization have created new opportunities but have also imposed challenges for the science of thermal transport and the technology of thermal management [24]. Advances in synthesis, processing and microanalysis are enabling the production of well-characterized materials with structural features ranging in size from micrometers down to nanometers. Enormous attention has been paid to functionalized structures such as superlattices, multilayer coatings, nanowire arrays and polymer nanocomposite materials [25–31]. Thermal interface materials and thermal fluids have drawn growing attention in both industry and the military for cooling of electronics [32–34]. Emerging two-dimensional (2D) materials such as graphene, hexagonal boron nitride (h-BN), black phosphorus (BP) and MoS_2, as well as bulk β-Ga_2O_3, have been considered as building blocks for future electronics [35–38]. Their anisotropic thermal transport properties, induced by the materials' structures, need to be understood more comprehensively to advance device performance.

9.1.2 Ultrafast-laser-based metrology for transport studies

For these emerging materials and technologies, ultrafast laser-based pump–probe techniques provide sensitive, powerful and high-throughput capabilities for the study of transport in materials. The high-temporal resolution of the ultrafast pump–probe method makes it suitable for studying dynamics occurring on time scales from hundreds of femtoseconds (fs, 10^{-15} s) to several nanoseconds (ns, 10^{-9} s), for both thermal and magnetic transport processes. The basic pump–probe configuration is time-domain thermoreflectance (TDTR) [39–41], which detects the temperature-dependent reflectance of the sample to extract the materials' thermal properties [41–53] or to quantify thermal transport across interfaces [54–58]. When integrated with the 'beam-offset' approach, TDTR can probe thermal transport, along both the through-plane and in-plane directions, to reconstruct the three-dimensional (3D) thermal conductivity tensor of thermally anisotropic materials [59, 60]. This method has been successfully demonstrated for certain materials [59–62]; however, it has also been proven to produce low measurement sensitivity to the in-plane thermal transport in materials with low thermal conductivity [63]. This is mainly attributed to the significant lateral heat spreading in the transducer layer during TDTR measurements, which smears out the thermal information of the underlying sample. Therefore, the transducer layer for TDTR measurements is preferred to have a low thermal conductivity to improve the measurement sensitivity to the in-plane thermal transport properties of the sample materials.

The TR-MOKE, a system initially invented to probe magnetization dynamics [64, 65], can be extended to thermal measurements by taking advantage of the temperature-dependent magnetization of the transducer. Therefore, TR-MOKE can be essentially treated as an upgraded version of a standard TDTR system [63, 66]. TR-MOKE uses optically thin magnetic transducers that are immune to contamination by thermoreflectance signals from the sample beneath and thus provides greatly enhanced measurement sensitivity to in-plane thermal transport. For magnetic transport studies, TR-MOKE can detect magnetization dynamics of materials with superb spatial (diffraction-limited beam spots) and temporal (sub-picosecond) resolutions. Particularly, the use of optical pumping and detection in TR-MOKE allows it to capture the ultrafast magnetization of 'hard' materials (with large magnetic anisotropy) that are not detectable using conventional FMR methods.

In this way, TR-MOKE provides a unique capability of studying the transport properties (both thermal and magnetic) of engineered materials, and it can enable exciting new technologies. Within this chapter, we aim to provide the foundation of the TR-MOKE technique, detail the information about data reduction for magnetization dynamics and thermal transport, and discuss several representative applications enabled by this promising technique.

9.2 TR-MOKE measurement technique

MOKE allows for direct optical measurements of the magnetic state of a material. To reveal the correlation between the optical response and the magnetism of the material, this section will discuss the physical foundation of MOKE measurements, the relationship between TR-MOKE signals and the thermal and magnetic transport properties of thin-film samples, and the typical optical set-up of TR-MOKE in the pump–probe configuration.

9.2.1 The physical foundation

As first described by Kerr [67], MOKE is a process that alters the polarization state of light reflected by a magnetic material. Fundamentally, MOKE stems from the different interactions of left- and right-circularly polarized light within a magnetized material. Linearly polarized light can be represented as the sum of equal proportions of left- and right-circularly polarized light, and each type of circular polarization will experience a different phase shift and absorption when interacting with a magnetic material [68]. The result of this process is the transformation from a linear polarization to an elliptical polarization upon the reflection of light (or transmission of light, for the analogous Faraday effect), as shown by figure 9.1.

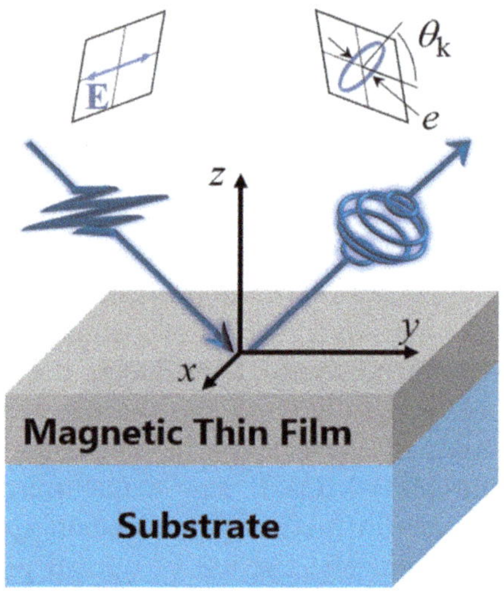

Figure 9.1. An illustration of the complex polarization rotation of reflected light from a magnetic material known as MOKE. The rotation of the polarization, from a linear polarization to an elliptical polarization, is denoted by the Kerr angle (θ_k). The ellipticity is denoted by e.

The rotation of linearly polarized light can be described by the response of an electric field vector to the dielectric tensor $\bar{\varepsilon}$, which is given by

$$\bar{\varepsilon} = \begin{pmatrix} \varepsilon_{xx} & \varepsilon_{xy} & \varepsilon_{xz} \\ -\varepsilon_{xy} & \varepsilon_{yy} & \varepsilon_{yz} \\ -\varepsilon_{xz} & -\varepsilon_{yz} & \varepsilon_{zz} \end{pmatrix}. \tag{9.1}$$

For an isotropic, non-magnetic material, the diagonal components of this tensor are equal ($\varepsilon_{xx} = \varepsilon_{yy} = \varepsilon_{zz}$), and the off-diagonal components are 0. For isotropic magnetic materials, however, the off-diagonal terms are related to the magnetization vector (**M**) through

$$\bar{\varepsilon} = \varepsilon_{xx} \begin{pmatrix} 1 & -\mathrm{i}Qm_z & Qm_y \\ \mathrm{i}Qm_z & 1 & -\mathrm{i}Qm_x \\ -\mathrm{i}Qm_y & \mathrm{i}Qm_x & 1 \end{pmatrix}, \tag{9.2}$$

where Q is the magneto-optical constant and $m_i = M_i/\|\mathbf{M}\|$ [69–72]. These nonzero off-diagonal terms cause different polarization changes to the opposing circular polarizations. This leads to a complex rotation angle of the polarization, given by $\tilde{\theta} = \theta_k + \mathrm{i}e$, where the real part of $\tilde{\theta}$ is the Kerr rotation and the imaginary part is the ellipticity [70, 71]. $\tilde{\theta}$ is also sometimes presented as components of a complex permittivity tensor [73]. Our discussion in this chapter will be limited to a discussion of the real component of the Kerr rotation, θ_k, the real rotation of the major axis of polarization upon reflection of linearly polarized light.

At equilibrium, θ_k contains information about the magnetization state in magnetic materials. It is therefore adopted as an alternative method for measuring magnetic hysteresis loops [74], in addition to vibrating sample magnetometry, alternating gradient magnetometry, and superconducting quantum interference device measurements. The MOKE response has proven to be powerful for measuring the magnetic properties of nanomaterials, including ferromagnetic monolayers [75]. MOKE microscopy has also been utilized to sense domains in magnetic materials, owing to the large contrast resulting from the opposite Kerr rotation of antiparallel magnetization between domains [76, 77]. These optical studies demonstrate well the use of MOKE for investigating magnetostatics (i.e. the magnetization of the sample is not changing in time). In the following sections, we will focus on transient magnetization dynamics induced by ultrafast laser pulses.

9.2.1.1 Ultrafast demagnetization induced by laser heating

The application of time-resolved Kerr rotations for ultrafast metrology began as a method to determine the non-equilibrium processes initiated by ultrafast laser excitation in ferromagnetic nickel [78]. Through MOKE, the magnetization within the sample can be measured, providing a window into the temperatures of various energy carriers, including electrons, phonons and magnons (wave-like variations in the magnetization). Due to the limitation of using lasers with pulse durations on the order of tens of picoseconds, early TR-MOKE measurements of ferromagnetic materials were unable to directly show these temperatures of carrier populations out of equilibrium with each other (the non-equilibrium regime) [79]. With the new application of femtosecond laser pulses (~60 fs), Beaurepaire *et al* were able to capture a sub-picosecond reduction in magnetization (demagnetization) resulting from the laser induced heating [80]. After several picoseconds to tens of picoseconds following laser excitation, the energy carriers approach thermal equilibrium, and the energy transfer will then be dominated by thermal transport via heat conduction. The temperature decay in the sample system can then be described by heat diffusion, which depends on the thermal conductivity (Λ), volumetric heat capacity (C) and the interfacial conductance (G) of the multi-layers and interfaces within the sample. The discussion of extracting these thermal parameters from the TR-MOKE signal is detailed in section 9.3.

9.2.1.2 Precessional magnetization dynamics

In addition to thermal information, this ultrafast demagnetization from laser pulses also initiates magnetization dynamics governed by the Landau–Lifshitz–Gilbert (LLG) equation, specifically magnetization (spin) precession [81, 82]. Further research into this all-optical, pump–probe technique showed that the frequency of magnetic precession extracted from TR-MOKE is consistent with frequency-domain FMR results [64]. The working principle for spin precession measured with TR-MOKE consists generally of three distinct regions, as illustrated in figure 9.2 [40, 64]. Initially, the magnetization **M** is in equilibrium and is parallel to the effective field ($\mathbf{H}_{\mathrm{eff}} = -\nabla_{\mathbf{M}} F$, with F being the magnetic free energy density), which is the minimum energy direction for the magnetization. Then, the pump beam deposits energy into the magnetic material, heating it up and inducing thermal demagnetization. Because of the heating, both the material's saturation magnetization (M_s)

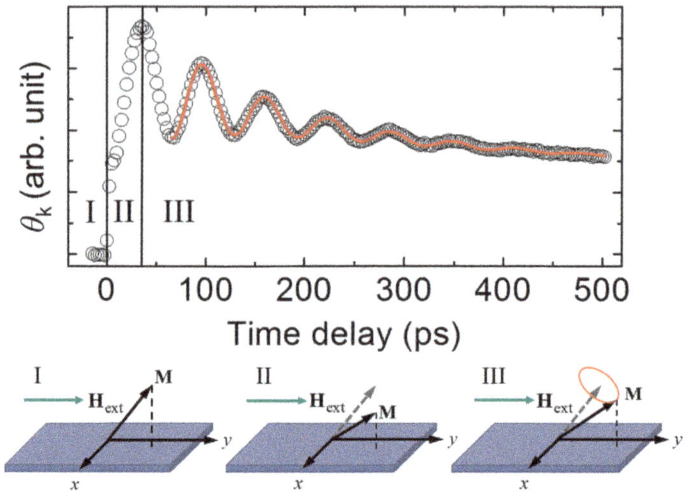

Figure 9.2. The typical signal of magnetic precession from polar TR-MOKE (open symbols). In region I, the system is in equilibrium with the magnetization (**M**) canted by an external field ($\mathbf{H}_{ext}$) to be along $\mathbf{H}_{eff}$. Following the laser pulse heating, both the saturation magnetization and magnetic anisotropy will decrease, which results in a change in the minimum energy direction in region II. After some amount of time, M_s will recover, but the angle between **M** and $\mathbf{H}_{eff}$ will result in precession (region III). The solid line indicates the fit of the data to a decaying sinusoid as expressed by equation (9.13). In regions I and III, $\mathbf{H}_{eff}$ is pointing along the equilibrium direction as denoted by the gray dashed line.

and magnetic anisotropy decrease, resulting in a change in $\mathbf{H}_{eff}$. Next, as the magnetic material cools down, M_s and magnetic anisotropy begin to recover to their initial values, restoring the minimum energy direction back to the original equilibrium direction. At this point, **M** does not align with $\mathbf{H}_{eff}$, resulting in a torque that acts on **M**. This torque causes damped precessional motion around the equilibrium direction, as described by the LLG equation with a damping parameter (α) and the gyromagnetic ratio (γ) [83]:

$$\frac{d\mathbf{M}}{dt} = -\gamma \mathbf{M} \times \mathbf{H}_{eff} + \frac{\alpha}{M_s}\left(\mathbf{M} \times \frac{d\mathbf{M}}{dt}\right). \tag{9.3}$$

After some mathematical manipulation, equation (9.3) will provide the theoretical foundation to analyze TR-MOKE measurement data for extracting both the spin precession frequency (f) and (α), which will be discussed further in section 9.4.

9.2.2 Optical setup of time-resolved magneto-optical Kerr effect

The TR-MOKE metrology belongs to the ultrafast pump–probe technique that uses a femtosecond laser to first pump energy into a material and then to probe the material response. The major difference between the TR-MOKE and TDTR techniques is the type of signals collected by the probe beam. In TR-MOKE measurements, the polarization state of the probe beam reflected from the sample is monitored [84], while in TDTR measurements, the reflectivity from the sample

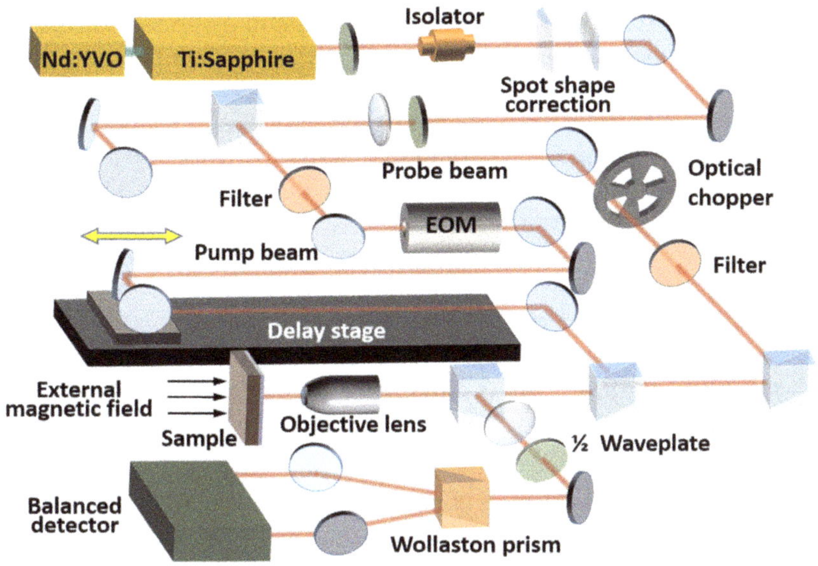

Figure 9.3. A schematic for the TR-MOKE measurement system. Reproduced with permission from [85]. Copyright 2016 the American Chemical Society.

surface is collected. For a small temperature rise, both TR-MOKE and TDTR signals can be treated as linearly proportional to the temperature variation of the sample.

Figure 9.3 depicts the optical layout of an example TR-MOKE set-up at the University of Minnesota, Twin Cities [40], which is upgraded from the basic two-tint time-domain thermoreflectance setup [41]. In TR-MOKE, a mode-locked Ti: Sapphire laser generates a train of pulses (typically ~100 fs in duration) at a repetition rate of 80 MHz (12.5 ns between pulses). An isolator placed right after the laser output prevents the back reflection of light into the laser cavity. The beam shape is corrected by a pair of cylindrical lenses to produce a circular beam spot (preferred for in-plane thermal transport measurements). A polarizing beam splitter separates the laser into pump and probe beams with orthogonal polarizations. The pump beam is modulated with an electro-optical modulator synchronized to a function generator, typically operated at a tunable frequency in the range of 0.1–20 MHz. The probe beam is modulated by a mechanical chopper (~200 Hz). The optical path of the pump beam can be adjusted by a delay stage, which produces a time separation of up to 4 ns between pump heating and probe sensing. The diffraction-limited beam spot size ($1/e^2$ radius) at the sample surface ranges from one to a few tens of micrometers, depending on the magnification of the objective lens [60]. A set of optical filters is exploited to create a spectral separation between pump and probe to suppress pump light that might otherwise leak into the detector. An electromagnet is placed near the sample to provide external magnetic fields (H_{ext}) for the sample.

To facilitate TR-MOKE measurements, a Wollaston prism in conjunction with a balanced detector (photodiode) are used to capture the Kerr rotation angle of the probe beam reflected from the sample. Caution should be exercised to carefully balance the photodiode prior to conducting a measurement, to suppress non-MOKE signals. Additional steps such as differential measurements can also be taken to reduce the non-MOKE components received by an imperfectly balanced detector [63, 66]. The output signal from the balanced detector is sent to a radio-frequency (RF) lock-in amplifier and then to a computer for signal processing with a digital audio-frequency (AF) lock-in. This double-modulation and double lock-in technique allows for the detection of low-level Kerr rotation signals. A more detailed description of the signal analysis for thermal and magnetic transport studies will be discussed, respectively, in sections 9.3 and 9.4.

9.3 Thermal measurements

At a fundamental level, TR-MOKE can be applied for thermal transport studies by correlating the magnetization variation of the material to its temperature excursion. In this section, we derive the relationship between MOKE signals and the temperature of a magnetic material and provide the detailed procedures for the data reduction of thermal measurements. Following that, we present examples of TR-MOKE measurements of several representative materials that are thermally anisotropic. These examples demonstrate the improved measurement sensitivity of TR-MOKE, compared to TDTR, to the in-plane thermal transport within materials.

9.3.1 Temperature information from TR-MOKE signals

Unlike TDTR which uses optically opaque and non-magnetic metallic films as the transducer to absorb light and to probe temperature, TR-MOKE can incorporate magnetic films that are optically semitransparent. When the magnetic transducer is pumped by a laser pulse, it undergoes a process of angular momentum transfer at short time scales, known as ultrafast demagnetization. The thermally induced demagnetization significantly alters the thermodynamic equilibrium among electrons, phonons and magnons and is then followed by a re-magnetization (or recovery) process which happens over times of 1–100 ps [86–91]. The polarization state change of the probe beam reflected from the magnetic transducer, or Kerr rotation change ($d\theta_k$), is temperature-dependent. When the temperature rise (ΔT) of the material is small, the signal (S) collected from TR-MOKE measurements is linearly proportional to ΔT:

$$S \approx \frac{dS}{dT}\Delta T = \gamma R \frac{d\theta_k}{dT}\Delta T, \tag{9.4}$$

where R is the reflectivity and γ is a conversion coefficient from the optical signal to electrical signal, taking into account the lock-in amplification and gain factors of other electronic devices. The temperature dependence of the Kerr rotation angle, $d\theta_k/dT$, is defined as the thermo-magneto-optical coefficient, which is analogous to the thermoreflectance coefficient ($\mathrm{d}R/\mathrm{d}T$) in TDTR. Both coefficients represent the responsivity (via either the Kerr rotation angle or the reflectivity) of a transducer to a temperature change. With a first approximation, $\mathrm{d}\theta_\mathrm{k}/\mathrm{d}T$ can be related to the magnetization of the magnetic transducer film:

$$\frac{\mathrm{d}\theta_\mathrm{k}}{\mathrm{d}T} = \frac{\mathrm{d}\theta_\mathrm{k}}{\mathrm{d}M}\frac{\mathrm{d}M}{\mathrm{d}T} = \frac{\mathrm{d}\theta_\mathrm{k}}{\mathrm{d}M/M_\mathrm{s}}\frac{\mathrm{d}M/M_\mathrm{s}}{\mathrm{d}T}. \tag{9.5}$$

Since θ_k is linearly proportional to M and is equal to zero when M equals zero, $\mathrm{d}\theta_\mathrm{k}/\mathrm{d}M$ can then be written as

$$\frac{\mathrm{d}\theta_\mathrm{k}}{\mathrm{d}M} \approx \frac{\theta_\mathrm{ks}}{M_\mathrm{s}}, \tag{9.6}$$

where θ_ks is the Kerr rotation angle at the saturated magnetization (M_s) state at room temperature [74]. By substituting equations (9.5) and (9.6) into equation (9.4), the MOKE signal (S) is linearly proportional to the product of several original factors:

$$S \propto R\theta_\mathrm{ks}\frac{\mathrm{d}M/M_\mathrm{s}}{\mathrm{d}T}\Delta T = p\Delta T. \tag{9.7}$$

The parameters R, θ_ks and $[\mathrm{d}M/M_\mathrm{s}]/\mathrm{d}T$ construct, together, the linear temperature dependence of the TR-MOKE signal. Similar to $\mathrm{d}\theta_\mathrm{K}/\mathrm{d}T$, the product ($p$) of these parameters remains constant for a certain magnetic material, provided that the temperature rise (ΔT) of the material is sufficiently small during TR-MOKE measurements to prevent any nonlinear effects in the $T \sim M$ relation. The product p before ΔT on the right-hand side of equation (9.7) is material dependent and can be optimized for enhancing TR-MOKE signals during thermal measurements. For this purpose, several magnetic materials have been explored as TR-MOKE transducers in the literature, including the 20 nm Co/Pt multilayer studied by Liu *et al* [63] and the 5 nm FePt:Cu ultra-thin alloy films prepared by Kimling *et al* [92]. To systematically investigate and optimize the thermo-magneto-optical coefficient (also the product p), Chen *et al* synthesized magnetic thin films of rare-earth transition metal (RE-TM) alloys and multilayer structures. They demonstrated that among the materials they tested, RE-TM alloys (TbFe) with an optimal thickness of ~20 nm provided the highest signal-to-noise ratio (SNR) for TR-MOKE measurements of thermal properties [85]. They further revealed the origin of TR-MOKE signals and attributed the SNR enhancement to the lower Curie temperatures and larger θ_ks at the laser operating wavelengths of RE-TM alloys. Since TR-MOKE is an emerging technique, there have been limited studies of the material selection and characterization for magnetic transducers. Further efforts should be devoted to identifying magnetic transducers for TR-MOKE thermal measurements under different conditions, such as varied temperature, pressure and laser wavelength.

9.3.2 Measurement process and data analysis of TR-MOKE

Similar to other optical pump–probe methods applied for thermal measurements, in TR-MOKE, the sample is coated by a transducer film (magnetic) with known properties and structural parameters, including thermal conductivity, specific heat and film thickness. These parameters will be used as inputs for the heat diffusion model for data analysis. For the ease of optical alignment and experimental operation, the polar MOKE configuration is preferred, in which the laser beam is normally incident onto to the sample surface. For this polar MOKE configuration, a magnetic transducer with PMA (the magnetization is along the through-plane direction parallel to the film surface normal) can be used to optimize the Kerr rotation signal. In addition, normal incidence in polar MOKE can also produce a circular beam spot shape that is preferred for in-plane thermal measurements. The circular beam shape can simplify the thermal analysis based on the heat diffusion model.

The in-phase (V_{in}) and out-of-phase (V_{out}) signals are collected by the RF lock-in amplifier in TR-MOKE experiments. Prior to the acquisition of each signal, the magnetic transducer is magnetized to saturation with an external magnet. The magnetization orientation of the magnetic layer can be flipped by reversing the polarity of the external magnet. Each sample is measured twice: with the initial magnetization orientation, and its reverse, referred to as $M+$ and $M-$. As shown in figure 9.4(A), both V_{in} and V_{out} change sign when the initial magnetization is flipped from $M+$ to $M-$. The actual signal used for thermal analysis is the corrected signal taken as the difference between the signals of two measurements with opposite magnetization orientations, which excludes contributions from all non-MOKE components (e.g. thermoreflectance signals). By fitting the time-resolved ratio ($-V_{in}/V_{out}$) from TR-MOKE measurements (figure 9.4(B)) to the heat diffusion model (similar to what is done in TDTR), the through-plane thermal transport properties of the sample underlying the magnetic transducer can be extracted [39, 66].

9.3.3 High-sensitivity thermal measurements enabled by TR-MOKE

Higher measurement sensitivity and therefore better measurement precision are always of critical importance for metrology advancement. For thermal transport studies based on the pump–probe technique, the measurement sensitivity to thermal properties of the sample can be strongly influenced by the transducer's properties and structural parameters. Generally speaking, thinner transducers with low thermal conductivities are beneficial for improving the measurement sensitivity to the thermal properties of the underlying sample material. In TDTR, thin films of aluminum (Al) are commonly chosen as metal transducers owing to the large thermoreflectance coefficient of Al at the laser operating wavelength (near-infrared) and strong adhesion between Al and the sample. This transducer film needs to be optically opaque (typically 70–100 nm) such that thermoreflectance signals of the underlying material will not contribute to the transient reflectivity [41]. TR-MOKE can instead work with optically thin magnetic transducers, which are immune to

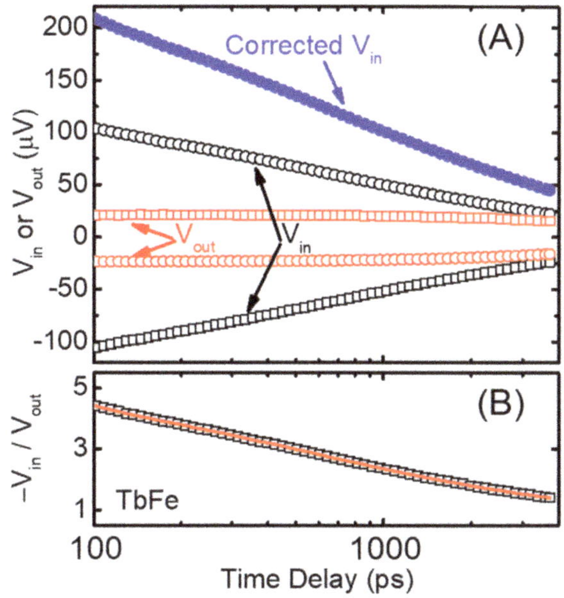

Figure 9.4. Example of the through-plane thermal measurements signal from TR-MOKE on a reference sample of a 300 nm SiO_2 film with an 11.5 nm TbFe transducer layer. (A) In-phase (V_{in}) or out-out-phase (V_{out}) voltage measured via TR-MOKE will change signs depending on the initial magnetization state. Prior to the measurement, the sample is magnetized out-of-plane along either the positive ($M+$) or negative ($M-$) direction resulting in flipped signs for V_{in} and V_{out}. The corrected V_{in} signal is the difference between the V_{in} from $M+$ and $M-$ to exclude non-MOKE signals. (B) The ratio with the corrected V_{in} is fitted to the thermal model to extract the through-plane thermal properties of the sample. Reproduced with permission from [85]. Copyright 2016 the American Chemical Society.

contamination by thermoreflectance signals from the sample itself. For this reason, the thickness of the magnetic transducer can be significantly reduced to enhance the measurement sensitivity to the thermal conductivity of the materials.

For measurements of thermal transport in the through-plane direction, TR-MOKE also provides a better sensitivity to interfacial thermal conductance (G). Using the standard TDTR method, it can be challenging to probe G between dissimilar materials (i.e. materials with large contrasts in stiffness or in the spectra of heat carriers' density of states), such as the interface between metals and oxides (e.g. Al/SiO_2). Using TR-MOKE, Kimling *et al* deposited ultra-thin Co/Pt magnetic layers (4.2–8.2 nm) as magnetic transducers to measure the G values between SiO_2 thin films (with thicknesses of 26–440 nm) and Si substrates [93]. They obtained an unexpected high value of G that is approximately 1.4 GW m^{-2} K^{-1}, which is challenging to detect using TDTR with optically opaque non-magnetic transducers.

As for in-plane thermal measurements, TR-MOKE has even more critical advantages to achieve higher sensitivities, and thus it is well aligned for studying the anisotropic thermal transport properties of 2D materials, such as graphene, h-BN, BP and MoS_2 [35–38]. To measure in-plane thermal transport, TR-MOKE also adopts the 'beam-offset' approach [59, 94]. The full-width at half-maximum

(FWHM) of V_{out} at a negative time delay is recorded as a function of the in-plane offset distance between the pump and probe beams. A heat conduction model taking into account the anisotropy in the materials' transport properties and heat source intensity profiles can be used to extract the sample's thermal conductivity tensor from the measured beam-offset data, given that the beam spot size is comparable to, or smaller than, the in-plane thermal penetration length (related to the heating modulation frequency) [60]. In addition to the small film thickness (<20 nm), the magnetic transducer typically has a lower thermal conductivity (<20 W m^{-1} K^{-1}) resulting from the multilayer or alloy structures. This can suppress the heat spreading in the transducer, and thus the signal contains more information about thermal transport along the in-plane direction within the sample beneath. Therefore, by using optically thin magnetic transducers with low thermal conductivities, TR-MOKE provides reduced systematic errors for measuring in-plane thermal transport [60].

Figure 9.4 summarizes several representative examples of TR-MOKE measurements of in-plane thermal transport from the literature. Zhu *et al* investigated the 3D thermal conductivity of single-crystal BP flakes along the three primary crystalline orientations using TR-MOKE with a 28 nm film of the TbFe alloy as the magnetic transducer [66]. The beam-offset approach was adopted to generate a 2D contour plot of V_{out}, by scanning the pump beam both vertically and horizontally, as shown in figure 9.5(A). The elliptical (instead of circular) shape of the 2D contour plot indicates the anisotropic in-plane thermal transport in BP. A line-cut can be made along any direction (preferably the primary crystalline direction, along the major and minor axes of the ellipse) from the 2D contour to generate the FWHM of V_{out} versus offset distance between pump and probe beams. Therefore, the 2D-scanning beam-offset method does not require a precise sample alignment with a specific crystalline orientation for sample loading, which greatly reduces the difficulties in thermal characterization of anisotropic samples that are a few micrometers in size. For example, in the measurements conducted by Zhu *et al*, the X-axis of the beam-offset direction (white solid line) was pre-aligned roughly along the zigzag orientation of the BP flake (the major axis of the ellipse, white dashed line). This 7° difference can be corrected in the data analysis.

Figure 9.5(B) shows the beam-offset signals taken as the line-cut from the 2D contour plot along the two primary crystal orientations indicated by the white dashed lines in figure 9.5(A). The values of the FWHM of measured V_{out} were compared with those predicted from the thermal model to obtain the in-plane thermal conductivity of BP along either the zigzag or armchair direction. The in-plane thermal conductivities of BP fitted from measurement data were 91 ± 10 W m^{-1} K^{-1} along the zigzag direction and 26 ± 3 W m^{-1} K^{-1} along the armchair direction. Rotating the sample by 90° and taking another beam-offset measurement of the same BP flake resolves the same thermal properties of BP, as demonstrated in [66]. The small difference (less than 1% in the FWHM of V_{out}) between the two measurements with different sample loading orientations was within the expected experimental uncertainty, justifying the effectiveness and reliability of this beam-offset TR-MOKE method.

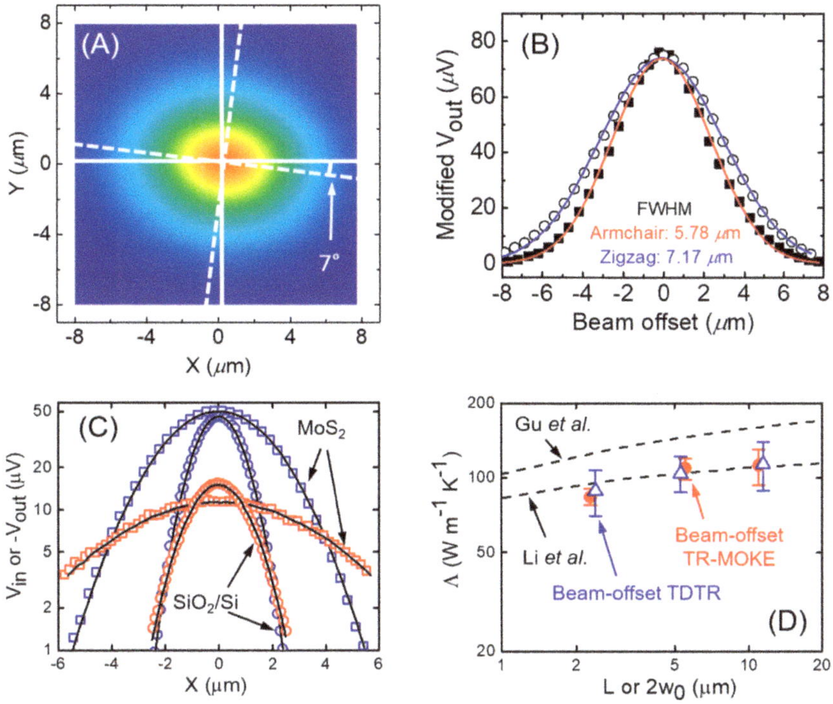

Figure 9.5. In-plane beam-offset TR-MOKE measurement examples. (A) 2D contour measured on a BP flake coated with 28 nm TbFe transducer, taken at a negative time delay ($t = -50$ ps) with a 20× objective lens and 1.6 MHz modulation frequency. The beam-offset signals in (B) are extracted from (a) along the dashed lines denoting the major and minor axes of the ellipse, which correspond to the zigzag and armchair directions, respectively [66]. (C) Beam-offset TR-MOKE data for SiO_2 film (open circles) and MoS_2 (open squares) coated with a 20 nm Co/Pt transducer. V_{out} signals at negative time delay ($t = -100$ ps) with a 20× objective lens and 1 MHz modulation frequency on MoS_2 were used to extract the FWHM [63]. (D) The beam-offset TR-MOKE (red circles) and beam-offset TDTR (blue triangles) measurement results of the MoS_2 in-plane thermal conductivity measured with three different beam spot sizes (w_0). The *ab initio* calculations of thermal conductivity for monolayer MoS_2 by Li *et al* [95] and by Gu and Yang [96] are included for comparison. Figures 9.5(C) and 9.5(D) are adapted from [63] with the permission of AIP Publishing.

The in-plane thermal conductivity of another 2D material, single-crystal MoS_2, was investigated by Liu *et al* [63] using beam-offset TR-MOKE. Their perpendicular magnetic transducer was a Co/Pt multilayer stack with a total thickness of 20 nm. As shown in figure 9.5(C), the beam-offset signals were collected and fitted by Gaussian functions to obtain the FWHM, which was then compared with thermal model predictions, similar to the procedures done for BP. The main difference between these two studies lies in the treatment of the in-plane thermal anisotropy. Liu *et al* did not take a full TR-MOKE 2D contour in [63] since it was assumed that crystalline MoS_2 is thermally isotropic in plane. Figure 9.5(D) shows the in-plane thermal conductivity of the MoS_2 crystal from beam-offset TR-MOKE measurements using three beam spot sizes. For comparison, Liu *et al* also performed beam-offset TDTR measurements of a MoS_2 crystal coated with a 65 nm NbV transducer.

They found that the thermal conductivity data obtained from TR-MOKE and TDTR were consistent, and TR-MOKE provided much smaller error bars. For example, when a 50× objective lens was used with the 20 nm Co/Pt transducer, the uncertainty for TR-MOKE measurements was reduced by nearly a factor of 3, compared to that of TDTR measurements with the 65 nm NbV transducer.

The difference in uncertainty between TDTR and TR-MOKE can be revealed through a sensitivity analysis, which further illustrates how much the TR-MOKE method with a thinner transducer will improve the measurement sensitivity to the in-plane thermal conductivity. For the through-plane thermal conductivity measurements using either TR-MOKE or TDTR, the sensitivity of the ratio signal $(-V_{\text{in}}/V_{\text{out}})$ to a nominal parameter 'σ' is defined as

$$S_z(\sigma) = \frac{\partial[\ln(-V_{\text{in}}/V_{\text{out}})]}{\partial[\ln(\sigma)]}, \tag{9.8}$$

where σ represents one of the geometrical parameters or material's thermal properties and the subscript 'z' denotes the through-plane direction. For in-plane beam-offset measurements, the FWHM of the V_{out} signals at negative time delay is analyzed; thus, the sensitivity of in-plane thermal transport measurements is defined as

$$S_r(\sigma) = \frac{\partial[\ln(\text{FWHM})]}{\partial[\ln(\sigma)]}, \tag{9.9}$$

where the subscript 'r' denotes the in-plane radial direction for general thermal properties.

Figure 9.6 depicts the absolute sensitivity plots of the through-plane TR-MOKE measurements with a 27 nm TbFe transducer, and in-plane beam-offset measurements with both the 27 nm TbFe (used in TR-MOKE) transducer and an 81 nm Al transducer (used in TDTR). The subscript 'm' refers to transducer parameters. The sensitivity analysis of through-plane thermal measurements is calculated for a reference sample of 300 nm SiO_2 on a Si substrate. At long time delay (>100 ps), the heat capacity of the TbFe transducer (C_{m}) is the dominant parameter with the largest measurement sensitivity. Unlike C_{m}, the through-plane measurements are nearly insensitive to the thermal conductivity of the TbFe transducer. This allows the heat capacity of the TbFe transducer to be uniquely determined (figure 9.6(A)). Figure 9.6(B) shows the sensitivities of beam-offset measurements using the same 27 nm TbFe transducer, as a function of the sample in-plane thermal conductivity. The analysis is conducted for the case of a beam spot size of $w_0 = 3$ μm, a laser modulation frequency of $f = 1.6$ MHz, and at a negative time delay of -50 ps. When the Al transducer is used for in-plane TDTR, the large in-plane thermal conductivity of Al ($\Lambda_{r,\text{m}} = 180$ W m^{-1} K^{-1}) causes significant heat spreading in the Al transducer layer, reducing the in-plane measurement sensitivity to Λ_r (figure 9.6(C)). Comparing the sensitivities of TR-MOKE and TDTR depicted in figure 9.6(B) and (C), the sensitivity of TR-MOKE to Λ_r is two to three times higher than that of TDTR, over the entire calculation range. On the other hand, the sensitivity of

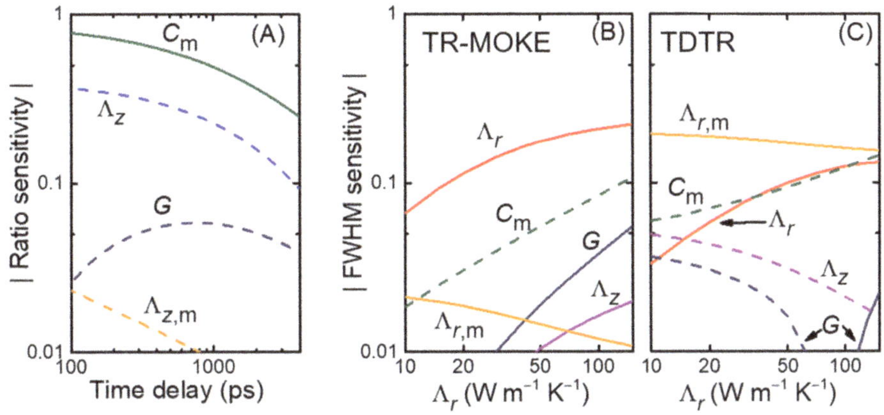

Figure 9.6. Sensitivity analysis of the through-plane and in-plane measurements to different parameters. (A) Absolute ratio sensitivities of through-plane TR-MOKE measurements to the thermal conductivity and heat capacity of a 27 nm TbFe transducer film on 300 nm SiO_2 reference. The measurement conditions are $w_0 = 12\ \mu$m and $f = 9$ MHz. Beam-offset FWHM sensitivities for (B) TR-MOKE with a 27 nm TbFe transducer and (C) TDTR with an 81 nm Al metal transducer as a function of the in-plane thermal conductivity Λ_r measured with $w_0 = 3\ \mu$m and $f = 1.6$ MHz. The time delay is set to −50 ps in both (B) and (C). Solid lines represent the positive values of sensitivities, while dashed lines represent negative sensitivities.

TR-MOKE to $\Lambda_{r,\mathrm{m}}$ is one order of magnitude lower than that of TDTR; thus, the use of magnetic transducers in TR-MOKE will also make the measurement results less influenced by the uncertainties of the transducer properties or geometric parameters. Both effects are favored for reducing the overall uncertainty of the in-plane TR-MOKE measurements.

9.4 Ultrafast magnetization dynamics

As pointed out in section 9.1, TR-MOKE is a technique that can be used to study the magnetization dynamics, in addition to thermal transport in materials. In this section, we will focus on detailing the measurement procedures and data analysis for probing the magnetization damping parameters of PMA materials that are of technological importance.

Relating the dynamic magnetization direction to a Kerr rotation (θ_k) requires the consideration of the optical incidence angle. Solving for the Kerr rotation through Fresnel coefficients [70, 71], it can be shown that, for the polar MOKE configuration with normal incidence, θ_k will only contain information from the magnetization component that is along the surface normal (M_z). For oblique incidence, θ_k will contain information from other magnetization components, such as longitudinal MOKE (where **M** is in the plane of incidence and perpendicular to the surface normal), or transverse MOKE (where **M** is perpendicular to both the plane of incidence and the surface normal) [97]. For thermal measurements using TR-MOKE, this distinction between different MOKE metrology is often neglected because the transducer magnetization is saturated along the through-plane direction ($\mathbf{M} = M_z\hat{z}$), and the change in **M** (through temperature variation) can be directly

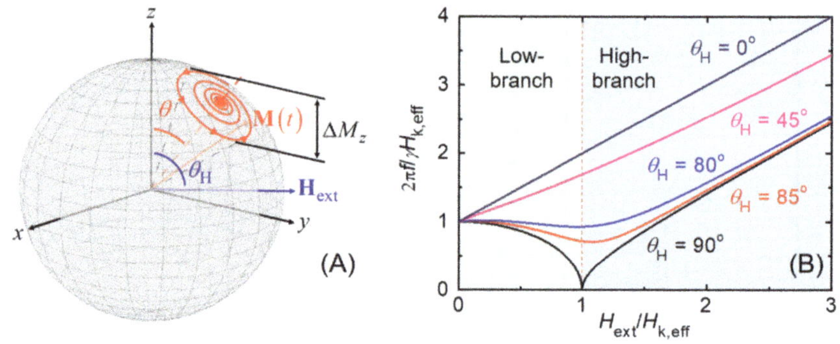

Figure 9.7. Magnetization precession and related key parameters. (A) A 3D representation of the magnetization vector (**M**) precessing around the equilibrium direction (θ) displayed on the surface of a sphere with a radius of M_s. The equilibrium direction is controlled by the magnitude (H_{ext}) and direction (θ_H) of the external magnetic field vector ($\mathbf{H}_{ext}$). The change in the z component of magnetization (ΔM_z) is proportional to the TR-MOKE signal. Reproduced with permission from [117]. Copyright 2018 AIP Publishing. (B) A plot of the resonance frequency (f) normalized to $(\gamma/2\pi)H_{k,eff}$ as a function of H_{ext} normalized to $H_{k,eff}$ for various θ_H. The transition between the low branch ($H_{ext} < H_{k,eff}$) and high branch ($H_{ext} \geqslant H_{k,eff}$) is indicated by different background colors for the $\theta_H = 0°$ case, where the minimum frequency reduces to zero.

related to θ_k without the presence of an external field (H_{ext}). TR-MOKE measurements of magnetization dynamics, on the other hand, utilize an optically induced magnetic torque via thermal demagnetization. Thus, the magnetization response of a material upon optical excitation depends on the direction of the incoming light. While MOKE can be used to determine a 3D profile of the magnetization precession [98], it is often beneficial to reduce the model complexity by focusing on a single component of **M** (such as M_z here).

9.4.1 Magnetization information from TR-MOKE signals

As introduced in section 9.2, the magnetization dynamics in a ferromagnetic film can be described with the LLG equation. Because TR-MOKE operates in the time domain, it is helpful to understand the time-domain response of materials following this differential equation. For this purpose, it is beneficial to transform the LLG equation (equation (9.3)) to a form closer to the original Landau–Lifshitz equation with relaxation [99]:

$$\frac{\mathrm{d}\mathbf{M}}{\mathrm{d}t} = -\frac{\gamma}{1+\alpha^2}\mathbf{M}\times\mathbf{H}_{\mathrm{eff}} - \frac{\gamma}{1+\alpha^2}\frac{\alpha}{M_s}\mathbf{M}\times(\mathbf{M}\times\mathbf{H}_{\mathrm{eff}}). \tag{9.10}$$

With this transformation, the use of a spherical coordinate system (as shown in figure 9.7(A)), the assumption that M_s is not changing in time and the small angle approximation, equation (9.10) will result in an eigenvalue problem akin to a damped oscillator system. Solving for the eigenvalue provides a complex resonance frequency (ω) with the real part of this resonance frequency (for $\alpha \ll 1$) described by the Smit–Suhl equation [100, 101]:

$$\omega = \frac{\gamma}{M_s \sin(\theta)} \sqrt{\frac{\partial^2 F}{\partial \theta^2} \frac{\partial^2 F}{\partial \varphi^2} - \left(\frac{\partial^2 F}{\partial \varphi \partial \theta}\right)^2}, \tag{9.11}$$

which relates the angular resonance frequency (ω) to the curvature of magnetic free energy density (F) with respect to the polar angle (θ), the azimuthal angle (φ) and the gyromagnetic ratio (γ). The Smit–Suhl equation is a generalized form of the well-known Kittel dispersion of FMR [20]. The imaginary part of the resonance frequency is the damping rate (inverse of relaxation time, $1/\tau$):

$$\frac{1}{\tau} = \frac{1}{2} \frac{\alpha \gamma}{M_s} \left(\frac{\partial^2 F}{\partial \theta^2} + \frac{1}{\sin^2(\theta)} \frac{\partial^2 F}{\partial \varphi^2} \right). \tag{9.12}$$

The resulting magnetization dynamics can then be represented by a decaying sinusoidal signal, with the frequency being the FMR frequency and the relaxation time which depends on α. For a typical TR-MOKE measurement, the time-dependent signal (S) can be described by

$$S(t) = A + B \exp(-t/C) + D \sin(\omega t + E) \exp(-t/\tau), \tag{9.13}$$

where variables A, B and C relate to the thermal information retained in the measurement (e.g. laser heating induced thermal demagnetization), D is the amplitude of the Kerr rotation resulting from the magnetization precession and E is a phase offset [102]. This measured signal can be treated as proportional to θ_k, given the small temperature rise induced by laser heating during measurements, detector linear responsivity and constant conversion factors of electronic devices. In this way, fitting the TR-MOKE signal to equation (9.13) will result in the FMR frequency and relaxation rate for the measured material. We will next discuss how to relate the measured frequencies and relaxation times to the extraction of important magnetic properties.

9.4.2 Magnetic anisotropy and damping

Magnetic materials with PMA are promising candidates to reduce the switching energy for many technologically important applications [103, 104]. This has led to a large amount of research focused on the magnetic characterization of these materials with TR-MOKE [105–115]. For materials with PMA (and all magnetic systems of interest), we start by defining the magnetic free energy density F, which contains contributions from the Zeeman energy, magnetocrystalline anisotropy (K_u), interfacial anisotropy (K_i) and shape-induced demagnetization energy. The macrospin approximation (treating the material as uniform in space represented by a single-spin mode) for a thin-film material results in

$$F = F_0 - \mathbf{M} \cdot \mathbf{H}_{ext} - K_u(\hat{u} \cdot \mathbf{m})^2 - \left(\frac{K_i}{h} - 2\pi M_s^2 \right) (\hat{z} \cdot \mathbf{m})^2, \tag{9.14}$$

where F_0 is the initial free energy density, H_{ext} is the externally applied magnetic field, $\hat{u}$ is the easy axis of the material (low-energy direction), $\mathbf{m}$ is a unit vector along the direction of the magnetization and h is the thickness of the magnetic film. The term F_0 is independent of magnetization or field and thus does not contribute to dynamics, so is often omitted in the discussions of free energy density. For samples with PMA, F can be expressed in terms of previously defined angles:

$$F = -M_s H_{ext}[\sin(\theta)\sin(\theta_H)\cos(\varphi) + \cos(\theta)\cos(\theta_H)] - K_{eff}\cos^2(\theta), \quad (9.15)$$

where K_{eff} contains the sum of K_u, K_i and the shape (demagnetization) anisotropy, θ_H is the polar angle of the external applied field, and θ and φ correspond to the polar and azimuthal angles of the magnetization, respectively. For magnetic materials with uniaxial perpendicular anisotropy, the azimuthal component of the external field (φ_H) is not considered because the azimuthal angle φ always follows φ_H. The first derivatives of this energy with respect to θ and φ indicate the energy minima, thus providing the equilibrium direction of magnetization:

$$\varphi = 0, \quad (9.16)$$

$$2H_{ext}\sin(\theta_H - \theta) = H_{k,eff}\sin(2\theta), \quad (9.17)$$

where $H_{k,eff}$ is the effective anisotropy field ($H_{k,eff} = 2K_{eff}/M_s - 4\pi M_s$), which indicates the external field required to change the magnetization direction from $\theta = 0°$ to $\theta = 90°$.

Following the Smit–Suhl equation (9.11), the dynamics of the material system can be described by the second derivatives (curvature) of the free energy density F in terms of θ and φ. The resulting equations to describe the resonance frequency (f) are [102]

$$f = \frac{\gamma}{2\pi}\sqrt{H_1 H_2}, \quad (9.18)$$

$$H_1 = H_{ext}\cos(\theta - \theta_H) + H_{k,eff}\cos^2(\theta), \quad (9.19)$$

$$H_2 = H_{ext}\cos(\theta - \theta_H) + H_{k,eff}\cos(2\theta). \quad (9.20)$$

These equations predict the trend of frequency as a function of the external field for a PMA thin film, which is depicted in figure 9.7(B) (the normalized frequency, $f/[(\gamma/2\pi)H_{k,eff}]$, versus normalized field, $H_{ext}/H_{k,eff}$). For the cases of field being aligned close to the in-plane direction, f decreases with increasing H_{ext} (low branch) until H_{ext} surpasses a critical field. After this point, f begins to increase with increasing field (high branch), saturating at a slope of $\gamma/2\pi$. For extreme cases, f increases monotonically with H_{ext} at the slope of $\gamma/2\pi$ when H_{ext} is along the surface normal ($\theta_H = 0°$). When H_{ext} is applied along the in-plane direction ($\theta_H = 90°$), these equations result in a singularity with $f = 0$ at $H_{ext} = H_{k,eff}$, which provides the well-known FMR frequency equations for field applied along the hard axis of a magnetic material [100, 116]:

$$f = \frac{\gamma}{2\pi}\sqrt{H_{k,eff}^2 - H_{ext}^2} \quad \text{for } H_{ext} < H_{k,eff} \text{ (low branch)}, \quad (9.21)$$

$$f = \frac{\gamma}{2\pi}\sqrt{H_{\text{ext}}(H_{\text{ext}} - H_{\text{k,eff}})} \quad \text{for } H_{\text{ext}} \geqslant H_{\text{k,eff}} \text{ (high branch).} \tag{9.22}$$

The measured relaxation time also depends on both θ_H and H_{ext}. Based on equation (9.12), the relaxation rate of a PMA thin film is given by

$$\frac{1}{\tau} = \frac{1}{2}\alpha\gamma(H_1 + H_2) + \frac{1}{2}\left|\frac{\mathrm{d}\omega}{\mathrm{d}H_{\text{k,eff}}}\right|\Delta H_{\text{k,eff}}, \tag{9.23}$$

where the second term on the right side of equation (9.23) incorporates the inhomogeneous broadening effect (apparent damping resulting from variation in $H_{k,eff}$ throughout the sample). TR-MOKE results often lump the inhomogeneous broadening (extrinsic) and intrinsic damping contribution into an effective damping [107–109] through the simplified relationship $\alpha_{eff} = 1/2\pi f\tau$ [109–111, 118]. At sufficiently high fields (relative to $H_{k,eff}$), the effect of inhomogeneous broadening can be minimized, resulting in $\alpha_{eff} \approx \alpha$. For several technologically relevant materials with large PMA (such as CoFeB and $L1_0$–FePt), it is challenging to reach high enough fields in experiments. Thus, the determination of α from the measured relaxation time often requires knowledge of the material properties (e.g. $H_{k,eff}$ derived from frequency measurements), numerical fittings [105, 106, 114] and precisely controlled field angles [115]. Figure 9.8 shows a representative study conducted by Lattery *et al*, exploring the angle and field dependence of the resonance frequency and damping for a PMA CoFeB thin-film sample [106].

Fundamentally, other physical phenomena also contribute to damping, including the scattering of magnons (quantized spin waves) within the material. In particular, two-magnon scattering, an elastic collision causing a uniform precession magnon ($\mathbf{k} = 0$) to scatter into a degenerate magnon ($\mathbf{k} \neq 0$), has been widely studied as an extrinsic form of magnetization damping [119–121]. While a theory exists to predict the magnon dispersion within PMA materials [122], incorporation of this theory into the understanding of TR-MOKE results is still a developing field [112].

With the combination of extrinsic sources of damping, inhomogeneous broadening and thickness dependent damping [123, 124], the measured effective damping in PMA materials can be significantly large [113]. For cases with a large effective damping, it is often necessary to consider the measurement conditions to maximize the precessional signal measured by TR-MOKE for reliable determination of α. Lattery *et al* have shown that, in addition to the frequency and effective damping, the TR-MOKE signal (proportional to ΔM_z) also depends on H_{ext} and θ_H [117]. An example is illustrated in figure 9.9(A), in which a contour plot shows the amplitude of TR-MOKE signals as a function of θ_H and H_{ext} (normalized to $H_{k,eff}$). The dashed line in figure 9.9(A) provides guidance on choosing field conditions to maximize the SNR of TR-MOKE measurements for complex samples [117]. Using a macromagnetic simulation, a simpler form of the precessional amplitude can be shown to follow

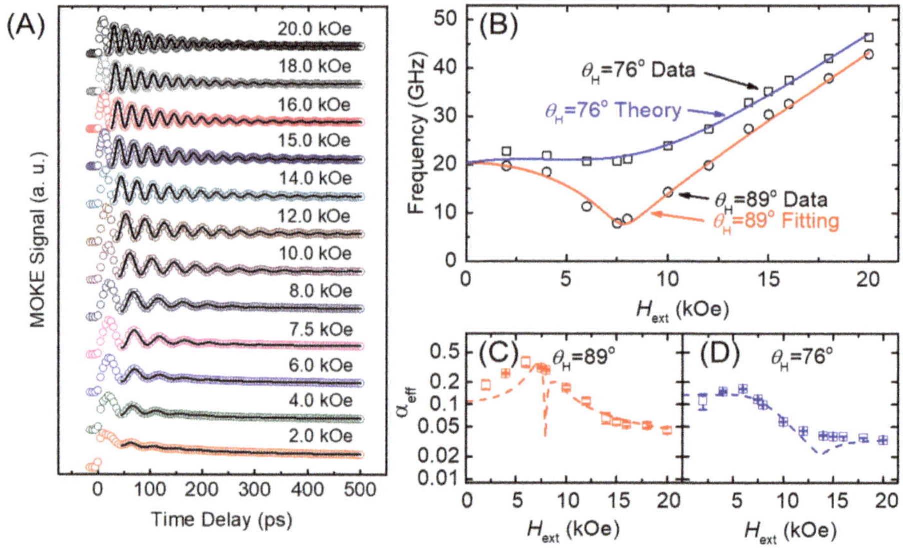

Figure 9.8. The process of data reduction in TR-MOKE measurements of magnetization precession. (A) The raw TR-MOKE signal is fitted to extract f and τ as a function of H_{ext}. (B) The resonance f (calculated from equation (9.18)) is plotted as a function of H_{ext} and fitted to measurement data to extract $H_{\text{k,eff}}$ and γ. The circle and square symbols are the measured data with $\theta_{\text{H}} = 89°$ and 76°, respectively. The red line shows the fitting of $\theta_{\text{H}} = 89°$, while the blue line corresponds to the resulting curve generated by using the fitting results from $\theta_{\text{H}} = 89°$ in equation (9.18) for $\theta_{\text{H}} = 76°$. (C) and (D) The inhomogeneous broadening is determined to extract a unique value for α. This α and the inhomogeneous broadening result in a field-dependent α_{eff} (dotted lines), agreeing well with the measured α_{eff} (open squares) for both field angles. Reproduced with permission from [106]. Creative Commons Attribution 4.0 International License.

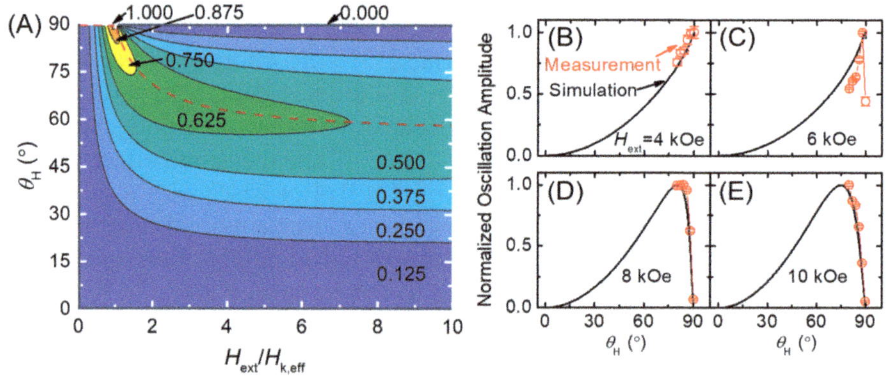

Figure 9.9. Simulation and measurements of TR-MOKE signal optimization for magnetization precession. (A) A contour plot of the normalized ΔM_z signal as a function of the field ratio ($H_{\text{ext}}/H_{\text{k,eff}}$) and θ_{H}, where the value of '1' indicates the maximum possible signal. The dotted red line corresponds to $\theta_{\text{H,MAX}}$ where the signal is maximized for a specific field ratio. (B)–(E) The measurement results from a W/CoFeB sample (open symbols) compared to a macromagnetic simulation (lines) for various values of H_{ext} normalized to $H_{\text{k,eff}}$ (6.1 kOe). Reproduced with permission from [117]. Copyright 2018 AIP Publishing.

$$\frac{\Delta M_z}{M_s} \propto \sin(\theta - \theta_H)\sin(\theta). \tag{9.24}$$

As a demonstration, Lattery *et al* extracted the TR-MOKE signal amplitudes from measurements of a PMA CoFeB thin film seeded with tungsten (W). The results are shown in figure 9.9(B)–(E) as a function of θ_H and H_{ext}. The dependence of TR-MOKE signal amplitudes on these two parameters are in excellent agreement with their model prediction.

This fundamental study of the optimal measurement conditions for higher TR-MOKE SNRs augments the research that has already been conducted with the TR-MOKE metrology for ultrafast magnetization dynamics, showing the versatility of this technique. Building upon an understanding of the FMR, the precession frequency and relaxation in these time-resolved measurements can be processed to extract magnetic properties, including $H_{k,eff}$ and α. As magnetic materials continue to advance, TR-MOKE has already served as a critical tool for understanding the magnetic transport properties and dynamics of these materials and for optimizing their usage and integration into advanced technologies.

9.5 Advanced capabilities for broader research directions

In addition to the studies of thermal properties and Gilbert damping with TR-MOKE, the ultrafast time scale is host to a plethora of physical processes that warrant further research. In this section, we will briefly highlight the potential of extending TR-MOKE capabilities to several other promising research areas.

9.5.1 Propagating spin waves

In analogy to lattice vibrations (phonons), spin waves (magnons) can be treated as oscillations in the magnetization of a material. Because the magnetization in the material is coupled together, when a single spin precesses, it will cause neighboring spins to also move accordingly and thus distribute the precession as waves. These waves lead to a difference in magnetization throughout the sample [125], and thus they can be imaged through interpretation of FMR data [126], or directly through MOKE. These spin waves within ferromagnetic materials are important for the future study of the transfer of spin angular momentum, even in ferrimagnetic insulators [127].

Research in the imaging of spin waves started with the study of standing spin waves in thick metallic films [64]. If the film thickness matches an integer number of magnon half-wavelengths, a standing spin wave will have a resonant frequency that is related to the magnon dispersion (specifically the spin wave exchange stiffness, *D*) within the material [125, 126]. Recent interests have been focused on studies of ultra-thin ferromagnetic films (less than 10 nm), which do not typically show the standing spin waves [64]. The perturbation from optical excitation can create spin waves that propagate along the in-plane direction of the film [128]. By varying the time delay between the excitation and probe beams (and potentially the distance between them as well), the resulting temporal and spatial map of the magnetization at the sample surface can detect propagating wave packets of the so-called 'magnetostatic spin waves' [128–130].

9.5.2 Ultrafast energy carrier coupling

As mentioned in section 9.2, due to the ultrafast optical pulses used in TR-MOKE (~100 fs), this technique can capture information about the coupling between energy carriers that is prevalent before the magnetization equilibrates with the electrons and phonons in the system [80]. For non-magnetic materials, the two-temperature model (2TM) has been used to describe the electron–phonon coupling (G_{ep}) [131], but the lack of understanding regarding the transfer and dissipation of angular momentum (and thus magnetization) in magnetic materials requires more complex models to interpret the signals from ultrafast demagnetization processes.

The most often discussed model is the phenomenological three-temperature model (3TM). In 3TM, electrons are heated by the pump pulse and the three thermal reservoirs assigned to lattice (phonons), electron and spin (magnons), respond via inter-reservoir coupling [80]. While useful, 3TM is almost purely phenomenological and reveals limited information about the physical processes that induce demagnetization. Expanding on the concept of this model, other models (such as the microscopic 3TM or M3TM) have incorporated aspects of Elliott–Yafet scattering (magnon–phonon scattering, i.e. a loss of magnon energy to atomic lattice vibrations) to explain the loss of magnetization as a spin-flip scattering event [89, 132]. Although further publications have expanded this model to include spin accumulation and spin currents [133], there have been very few modifications to this model since it was originally proposed.

In parallel with this, there has been an effort to interpret the ultrafast demagnetization through the lens of magnetization dynamics. These models utilize the high-temperature of electrons as an input into atomistic simulations of spin [134]. As such, these models would still require the 2TM and G_{ep} as an input. Where these atomistic spin models differ from the 3TM approach is in the treatment of spin states not as a thermal reservoir, but as individual components of a larger system with their own unique magnetic moments [134]. These temperature-dependent atomistic simulations require a large amount of computational resources to simulate ultrafast magnetization within systems of interest. Therefore, it is often of interest to utilize single-spin (macrospin) approaches, such as the LLB equation, as an approximation [135, 136]. In fact, there are literature studies showing that the LLB approach is approximately equivalent to the M3TM in understanding ultrafast demagnetization processes [137]. As a system-averaged approach to understand high-temperature magnetization, LLB has been argued to fail to consider the change in magnon heat capacity at high temperatures [92], it still offers a useful tool for applications such as HAMR and all-optical switching, where temperature dependence is crucial.

9.5.3 Straintronics (coupling between spin and strain)

Magnetostriction is a well-known phenomenon in which the magnetization within a magnetic material causes a structural deformation and can even launch acoustic strains [138]. The inverse effect is also important, particularly in pump–probe measurements that create longitudinal strain waves via thermal expansion after optical excitation [139, 140]. This strain will create picosecond acoustic signals in TDTR measurements that can be used to determine the thickness of thin films. Through the inverse magnetostriction effect, strain waves can also influence magnetization dynamics, as shown in the study of ferromagnetic Ni films by Kim *et al* [141]. Further research has also shown how this effect can be achieved by using materials with large magnetostriction, such as Galfenol (an alloy of Fe and Ga), [142] or controlling the strain in the material through the use of acoustic Bragg mirrors [143]. Utilizing the coupling between strain and magnetization can potentially offer a unique non-thermal approach to excite magnetization precession for realizing high-speed and low-energy switching in spintronics.

9.5.4 Spin caloritronics

The recent discovery of the spin Seebeck effect [144, 145] has resulted in a large amount of research into the field of spin caloritronics [146, 147]. Analogous to the standard Seebeck effect, the spin Seebeck effect results in a voltage difference when a temperature gradient is present in a magnetic material. While many measurements of the spin Seebeck constant in magnetic materials occur at steady state [144, 145, 148, 149], the ultrafast temperature rise from a pump–probe measurement can lead to the existence of spin currents and spin accumulation (e.g. the accumulation of electron angular momentum at interfaces) at an ultrafast time scale, which can be measured by TR-MOKE [132, 133]. Due to a difference in the density of states of spin up/down electrons in a magnetic material, an ultrafast temperature gradient can also lead to a spin current from the spin-dependent Seebeck effect (not to be confused with the spin Seebeck effect). This spin current then creates a spin-transfer torque that can initiate precession in a ferromagnetic detector layer [132, 150]. These initial measurements have shown the efficacy of measuring spin caloritronics with TR-MOKE, but most of the advances in this field have been determined through other techniques [147].

9.6 Summary and outlook

As technology continues to shrink in scale and technological advancements such as the field of spintronics come to life, it becomes increasingly important to study materials' transport behaviors in the ultrafast time scale. In this chapter, we have discussed a number of applications for TR-MOKE measurements including measurements of thermal properties, magnetization dynamics and the coupling between the two. As we have shown, TR-MOKE has been applied to understand complex materials with high anisotropy and to increase the measurement sensitivity to the thermal transport within materials and across interfaces. The results of these experimental studies can improve the design of future devices and interfaces for better thermal management. This powerful technique can also measure the dynamic magnetization within ferromagnetic materials, which has already become a crucial demand for spintronic materials and devices. Exciting new research has already utilized TR-MOKE to probe deeper into unique physics such as the coupling between energy carriers, and the phenomena of spin caloritronics.

Acknowledgements

The work described in this chapter was supported by the following funding agencies: the National Science Foundation (NSF) Award No. 1804840, the University of Minnesota MRSEC under the NSF Award No. DMR-1420013, C-SPIN (one of six centers of STARnet, a Semiconductor Research Corporation program, sponsored by MARCO and DARPA), the Legislative-Citizen Commission on Minnesota Resources (LCCMR), the Advanced Storage Research Consortium, 3M, the Institute on the Environment at the University of Minnesota and the Office of the Vice President for Research at the University of Minnesota. DL also appreciates the support from the university's 2019–2020 Doctoral Dissertation Fellowship. Portions of the work related to sample fabrication and characterization were conducted in the Minnesota Nano Center, supported by the NSF through the National Nano Coordinated Infrastructure Network (NNCI) under Award No. ECCS-1542202.

References

[1] Moore G E 2006 *IEEE Solid-State Circuits Soc. Newslett.* **11** 36–7

[2] Theis T N and Wong H P 2017 *Comput. Sci. Eng.* **19** 41–50

[3] Baibich M N, Broto J M, Fert A, Van Dau F N, Petroff F, Etienne P, Creuzet G, Friederich A and Chazelas J 1988 *Phys. Rev. Lett.* **61** 2472–5

[4] Binasch G, Grünberg P, Saurenbach F and Zinn W 1989 *Phys. Rev.* B **39** 4828–30

[5] Tehrani S, Slaughter J M, Chen E, Durlam M, Shi J and DeHerren M 1999 *IEEE Trans. Magn.* **35** 2814–9

[6] Slonczewski J C 1996 *J. Magn. Magn. Mater.* **159** L1–7

[7] Berger L 1996 *Phys. Rev.* B **54** 9353–8

[8] Liu L, Pai C-F, Li Y, Tseng H W, Ralph D C and Buhrman R A 2012 *Science* **336** 555–8

[9] Cubukcu M *et al* 2018 *IEEE Trans. Magn.* **54** 1–4

[10] Behin-Aein B, Datta D, Salahuddin S and Datta S 2010 *Nat. Nanotechnol.* **5** 266

[11] Kim J, Paul A, Crowell P A, Koester S J, Sapatnekar S S, Wang J and Kim C H 2015 *Proc. IEEE* **103** 106–30

[12] Camsari K Y, Faria R, Sutton B M and Datta S 2017 *Phys. Rev.* X **7** 031014

[13] Kiselev S I, Sankey J C, Krivorotov I N, Emley N C, Schoelkopf R J, Buhrman R A and Ralph D C 2003 *Nature* **425** 380

[14] Kaka S, Pufall M R, Rippard W H, Silva T J, Russek S E and Katine J A 2005 *Nature* **437** 389

[15] Houssameddine D *et al* 2007 *Nat. Mater.* **6** 447

[16] Mizushima K, Kudo K and Sato R 2007 *J. Magn. Magn. Mater.* **316** e960–2

[17] Kryder M H, Gage E C, McDaniel T W, Challener W A, Rottmayer R E, Ju G, Hsia Y and Erden M F 2008 *Proc. IEEE* **96** 1810–35

[18] Challener W A *et al* 2009 *Nat. Photonics* **3** 220
[19] Ju G *et al* 2015 *IEEE Trans. Magn.* **51** 1–9
[20] Kittel C 1948 *Phys. Rev.* **73** 155–61
[21] Kittel C 1947 *Phys. Rev.* **71** 270–1
[22] Rossing T D 1963 *J. Appl. Phys.* **34** 995
[23] Farle M, Silva T and Woltersdorf G 2013 *Magnetic Nanostructures: Spin Dynamics and Spin Transport* ed H Zabel and M Farle (Berlin: Springer), pp 37–83
[24] Cahill D G *et al* 2014 *Appl. Phys. Rev.* **1** 011305
[25] Nan C-W, Birringer R, Clarke D R and Gleiter H 1997 *J. Appl. Phys.* **81** 6692–9
[26] Goodson K E and Ju Y S 1999 *Annu. Rev. Mater. Sci.* **29** 261–93
[27] Cahill D G, Bullen A and Seung-Min L 2000 *High Temp. High Press.* **32** 135–42
[28] Soyez G, Eastman J A, Thompson L J, Bai G-R, Baldo P M, McCormick A W, DiMelfi R J, Elmustafa A A, Tambwe M F and Stone D S 2000 *Appl. Phys. Lett.* **77** 1155–7
[29] Hung M-T, Choi O, Ju Y S and Hahn H T 2006 *Appl. Phys. Lett.* **89** 023117
[30] Chiritescu C, Cahill D G, Nguyen N, Johnson D, Bodapati A, Keblinski P and Zschack P 2007 *Science* **315** 351–3
[31] Losego M D, Moh L, Arpin K A, Cahill D G and Braun P V 2010 *Appl. Phys. Lett.* **97** 011908
[32] Gundrum B C, Cahill D G and Averback R S 2005 *Phys. Rev.* B **72** 245426
[33] Hamad-Schifferli K, Schwartz J J, Santos A T, Zhang S and Jacobson J M 2002 *Nature* **415** 152
[34] Vallabhaneni A K, Qiu B, Hu J, Chen Y P, Roy A K and Ruan X 2013 *J. Appl. Phys.* **113** 064311
[35] Novoselov K S, Geim A K, Morozov S V, Jiang D, Katsnelson M I, Grigorieva I V, Dubonos S V and Firsov A A 2005 *Nature* **438** 197
[36] Ferrari A C and Basko D M 2013 *Nat. Nanotechnol.* **8** 235
[37] Rodin A S, Carvalho A and Castro Neto A H 2014 *Phys. Rev. Lett.* **112** 176801
[38] Cai Y, Zhang G and Zhang Y-W 2014 *JACS* **136** 6269–75
[39] Cahill D G 2004 *Rev. Sci. Instrum.* **75** 5119–22
[40] Zhu J, Wu X, Lattery D M, Zheng W and Wang X 2017 *Nanoscale Microscale Thermophys. Eng.* **21** 177–98
[41] Kang K, Koh Y K, Chiritescu C, Zheng X and Cahill D G 2008 *Rev. Sci. Instrum.* **79** 114901
[42] Persson A I, Koh Y K, Cahill D G, Samuelson L and Linke H 2009 *Nano Lett.* **9** 4484–8
[43] Koh Y K, Bae M-H, Cahill D G and Pop E 2011 *ACS Nano* **5** 269–74
[44] Monachon C and Weber L 2012 *Emerg. Mater. Res.* **1** 89–98
[45] Cheaito R, Duda J C, Beechem T E, Hattar K, Ihlefeld J F, Medlin D L, Rodriguez M A, Campion M J, Piekos E S and Hopkins P E 2012 *Phys. Rev. Lett.* **109** 195901
[46] Wang X, Liman C D, Treat N D, Chabinyc M L and Cahill D G 2013 *Phys. Rev.* B **88** 075310
[47] Zhu J, Zhu Y, Wu X, Song H, Zhang Y and Wang X 2016 *Appl. Phys. Lett.* **108** 231903
[48] Zhu J, Feng T, Mills S, Wang P, Wu X, Zhang L, Pantelides S T, Du X and Wang X 2018 *ACS Appl. Mater. Interfaces* **10** 40740–7
[49] Prakash A, Xu P, Wu X, Haugstad G, Wang X and Jalan B 2017 *J. Mater. Chem.* C **5** 5730–6

[50] Liu J, Wang X, Li D, Coates N E, Segalman R A and Cahill D G 2015 *Macromolecules* **48** 585–91
[51] Wang X, Ho V, Segalman R A and Cahill D G 2013 *Macromolecules* **46** 4937–43
[52] Mai C-K, Schlitz R A, Su G M, Spitzer D, Wang X, Fronk S L, Cahill D G, Chabinyc M L and Bazan G C 2014 *JACS* **136** 13478–81
[53] Xu D, Wang Q, Wu X, Zhu J, Zhao H, Xiao B, Wang X, Wang X and Hao Q 2018 *Front. Energy* **12** 127–36
[54] Hopkins P E, Beechem T, Duda J C, Hattar K, Ihlefeld J F, Rodriguez M A and Piekos E S 2011 *Phys. Rev.* B **84** 125408
[55] Losego M D, Grady M E, Sottos N R, Cahill D G and Braun P V 2012 *Nat. Mater.* **11** 502
[56] Hohensee G T, Wilson R B and Cahill D G 2015 *Nat. Commun.* **6** 6578
[57] Zheng K, Sun F, Zhu J, Ma Y, Li X, Tang D, Wang F and Wang X 2016 *ACS Nano* **10** 7792–8
[58] Wu X, Ni Y, Zhu J, Burrows N D, Murphy C J, Dumitrica T and Wang X 2016 *ACS Appl. Mater. Interfaces* **8** 10581–9
[59] Feser J P and Cahill D G 2012 *Rev. Sci. Instrum.* **83** 104901
[60] Feser J P, Liu J and Cahill D G 2014 *Rev. Sci. Instrum.* **85** 104903
[61] Jang H, Ryder C R, Wood J D, Hersam M C and Cahill D G 2017 *Adv. Mater.* **29** 1700650
[62] Jang H, Wood J D, Ryder C R, Hersam M C and Cahill D G 2015 *Adv. Mater.* **27** 8017–22
[63] Liu J, Choi G-M and Cahill D G 2014 *J. Appl. Phys.* **116** 233107
[64] van Kampen M, Jozsa C, Kohlhepp J T, LeClair P, Lagae L, de Jonge W J M and Koopmans B 2002 *Phys. Rev. Lett.* **88** 227201
[65] Barman A, Kimura T, Otani Y, Fukuma Y, Akahane K and Meguro S 2008 *Rev. Sci. Instrum.* **79** 123905
[66] Zhu J *et al* 2016 *Adv. Electron. Mater.* **2** 1600040
[67] Kerr J 1900 *The Effects of a Magnetic Field on Radiation* ed E P Lewis (New York: American Book), pp 27–52
[68] Pedrotti F L, Pedrotti L M and Pedrotti L S 2007 *Introduction to Optics* (London: Pearson Education), p 562
[69] Yang Z J and Scheinfein M R 1993 *J. Appl. Phys.* **74** 6810–23
[70] You C Y and Shin S C 1996 *Appl. Phys. Lett.* **69** 1315–7
[71] You C-Y and Shin S-C 1998 *J. Appl. Phys.* **84** 541–6
[72] Landau L D and Lifshitz E M 1984 *Electrodynamics of Continuous Media* ed L D Landau and E M Lifshitz (Amsterdam: Pergamon), pp 257–89
[73] Krinchik G S and Arem'ev V A 1968 *Sov. Phys. JETP* **26** 1080–5
[74] Qiu Z Q and Bader S D 2000 *Rev. Sci. Instrum.* **71** 1243–55
[75] Huang B *et al* 2017 *Nature* **546** 270
[76] Schmidt F, Rave W and Hubert A 1985 *IEEE Trans. Magn.* **21** 1596–8
[77] Rave W, Schäfer R and Hubert A 1987 *J. Magn. Magn. Mater.* **65** 7–14
[78] Agranat M B, Ashitkov S I, Granovskii A B and Rukman G I 1984 *Sov. Phys. JETP* **59** 804–6
[79] Vaterlaus A, Beutler T, Guarisco D, Lutz M and Meier F 1992 *Phys. Rev.* B **46** 5280–6
[80] Beaurepaire E, Merle J C, Daunois A and Bigot J Y 1996 *Phys. Rev. Lett.* **76** 4250–3
[81] Ju G, Nurmikko A V, Farrow R F C, Marks R F, Carey M J and Gurney B A 1999 *Phys. Rev. Lett.* **82** 3705–8

[82] Ju G, Chen L, Nurmikko A V, Farrow R F C, Marks R F, Carey M J and Gurney B A 2000 *Phys. Rev.* B **62** 1171–7
[83] Gilbert T L 2004 *IEEE Trans. Magn.* **40** 3443–9
[84] Koopmans B 2003 *Spin Dynamics in Confined Magnetic Structures II* ed B Hillebrands and K Ounadjela (Berlin: Springer), pp 253–316
[85] Chen J-Y, Zhu J, Zhang D, Lattery D M, Li M, Wang J-P and Wang X 2016 *J. Phys. Chem. Lett.* **7** 2328–32
[86] Güdde J, Conrad U, Jähnke V, Hohlfeld J and Matthias E 1999 *Phys. Rev.* B **59** R6608–11
[87] Melnikov A, Bovensiepen U, Radu I, Krupin O, Starke K, Matthias E and Wolf M 2004 *J. Magn. Magn. Mater.* **272–276** 1001–2
[88] Wietstruk M, Melnikov A, Stamm C, Kachel T, Pontius N, Sultan M, Gahl C, Weinelt M, Dürr H A and Bovensiepen U 2011 *Phys. Rev. Lett.* **106** 127401
[89] Koopmans B, Malinowski G, Dalla Longa F, Steiauf D, Fähnle M, Roth T, Cinchetti M and Aeschlimann M 2010 *Nat. Mater.* **9** 259–65
[90] Roth T, Schellekens A J, Alebrand S, Schmitt O, Steil D, Koopmans B, Cinchetti M and Aeschlimann M 2012 *Phys. Rev.* X **2** 021006
[91] Hohensee G T, Wilson R B, Feser J P and Cahill D G 2014 *Phys. Rev.* B **89** 024422
[92] Kimling J, Kimling J, Wilson R B, Hebler B, Albrecht M and Cahill D G 2014 *Phys. Rev.* B **90** 224408
[93] Kimling J, Philippi-Kobs A, Jacobsohn J, Oepen H P and Cahill D G 2017 *Phys. Rev.* B **95** 184305
[94] Wang X J, Mori T, Kuzmych-Ianchuk I, Michiue Y, Yubuta K, Shishido T, Grin Y, Okada S and Cahill D G 2014 *APL Mater.* **2** 046113
[95] Li W, Carrete J and Mingo N 2013 *Appl. Phys. Lett.* **103** 253103
[96] Gu X and Yang R 2014 *Appl. Phys. Lett.* **105** 131903
[97] Mansuripur M 1995 *The Physical Principles of Magneto-optical Recording* (Cambridge: Cambridge University Press), pp 128–79
[98] Acremann Y, Back C H, Buess M, Portmann O, Vaterlaus A, Pescia D and Melchior H 2000 *Science* **290** 492–5
[99] Iida S 1963 *J. Phys. Chem. Solids* **24** 625–30
[100] Smit J and Beljers H G 1955 *Philips Res. Rep.* **10** 113–30
[101] Suhl H 1955 *Phys. Rev.* **97** 555–7
[102] Mizukami S 2015 *J. Magn. Soc. Jpn.* **39** 1–7
[103] Mangin S, Ravelosona D, Katine J A, Carey M J, Terris B D and Fullerton E E 2006 *Nat. Mater.* **5** 210
[104] Meng H and Wang J-P 2006 *Appl. Phys. Lett.* **88** 172506
[105] Iihama S, Mizukami S, Naganuma H, Oogane M, Ando Y and Miyazaki T 2014 *Phys. Rev.* B **89** 174416
[106] Lattery D M, Zhang D, Zhu J, Hang X, Wang J-P and Wang X 2018 *Sci. Rep.* **8** 13395
[107] Zhang B *et al* 2017 *Appl. Phys. Lett.* **110** 012405
[108] Song H-S, Lee K-D, Sohn J-W, Yang S-H, Parkin S S P, You C-Y and Shin S-C 2013 *Appl. Phys. Lett.* **103** 022406
[109] Mizukami S, Watanabe D, Kubota T, Zhang X, Naganuma H, Oogane M, Ando Y and Miyazaki T 2010 *Appl. Phys. Exp.* **3** 123001
[110] Iihama S, Sakuma A, Naganuma H, Oogane M, Miyazaki T, Mizukami S and Ando Y 2014 *Appl. Phys. Lett.* **105** 142403

[111] Wu D, Li W, Tang M, Zhang Z, Lou S and Jin Q Y 2016 *J. Magn. Magn. Mater.* **409** 143–7
[112] Iihama S, Sakuma A, Naganuma H, Oogane M, Mizukami S and Ando Y 2016 *Phys. Rev.* B **94** 174425
[113] Becker J, Mosendz O, Weller D, Kirilyuk A, Maan J C, Christianen P C M, Rasing T and Kimel A 2014 *Appl. Phys. Lett.* **104** 152412
[114] Takahashi Y K *et al* 2017 *Appl. Phys. Lett.* **110** 252409
[115] Capua A, Yang S-h, Phung T and Parkin S S P 2015 *Phys. Rev.* B **92** 224402
[116] Morrish A H 1965 *The Physical Principles of Magnetism* (New York: Wiley), p 555
[117] Lattery D M, Zhu J, Zhang D, Wang J-P, Crowell P A and Wang X 2018 *Appl. Phys. Lett.* **113** 162405
[118] Qiao S, Nie S, Zhao J, Huo Y, Wu Y and Zhang X 2013 *Appl. Phys. Lett.* **103** 152402
[119] Sparks M 1964 *Ferromagnetic-Relaxation Theory* (New York: McGraw-Hill)
[120] McMichael R D and Krivosik P 2004 *IEEE Trans. Magn.* **40** 2–11
[121] Landeros P, Arias R E and Mills D L 2008 *Phys. Rev.* B **77** 214405
[122] Beaujour J M, Ravelosona D, Tudosa I, Fullerton E E and Kent A D 2009 *Phys. Rev.* B **80** 180415
[123] Tserkovnyak Y, Brataas A and Bauer G E W 2002 *Phys. Rev. Lett.* **88** 117601
[124] Ikeda S, Miura K, Yamamoto H, Mizunuma K, Gan H D, Endo M, Kanai S, Hayakawa J, Matsukura F and Ohno H 2010 *Nat. Mater.* **9** 721
[125] Kittel C 1958 *Phys. Rev.* **110** 1295–7
[126] Weber R 1968 *IEEE Trans. Magn.* **4** 28–31
[127] Kajiwara Y *et al* 2010 *Nature* **464** 262
[128] Iihama S, Sasaki Y, Sugihara A, Kamimaki A, Ando Y and Mizukami S 2016 *Phys. Rev.* B **94** 020401
[129] Yun S-J, Cho C-G and Choe S-B 2015 *Appl. Phys. Exp.* **8** 063009
[130] Wessels P, Vogel A, Tödt J-N, Wieland M, Meier G and Drescher M 2016 *Sci. Rep.* **6** 22117
[131] Qiu T Q, Juhasz T, Suarez C, Bron W E and Tien C L 1994 *Int. J. Heat Mass Transfer* **37** 2799–808
[132] Choi G-M, Min B-C, Lee K-J and Cahill D G 2014 *Nat. Commun.* **5** 4334
[133] Kimling J and Cahill D G 2017 *Phys. Rev.* B **95** 014402
[134] Kazantseva N, Nowak U, Chantrell R W, Hohlfeld J and Rebei A 2008 *Europhys. Lett.* **81** 27004
[135] Chubykalo-Fesenko O, Nowak U, Chantrell R W and Garanin D 2006 *Phys. Rev.* B **74** 094436
[136] Kazantseva N, Hinzke D, Nowak U, Chantrell R W, Atxitia U and Chubykalo-Fesenko O 2008 *Phys. Rev.* B **77** 184428
[137] Atxitia U and Chubykalo-Fesenko O 2011 *Phys. Rev.* B **84** 144414
[138] Joule J 1842 *Ann. Electr. Magn. Chem* **8** 219–24
[139] O'Hara K E, Hu X and Cahill D G 2001 *J. Appl. Phys.* **90** 4852–8
[140] Hohensee G T, Hsieh W-P, Losego M D and Cahill D G 2012 *Rev. Sci. Instrum.* **83** 114902
[141] Kim J-W, Vomir M and Bigot J-Y 2012 *Phys. Rev. Lett.* **109** 166601
[142] Jäger J V *et al* 2013 *Appl. Phys. Lett.* **103** 032409
[143] Jäger J V, Scherbakov A V, Glavin B A, Salasyuk A S, Campion R P, Rushforth A W, Yakovlev D R, Akimov A V and Bayer M 2015 *Phys. Rev.* B **92** 020404

[144] Uchida K, Takahashi S, Harii K, Ieda J, Koshibae W, Ando K, Maekawa S and Saitoh E 2008 *Nature* **455** 778

[145] Uchida K *et al* 2010 *Nat. Mater.* **9** 894

[146] Bauer G E W, Saitoh E and van Wees B J 2012 *Nat. Mater.* **11** 391

[147] Boona S R, Myers R C and Heremans J P 2014 *Energy Env. Sci.* **7** 885–910

[148] Jaworski C M, Yang J, Mack S, Awschalom D D, Heremans J P and Myers R C 2010 *Nat. Mater.* **9** 898

[149] Jaworski C M, Yang J, Mack S, Awschalom D D, Myers R C and Heremans J P 2011 *Phys. Rev. Lett.* **106** 186601

[150] Choi G-M, Moon C-H, Min B-C, Lee K-J and Cahill D G 2015 *Nat. Phys.* **11** 576

Chapter 10

Investigation of nanoscale energy transport with time-resolved photoemission electron microscopy

Rebecca Wong, Michael K L Man and Keshav M Dani

Electronics are becoming increasingly more compact and embedded with more functionality. Miniaturization of electronics leads to the emergence of new quantum phenomena, and the study of ultrafast electronic processes in these ultrasmall devices requires techniques that can simultaneously provide high spatial and temporal resolution. In response to this need, techniques such as ultrafast transient absorption microscopy, ultrafast electron microscopy, scanning ultrafast electron microscopy and time-resolved photoemission electron microscopy (TR-PEEM), have been developed. In this chapter, we will focus mainly on the development and perspective of TR-PEEM techniques. We will go through examples demonstrating the versatility of TR-PEEM in the study of carrier dynamics and charge transport in semiconductors, and we will look into the latest developments that promise to expand its capability, allowing TR-PEEM to explore electron dynamics and obtain detailed information, such as carrier scattering mechanisms and resolving the electronics spin texture.

10.1 Introduction

The constant discovery of new materials and semiconductors exhibiting properties beneficial for technologies ranging from transistors to solar cells has commenced a race to better understand and control their physical and chemical properties. For many of these materials, processes of interest, including those involving the materials' electronic, magnetic and photonic capabilities, occur on ultrasmall (micron to nanometer) and ultrafast (picosecond or faster) scales. The understanding of these ultrafast processes will be crucial to realizing the full potential of current and newly discovered materials, as well as for understanding how to manipulate material properties to further efficiency and economic viability. However, to access this

doi:10.1088/978-0-7503-1738-2ch10

information, the experimental techniques must be able to resolve processes on the nanometer and sub-picosecond scales simultaneously, which is a non-trivial task. Since the late twentieth century, several groups have developed methods that have enabled such studies and are revolutionizing the understanding and visualization of charge carrier dynamics through high-resolution, rapid imaging techniques. These images can be stitched together to form movies, enabling processes to be detailed in an unprecedented, 'seeing is believing' fashion at their fundamental length and time scales.

The objective of this chapter is to elucidate electron microscopy techniques that have both high spatial and temporal resolution—with emphasis on the technique of time-resolved photoemission electron microscopy—and to investigate their usage in the observation and understanding of charge dynamics, specifically in semiconductors. This chapter will first illustrate the necessity of semiconductors and methods for manipulation of semiconductors' properties followed by a brief mention of the development of other ultrafast techniques, in particular those providing high spatial resolution, including ultrafast transient absorption microscopy, ultrafast electron microscopy and scanning ultrafast electron microscopy. The main focus will be on the technique of time-resolved photoemission electron microscopy, including highlights from key studies on semiconductors using this technique and future perspectives of this technology.

10.1.1 The era of semiconductor technologies

The importance of semiconductors cannot be overstated—they have radically revolutionized the everyday lives of humans from cars and home appliances to computers. Their compactness, power efficiency and low cost have allowed them to be widely adopted in numerous devices spanning sensors and integrated circuits, and make them indispensable for future applications. In particular, the properties of semiconductors can be manipulated using band engineering methods that allow for nearly arbitrary and continuous spatial bandgap variations to suit different optoelectronic purposes, such as light-emission and energy conversion [1–4]. For example, the conductivity of semiconductors can be modified using pressure, as demonstrated on graphene by Yankowitz *et al* [5], as well as through thermal manipulation, as shown on a variety of semiconductors by Zheng *et al* [6]. Other methods, such as doping, can increase free-carrier concentration with doping of just parts per million strongly influencing conductivity. For example, intrinsic silicon has a bandgap energy of 1.1 eV. However, in n-type silicon doped with phosphorus, electrons residing in a higher donor level than in intrinsic silicon require only 0.045 eV to be excited to the conduction band, meaning that the thermal energy within the crystal is enough to create free-electron carriers [7]. Depending on what impurities are added, the subsequent material is then referred to as n-type (the addition of donor impurities with electrons as the majority carrier) or p-type (the addition of acceptor impurities with holes as the majority carrier) [8]. When an n-type and p-type material are put together, electrons near the interface of the n-type material diffuse toward the p region and leave behind positively charged holes.

Similarly, holes from the p-type material interface move toward the n region and leave negatively charged ions behind. The charged ions form a depletion region that has a built-in potential that prevents electrons or holes from crossing to the other side. This interface, known as the p–n junction, was first discovered by Ohl with its theory developed by Shockley, Sah, Noyce and Moll [8]. The p–n junction is commonly used as a diode that allows the flow of current in only a singular direction, which can be manipulated by applying a positive or negative bias voltage. Thus, the properties of these devices can be manipulated, such as through voltage to control the resistivity of the device, and they have led to the development of transistors and therefore most electronic technologies [7].

10.1.2 The importance of reaching the ultrafast frontier in semiconductor research

In semiconductor microelectronics, size and speed are closely related to shorter effective diffusion lengths entailing faster processing speeds [9, 10]. Therefore, industry continues to realize smaller semiconductor structures—the size of these components has been reduced one million times at the production level over the past 100 years [11]—with structures projected to eventually be as small as a few nanometers [12]. At this length scale, the transit times of electrons happen at sub-picoseocond or faster time scales. The development of these high-speed devices requires better understanding of the various dynamic processes happening to the carriers at ultrafast time scales. Ultrafast optical techniques developed during the twentieth century gave us such an ability to observe the temporal evolution of various fundamental processes involved in microscopic carrier dynamics [10]. These techniques enabled the study of physical phenomena such as excitation and relaxation of hot carrier distributions as well as carrier–carrier scattering and electron–phonon interactions. These techniques have shown that Auger recombination and surface trapping processes limit the efficiencies of semiconductor devices [13–16]. Additionally, optical techniques have also been used to investigate electron–photon coupling in nanodevices and ultrafast nanoplasmonic dynamics for new materials such as van der Waals heterostructures and high-speed optoelectronics [18, 20]. Nonetheless, as semiconductor devices approach physical limits in size, technological breakthroughs will be needed to study these increasingly small devices [12], including following the carrier movement within these devices as well as to investigating the carrier dynamics, particularly at the interfaces. Therefore, tools are required that can spatially resolve individual charge carrier dynamics in semiconductors with nanometer resolution while simultaneously maintaining temporal resolutions on a faster than picosecond timescale [17–19].

10.1.3 The grand unification of electron microscopy and femtosecond spectroscopy

Understanding ultrafast processes, particularly for devices with characteristic lengths on the order of nanometers, requires techniques that can temporally and spatially resolve the electron dynamics of individual nanostructures, a non-trivial task requiring the maturation of two important technologies: electron microscopy to provide spatial resolution and optical pump–probe techniques to provide temporal resolution [20]. Over the past few decades, great strides have been made regarding the resolution of microscopes and the laser pulse duration, including the development of electron microscopes (EMs) with spatial resolutions of sub-nanometers and Ti:sapphire lasers capable of detailing processes on the order of sub 10 fs. Until recently, the two techniques were largely separated—electron microscopy techniques, with their excellent spatial resolution, were suited to image structures and surfaces while optical techniques, with their excellent temporal resolution, were used to study ultrafast dynamic processes.

The first EM, a simple two-lens system, was invented by Max Knoll and Ernst Ruska in 1931 when Ruska noticed that the electromagnetic field could be used as a sort of lens for the electron beam [21]. Although this first system achieved magnification of only 17.4 times, the abilities of electron microscopy would soon rapidly increase. The advent of modern computers enabled more accurate focusing and proper adjustment of image astigmatism, leading to the advent of high-resolution electron microscopy (HREM) in the mid-twentieth century with a spatial resolution of up to 1 nm. Currently, EMs with the ability to correct for spherical and chromatic aberrations can easily achieve sub-angstrom spatial resolutions [22].

Traditional EM techniques, such as those utilizing scanning electron microscopy (SEM), transmission electron microscopy (TEM) and low energy/photoemission electron microscopy (LEEM/PEEM), can detail processes on the order of milliseconds in real time and are mainly limited by the speed of the imaging cameras. However, these microscopes employ electron sources generated thermally by heating the cathode or via field emission, meaning the electron beam is composed of random, single-electron bursts that do not afford temporal control for imaging [23]. More recently, high-speed microscopy (HSM) techniques were used to image details of interfacial motion, crystal formation, twinning, phase transition changes in metastable morphologies and other fundamental material processes on the order of tens to hundreds of nanoseconds [24, 25]. Yet these traditional EM techniques are still far from reaching picosecond or better temporal resolution.

Since the early 1990s, time-resolved femtosecond optical spectroscopy techniques such as pump–probe spectroscopy, four-wave-mixing spectroscopy, ultrafast luminescence and terahertz spectroscopy, among others, have been utilized to understand ultrafast dynamics in nanomaterials [26]. Beginning in 1864, Toepler used short sparks to image sound-wave phenomena, which is regarded as one of the first instances of a pump–probe technique being used to capture microscopic dynamics [27]. In 1899, Abraham and Lemoine improved upon the temporal resolution of this system. Since that time, pump–probe spectroscopy has been further developed to allow for direct probing of the dynamics themselves, rather than of the energy levels, with increased spatial and temporal resolution [28]. By 1960, the first pulsed lasers were used in time-resolved studies with mode-locking techniques (discovered in 1964) allowing for advances into even shorter time regimes [1, 29]. The ability to probe ultrafast phenomena has continued to improve with the first attosecond pulse trains realized in 2001 [30]. This reduction of pulse duration from femtosecond to attosecond was possible with advances in dispersion control and Kerr nonlinearities as well as improvements in material manufacturing [29, 31]. Such speeds have allowed for the study of atomic ionization and structural and electronic wave-packet dynamics in simple molecules; in the future, studies of real-time charge transfer in molecular systems might be feasible, which may push the speed limits of digital electronics and signal processing [29, 30, 32].

During pump–probe experiments, the pump pulse excites carriers from the ground state to unoccupied excited energy states in the conduction band, sending the system out of equilibrium. After excitation, equilibrium is restored through several different relaxation processes, such as hot carrier thermalization, carrier–carrier scattering, carrier–phonon scattering and carrier migration. The dynamics of these processes are followed using probe pulses sent at different, but carefully controlled, time delays relative to the pump [8, 24, 33]. Using this technique, the design, synthesis and implementation of nanomaterials in various photochemical and photophysical applications, such as for photocatalysis and photoelectrochemistry, can be better optimized and guided [26].

Although optical studies provide excellent temporal resolution that is limited only by pulse duration, the use of photons as a probe results in spatial resolutions that cannot exceed the diffraction limit (hundreds of nanometers). Most optical studies also focus on interband excitation at photon energies close to or greater than the bandgap, meaning the transient optical properties of electron–hole pairs are investigated simultaneously, making it difficult to distinguish between the two types of charge carriers [34]. Additionally, converting measured optical quantities, such as transient reflection, into an understanding of charge carrier dynamics can often be non-trivial and makes quantitative analysis of these experiments challenging.

As can be seen, both electron microscopy techniques and optical techniques alone show unique advantages and limitations. While traditional EM techniques provide nanometer spatial resolution, their millisecond temporal resolutions would prove problematic for observing ultrafast phenomena; even HSM techniques are orders of magnitude too slow to image electron dynamics at the pico- to femtosecond scale [24]. To overcome the spatial *and* temporal barrier, several groups began experimenting with combining optical and EM techniques, marking the beginning of a new realm of science that would enable a new way of seeing and understanding ultrafast, sub-micron processes. With this combination of EMs and ultrafast lasers, techniques such as ultrafast electron microscopy, scanning ultrafast electron microscopy and time-resolved photoemission electron microscopy can provide simultaneously the spatial and temporal resolution necessary for imaging ultrafast dynamics at nanometer length scales.

10.2 Unlocking high spatial–temporal resolution in studies of ultrafast dynamics in semiconductors

10.2.1 Ultrafast transient absorption microscope (ultrafast TAM)

Before detailing the development of ultrafast EMs, it is important to note a purely optical technique that also brings ultrasmall resolution into the ultrafast world. Ultrafast TAM, despite being purely optical, overcomes the diffraction limit through nonlinear optics that enable the characterization of electronic excited-states with high spatial and temporal resolution. This process allows one to create maps of excited-state processes, such as photoexcitation energy flow [35]. Normally, when a single photon transitions at the pump and probe frequencies that are used to generate the signal, the resolution is determined by the overlap between the intensity point spread function between the pump and probe. The nonlinear nature of TAM allows for two photons to be used, which increases the imaging spatial resolution by up to a factor of $\sqrt{2}$ compared to linear optical microscopy at the same wavelength [36]. Ultrafast TAM uses a femtosecond laser in an optical pump–probe set-up to study nonlinear dynamics in materials, where pump-induced change in the material is monitored by the probe (figure 10.1). The pump is modulated with an acousto-optic modulator (AOM) before entering the microscope objective and hitting the sample, which is scanned in a raster pattern utilizing a *X*–*Y*–*Z* piezo stage. The transmitted light is then collected by another objective, and the transmitted probe intensity is detected by an avalanche photodiode (APD). By varying the delay time between the pump and the probe, photoinduced absorption changes and TA traces can be collected as a contrast mechanism for imaging [35].

Ultrafast TAM has been used in a variety of studies, including: determining charge carrier diffusion coefficient and length [37]; estimating the contributions of the excitons and free charge carriers in the observed transient absorption response [35]; observing the complex local structure and charge separation in CdSe nanocrystals (NCs) [38]; determining the carrier cooling dependence on carrier density and understanding the fast and slow lifetimes in graphene (figures 10.2(a–c)) [39, 40]; and investigating hot carrier migration in $CH_3NH_3PbI_3$ thin film hybrid perovskites (figures 10.2(d) and (e)) [41]. Ultrafast TAM has also been used to image the excited-state dynamics in a variety of nanostructures such as single-walled carbon nanotubes (SWNT) [42, 43], ZnO nanorods [44] and Si nanowires [45]. Although ultrafast

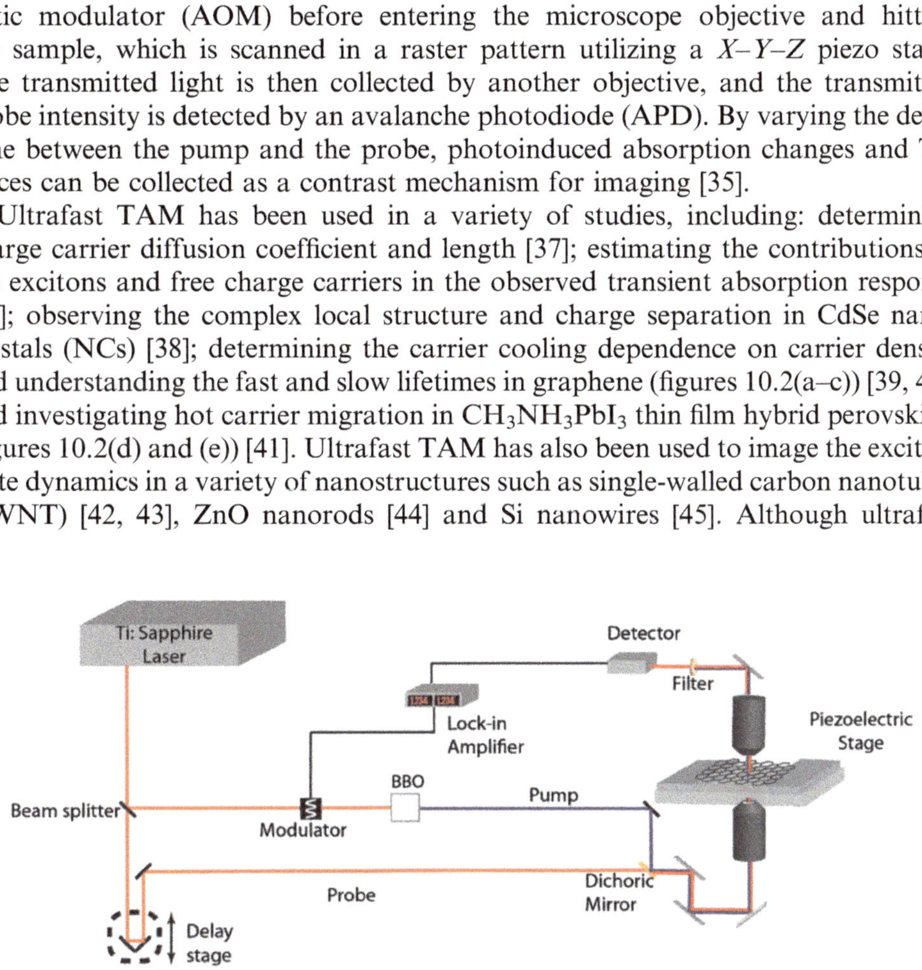

Figure 10.1. Schematic diagram of an ultrafast TAM set-up. The piezoelectric stage allows for raster-scanning of the sample in the *X*-, *Y*- and *Z*-directions. An objective lens with sub-micrometer spot size is used to focus the pump and probe beams on the sample with a second lens used to collect the transmitted light (or collected by the same objective lens for reflected light). The probe is then detected by an avalanche photodiode. Reproduced with permission from [148]. Copyright 2013 Institute of Physics and IOP Publishing.

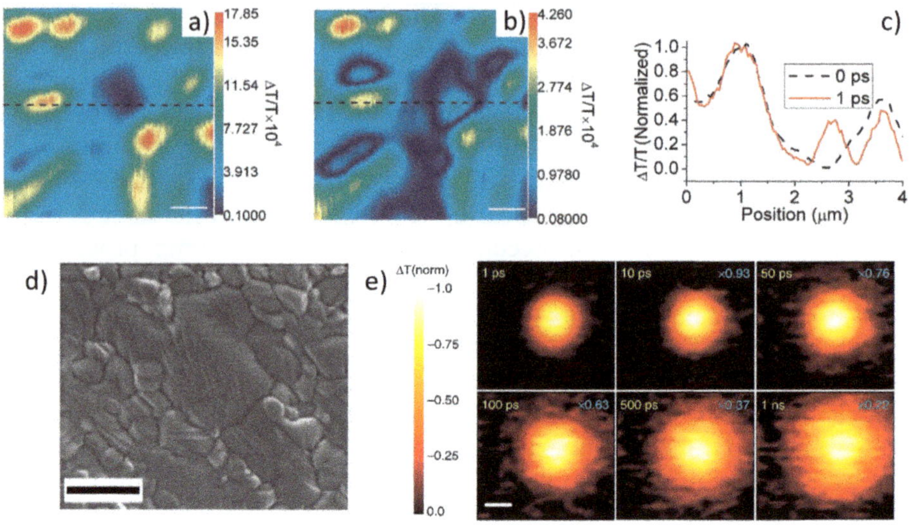

Figure 10.2. (a) 2D image carrier density profiles from an ultrafast TAM experiment on graphene at (a) 0 ps and (b) 1 ps delay times. The scale bar is 0.5 μm. The differences in thicknesses within the graphene sample result in observed differences in slow relaxation time between (a) and (b). (c) Line profiles are taken along the dashed lines as indicated in figures (a) and (b). Adapted with permission from [39]. Copyright 2010 American Chemical Society. (d) SEM micrograph of a perovskite flake. The scale bar is 0.5 μm. (e) 2D image carrier density profiles from an ultrafast TAM experiment at various pump–probe delay times on $CH_3NH_3PbI_3$ perovskite film. Each image was normalized to the signal maximum at 1 ps. The scale bar is 300 nm. Adapted with permission from [37]. Creative Commons Attribution 4.0 International License.

TAM gives detailed morphological information using label-free molecular contrast with high spatial and temporal resolution [36], its resolution is more limited compared to ultrafast EM techniques due to its lower signal-to-noise ratio (SNR), meaning it cannot resolve photocarrier dynamics with sub-wavelength spatial features (tens of nanometers spatial resolution rather than several nanometers) [17]. These limitations can be seen in figure 10.2(a–c), from Huang *et al*'s work, where they studied the carrier dynamics in graphene. Although the study had a high spatial resolution of 0.3 μm, even higher spatial resolution would be necessary to visualize the inhomogeneous response between graphene and the substrate.

Besides ultrafast TAM, it is worth noting other high spatial resolution, ultrafast techniques, such as time-resolved scanning tunneling microscopy (TR-STM) [46–53], time resolved near-field scanning optical microscopy (TR-NSOM/SNOM) [54–59] and ultrafast scanning probe microscopy techniques such as pump–probe atomic force microscopy (AFM) [60] and pump–probe Kelvin-probe force microscopy [61, 62], all of which are capable of imaging with several nanometer resolution and detecting femtosecond charge carrier dynamics among other phenomena. However, these techniques differ fundamentally from time-resolved EM techniques and will thus not be discussed in-depth in this chapter. For more information regarding these topics, please see the above-cited articles.

10.2.2 Ultrafast techniques utilizing electron microscopes

To obtain high spatial and temporal resolution, another approach is to combine femtosecond lasers with various EMs, which gave birth to new techniques including ultrafast electron microscopy (UEM), scanning ultrafast electron microscopy (S-UEM or SUEM) and time-resolved photoemission electron microscopy (TR-PEEM). Like the pump–probe technique described previously, these techniques employ stroboscopic measurements, where a pulsed source is used to generate many rapidly repeating pump–probe cycles to achieve a high SNR.

The primary difference between optical pump–probe microscopy techniques and ultrafast EM techniques is that electrons (in UEM and SUEM) and photoemitted electrons (in TR-PEEM), rather than photons, are used to provide the spatial resolution. By using electrons, these techniques bypass the diffraction limit of light, allowing for the study of ultrafast phenomena at high spatial resolution [17]. Low energy electrons with kinetic energies of 3.1 eV can achieve an approximately 1 nm spatial resolution, and the resolving power of electrons improves with higher kinetic energies. Although in principle the spatial resolution of these EM techniques is limited only by the wavelength of electrons, in practice, the spatial resolution is limited by chromatic and spherical aberrations in electron optics [63].

10.2.2.1 Ultrafast electron microscopy (UEM)

The idea of UEM began when Zewail and Williamson determined in 1991 that replacing optical pulses with electron pulses would resolve structural bond dynamics in chemical systems. Although the creation of the first UEM by Zewail would take more than a decade, it has proven its significance in various studies and is being pursued at many institutions [64]. UEM utilizes a TEM in a pump–probe stroboscopic setting. An ultrafast optical pump is used for excitation and a synchronized pulsed electron beam, driven by another ultrafast laser pulse, probes the changes in the sample. Since the temporal resolution only depends upon the pulse duration of the pump and probe pulses, it is independent of the speed of the detector, such as a CCD's read-out time or the electronic response of a video recorder. In conventional TEMs, electron beams are generated by a typical LaB_6 cathode or field emission electron gun that consists of spatially and temporally randomly distributed electrons; in contrast, UEM utilizes ultrafast laser pulses to trigger electron pulses that are temporally and spatially coherent [65–67]. A typical UEM set-up is shown in figure 10.3.

Ultrashort optical pulses produce photoelectron packets through the photoelectric effect upon impact with the cathode tip. With few electrons per packet in each probe pulse (usually only tens of electrons), Coulombic space-charge effects are minimized, making high temporal precision possible [68]. These pulses can be further collimated, focused and compressed, allowing for precise control over pulse broadening and therefore pulse resolution of the electron packets (figure 10.3) [20, 68, 69]. Since beam coherency is preserved, spatial resolution intrinsic to a conventional TEM instrument can be obtained [65, 70–73]. However, because UEM

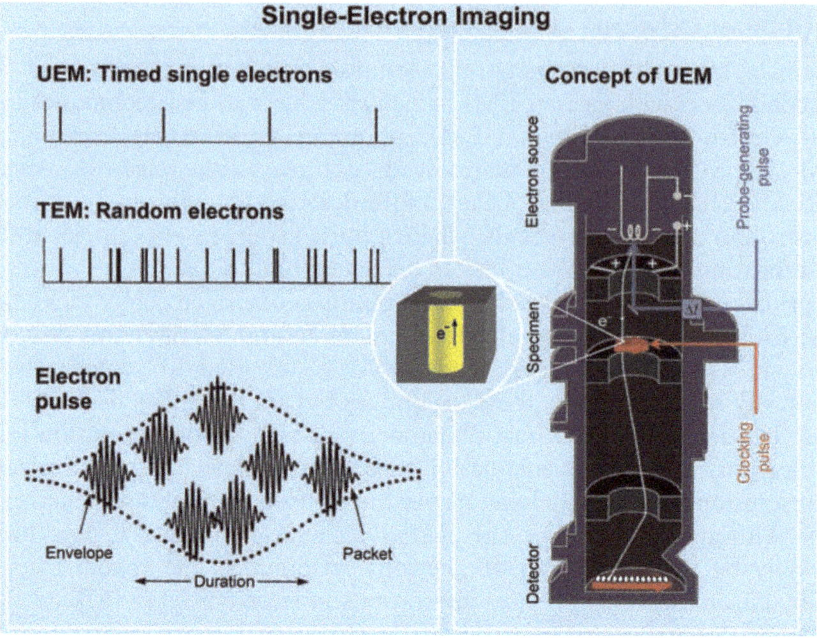

Figure 10.3. Schematic depicting a UEM. Unlike in TEM, timed electron packets utilized in UEM allow for high temporal precision with the pulse duration dictating the overall temporal resolution. Adapted with permission from [65]. Copyright 2010 the American Association for the Advancement of Science.

utilizes so few electrons per pulse, the signal from each pulse is low, and experiments must be aggregated over more independent shots compared to other similar techniques [68, 74].

Inheriting the great versatility of TEM, UEM can operate in several modes: real (imaging), reciprocal (diffraction) and energy (spectroscopy) space [67]. In imaging and diffraction modes, UEM can investigate atomic and structural changes [20]. By exploiting the energy-filtering capability of TEM, UEM can perform electron energy loss spectroscopy (EELS), which provides information about the composition, chemical state and collective excitation modes supported by the sample. Additionally, a method called photon-induced near-field electron microscopy (PINEM) enables mapping and imaging of localized electric fields at interfaces (figure 10.4) [20, 23, 24]. Many investigations of semiconductor materials have been performed using UEM techniques, including: imaging of coherent transients during morphology changes of graphite [73]; visualizing structural changes and their time scales in crystalline Si [75]; imaging the role of electronic and vibrational excitations of axial and radial expansions in carbon nanotubes (CNT) (figure 10.4(a)) [76]; understanding the nanoscopic crystallization of amorphous silicon nitride (Si_3N_4) [52]; investigating the ultrashort optomechanical motion of the Si_3N_4 and CNT network (figures 10.4(b–d)) [77]; and imaging the photoinduced structural dynamics in MoS_2 [78].

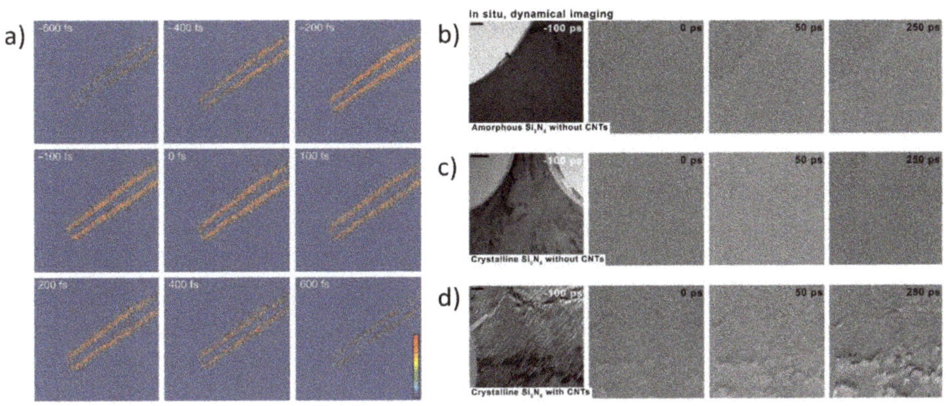

Figure 10.4. (a) PINEM (energy-filtered imaging) of an individual nanotube. Blue indicates regions of no counts while red indicates the fields created by the femtosecond pulse around the nanostructure and their subsequent relaxation at various times. Adapted with permission from [23]. Copyright 2009 Macmillan Publishers. UEM stroboscopic imaging of (b) amorphous Si_3N_4 without an adhering CNT network (scale bar: 200 nm), (c) crystalline Si_3N_4 devoid of CNTs (scale bar: 500 nm) and (d) crystalline Si_3N_4 with an adhering CNT network (scale bar: 200 nm). Each image was obtained by subtracting the reference image at −100 ps. Adapted with permission from [77]. Copyright 2010 the American Chemical Society.

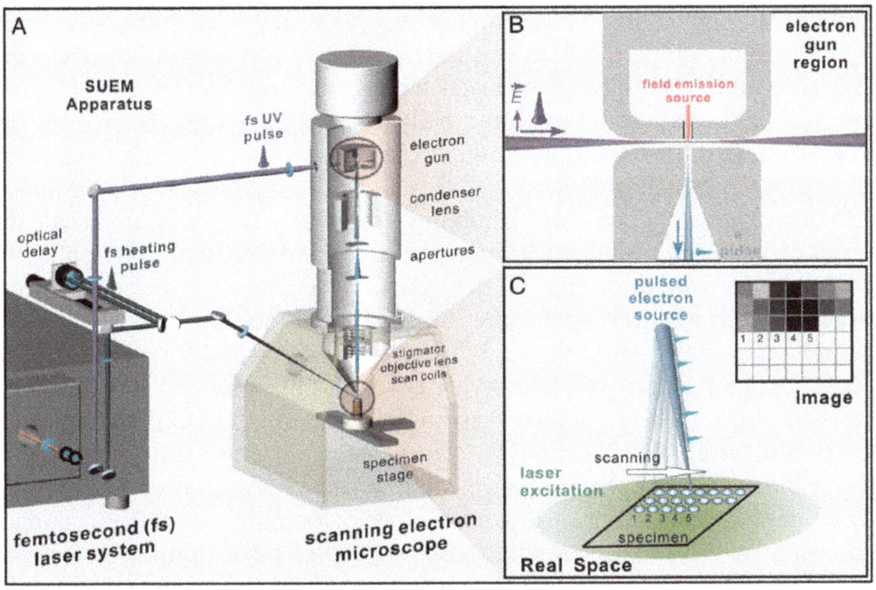

Figure 10.5. (a) Schematic for an SUEM set-up including the ultrafast pump–probe system. (b) and (c) Illustrations of the electron source as it scans across the sample to form the image. Used with permission from [24]. Copyright 2010 National Academy of Sciences.

10.2.2.2 Scanning ultrafast electron microscopy (SUEM)

SUEM operates similarly to UEM but utilizes an SEM rather than a TEM (figure 10.5(a)). In SUEM, photoinduced ultrashort electron packets from a sharp field-emitter tip are used as a probe (figure 10.5(b)) and like conventional SEM, images are formed by scanning a focused electron beam across the sample surface (figure 10.5(c)) and measuring the number of emitted secondary electrons and primary backscattered electrons at each position with an Everhart–Thornley scintillator detector. Since the signal obtained from each pulse is quite weak, millions of electrons pulses are averaged to generate each pixel in an image [20]. The conventional field-emitter tip used for SUEM is tens to hundreds of nanometers in size while in UEM, the typical LaB_6 photocathode active area size is on the order of tens of micrometers, making the SUEM's electron beam in this configuration brighter than the source for the UEM [79, 80]. Spatial resolutions as high as approximately 10 nm with controlled electron packet emission were achievable at a 30 kV acceleration voltage [17, 79].

In SUEM, the energy of the primary pulsed electrons can be varied through adjustment of the accelerating voltage in the EM column. This allows for control of the penetration depth of the primary electrons and subsequently the sampling depth from which secondary electrons are generated and detected [70]. With this versatility, SUEM is an effective technique not only for investigations of surfaces and interfaces but also of the bulk. There are several merits of bulk sensitivity. Unlike UEM, where an ultrathin sample is needed, SUEM requires no special sample preparation as the technique is not sensitive to surface impurities and roughness [17]. The ability to examine bulk dynamics also means that one can investigate thicker samples, which can tolerate larger temperature fluctuations and sustain greater radiation damage by lasers when compared to ultrathin samples [81]. However, SUEM cannot study electrically insulating materials due to sample charging issues, and the technique suffers from a low SNR, meaning long averaging times for image acquisition [17].

SUEM is particularly useful for resolving electronic changes in nanoscale devices using its contrast mechanisms, including secondary and backscattered electrons [24]. It is capable of studying charge distributions, chemical reactions, molecular interfaces, phase transitions and more. Some SUEMs are equipped with environmental and low-voltage imaging capabilities, which are useful for studies of reactive surfaces [20, 81]. In a study of carrier–carrier scatterings in GaAs, it was shown that secondary electron yield is influenced strongly by doping and carrier concentration (figures 10.6(a) and (b)) [82], and the measured transients can be related to the underlying electronic structural changes. The spatial resolution of SUEM has also been exploited to image the anisotropic diffusion of photoexcited hot carriers on black phosphorus (BP) in the armchair and zigzag directions (figures 10.6(c) and (d)) [83]. SUEM has also been utilized to study many different semiconductor samples including Si p–n junctions [84], hydrogenated amorphous Si [85], InGaN NWs [86, 87], CdSe thin films [88], CdSe single crystal and its powder film [89], shelled copper indium gallium selenide (CIGSe) film [90] and ZnS passivated CIGSe NCs [91].

10.2.2.3 Time-resolved photoemission electron microscopy (TR-PEEM)

PEEM is a versatile EM technique that is used widely in studies of electronic, chemical and magnetic structures at surfaces and buried interfaces with nanometer spatial

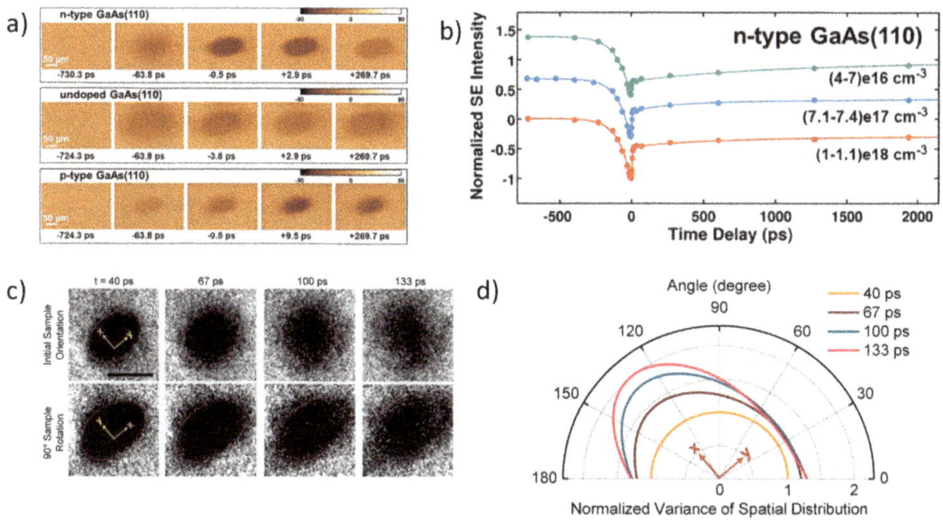

Figure 10.6. An application of SUEM demonstrating (a) the change in photoemission intensities for various doping conditions, including n-type, undoped and p-type GaAs (110) and (b) the temporal evolution of the average intensity with lifetimes for photocarrier generation and recombination. Adapted from [82]. (c) SUEM images of the distribution of hot-holes on the surface of a black phosphorus sample with two orthogonal orientations, where 'X' is the armchair direction and 'Y' is the zigzag direction. The scale bar is 60 μm. (d) The photogenerated holes in both cases diffused preferentially along the armchair direction in this experiment. Reproduced with permission from [83]. Copyright 2017 the American Chemical Society.

resolution. Its combination with ultrafast lasers led to the development of TR-PEEM, which was first used to investigate the formation, propagation and nonlinear effects in surface plasmon polaritons on metal surfaces [92–98]. More recently, several groups have pioneered the use of TR-PEEM in the investigation of semiconductor interfaces and nanostructures [99–107]. The focus of this section will be on a few aspects of TR-PEEM that are fundamentally different from the other electron microscopy techniques discussed above. For a more comprehensive review of PEEM techniques, please see these publications and the references in [108–110].

PEEM images electrons based on the photoelectric effect, a phenomenon discovered by Hertz in 1887 and later explained by Einstein in 1905 [111]. Photoelectrons can be emitted from materials when the energy of the photons exceeds the work function of the material, which is around 4 to 6 eV for most solids. Generally, UV lamps, UV lasers or x-ray sources are used to illuminate and generate photoemitted electrons from the sample. These electrons are accelerated and collected by an objective lens sitting in front of the sample at high electric potential (figure 10.7). The image formed at the back objective is then transferred and magnified by a series of lenses and projectors. At the end, the magnified image is intensified by the microchannel plane and imaged on a phosphor screen by a CCD camera. In addition to the ability to collect real space images (figure 10.7(a)), reciprocal space images can also be collected by varying the setting of the lenses (figure 10.7(b)). Since all of the photoemitted electrons are collected simultaneously

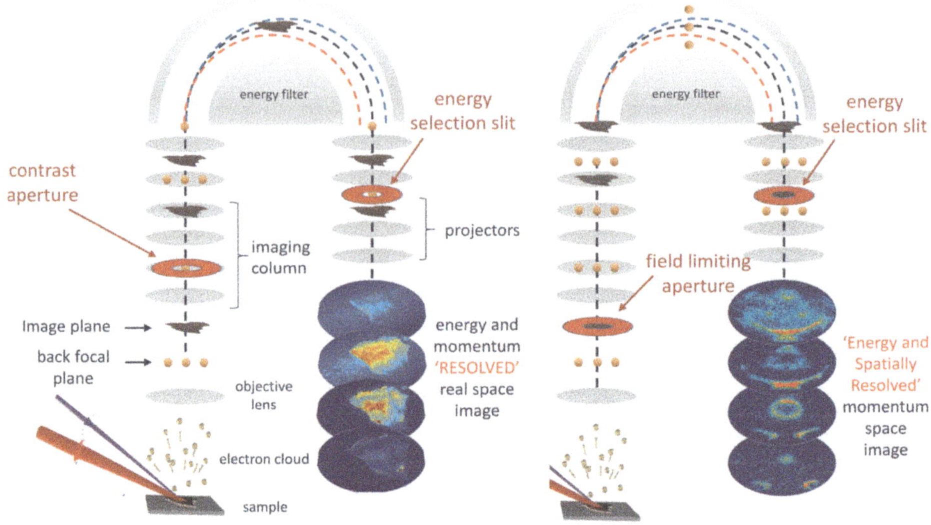

Figure 10.7. TR-PEEM system working in (a) real space imaging and (b) momentum space imaging mode. In (a) real space imaging mode, photoemitted electrons are collected by the objective lens in front of the sample and a contrast aperture can be inserted into the back focal plane of the electron optics to select electrons emitted at a certain angle or momentum. An energy selection slit inserted after the hemispherical energy filter provides further spectroscopy capability to selectivity image electrons of certain kinetic energies. In (b) momentum space imaging mode, a field limiting aperture is inserted at the image plane to select a small region of interest. Together with the energy filter, it provides the means to perform energy- and spatially resolved imaging in momentum space.

by the objective lens without scanning the photoexcitation beam, PEEM can capture images from a large area in the relatively short time of 100 ms or less [112, 113]. The spatial resolution of the instrument does not depend on the size of the optical beam; it is mainly limited by aberrations in the electron optics, which are around several and few tens of nanometers for photoelectrons.

Photoelectron spectroscopy by itself is a powerful technique that allows for the complete determination of the electronic structures and chemical compositions of materials. In the last few decades, it has developed into one of the most popular tools for studying band dispersions as well as mapping out the Fermi surfaces of complex quantum systems [114, 115]. In PEEM, an energy filter can be added to allow simultaneous determination of both the momentum and energy of the photoemitted electrons. As shown in figure 10.7, an electrostatic concentric hemispherical deflection analyzer lets electrons traveling at different kinetic energies disperse, and electrons with different energies can be selected by a slit located at the exit of the analyzer. While a traditional angle-resolved photoemission spectroscopy (ARPES) apparatus using a similar energy analyzer has a limited acceptance angle for electron collection, in PEEM the objective lens allows for simultaneous collection of photoelectrons at different emission angles. With high enough photon energy, full or multiple Brillouin zones can be recorded readily within the same image, allowing for fast data acquisition of band structures. In addition to the energy filter, an aperture can be inserted at the image plane of the electron beam path (figure 10.7(b)). This apparatus limits the field of view of the image transmitted through the lenses. Various combinations of the energy filters and field limiting apertures makes it possible to perform selected area ARPES measurements for micron-sized sample areas, which is useful in studies of inhomogeneous samples [116]. In addition to adding apertures to the image plane, it is also possible to add apertures to the back focal plane (figure 10.7(a)); the so-called contrast aperture selects photoelectrons passing at certain angles. The combination of energy filters with the contrast aperture allows for selective imaging of electrons residing within a specific position in momentum space, facilitating mapping of charge and energy transfer in materials [105].

In TR-PEEM, PEEM is combined with ultrafast lasers to perform pump–probe stroboscopic measurements and ultrafast imaging. However, unlike in other EM techniques, such as UEM and SUEM where an optical beam is used as a pump and ultrafast electron beam as a probe, TR-PEEM uses ultrafast optical beams for both the pump and probe, just like in a traditional all-optical pump–probe measurement. This provides TR-PEEM with the full capabilities of ultrafast temporal resolution afforded by traditional all-optical pump–probe techniques. However, unlike the traditional all-optical pump–probe techniques, which measure the properties of the transmitted and reflected probe beam, TR-PEEM measures the properties of the photoelectrons emitted by the probe beam. In particular, images are acquired with the photoelectrons generated by the probe beam, providing spatial resolution as described for PEEM techniques above, well beyond the resolution obtained by all-optical pump–probe techniques.

The very first attempt at time resolved PEEM dates back to 1997 when a Nd:YAG laser was used to study the time dependence melting and recrystallization of Al films at a temporal resolution of a few nanoseconds [117]. Later, Ti:sapphire lasers with frequency doubled pump and probe beams of 3 to 3.4 eV in an interferometric geometry were used to study the decay of excited electrons in patterned Ag films on Si and GaAs [92]. Faster electron decays in metals in comparison to the GaAs substrate were observed in this experiment. Notice that the Ag film here could only be imaged via two-photon photoemission (2PPE) using frequency doubled laser light or by multiphoton photoemission (*n*PPE) at the fundamental laser light (1.55 eV). The observation of hot spots on the Ag film has triggered many studies in surface plasmon and polarity dynamics on Ag and Au nanostructures using the TR-PEEM technique [92–98].

Notice that in 2PPE the pump and probe pulses are of the same color and that they are indistinguishable. Although this set-up works well with imaging plasmon dynamics, it does not provide the versatility to study excited-state dynamics, particularly in semiconductor materials, where a two color pump–probe set-up allows for different photon energies to be used; namely where one photon selectively excites while another photon probes electrons at different energy levels. In general, photoemitting electrons from the sample require the probe pulse in TR-PEEM to typically be in the UV range in order to overcome the work function of the sample.

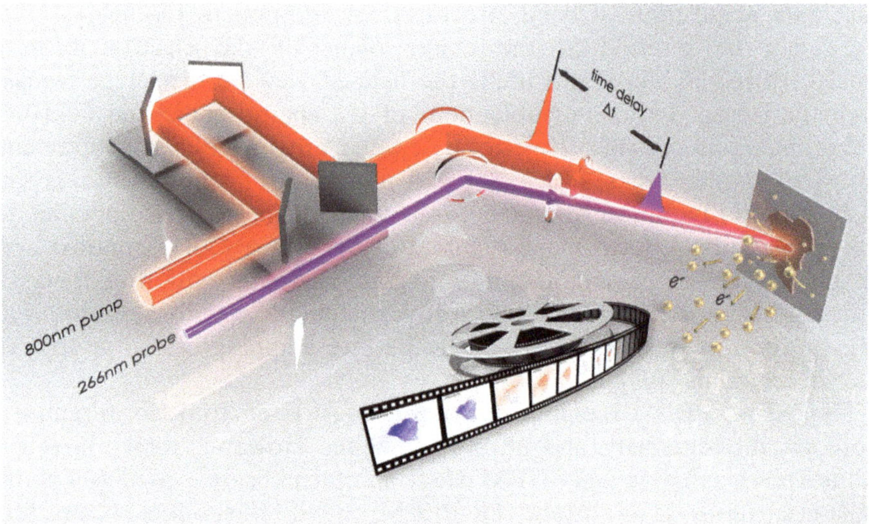

Figure 10.8. Illustration of a TR-PEEM set-up. An 800 nm pump beam (in red) excites the sample, and the electronic dynamics of the photoexcited carriers are imaged by the photoemitted electrons using a 266 nm pump beam (in purple) that arrives at a delayed time. Adapted with permission from [105]. Copyright 2016 Macmillan Publishers.

Figure 10.8 shows a TR-PEEM set-up that utilizes a Ti:sapphire oscillator laser running at 4 MHz with wavelength centered at around 800 nm (approximately 1.55 eV) and pulse width of around 45 fs [105]. The laser beam is split into two paths with a beam splitter: one path is used for the pump and the other path for generating the UV probe. The pump can use the fundamental wavelength of the laser or, depending on the experiment, it can also be changed to a different wavelength using an optical parametric amplifier or frequency doubling technique. To access the electronic bands close to the Fermi level, the photons in the probe beam need enough energy to overcome the work function of a material. Third and fourth harmonics are accomplished utilizing various nonlinear crystals that can generate photons with the necessary energies ranging from 4.7 eV to 6.2 eV in the deep-UV. Alternatively, it is also possible to use a very intense, ultrafast laser beam to initiate nonlinear multiphoton photoemission (nPPE) processes in the material directly to reach the photoemission threshold. In TR-PEEM, the dynamics of these carriers are followed by photoemission of electrons by the probe pulses.

10.3 Studies of semiconductors utilizing TR-PEEM

As mentioned in the opening of the previous section, most early developments in the TR-PEEM technique were in the study of surface plasmon polaritons of metal surfaces, and until very recently, surprisingly little research had been conducted on semiconductors using the technique. This section will summarize the works from several groups that led to the development of the TR-PEEM technique for use in studies of electron dynamics in semiconductors. The next section will then give a brief perspective about the future of this technique.

In 2014, Fukumoto *et al* [99, 100] demonstrated TR-PEEM's capabilities in visualizing the motion of charge carriers in semiconductors. In this work, they investigated the recombination dynamics and lateral motion of hot electrons on a GaAs surface under an external electric field (figure 10.9). A pump of 2.4 eV was

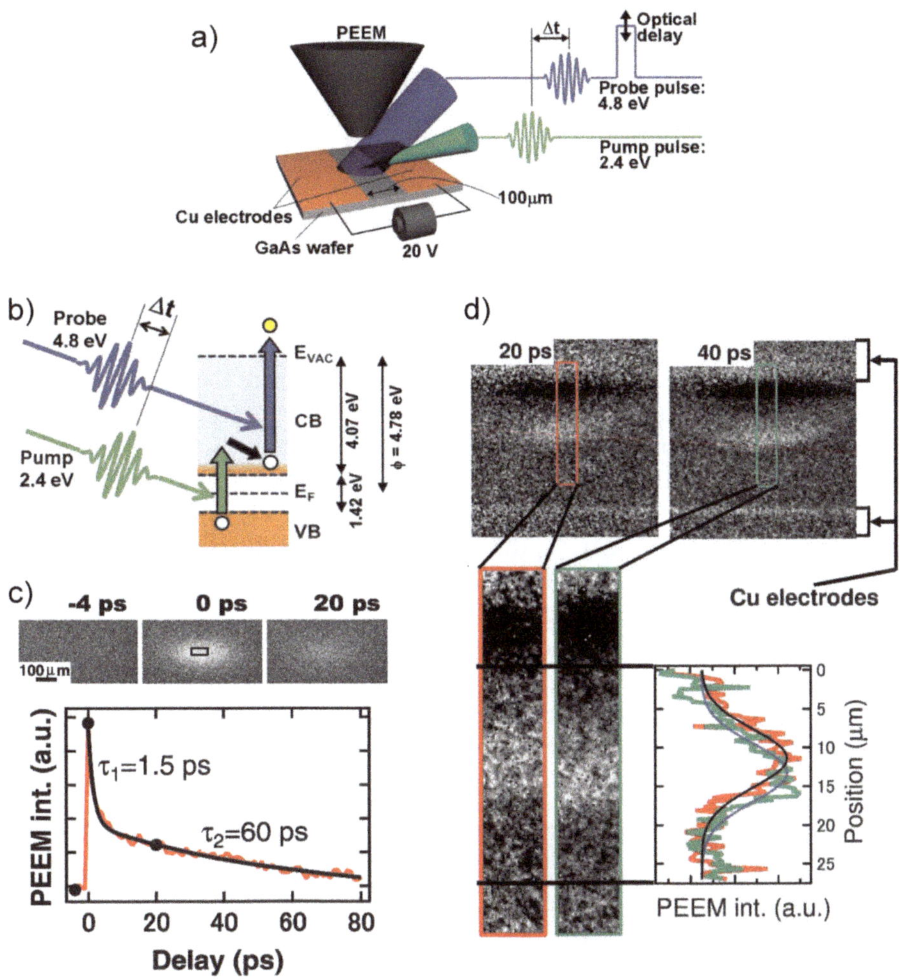

Figure 10.9. (a) Schematic of the experimental set-up showing copper electrodes deposited on the GaAs surface, with a potential difference applied between them to drive lateral motion of hot electrons. (b) Schematic depicting charge transport with the pump (2.4 eV) exciting the electrons from the valence band to the conduction band and the probe (4.8 eV) exciting the electrons to the vacuum (E_{VAC}). (c) Carrier lifetime estimates averaged over the rectangular region marked in the TR-PEEM image and taken at −4, 0 and 20 ps. (d) TR-PEEM images taken at 20 and 40 ps with the subsequent intensity profiles plotted (red and green marked areas) with fits (solid black lines) showing the external electric field driven hot electron motion. Adapted with permission from [100]. Copyright 2014 the AIP Publishing.

used to excite photocarriers over the bandgap of GaAs; with a probe at 4.8 eV, they monitored the population of photoexcited carriers within the conduction band by measuring the changes of photoemission intensities after photoexcitation (figures 10.9(a) and 10.10(b)). The decay of PEEM intensity was fitted by two exponentials (figure 10.9(c)). The fast component was attributed to annihilation of excited electrons on the GaAs surface through carrier re-trapping by mid-gap defect states, inter-valley scattering and intra-band relaxation via electron–phonon scattering. The second, slower decay component was attributed to the surface recombination of photogenerated carriers. By applying a potential between two Cu electrodes on the GaAs surface, they observed a drift of hot electrons under the applied electric field (figure 10.9(d)). A drift velocity of 7.4×10^6 cm s^{-1} to 2.4×10^6 cm s^{-1} under the field gradient from 2000 to 1000 V cm^{-1} was extracted from the data. The measured electron mobility agreed well with known values of electron mobility in bulk GaAs, showing that TR-PEEM is capable of measuring the transport of photoexcited carriers in materials with sub-picosecond temporal resolution and sub-micron spatial resolution.

Instead of using an external electric field to drive electrons, Wong *et al* [108] showed that it is also possible to manipulate the movement of electrons using an ultrafast optical pulse. Here, by using the non-uniform profile of an intense laser pulse, Wong *et al* showed that they can control and separate the charge carriers on a GaAs surface. In this study, they used a 1.55 eV, 45 fs pump pulse and frequency-tripled, 4.6 eV probe pulse. For low excitation intensity (1.4×10^{18} cm^{-3}), the distribution profile of the photoexcited carriers retained a Gaussian shape with carriers diffusing and spreading out normally. However, for high photoexcited carrier density (2.1×10^{19} cm^{-3}), the evolution of the distribution of the photoexcited electrons became non-trivial: the photoexcitation profile started as a simple Gaussian, but later evolved and split into two Gaussians (figure 10.10). This separation was attributed to an induced inhomogeneous surface electric field caused by changes in the surface potential (or surface photovoltage), which is a manifest of the irregular laser excitation profile—in this case, an elliptical beam shape. The degree of separation and the velocity of the two separating Gaussians can be

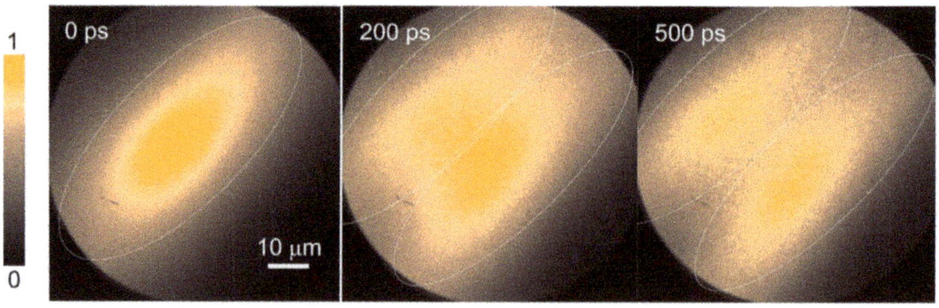

Figure 10.10. TR-PEEM images taken at 0, 200 and 500 ps of the normalized spatial distribution of the photoexcited electrons. At a photoexcitation density of 2.1×10^{19} cm^{-3}, the initial Gaussian profile at 0 ps starts to separate and eventually splits into two distinct peaks at 500 ps. Reproduced with permission from [108]. Copyright 2018 the American Association for the Advancement of Science.

controlled by varying the intensity of the laser. This work showed the ability to manipulate and redirect the flow of charges, which is promising for building new nanoscale optoelectronic devices.

With high spatial resolution, Fukumoto *et al* [102] demonstrated the use of TR-PEEM in their study on the carrier lifetimes of nano-sized structural defects on a GaAs surface (figure 10.11). In this work, they found that photoexcited carriers had shorter lifetimes at the defect sites compared to the rest of the GaAs surface and that the carrier lifetimes were inversely proportional to the photoemission intensity at each defect. To explain these findings, they argued that there are two different contributions to the photoemission intensity: one from the excitation of electrons from the valence band into the conduction band and another from the excitation of electrons from mid-gap defect states to the conduction band. They also proposed that the observable carrier dynamics mainly consisted of electrons relaxing from the

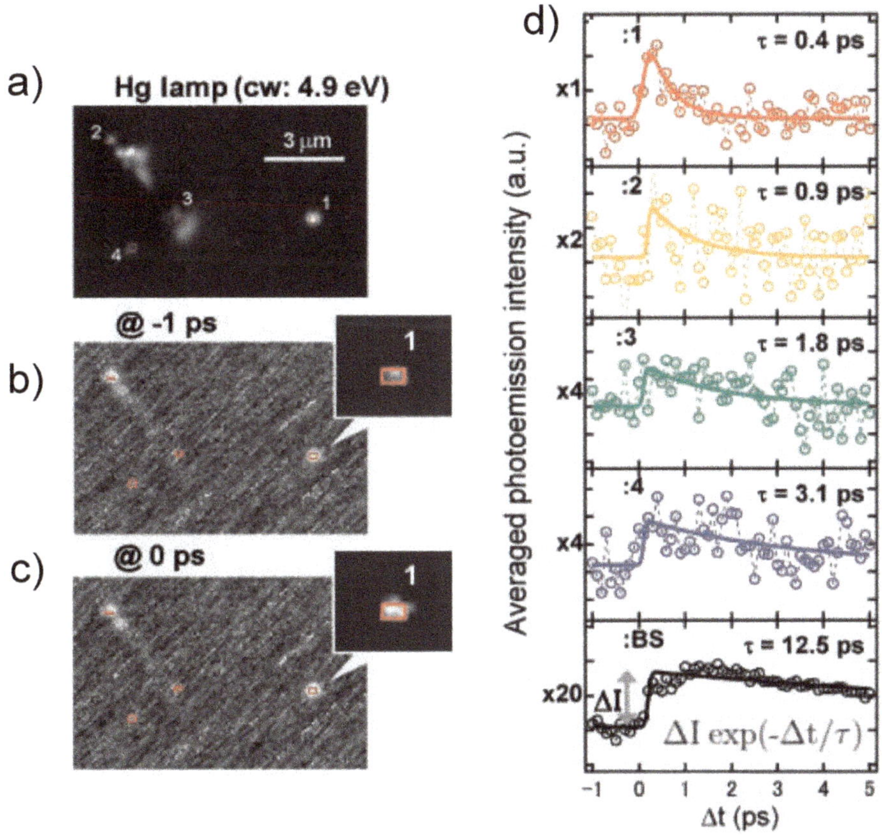

Figure 10.11. Time dependence photoemission intensity on GaAs defects. (a) PEEM image obtained using a Hg discharge lamp with defects shown as bright spots. (b) and (c) TR-PEEM images of the defects taken at −1 ps and 0 ps delays. (d) Time dependence photoemission intensity curves of different defects marked in (a) show that bright defects have shorter lifetimes. Adapted with permission from [102]. Copyright 2015 The Japan Society of Applied Physics.

conduction band back to the mid-gap defect states. In their model, the decay rate of photoemission intensity was proportional to the defect density, and from that they estimated the defect density within each individual defect. However, this study used a probe pulse of 4.8 eV, which is only high enough to probe electrons in the conduction band. This energy is insufficient to reach the energy level of the mid-gap defect states where all the proposed dynamics are happening. Thus, this experiment only showed indirect evidence of the relaxation pathways of the photoexcited elections.

Answering questions about charge transfer and relaxation of carriers among different energy levels requires the use of a probe beam with enough energy to access the energy states of interest. In addition, it is also beneficial to perform photoelectron spectromicroscopy to resolve the electron populations within each electronic state. In the work by Man *et al* [105], they took advantage of the idea of spectromicroscopy and captured a movie of electrons going from high to low potential states in a type-II InSe/GaAs heterostructure, a typical photovoltaic device structure. In this study, they followed the dynamics of the photoexcited carriers in the heterostructure with TR-PEEM using femtosecond, near-infrared (1.55 eV) pump pulses and deep-UV (4.66 eV) probe pulses. The heterostructure consisted of a thin InSe flake mechanically exfoliated onto a GaAs wafer. Since both InSe and GaAs have bandgaps smaller than the pump pulse, the photoexcited electrons populated the conduction band of both InSe and GaAs. Here, with an energy filter, they monitored the spatial and energy distribution of the photoexcited electrons simultaneously (figure 10.12). By stitching together images taken at various pump–probe time delays, they produced a movie showing the migration of electrons within the heterostructure following the energy landscape, where electrons from GaAs at higher potential flowed into InSe at a lower potential (figure 10.13). Within the InSe sample, they also observed a flow of charges between different regions where there was an offset in the band alignment due to the inhomogeneous sample thickness. This experiment was the first time that the non-equilibrium distribution of electrons in space at the instant of photoexcitation was imaged, and the subsequent electron motion within the semiconductor heterostructure was directly imaged.

In addition to resolving and following carrier dynamics in the energy landscape and in real space, TR-PEEM can elucidate the recombination dynamics and scattering pathways in inhomogeneous samples in momentum space, which is not possible with any other currently available technique. For example, Fukumoto *et al* [104] studied multilayer graphene structures that were grown using the chemical vapor deposition (CVD) method. This kind of sample, namely graphene composed of many different thicknesses and twist angles, is impractical to study with spatial averaging techniques where responses from different sample regions will be averaged out. With TR-PEEM, Fukumoto *et al* have succeeded in resolving the recombination dynamics of different sample regions. In this study, they investigated the decay rate of photocarriers with a 1.2 eV pump and 4.8 eV probe. They categorized their sample into regions with different twist angles and found that all regions of multilayer graphene exhibit higher work functions as well as longer lifetimes than monolayer graphene. In particular, superstructures with twist angles greater or less

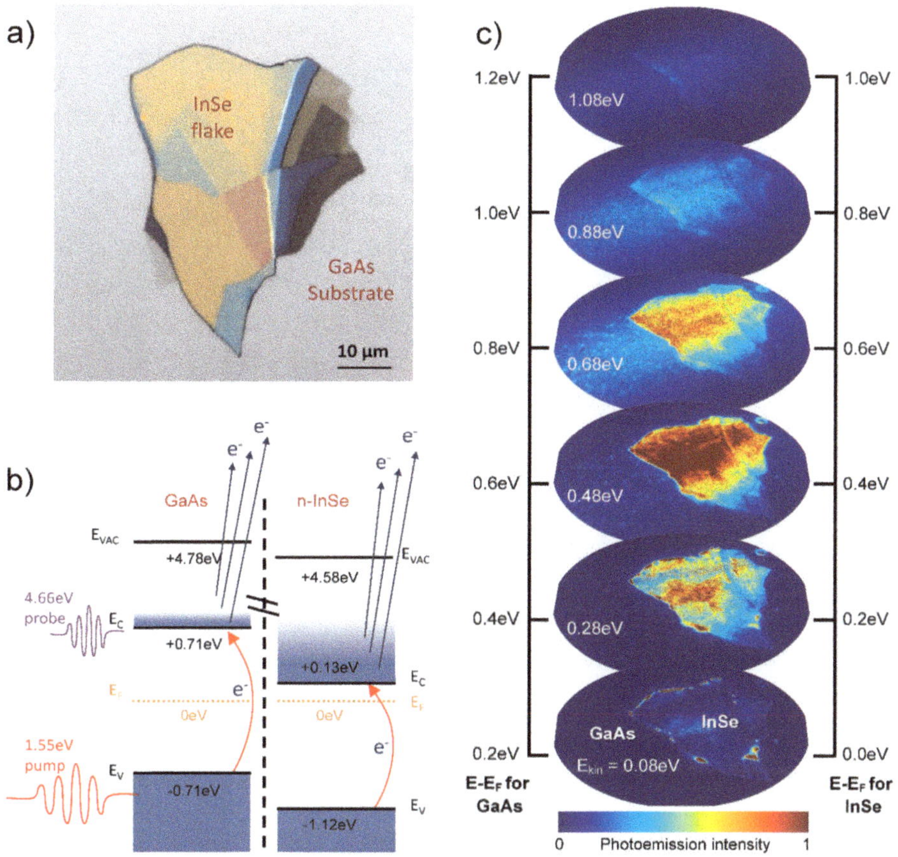

Figure 10.12. Band alignment of the InSe/GaAs heterostructure. (a) Optical image of an InSe flake sitting on top of a GaAs substrate. (b) Inferred band diagram of the InSe/GaAs heterostructure showing a type-II band alignment. The 1.55 eV pump excites carriers into the conduction bands and with the 4.66 eV probe, the photoexcited carriers are photoemitted and imaged by the PEEM. (c) With an energy analyzer, one can visualize the non-equilibrium distribution of electrons in energy and space during the temporal overlap between the pump and probe pulses. Adapted with permission from [105]. Copyright 2016 Macmillan Publishers.

than 12° show longer lifetimes than regions with twist angles of exactly 12°. They attributed this difference to complex band structure formation in multilayer graphene structures with different twist angles.

As another example, Wang *et al* [106] showed that they could resolve very pronounced spatial heterogeneity in the carrier dynamics within the flakes of a single layer WSe_2 sample. In their work, with spatial resolution better than 80 nm, they observed different decay dynamics along the peripheral edges of the flakes as well as on some puddles within the interior of these flakes (figure 10.14). The differences in lifetimes are attributed to structural disorder in the WSe_2 flakes and charge puddles induced by the substrate. One thing to note in this study is since they employed a 3.61 eV pump, they excited carriers into the conduction band via many different

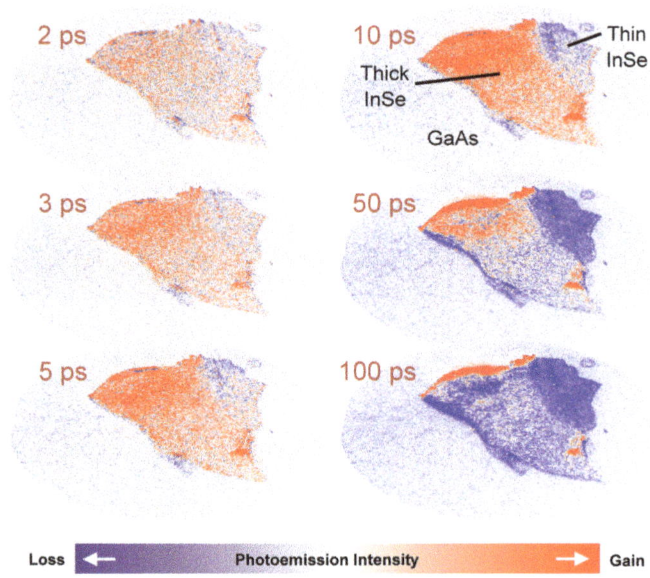

Figure 10.13. PEEM images showing electron transport in the InSe/GaAs heterostructure after photoexcitation. Images at different time delays show the accumulation (red) and depletion (blue) of photoexcited electrons. A reference image taken at 500 fs is subtracted to enhance the visualization of the redistribution of electrons. In general, electrons accumulate everywhere in InSe at an early time due to a transfer of electrons from the GaAs. After 10 ps, depletion occurs in the thin layer, where electrons migrate from the thin to thick regions due to differences in band alignment. Finally, at longer delays, electrons are depleted everywhere, and the sample returns to ground state. Reproduced with permission from [105]. Copyright 2016 Macmillan Publishers.

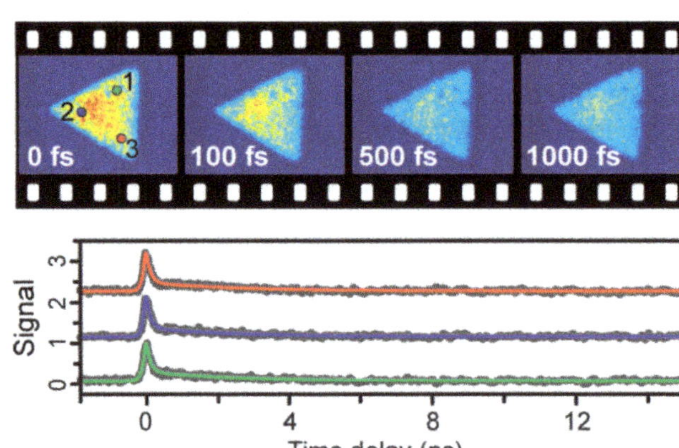

Figure 10.14. TR-PEEM is used to spatially resolve the ultrafast photocarrier response on inhomogeneous micron-sized WSe_2 flakes. Adapted with permission from [106]. Copyright 2018 the American Chemical Society.

regions in the Brillouin zone. However, with their 2.41 eV probe, they could only analyze electrons in the vicinity of the Γ point of the WSe_2, and with this limited photon energy of the probe, they could not probe electrons at the edge of the Brillouin zone (the high symmetry K valley), where the direct bandgap of WSe_2 is located. In general, in most 2D materials such as WSe_2 or graphene, the interesting electron dynamics are happening at the edge of the Brillouin zone, and it is desirable to utilize probe pulses of higher photon energy at the extreme ultraviolet regime to reach the zone boundaries. These higher photon energies cannot be generated with nonlinear crystals due to their limited bandwidth in the ultraviolet region. To reach the extreme ultraviolet regime, one needs to use higher harmonic generation (HHG) techniques [118–121] or use a free-electron-laser facility [122, 123]. For many years, with limited spatial resolution, the successful combination of HHG and ARPES has produced many works unveiling the electron dynamics in a variety of 2D materials [124–127]. By combining HHG with TR-PEEM, one can potentially explore electron dynamics in 2D materials with high spatial resolution.

10.4 Outlook and perspective of TR-PEEM technique

The TR-PEEM technique started off with studies of surface plasmon polaritons on metal surfaces and has shifted recently towards studies of electron dynamics in semiconductors and 2D materials. There is no doubt that TR-PEEM will continue to be developed and utilized in novel applications. The demands of new physics problems will drive the development of ultrafast laser technology and new ideas in PEEM instrumentation that will further enhance the capabilities of the TR-PEEM technique. A few examples are listed below of the ongoing developments in TR-PEEM techniques.

10.4.1 Ultrafast light sources with optimal repetition rate, peak power, pulse duration and energy bandwidth depending on application

In TR-PEEM, the laser source is one of the most critical parts of the system. The pulse duration of the laser defines the temporal resolution of the measurement, while the energy bandwidth of the laser pulses limits the energy resolution of all spectroscopy measurements. Additionally, as in all stroboscopic pump–probe measurements, the repetition rate of the laser pulses plays a critical role in the improvement of the SNR. In general, oscillator laser systems running at 80 MHz provide a good repetition rate, with the potential for a decent SNR. However, these 80 MHz oscillator systems have limited pulse energy. With low pulse energy, there is limited access to deep-UV light using nonlinear frequency conversion in different nonlinear crystals [128–130], and one can reach only the valence electrons close to the Fermi level and with low momentum in the reciprocal space. On the other hand, Ti:sapphire amplifier systems provide millijoule-class or higher pulse energies enabling high-order harmonic generation [118–123], where the high peak powers in the laser pulses drive nonlinear processes in noble gasses to reach photon energies from tens of electron-volts [118–121] to a few kilo-electron-volts [68, 130]. High photon energies give access to all the valence band electrons, even reaching the core-level electrons, and it also provide access to the full Brillouin zone in momentum space. However, these millijoule-class amplifier systems have low repetition rates, making data acquisition very time-consuming and challenging. In general, laser systems display trade-offs between repetition rate and pulse energy. Naturally, there is demand for systems with high energy photon sources running at high repetition rates [119, 120, 132, 133]. More recently, there have been ytterbium-based amplifier systems that provide similar pulse energies as Ti:sapphire amplifier systems but with higher repetition rates. These laser sources should dramatically improve the SNR and shorten the experimental time for all experiments that require high photon energies.

Additionally, high-repetition systems are also useful in minimizing space-charge issues in EM electron pulses. Unlike UEM or SUEM, where electrons are traveling at high speeds with kilos of electron-volts and therefore are not packed so closely in space, in TR-PEEM low energy pulses mean electrons leave the sample with low kinetic energy. These electrons travel together slowly, giving them sufficient time to interact. Subsequently, space-charge issues in TR-PEEM can be severe and can lead to broadening of the energy spectrum of the photoemitted electrons and blurring of the image [134–137]. Many parameters, such as the number of electrons per pulse, the pulse duration, photon energy, kinetic energy, spot size and electron optics of the microscope must be considered in space-charge issues. In general, one should always consider the possibility of space charge and limit the pulse energy and number of photoemitted electrons in the system. Improving the signal therefore requires a higher repetition rate laser system.

In parallel to the development of laser systems with high repetition rates is the development of optical parametric amplifiers (OPAs) that can work at high repetition rates [138, 139]. OPAs previously only worked with low repetition rate, high powered lasers, and without the OPA one must either use nonlinear crystals or is limited to the fundamental wavelength or the harmonics of it. OPAs provide a wide and continuously tunable laser wavelength. It would be particularly useful in the study of electron dynamics as an OPA will provide the ability to selectivity pump on- or off-resonance to the bandgap of the semiconductor, and a high repetition rate OPA is essential to TR-PEEM for the reasons mentioned above.

10.4.2 Parallel data acquisition for multidimensional data

TR-PEEM can simultaneously record energies, spatial distributions and emission angles of the photoemitted electrons, making it a strategic tool for spatially resolving electronic and chemical states in a system. However, due to the design of PEEMs and the detector system, TR-PEEM is limited to dissecting and looking at a small section of this multidimensional data. For example, to spatially resolve the kinetic energy of the photoemitted electrons in TR-PEEM systems equipped with a hemispherical energy analyzer and CCD camera, an energy selection slit is inserted into the energy analyzer, allowing selection and imaging of electrons at only one particular energy at a time. To improve the efficiency of the system, a different design for both the energy analyzer and the detector is needed to enable acquisition

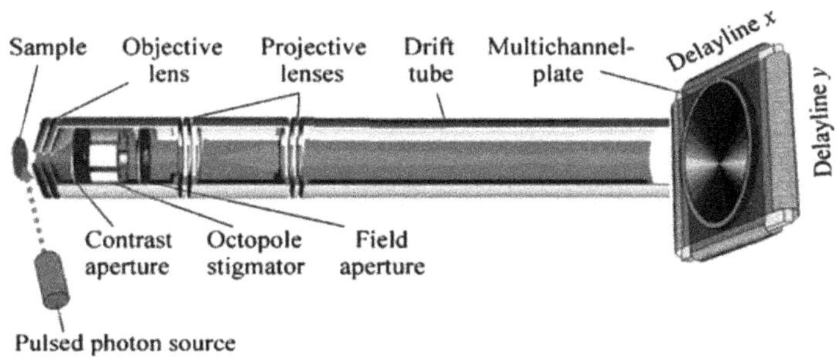

Figure 10.15. TOF-PEEM is a photoemission electron microscope equipped with a drift tube and a delay line detector. With low drift energy in the drift tube, electrons traveling at different kinetic energies are separated in time and spatially resolved using the delay line detector. Reproduced with permission from [141]. Copyright 2010 Elsevier.

of images at different energies simultaneously. One way to do this is to replace the hemispherical energy analyzer with time-of-flight (TOF) optics coupled with an ultrafast gated CCD camera [140] or a delay line detector (DLD) [141] (figure 10.15). The TOF optics includes a drift tube in which electrons traveling at different kinetic energies are spread out in time. The DLD detector at the end discriminates the arrival times and hence the kinetic energies of these electrons. Using TOF optics, there is no need to block off and discard any electrons, and 3D datasets of electron distributions created by each laser pulse in energy and real/momentum space can be obtained. This kind of energy filter dramatically increases the overall efficiency of the microscope. Additionally, the electronics of the detector currently constrains the number of electrons and the maximum repetition rate one can use in the laser system. At the moment, DLDs have a time resolution around 150 ps and can count up to several million events per second. In the near future, new DLD architectures, such as multi-segment designs, can potentially bring the count rate up to 100 million.

10.4.3 Resolving electron spin in TR-PEEM

The spin state of photoemitted electrons is one of the final pieces of information needed to fully determine the electronic structure of materials. Spin information has become increasingly important since the discovery of new topological quantum materials and in the development of spintronic applications [142–144]. Kronast *et al* [145] developed a spin filter for PEEM based on the Mott scattering mechanism. As illustrated in figure 10.16, two Mott detectors and two electrostatic deflector lenses are added to both ends of the microscope. In spin filter mode, the two deflectors are independently activated, allowing for determination of all three spin components of the photoemitted electrons from a small sample area defined by a field selecting aperture. One issue with the Mott detector is that it has a low efficiency; namely, it can only measure one point in energy, momentum and real space at any given time. To overcome this problem, Tusche *et al* [146, 147] came up with another spin filter design based on spin dependence diffraction of low energy electrons on single crystals (figure 10.17). A W(100) single crystal is set at an angle of 45° at the exit of

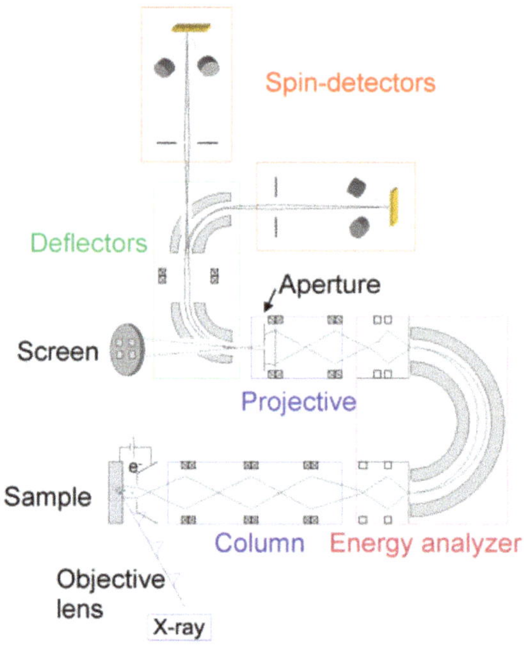

Figure 10.16. Schematic of a spin resolved PEEM with a spin filter using two Mott detectors. Reproduced with permission from [145]. Copyright 2010 Wiley.

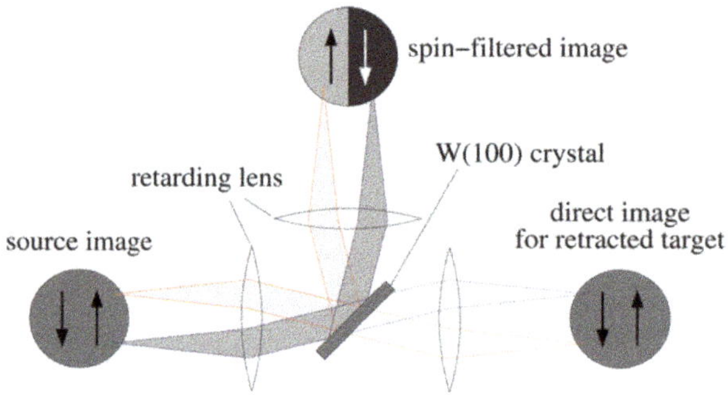

Figure 10.17. Schematic of a 2D spin filter using spin dependence reflection in low energy electron diffraction on a W(100) single crystal surface. Reproduced with permission from [146]. Copyright 2011 American Institute of Physics.

the energy analyzer of the microscope. Lenses are added before the crystal to optimize the electron energy and to obtain the diffraction conditions necessary for maximum spin selected scattering. Two-dimensional images of real space or reciprocal space spin-filtered images are obtained after 90° reflection by the single crystal. By using this design, several thousands of data points can be taken simultaneously, giving it an enormous boost in data acquisition efficiency over traditional spin detectors, where only one channel can be processed at a time.

10.5 Final remarks

The development of ultrafast lasers and EM techniques has created a new realm of studies within the field of semiconductors. Investigations of charge carriers in semiconductors using UEM, SUEM and TR-PEEM have only begun recently, but as mentioned above, the continued improvement of existing technologies, as well as the development of new ones, will allow for expanded instrument capabilities as well as the availability of these techniques. With better optical components to minimize image aberration, the use of OPAs that are compatible with high-repetition laser systems and a new generation of ultrafast lasers providing atto-second temporal resolution, greater stability and ease of operation, the TR-PEEM technique will continue to become more readily accessible and effective in observing and explaining charge carrier phenomena in semiconductors with high spatial and temporal resolution. With these new developments in laser technologies and advancements of PEEM instrumentation, TR-PEEM will play an important role in studies of ultrafast electronic dynamics and novel phenomena at the ultrasmall scale in semiconductor electronics and other quantum materials in the future.

References

[1] Axt V M and Kuhn T 2004 Femtosecond spectroscopy in semiconductors: a key to coherences, correlations and quantum kinetics *Rep. Prog. Phys.* **67** 433

[2] Kim G H, Shao L, Zhang K and Pipe K P 2013 Engineered doping of organic semiconductors for enhanced thermoelectric efficiency *Nat. Mater.* **12** 719–23

[3] López N, Reichertz L A, Yu K M, Campman K and Walukiewicz W 2011 Engineering the electronic band structure for multiband solar cells *Phys. Rev. Lett.* **106** 028701

[4] Blochwitz J *et al* 2001 Interface electronic structure of organic semiconductors with controlled doping levels *Org. Electron.* **2** 97–104

[5] Yankowitz M *et al* 2018 Dynamic band-structure tuning of graphene moiré superlattices with pressure *Nature* **557** 404–8

[6] Zheng R, Gao J, Wang J and Chen G 2011 Reversible temperature regulation of electrical and thermal conductivity using liquid–solid phase transitions *Nat. Commun.* **2** 289

[7] Largent R and Wenham S 2003 *Solar Cells Resources for the Secondary Science Teacher* (Sydney: UNSW Photovoltaics)

[8] Sze S M and Ng K K 2006 *Physics of Semiconductor Devices* 3rd edn (Hoboken, NJ: Wiley)

[9] Othonos A 1998 Probing ultrafast carrier and phonon dynamics in semiconductors *J. Appl. Phys.* **83** 1789

[10] Rossi F and Kuhn T 2002 Theory of ultrafast phenomena in photoexcited semiconductors *Rev. Mod. Phys.* **74** 895

[11] Iwai H, Kakushima K and Wong H 2006 Challenges for future semiconductor manufacturing *Int. J. High Speed Electron. Syst.* **16** 43–81

[12] Peercy P S 2000 The drive to miniaturization *Nature* **406** 1023–6

[13] Zhai X, Wu S, Shang A and Li X 2015 Limiting efficiency calculation of silicon single-nanowire solar cells with considering Auger recombination *Appl. Phys. Lett.* **106** 063904

[14] García-Santamaría F *et al* 2009 Suppressed Auger recombination in 'giant' nanocrystals boosts optical gain performance *Nano Lett.* **9** 3482–8

[15] Bulashevich K A and Karpov S Y 2008 Is Auger recombination responsible for the efficiency rollover in III-nitride light-emitting diodes? *Phys. Status Solidi* **5** 2066–9
[16] Saari J I *et al* 2013 Ultrafast electron trapping at the surface of semiconductor nanocrystals: excitonic and biexcitonic processes *J. Phys. Chem.* **B 117** 4412–21
[17] Liao B and Najafi E 2017 Scanning ultrafast electron microscopy: a novel technique to probe photocarrier dynamics with high spatial and temporal resolutions *Mater. Today Phys.* **2** 46–53
[18] Jin C, Ma E Y, Karni O, Regan E C, Wang F and Heinz T F 2018 Ultrafast dynamics in van der Waals heterostructures *Nat. Nanotechnol.* **13** 994–1003
[19] Miller R J D 1995 Time scale issues for charge transfer and energy storage using semiconductor junctions *Sol. Energy Mater. Sol. Cells* **38** 331–3
[20] Vanacore G M, Fitzpatrick A W P and Zewail A H 2016 Four-dimensional electron microscopy: ultrafast imaging, diffraction and spectroscopy in materials science and biology *Nano Today* **11** 228–9
[21] Freundlich M M 1963 Origin of the electron microscope *Sci.* **142** 185–8
[22] Smith D J 2008 Development of aberration-corrected electron microscopy *Microsc. Microanal.* **14** 2–15
[23] Barwick B, Flannigan D J and Zewail A H 2009 Photon-induced near-field electron microscopy *Nature* **462** 902–6
[24] Barwick B and Zewail A H 2015 Photonics and plasmonics in 4D ultrafast electron microscopy *ACS Photonics* **2** 1391–402
[25] Kim J S *et al* 2008 Imaging of transient structures using nanosecond *in situ* TEM *Science* **321** 1472–5
[26] Zhang Q and Luo Y 2016 Probing the ultrafast dynamics in nanomaterial complex systems by femtosecond transient absorption spectroscopy *High Power Laser Sci. Eng.* **4** 1–10
[27] Krehl P and Engemann S 1995 August Toepler—the first who visualized shock waves *Shock Waves* **5** 1–18
[28] Dantus M and Gross P 1998 Ultrafast spectroscopy *Encycl. Appl. Phys.* **22** 431–56
[29] Corkum P B and Krausz F 2007 Attosecond science *Nat. Phys.* **3** 381–7
[30] Krausz F and Ivanov M 2009 Attosecond physics *Rev. Mod. Phys.* **81** 163
[31] Shah J 1999 *Ultrafast Spectroscopy of Semiconductors and Semiconductor Nanostructures* vol 115 (Berlin: Springer)
[32] Keller U 2003 Recent developments in compact ultrafast lasers *Nature* **424** 831–8
[33] Lundstrom M 2000 *Fundamentals of Carrier Transport* 2nd edn (New York: Cambridge University Press)
[34] Elsaesser T and Woerner M 1999 Femtosecond infrared spectroscopy of semiconductors and semiconductor nanostructures *Phys. Rep.* **321** 253–305
[35] Simpson M J, Doughty B, Yang B, Xiao K and Ma Y Z 2016 Separation of distinct photoexcitation species in femtosecond transient absorption microscopy *ACS Photonics* **3** 434–42
[36] Davydova D, de la Cadena A, Akimov D and Dietzek B 2016 Transient absorption microscopy: advances in chemical imaging of photoinduced dynamics *Laser Photon. Rev.* **10** 62–81
[37] Guo Z, Manser J S, Wan Y, Kamat P V and Huang L 2015 Spatial and temporal imaging of long-range charge transport in perovskite thin films by ultrafast microscopy *Nat. Commun.* **6** 7471

[38] Grancini G *et al* 2012 Dynamic microscopy study of ultrafast charge transfer in a hybrid P3HT/hyperbranched CdSe nanoparticle blend for photovoltaics *J. Phys. Chem. Lett.* **3** 517–23
[39] Huang L *et al* 2010 Ultrafast transient absorption microscopy studies of carrier dynamics in epitaxial graphene *Nano Lett.* **10** 1308–13
[40] Gao B *et al* 2011 Studies of intrinsic hot phonon dynamics in suspended graphene by transient absorption microscopy *Nano Lett.* **11** 3184–9
[41] Guo Z, Wan Y, Yang M, Snaider J, Zhu K and Huang L 2017 Long-range hot-carrier transport in hybrid perovskites visualized by ultrafast microscopy *Sci.* **356** 59–62
[42] Ruzicka B A, Wang R, Lohrman J, Ren S and Zhao H 2012 Exciton diffusion in semiconducting single-walled carbon nanotubes studied by transient absorption microscopy *Phys. Rev.* B **86** 205417
[43] Gao B, Hartland G V and Huang L 2012 Transient absorption spectroscopy and imaging of individual chirality-assigned single-walled carbon nanotubes *ACS Nano* **6** 5083–90
[44] Mehl B P, Kirschbrown J R, House R L and Papanikolas J M 2011 The end is different than the middle: spatially dependent dynamics in ZnO rods observed by femtosecond pump-probe microscopy *J. Phys. Chem. Lett.* **2** 1777–81
[45] Gabriel M M *et al* 2013 Direct imaging of free carrier and trap carrier motion in silicon nanowires by spatially-separated femtosecond pump–probe microscopy *Nano Lett.* **13** 1336–40
[46] Steeves G M, Elezzabi A Y, Freeman M R, Steeves G M, Elezzabi A Y and Freeman M R 2011 Nanometer-scale imaging with an ultrafast scanning tunneling microscope nanometer-scale imaging with an ultrafast scanning tunneling microscope **504** 1996–9
[47] Terada Y, Aoyama M and Kondo H 2007 Ultrafast photoinduced carrier dynamics in GaNAs probed using femtosecond time-resolved scanning tunnelling *Nanotechnology* **18** 04402
[48] Terada Y, Yoshida S, Takeuchi O and Shigekawa H 2010 Laser-combined scanning tunnelling microscopy for probing ultrafast transient dynamics *J. Phys. Condens. Matter* **22** 264008
[49] Cocker T L *et al* 2013 An ultrafast terahertz scanning tunnelling microscope *Nat. Photonics* **7** 620–5
[50] Shigekawa H, Takeuchi O and Aoyama M 2005 Development of femtosecond time-resolved scanning tunneling microscopy for nanoscale science and technology *Sci. Technol. Adv. Mater.* **6** 582–8
[51] Yoshida S *et al* 2014 Probing ultrafast spin dynamics with optical pump–probe scanning tunnelling microscopy *Nat. Nanotechnol.* **9** 588–93
[52] Gerstner V, Knoll A, Pfeiffer W, Thon A and Gerber G 2000 Femtosecond laser assisted scanning tunneling microscopy *J. Appl. Phys.* **88** 4851–9
[53] Loth S, Etzkorn M, Lutz C P, Eigler D M and Heinrich A J 2010 Measurement of fast electron spin relaxation times with atomic resolution *Sci.* **329** 1628–30
[54] Nechay B A, Siegner U, Morier-Genoud F, Schertel A and Keller U 1999 Femtosecond near-field optical spectroscopy of implantation patterned semiconductors *Appl. Phys. Lett.* **74** 61–3
[55] Nechay B A, Siegner U, Achermann M, Bielefeldt H and Keller U 1999 Femtosecond pump–probe near-field optical microscopy *Rev. Sci. Instrum.* **70** 2758–64

[56] Richter A *et al* 1999 Time-resolved near-field optics: exciton transport in semiconductor nanostructures *J. Microsc.* **194** 393–400
[57] Lienau C, Emiliani V, Günther T, Intonti F and Elsaesser T 1999 Picosecond and femtosecond near-field optical spectroscopy of carrier dynamics in semiconductor nanostructures *Physica* B **272** 96–100
[58] Guenther T *et al* 1999 Femtosecond near-field spectroscopy of a single GaAs quantum wire *Appl. Phys. Lett.* **75** 3500–2
[59] Huber M A *et al* 2017 Femtosecond photo-switching of interface polaritons in black phosphorus heterostructures *Nat. Nanotechnol.* **12** 207–11
[60] Schumacher Z, Spielhofer A, Miyahara Y and Grutter P 2017 The limit of time resolution in frequency modulation atomic force microscopy by a pump–probe approach *Appl. Phys. Lett.* **110** 053111
[61] Murawski J *et al* 2015 Tracking speed bumps in organic field-effect transistors via pump–probe Kelvin-probe force microscopy *J. Appl. Phys.* **118** 244502
[62] Liscio A, Palermo V and Samorì P 2010 Nanoscale quantitative measurement of the potential of charged nanostructures by electrostatic and Kelvin probe force microscopy: unraveling electronic processes in complex materials *Acc. Chem. Res.* **43** 541–50
[63] Haight R, Ross F M and Hannon J B 2011 *Handbook of Instrumentation and Techniques for Semiconductor Nanostructure Characterization* vol 1 (Singapore: World Scientific)
[64] Shorokhov D and Zewail A H 2016 Perspective: 4D ultrafast electron microscopy—evolutions and revolutions *J. Chem. Phys.* **144** 080901
[65] Zewail A H 2010 Four-dimensional electron microscopy *Sci.* **328** 187–93
[66] Zewail A H 2016 Ultrafast light and electrons: imaging the invisible *Optics in Our Time* (Berlin: Springer), pp 43–68
[67] Plemmons D A, Suri P K and Flannigan D J 2015 Probing structural and electronic dynamics with ultrafast electron microscopy *Chem. Mater.* **27** 3178–92
[68] Baskin J S and Zewail A H 2014 Seeing in 4D with electrons: development of ultrafast electron microscopy at Caltech *Compt. Rend. Phys.* **15** 176–89
[69] Seres E, Seres J and Spielmann C 2006 X-ray absorption spectroscopy in the keV range with laser generated high harmonic radiation *Appl. Phys. Lett.* **89** 1–4
[70] Adhikari A, Eliason J K, Sun J, Bose R, Flannigan D J and Mohammed O F 2017 Four-dimensional ultrafast electron microscopy: insights into an emerging technique *ACS Appl. Mater. Interfaces* **9** 3–16
[71] Park H S, Baskin J S, Kwon O H and Zewail A H 2007 Atomic-scale imaging in real and energy space developed in ultrafast electron microscopy *Nano Lett.* **7** 2545–51
[72] Bücker K *et al* 2016 Electron beam dynamics in an ultrafast transmission electron microscope with Wehnelt electrode *Ultramicroscopy* **171** 8–18
[73] Barwick B, Park H S, Kwon O-H, Baskin J S and Zewail A H 2008 4D imaging of transient structures and morphologies in ultrafast electron microscopy *Sci.* **322** 1227–31
[74] King W E *et al* 2005 Ultrafast electron microscopy in materials science, biology, and chemistry *J. Appl. Phys.* **97** 111101
[75] Yurtsever A and Zewail A H 2009 4D nanoscale diffraction observed by convergent-beam ultrafast electron microscopy *Sci.* **326** 708–12
[76] Vanacore G M, Van Der Veen R M and Zewail A H 2015 Origin of axial and radial expansions in carbon nanotubes revealed by ultrafast diffraction and spectroscopy *ACS Nano* **9** 1721–9

[77] Flannigan D J and Zewail A H 2010 Optomechanical and crystallization phenomena visualized with 4D electron microscopy: interfacial carbon nanotubes on silicon nitride *Nano Lett.* **10** 1892–9
[78] McKenna A J, Eliason J K and Flannigan D J 2016 Imaging photoinduced structural and morphological dynamics of a single MoS_2 flake with ultrafast electron microscopy *Proc. Microsc. Microanal.* **22** 1662–3
[79] Yang D-S, Mohammed O F and Zewail A H 2010 Scanning ultrafast electron microscopy *Proc. Natl Acad. Sci.* **107** 14993–8
[80] Hassan M T 2018 Attomicroscopy: from femtosecond to attosecond electron microscopy *J. Phys. B: At. Mol. Opt. Phys.* **51** 032005
[81] Mohammed O F, Yang D S, Pal S K and Zewail A H 2011 4D scanning ultrafast electron microscopy: visualization of materials surface dynamics *J. Am. Chem. Soc.* **133** 7708–11
[82] Cho J, Hwang T Y and Zewail A H 2014 Visualization of carrier dynamics in p(n)-type GaAs by scanning ultrafast electron microscopy *Proc. Natl Acad. Sci.* **111** 2094–9
[83] Liao B *et al* 2017 Spatial–temporal imaging of anisotropic photocarrier dynamics in black phosphorus *Nano Lett.* **17** 3675–80
[84] Najafi E, Scarborough T D, Tang J and Zewail A 2015 Four-dimensional imaging of carrier interface dynamics in p–n junctions *Sci.* **347** 164–7
[85] Liao B, Najafi E, Li H, Minnich A J and Zewail A H 2017 Photo-excited hot carrier dynamics in hydrogenated amorphous silicon imaged by 4D electron microscopy *Nat. Nanotechnol.* **12** 871–6
[86] Khan J I *et al* 2016 Enhanced optoelectronic performance of a passivated nanowire-based device: key information from real-space imaging using 4D electron microscopy *Small* **12** 2312
[87] Bose R *et al* 2018 Imaging localized energy states in silicon-doped InGaN nanowires using 4D electron microscopy *ACS Energy Lett.* **3** 476–81
[88] Shaheen B S, Sun J, Yang D S and Mohammed O F 2017 Spatiotemporal observation of electron-impact dynamics in photovoltaic materials using 4D electron microscopy *J. Phys. Chem. Lett.* **8** 2455–62
[89] Sun J, Melnikov V A, Khan J I and Mohammed O F 2015 Real-space imaging of carrier dynamics of materials surfaces by second-generation four-dimensional scanning ultrafast electron microscopy *J. Phys. Chem. Lett.* **6** 3884–90
[90] Meizyte G *et al* 2018 Imaging the reduction of electron trap states in shelled copper indium gallium selenide nanocrystals using ultrafast electron microscopy *J. Phys. Chem.* C **122** 15010–6
[91] Bose R *et al* 2016 Real-space mapping of surface trap states in CIGSe nanocrystals using 4D electron microscopy *Nano Lett.* **16** 4417–23
[92] Schmidt O *et al* 2002 Time-resolved two photon photoemission electron microscopy *Appl. Phys.* B **74** 223–7
[93] Kahl P *et al* 2018 Direct observation of surface plasmon polariton propagation and interference by time-resolved imaging in normal-incidence two photon photoemission microscopy *Plasmonics* **13** 239–46
[94] Mahro A K *et al* 2017 Revealing the subfemtosecond dynamics of orbital angular momentum in nanoplasmonic vortices *Sci.* **355** 1187–91

[95] Bayer D, Wiemann C, Gaier O, Bauer M and Aeschlimann M 2008 Time-resolved 2PPE and time-resolved PEEM as a probe of LSPs in silver nanoparticles *J. Nanomater.* **2008** 249514
[96] Petek H, Sametoglu V, Pontius N and Kubo A 2005 Imaging of surface plasmon dynamics in nanostructured silver films *IQEC, Int. Quantum Electronics Conf. Proc.*
[97] Mårsell E *et al* 2015 Nanoscale imaging of local few-femtosecond near-field dynamics within a single plasmonic nanoantenna *Nano Lett.* **15** 6601–8
[98] Bauer M, Wiemann C, Lange J, Bayer D, Rohmer M and Aeschlimann M 2007 Phase propagation of localized surface plasmons probed by time-resolved photoemission electron microscopy *Appl. Phys.* A **88** 473–80
[99] Fukumoto K *et al* 2014 Femtosecond time-resolved photoemission electron microscopy for spatiotemporal imaging of photogenerated carrier dynamics in semiconductors *Rev. Sci. Instrum.* **85** 083705
[100] Fukumoto K, Yamada Y, Onda K and Koshihara S Y 2014 Direct imaging of electron recombination and transport on a semiconductor surface by femtosecond time-resolved photoemission electron microscopy *Appl. Phys. Lett.* **104** 053117
[101] Fukumoto K *et al* 2014 Femtosecond time-resolved photoemission electron microscopy for spatiotemporal imaging of photogenerated carrier dynamics in semiconductors *Rev. Sci. Instrum.* **85** 083705
[102] Fukumoto K, Yamada Y, Koshihara S Y and Onda K 2015 Lifetimes of photogenerated electrons on a GaAs surface affected by nanostructural defects *Appl. Phys. Express* **8** 101201
[103] Shibuta M, Yamagiwa K, Eguchi T and Nakajima A 2016 Imaging and spectromicroscopy of photocarrier electron dynamics in C_{60} fullerene thin films *Appl. Phys. Lett.* **109** 203111
[104] Fukumoto K *et al* 2017 Ultrafast electron dynamics in twisted graphene by femtosecond photoemission electron microscopy *Carbon* **124** 49–56
[105] Man M K L *et al* 2017 Imaging the motion of electrons across semiconductor heterojunctions *Nat. Nanotechnol.* **12** 36–40
[106] Wang L, Xu C, Li M Y, Li L J and Loh Z H 2018 Unraveling spatially heterogeneous ultrafast carrier dynamics of single-layer WSe_2 by femtosecond time-resolved photoemission electron microscopy *Nano Lett.* **18** 5172–8
[107] Wong E L, Winchester A J, Pareek V, Madéo J, Man M K L and Dani K M 2018 Pulling apart photoexcited electrons by photoinducing an in-plane surface electric field *Sci. Adv.* **4** eaat9722
[108] Bauer E 2014 *Surface Microscopy with Low Energy Electrons* (New York: Springer)
[109] Man K L and Altman M S 2012 Low energy electron microscopy and photoemission electron microscopy investigation of graphene *J. Phys. Condens. Matter* **24** 314209
[110] Menteş T O, Zamborlini G, Sala A and Locatelli A 2014 Cathode lens spectromicroscopy: methodology and applications *Beilstein J. Nanotechnol.* **5** 1873–86
[111] Einstein A 1905 Über einen die Erzeugung und Verwandlung des Lichtes betreffenden heuristischen Gesichtspunkt *Ann. Phys.* **322** 132–48
[112] Mårsell E *et al* 2015 Nanoscale imaging of local few-femtosecond near-field dynamics within a single plasmonic nanoantenna *Nano Lett.* **15** 6601–8
[113] Renault O and Chabli A 2007 Energy-filtered photoelectron emission microscopy (EF-PEEM) for imaging nanoelectronic materials *AIP Conf. Proc.* **931** 502–6
[114] Hüfner S 2003 *Photoelectron Spectroscopy* vol 53 (Berlin: Springer)

[115] Suga S and Sekiyama A 2014 *Photoelectron Spectroscopy—Bulk and Surface Electronic Structures* vol 176 (Berlin: Springer)
[116] Amati M *et al* 2018 Photoelectron microscopy at Elettra: recent advances and perspectives *J. Electron Spectros. Relat. Phenomena* **224** 59–67
[117] Bostanjoglo O and Weingärtner M 1997 Pulsed photoelectron microscope for imaging laser-induced nanosecond processes *Rev. Sci. Instrum.* **68** 2456–60
[118] Schmitz C *et al* 2016 Compact extreme ultraviolet source for laboratory-based photoemission spectromicroscopy *Appl. Phys. Lett.* **108** 234101
[119] Chiang C T *et al* 2015 Boosting laboratory photoelectron spectroscopy by megahertz high-order harmonics *New J. Phys.* **17** 013035
[120] Wang H, Xu Y, Ulonska S, Robinson J S, Ranitovic P and Kaindl R A 2015 Bright high-repetition-rate source of narrowband extreme-ultraviolet harmonics beyond 22 eV *Nat. Commun.* **6** 1–7
[121] Heyl C M, Güdde J, L'Huillier A and Höfer U 2012 High-order harmonic generation with μJ laser pulses at high repetition rates *J. Phys. B: At. Mol. Opt. Phys.* **45** 074020
[122] Hellmann S *et al* 2012 Time-resolved x-ray photoelectron spectroscopy at FLASH *New J. Phys.* **14** 013062
[123] Pietzsch A *et al* 2008 Towards time resolved core level photoelectron spectroscopy with femtosecond x-ray free-electron lasers *New J. Phys.* **10** 033004
[124] Johannsen J C *et al* 2013 Direct view of hot carrier dynamics in graphene *Phys. Rev. Lett.* **111** 027403
[125] Grubišić Čabo A *et al* 2015 Observation of ultrafast free carrier dynamics in single layer MoS_2 *Nano Lett.* **15** 5883–7
[126] Wallauer R, Reimann J, Armbrust N, Güdde J and Höfer U 2016 Intervalley scattering in MoS_2 imaged by two-photon photoemission with a high-harmonic probe *Appl. Phys. Lett.* **109** 162102
[127] Ulstrup S *et al* 2017 Spin and valley control of free carriers in single-layer WS_2 *Phys. Rev.* B **95** 041405
[128] Nikogosyan D N 1991 Beta barium borate (BBO)—a review of its properties and applications *Appl. Phys.* A **52** 359–68
[129] Togashi T *et al* 2003 Generation of vacuum-ultraviolet light by an optically contacted, prism-coupled $KBe_2BO_3F_2$ crystal *Opt. Lett.* **28** 254
[130] Guo S *et al* 2016 $BaBe_2BO_3F_3$: a KBBF-type deep-ultraviolet nonlinear optical material with reinforced $[Be_2BO_3F_2]\infty$layers and short phase-matching wavelength *Chem. Mater.* **28** 8871–5
[131] Aliŝauskas S *et al* 2012 Bright coherent ultrahigh harmonics in the keV x-ray regime from mid-infrared femtosecond lasers *Sci.* **1287** 1287–92
[132] Pupeza I *et al* 2013 Compact high-repetition-rate source of coherent 100 eV radiation *Nat. Photonics* **7** 608–12
[133] Schmidt J *et al* 2017 Development of a 10 kHz high harmonic source up to 140 eV photon energy for ultrafast time-, angle-, and phase-resolved photoelectron emission spectroscopy on solid targets *Rev. Sci. Instrum.* **88** 083105
[134] Passlack S, Mathias S, Andreyev O, Mittnacht D, Aeschlimann M and Bauer M 2006 Space charge effects in photoemission with a low repetition, high intensity femtosecond laser source *J. Appl. Phys.* **100** 024912

[135] Buckanie N M, Göhre J, Zhou P, von der Linde D, H-von Hoegen M and Meyer Zu Heringdorf F-J 2009 Space charge effects in photoemission electron microscopy using amplified femtosecond laser pulses *J. Phys. Condens. Matter* **21** 314003

[136] Hellmann S, Rossnagel K, Marczynski-Bühlow M and Kipp L 2009 Vacuum space-charge effects in solid-state photoemission *Phys. Rev.* B **79** 1–12

[137] Schoenhense B *et al* 2018 Multidimensional photoemission spectroscopy—the space-charge limit *New J. Phys.* **20** 033004

[138] Rothhardt J *et al* 2010 High average and peak power few-cycle laser pulses delivered by fiber pumped OPCPA system *Opt. Express* **18** 12719

[139] Tavella F *et al* 2010 Fiber-amplifier pumped high average power few-cycle pulse non-collinear OPCPA *Opt. Express* **18** 4689

[140] Spiecker H *et al* 1998 Time-of-flight photoelectron emission microscopy TOF-PEEM: first results *Nucl. Instrum. Methods Phys. Res.* A **406** 499–506

[141] Oelsner A, Rohmer M, Schneider C, Bayer D, Schönhense G and Aeschlimann M 2010 Time- and energy resolved photoemission electron microscopy-imaging of photoelectron time-of-flight analysis by means of pulsed excitations *J. Electron Spectros. Relat. Phenomena* **178–179** 317–30

[142] Wolf S A *et al* 2001 Spintronics: a spin-based electronics vision for the future *Sci.* **294** 1488

[143] Hasan M Z and Kane C L 2010 Colloquium: topological insulators *Rev. Mod. Phys.* **82** 3045–67

[144] Qi X L and Zhang S C 2011 Topological insulators and superconductors *Rev. Mod. Phys.* **83** 1057

[145] Kronast F, Schlichting J, Radu F, Mishra S K, Noll T and Dürr H A 2010 Spin-resolved photoemission microscopy and magnetic imaging in applied magnetic fields *Surf. Interface Anal.* **42** 1532–6

[146] Tusche C *et al* 2011 Spin resolved photoelectron microscopy using a two-dimensional spin-polarizing electron mirror *Appl. Phys. Lett.* **99** 15–8

[147] Tusche C, Krasyuk A and Kirschner J 2015 Spin resolved bandstructure imaging with a high resolution momentum microscope *Ultramicroscopy* **159** 520–9

[148] Murphy S and Huang L 2013 Transient absorption microscopy studies of energy relaxation in graphene oxide thin film *J. Phys. Condens. Matter* **25** 144203

IOP Publishing

Nanoscale Energy Transport
Emerging phenomena, methods and applications
Bolin Liao

Chapter 11

Exploring nanoscale heat transport via neutron scattering

Qiyang Sun and Chen Li

Since its establishment as a powerful tool for structure and dynamics studies, neutron scattering has been widely applied in investigating heat transport in a variety of materials, particularly through the measurement of full phonon lattice dynamics. In this chapter, neutron scattering theory, instrumentation, data reduction and analysis related to applications in heat transport research are briefly reviewed. A few examples of inelastic neutron scattering studies of phonon dynamics and their relevance to nanoscale heat transport are discussed. Neutron scattering is not only a powerful tool for investigating the thermal transport mechanism in materials, but can also provide valuable insights in refining the microscopic thermal transport theory.

11.1 Introduction

11.1.1 A short history

Neutron scattering has been widely used to study phonon dynamics in nanoscale heat transport processes. Shortly after James Chadwick won the Nobel Prize in Physics for the discovery of the neutron in 1932, Enrico Fermi investigated the scattering cross-sections of cold and thermal neutrons and received the Nobel Prize in Physics in 1938. In 1955, the first measurements of phonons on a prototype triple-axis spectrometer were performed Bertram N Brockhouse and he later shared the Noble Prize in Physics with Clifford G Shull for their contribution to neutron scattering techniques. In the last eight decades, various neutron scattering instruments have been developed for structure determination and dynamics studies in a wide range of scientific disciplines from biology to chemistry, and from materials science to condensed matter physics. Phonon dynamics remains one of the central topics of neutron scattering and remains a field under active development.

doi:10.1088/978-0-7503-1738-2ch11

11.1.2 Neutron advantages

Neutron scattering provides information that is highly complementary to that from other diffraction and spectroscopy techniques due to their unique physical properties (table 11.1). The most important difference between neutron scattering and other experimental methods is the type of interaction studied. For example, photons in inelastic x-ray scattering interact with the materials mostly through Coulomb interactions. Neutrons are charge-neutral and interact with nuclei in materials via strong-force interactions, which have a very short range. As a result, neutrons have good penetration and are capable of probing bulk properties. Also, the neutron scattering cross-section is isotope-dependent and allows elementary contrast through isotopic substitution without much change to the materials' chemical properties. Due to their magnetic moment, neutrons are capable of interacting with magnetic moments, thus both magnetic structures and excitations (spin-waves) in materials can be measured.

Cold and thermal neutrons are particularly suitable for probing phonons due to their energy and wavelength scales. As can be seen in table 11.2, the wavelengths of cold and thermal neutrons are comparable to inter-atomic distances, allowing structure determination by diffraction. The energies of the cold and thermal neutrons are also comparable to the phonons and magnons, allowing inelastic scattering investigation of their dynamics. This is different from other scattering techniques, such as inelastic x-ray scattering.

As a probe, neutrons do have some drawbacks. First, neutron sources are costly to build and operate. As a result, neutron scattering experiments can only be conducted at a limited number of facilities. Second, generating neutrons is difficult and their flux is extremely weak compared to other techniques such as x-rays. Because of this, neutron experiments are usually slow and have low counting statistics and large samples, preferably single-crystal samples, are required. In addition, some elements such as boron (B), cadmium (Cd) and gadolinium (Gd) absorb neutrons strongly with their natural abundance. Neutron scattering

Table 11.1. The physical properties of the neutron.

Mass	Charge	Spin
1.675×10^{-27} kg	0	½

Table 11.2. The relations between the energy, temperature and wavelength of neutrons.

	Energy (meV)	Temperature (K)	Wavelength (nm)
Cold	0.1–10	1–120	0.4–3
Thermal	10–100	60–1000	0.1–0.4
Hot	100–500	1000–6000	0.04–0.1

experiments on materials that contain such elements usually require isotope substitution, adding to the expense and difficulty of sample preparation.

11.1.3 Neutron sources

Some representative neutron sources are listed in table 11.3. Neutrons for scattering are usually generated by nuclear fission or spallation. Fission is a chain reaction of radioactive isotopes, such as uranium-235, and thermal neutrons. In such processes, one incoming neutron interacts with a nucleus and produces multiple neutrons, which in turn react with additional nuclei. A fission reactor operating at steady state maintains a continuous-wave output of neutrons, which may be used for scattering experiments. A pulsed neutron beam from a reactor-based neutron source is also possible.

In a spallation-based neutron source, GeV protons from a linear accelerator (after the electrons are stripped from H^-) are used to bombard heavy metal targets (lead/mercury/tungsten) to generate neutrons. The spallation neutron sources produce pulsed neutron beams that are required by time-of-flight (TOF) neutron instruments. The neutrons directly generated by fission and spallation may have too much energy for scattering experiments. As a result, neutrons might need to be slowed down by interacting with moderators containing media such as light water, coupled or decoupled parahydrogen. From there, the cold or thermal neutrons are fed to the instruments via neutron beam guides.

11.1.4 Scattering theory

The neutron scattering formalism has been well developed and elaborated elsewhere [1, 2]. Only some major concepts will be briefly reviewed. The scattering is classified as elastic and inelastic processes. The energy of neutrons remains unchanged during elastic scattering, which is used for structure determination in diffraction. On the other hand, energy transfer occurs between neutrons and scatters in inelastic

Table 11.3. List of representative active neutron sources.

Country	Facility	Type	Power	Location	Beamlines	Operational since
US	SNS	Spallation	1.4 MW	Oak Ridge, TN	19	2006
US	HFIR	Fission	85 MW	Oak Ridge, TN	14	1966
US	NCNR	Fission	20 MW	Gaithersburg, MD	18	1969
Australia	OPAL	Fission	20 MW	Sydney	15	2006
UK	ISIS	Spallation	300 kW	Didcot	25	1984
Germany	BER-II	Fission	10 MW	Berlin	20	1972
France	ILL	Fission	58.3 MW	Grenoble	40	1967
Russia	IBR-2	Fission	2 MW	Dubna	13	1977
China	CSNS	Spallation	500 kW	Dongguan	3	2018
Japan	JRR-3	Fission	20 MW	Ibaraki	31	1962
Japan	J-PARC	Spallation	1 MW	Ibaraki	22	2009

scattering, which can be used to map the dynamics of excitations such as phonons and spin-waves.

Applying the Fermi golden rule within the first Born approximation, a scattering process can be described by

$$W_{i\to f} = \frac{2\pi}{\hbar} |\langle f|V|i\rangle|^2 \rho_f,$$

in which $W_{i\to f}$ is the rate of transition of the whole system from initial state i to final state f, V is the scattering potential and ρ_f is the density of final scattered states. Summing over the available final states in $\delta\Omega$ and normalizing the transition probability by the incident flux gives

$$\frac{\mathrm{d}\sigma}{\mathrm{d}\Omega} = \frac{k_f}{k_i}\left(\frac{m}{2\pi\hbar^2}\right)^2 |\langle f|V|i\rangle|^2.$$

The term $\frac{\mathrm{d}\sigma}{\mathrm{d}\Omega}$ is the scattering cross-section for neutrons scattered to solid angle $\delta\Omega$. Furthermore, a partial differential cross-section $\frac{\mathrm{d}^2\sigma}{\mathrm{d}\Omega\mathrm{d}E}$ can be written by considering conservation of energy.

The scattering of neutrons by a single fixed nucleus can be described by its neutron scattering cross-section σ. Neutron scattering by a large number of nuclei in a crystal generally have both coherent and incoherent components depending on the isotopes and their concentrations. This is usually described by coherent and incoherent scattering cross-sections. The coherent scattering arises from the correlation between the positions of the same nuclei at different times, and the correlations between the positions of different nuclei at different times. As a result, it gives interference effects. The incoherent scattering only depends on the correlation between the positions of the same nucleus at different times so there are no interference effects. The total partial differential cross-section is

$$\frac{\mathrm{d}^2\sigma}{\mathrm{d}\Omega\mathrm{d}E} = \frac{k_f}{k_i}\sigma_{\mathrm{coh}}\frac{N}{4\pi}S_{\mathrm{coh}}(\boldsymbol{Q}, \omega) + \frac{k_f}{k_i}\sigma_{\mathrm{incoh}}\frac{N}{4\pi}S_{\mathrm{incoh}}(\boldsymbol{Q}, \omega)$$

in which the term $\frac{\delta^2\sigma}{\delta\Omega\delta \mathrm{E}}$ represents the number of neutrons scattered per unit time into a small solid angle $\delta\Omega$ with final energy between $E + \delta E$. This is the measurable quantity in neutron scattering experiments. The term $\mathrm{S}(\boldsymbol{Q}, \omega)$, the dynamical structure factor, is the double Fourier transform of pair correlation function $G(r, t)$ in real space.

For coherent one-phonon creation processes, the partial differential cross-section is

$$\left(\frac{\mathrm{d}^2\sigma}{\mathrm{d}\Omega\mathrm{d}E}\right)_{\mathrm{coh}+1} = \frac{k_f}{k_i}\frac{(2\pi)^3}{2v_0}\sum_{\boldsymbol{q}}\sum_{j}\frac{1}{\omega_j}$$
$$\times\left|\sum_d \frac{\bar{b}_d}{\sqrt{M_d}}\exp(-W_d)\boldsymbol{Q}\cdot \boldsymbol{e}_{dj}\exp(\mathrm{i}\boldsymbol{Q}\cdot \boldsymbol{d})\right|^2.$$
$$\times(n_j + 1)\delta(\boldsymbol{Q} - \boldsymbol{q} - \boldsymbol{q}_0)\delta(\omega - \omega_j)$$

The sums are over reduced momentum transfer $\boldsymbol{q}$, phonon modes j and atom index d. $\boldsymbol{e}_{dj}$ are the phonon eigenvectors, $\boldsymbol{q}_0$ is the reciprocal lattice vector, ω_j is the phonon frequency, $\bar{b}_d$ is the neutron scattering length and W_d is the Debye–Waller factor. The scattering intensity is non-zero, where phonon wavevector $\boldsymbol{Q}$ and $\boldsymbol{e}_{dj}$ are not orthogonal. The expressions for phonon annihilation processes, incoherent processes and multi-phonon processes are similar.

The measured partial differential cross-section allows extraction of the information on phonon dynamics. From first-principles phonon calculations, it is possible to simulate the neutron scattering intensity as a function of different momentum and energy transfers based on the phonon eigenvalues and eigenvectors.

11.1.5 Neutron instruments

In the context of lattice thermal transport, neutron spectrometers are valuable in providing knowledge on phonon dynamics. There are two major types of neutron spectrometers: time-of-flight (TOF) and triple-axis instruments. TOF neutron spectrometers require pulsed neutron beams and usually use area detectors, and they can map out phonons throughout the reciprocal space efficiently. In a typical TOF instrument, the energy of the incident neutrons is selected by spinning choppers working as monochromators. For phonon measurements, the incident neutron energy is chosen to optimize energy/wavevector resolutions and ranges of coverage. To map out the overall phonon dispersion, an incident energy larger than the phonon cut-off energy is used. For high-resolution measurement of acoustic phonons, a lower incident energy is usually preferred. The final energy and momentum of the scattered neutrons are determined by their travel time and the position of illuminated pixels in the area detector. From the energy/momentum of the incident and scattered neutrons, the momentum and energy transfer between neutrons and phonons will be calculated. The reduced data provides the dynamical structure factor, $S(\boldsymbol{Q}, \omega)$, in four dimensions (one for energy transfer and three for momentum transfer).

Triple-axis spectrometers (TASs) are more common for reactor-based neutron sources. They have flux and resolution advantages in measuring the phonon energy spectra at specific momentum transfers. In a TAS instrument, the neutron beam first goes through a single-crystal monochromator and the sample. Then the scattered neutrons go through an analyser, where the neutrons with correct energy will be diffracted into the detector. Based on the energy/orientation of the incident and the scattered beam the momentum and energy transfer may be determined. The measurement scans are usually performed at constant momentum or energy transfer with fixed final neutron energy. TAS is a good complement to TOF spectrometers for studying phonons at specific wavevectors, particularly as a function of temperature, pressure or external fields.

It should be noted that even though neutron diffraction does not probe phonon dynamics directly, it is still valuable for understanding lattice vibrations. Due the unique physical properties of the particle, neutron diffraction is sensitive to small lattice distortions and the spatial distribution due to phonon dynamics. Both powder and single-crystal diffraction allow extracting atomic pair-distribution functions (PDF) and anharmonic components in the thermal atomic spatial distribution.

11.2 Inelastic neutron scattering and phonon transport

11.2.1 Thermal transport and measurable phonon properties

In most non-magnetic solid-state insulators and semiconductors, phonons are the primary heat carriers. Due to the anharmonicity of the inter-atomic potential, phonons can be scattered by each other. There are typically two types of phonon–phonon scattering processes: the normal process and Umklapp process. The normal process happens with an unchanged direction of momentum that does not necessarily hinder the propagation of the phonon. The Umklapp process changes the direction of momentum between scatterings, which gives resistance to thermal conductivity. Using the phonon mean-free-path, the thermal conductivity due to lattice vibrations gives

$$\kappa_{\text{lattice}} = \frac{1}{3}\sum C_v v_g l,$$

where C_v is the mode-specific heat, v_g is the mode group velocity of the wave packet and l is the phonon mean-free-path. In the context of the Debye model, above the Debye temperature where the specific heat C_v saturates, l and v_g would be responsible for the change of lattice thermal conductivity. With relaxation time approximation, the phonon mean-free-path for a certain phonon frequency ω could be described as the product of the phonon lifetime and group velocity:

$$l = v_g \tau.$$

So the lattice thermal conductivity may be written as

$$\kappa_{\text{lattice}} = \frac{1}{3}\sum C_v {v_g}^2 \tau.$$

For a quantum state with finite lifetime τ and energy distribution ΔE,

$$\Delta E \cdot \tau \geqslant \hbar.$$

As a result, the energy distribution, i.e. linewidth ΔE, is inversely proportional to the lifetime [3]. A reduction in phonon lifetime due to enhanced phonon scattering processes will be shown by the broadening of measured phonon modes, a quantity measurable by inelastic neutron scattering.

The phonon group velocity is determined by the slope of the dispersion relation at the specific phonon mode. By mapping out the overall dispersion relation, the phonon group velocity as a function of conditions can be measured by inelastic neutron scattering. Within the quasi-harmonic approximation (QHA), phonon frequencies depend solely on the dilation of the lattice due to thermal expansion. The change of phonon energy is usually described by Grüneisen parameter γ:

$$\gamma = \frac{-\text{dln}(E)}{\text{dln}(V)},$$

which is a mode-specific quantity. Typically, γ is positive—phonon energy decreases with increasing temperature. A broad distribution of Grüneisen parameters for different phonon modes is a sign of large phonon anharmonicity.

To investigate the phonon dynamics and its connection to the lattice thermal conductivity, inelastic neutron scattering (INS) allows direct access to all related phonon properties of the materials. The measured phonon dynamical structure factor by INS contains the information on both phonon lifetime and group velocity, and their changes under different conditions and external stimuli such as temperature, external magnetic and electric fields, and pressure. TOF neutron spectrometers are capable of mapping out the overall phonon dispersion or phonon density of states, while TAS neutron spectrometers, working via one-dimensional constant **Q**-transfer or **E**-transfer scans, often provide better resolution, necessary for determining the lifetime of long-lived acoustic phonons. Within each type of neutron spectrometer, different instrument implementations allow different operating energy ranges and resolutions on momentum and energy. As a result, complementary sets of scattering instruments are common used to address different aspects of one problem.

11.2.2 Data reduction and analysis

11.2.2.1 TOF data reduction and analysis

The reduction of raw detector signals from TOF spectrometers to dynamical structure factors of the neutron scattering intensity is instrument-dependent and beyond the scope of this chapter. The basic ideas of data reduction are converting the temporal and spatial information from neutron detectors to the four-dimensional energy–wavevector space. The literature on TOF neutron spectrometers at SNS provides some good examples [4, 5].

Visualizing four-dimensional data is difficult and the typical approach is projecting a subset of the data to two or one axes by making two-dimensional 'slices' or one-dimensional 'cuts'. Usually, integration is performed on the other axes over a short range near the region of interest to produce such slices and cuts. A two-dimensional slice may contain one Q-axis and one E-axis or two Q-axes. For example, a two Q-axis slice near zero energy transfer ($E = 0$ for elastic scattering/diffraction) gives a two-dimensional cross-section of the reciprocal space lattice and the position of the cross-section is determined by the two Q-axes. Elastic slices are useful to determine the quality of the single-crystal sample and its alignment. A two-dimensional slice of one Q-axis and one E-axis is commonly used to map out the dispersion relation along the Q-axis in the reciprocal space. Usually, multiple Brillouin zones are covered by the slice with varying structure factors and a higher scattering intensity of phonons is expected for the higher order zones. One-dimensional cuts can be generated at constant E or Q. This is useful in determining the energy, linewidth and line shape of the phonon modes at that point in reciprocal space.

A good understanding of the resolution function is necessary to understand the TOF data, in particular for extracting linewidth. This resolution function is usually quite complex and depends on Q, E and the experiment parameters, such as neutron chopper settings. Monte Carlo simulation of the neutron instrument, for example as implemented in McStas [6], is the common approach to obtain such information. Another unique challenge of working with TOF data is that the high dimensionality makes it difficult to extract all valuable information without missing important physical phenomena. This is particularly true if such phenomena happen in the low symmetry parts of the reciprocal space.

11.2.2.2 TAS data analysis

The design of typical triple-axis spectrometers determines that the scattering data are one-dimensional spectra of energy (E) or momentum transfer (Q), similar to the one-dimensional cuts of TOF data. While this greatly simplifies the data collection, careful advance planning of a TAS experiment is always necessary to locate the region of interest. Sometimes, phonon peaks in the energy spectra from constant-Q scans are fit to a Lorentzian function to extract peak center and width. However, this is non-ideal because the instrument resolution function of TAS is highly anisotropic with strong E and Q dependence [1]. For example, it is common for TAS phonon spectra to have a complicated line shape as a result of the behaviors of the resolution function near the measured point. To obtain accurate intrinsic phonon linewidths, knowledge of both the nearby phonon dispersion relation and the resolution function itself is necessary. The phonon linewidth then can be solved by optimizing the calculated spectra via the convolution of the phonon dispersion and the resolution functions, as implemented in the *Reslib* package [7]. Such a calculation is also crucial for TAS experiment planning because the instrument resolution can vary more than one order of magnitude depending on the focus condition.

11.2.3 Some examples

11.2.3.1 Determine phonon energy shift and change of lifetime

A combination of complementary triple-axis and TOF neutron scattering measurements allows complete and accurate mapping of the phonon dispersion in single-crystal samples, as demonstrated by the phonon work on SnSe, a thermoelectric material with low lattice thermal conductivity [8]. This inelastic neutron scattering work on SnSe utilized both TOF (CNCS) and TAS neutron spectrometers (HB-3, CTAX). The full Brillouin zone phonon measurements on CNCS pronounced softening of the transverse acoustic phonon mode near the zone center for certain phonon polarizations with increasing temperature. More detailed measurements are then carried out on TAS to investigate the temperature dependence of these phonon modes by constant-Q scans. As seen in figure 11.1, the TAS data agree fairly well

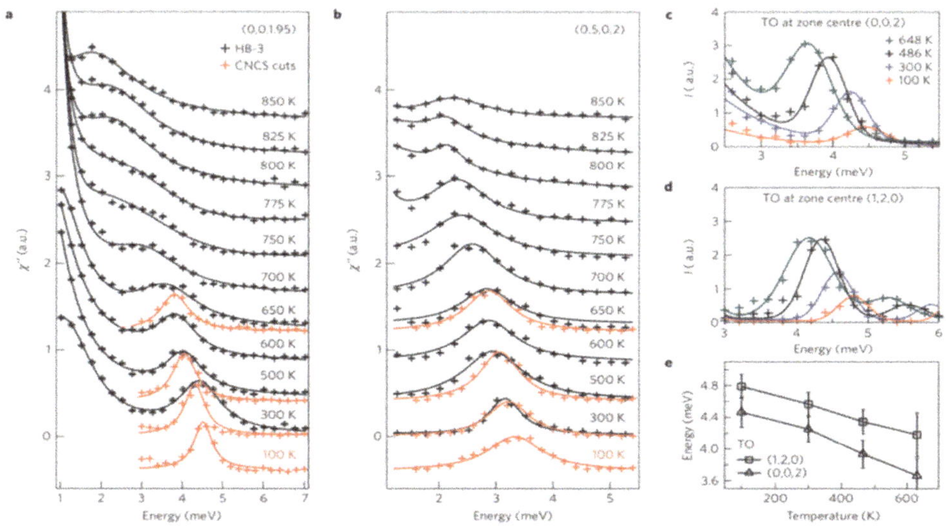

Figure 11.1. Strong softening of the TO phonon in SnSe with increasing temperature. Constant-Q spectra for several low-energy phonon modes as a function of temperature, showing the pronounced temperature softening. (a) Temperature dependence of spectra for the TO mode at Q = [0, 0, 1.95] from CNCS and HB-3. (b) The same as (a) but for Q = [0.5, 0, 2]. (c), (d) Spectra for c- and b-polarized TO modes at the zone centers [0,0,2] and [1,2,0], respectively, from CNCS data. (e) Temperature dependence of zone-center TO phonon energies at [0,0,2] and [1,2,0] with error bars indicating linewidths. Reproduced with permission from [8]. Copyright 2015 Macmillan Publishers.

with the TOF results for the same temperatures. Such temperature-dependent scans for given Q points are more efficient in triple-axis spectrometers that TOF.

From the fitting of the phonon spectra by Lorentzian functions, temperature-dependent phonon energy shift and change of the linewidth in the spectra indicate a giant phonon anharmonicity in this material. From 500 K to 750 K, the TO branch exhibits a marked softening. Crossing the phase transition near 810 K, the TO phonon mode is damped and overlaps with the zone-center low-energy TA mode. At even higher temperatures, the TO phonon shows a stiffening with increasing temperature, in particular at the zone boundary. Broadening of the phonon linewidth with increasing temperature is a result of the suppressed phonon lifetime from enhanced anharmonic phonon–phonon scattering processes. Together with first-principles simulation, it was found that the giant phonon scattering arises from an unstable electronic structure, with orbital interactions leading to a ferroelectric-like lattice instability.

11.2.3.2 Measure phonon density of states

Before the advent of the time-of-flight instruments, the phonon density of state (PDOS) was commonly obtained by first fitting the harmonic inter-atomic force constants to the phonon dispersion from the TAS measurements and then calculating the PDOS from these force constants [9]. There are several issues with such an approach: the inter-atomic forces are poor descriptors of the lattice dynamics when the anharmonicity is large; the process is tedious and not a direct measurement; and the results are heavily weighted on the phonons along the measured high symmetry directions.

TOF inelastic neutron scattering has a unique advantage when it comes to measuring the total PDOS. For such measurements, polycrystalline powder samples are usually used. Because of the random orientation distribution of the crystalline grains, the measurement provides a good average of contribution from a large number of Brillouin zones. The technique is well established and reliable, as shown by the work on a good number of materials [10, 11]. It should be noted that the measured PDOS is weighted by the total elementary neutron scattering cross-sections. To obtain the correctly normalized total PDOS, or projected partial PDOS, either first-principles calculations or isotope substitution is necessary. The measured PDOS has an energy-dependent resolution function—it is always favorable to work on the energy loss side and the energy resolution is better at high-energy transfer. Multi-phonon and multiple scattering contributions need to be subtracted carefully using iterative methods [12].

PDOS gives valuable information for understanding nanoscale thermal transport. A good example is the work on thermoelectric $Cu_{12}Sb_4S_{13}$ [13]. The PDOS measured by inelastic neutron scattering under various temperatures is shown in figure 11.2. For this sample, there exists a marked stiffening of the lowest-energy peak with increasing temperature. By fitting with Gaussian functions, plus a second-order polynomial background, the center and width of the features are extracted. Contrary to the quasi-harmonic (QH) model, the doped $Cu_{10}Zn_2Sb_4S_{13}$ exhibits a

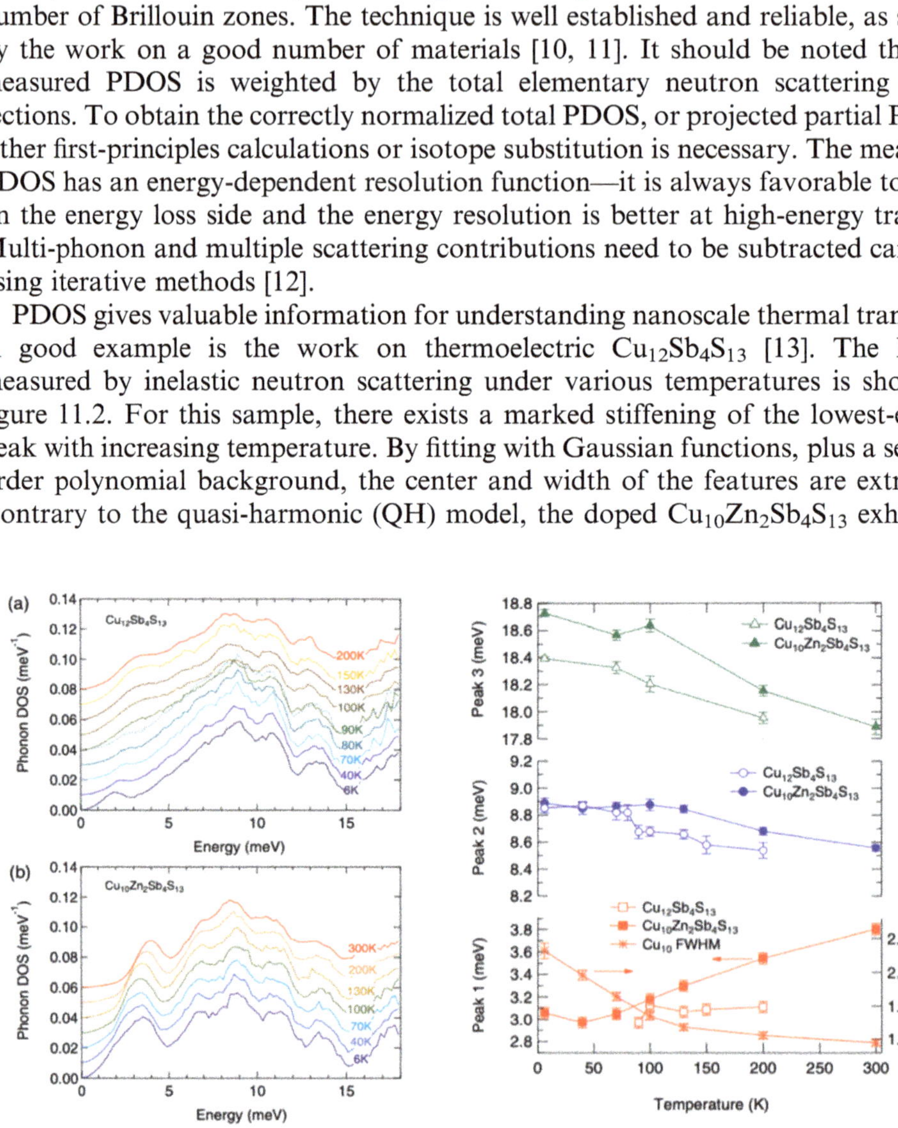

Figure 11.2. Temperature dependence of the phonon DOS for (a) $Cu_{12}Sb_4S_{13}$ and (b) $Cu_{10}Zn_2Sb_4S_{13}$. The fitting is shown one the right for peak 1 ($E \simeq 3$ meV), peak 2 ($E \simeq 9$ meV) and peak 3 ($E \simeq 18$ meV). The width of peak 1 is FWHM, corrected for instrument resolution. Reproduced with permission from [13]. Copyright 2016 the American Physical Society.

striking stiffening and narrowing of the low-energy peak with increasing temperature, more than its parent compound. Meanwhile, the higher energy feature near 9 meV and 18 meV follows the expected softening in the QH model. The lowest-energy feature mainly consists of acoustic modes and Einstein-like low-energy optical modes. This indicates that the doped material has strong anharmonicity at low temperatures, resulting in its very low thermal conductivity.

11.2.3.3 Determine branch specific thermal conductivity

Inelastic neutron scattering can directly measure phonon group velocity and lifetime along each phonon dispersion throughout the Brillouin zone. This allows quantifying the contribution from each phonon branch for heat transfer and the overall thermal conductivity. A good example is the work on investigating branch specific thermal conductivity in uranium dioxide (UO_2) [14]. As shown in figure 11.3, the TA and LA branches contribute to a major part of the total lattice thermal conductivity. However, it is surprising that the LO also plays a fairly important role. Even at 1200 K, the LO branch transport ~30% of heat in the lattice. The results are compared to the density functional theory calculations with generalized gradient approximation and Habbard-U correction (GGA+U). Although in both of them LO takes up a big part, it is clear that the simulation overestimates the overall thermal conductivity at low temperatures and underestimates it at high temperatures. The inelastic neutron scattering technique is a powerful tool to validate the thermal transport theory and guide the first-principles simulations and in the case of UO_2, help to improve the understanding of the role of strongly correlated 5*f* electrons in this material system.

11.2.3.4 Probe phonon scattering processes

The inelastic neutron scattering has direct access to individual phonon branches. This is valuable to identify specific phonon–phonon scattering processes, as shown by the work on the anomalous phonon features in thermoelectric PbTe with low thermal conductivity (figure 11.4) [15]. Comparing the phonon dispersion

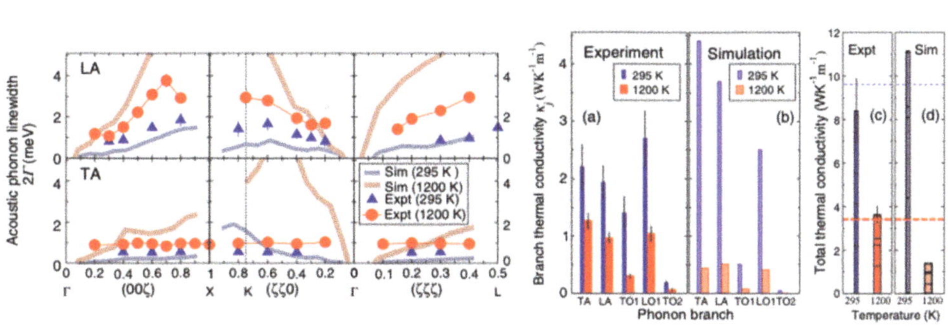

Figure 11.3. Left: measured and simulated intrinsic phonon linewidths for UO_2 for the TA and LA phonon branches. Right: measured and simulated branch specific thermal conductivity and total thermal conductivities. The thin black vertical lines in (a) and (c) are the experimental uncertainties; the thin and thick dashed horizontal lines in (c) and (d) denote the macroscopic thermal conductivities for UO_2 at 295 and 1200 K, respectively. Reproduced with permission from [14]. Copyright 2013 the American Physical Society.

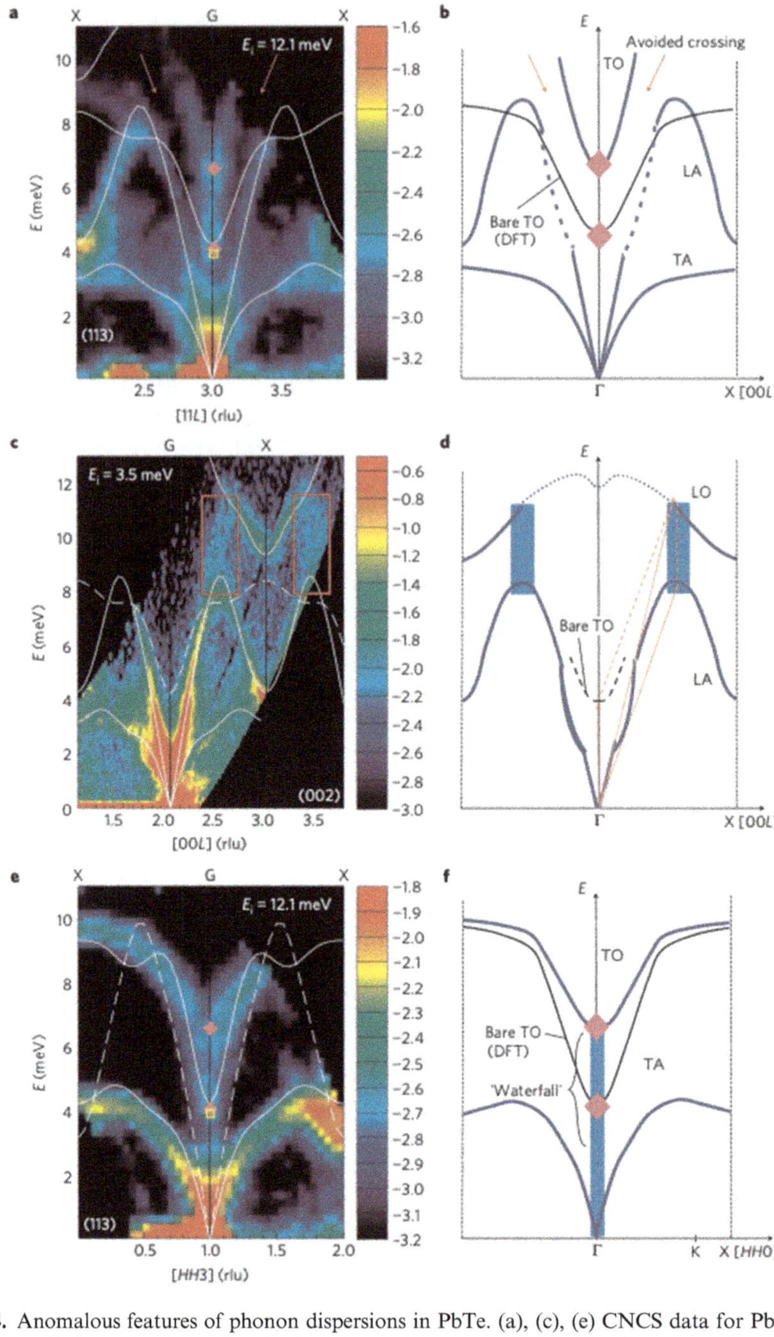

Figure 11.4. Anomalous features of phonon dispersions in PbTe. (a), (c), (e) CNCS data for PbTe at 300 K, showing the avoided-crossing behavior of LA and TO phonon branches in (a), the LA + TO → LO scattering in (c) and the 'waterfall' effect for the TO branch in (e). The solid and dashed white lines in (a), (c), (e) are harmonic dispersions calculated with DFT. (b), (d), (f) Schematic representations of the dispersions (blue lines). In all the panels, pink diamonds indicate the positions of the peaks in the TO obtained with HB-3. Reproduced with permission from [15]. Copyright 2011 Macmillan Publishers.

measurements at CNCS and HB-3, and the density functional theory calculation, shows an avoided-crossing behavior between longitudinal acoustical (LA) and transverse optical (TO) branches. Additionally, LA extinction in the $HKL = 113$ Brillouin zones near $q = 3.3$ rlu indicates anharmonic repulsion between the LA and TO modes which directly affects the phonon scattering process, causing low thermal conductivity in PbTe.

According to the scattering measurement, an extra intensity near $q = 2.5$ rlu is related to the three-phonon scattering process between LA, TO and LO. A 'waterfall' effect is observed for the zone-center TO branch. In comparison work between PbTe and SnTe, it is revealed that such an effect is a result of phonon nesting between the acoustic and optical phonon dispersion [8]. Such phonon nesting creates a resonance effect and an expanded phonon scattering phase space, and therefore a suppression of lattice thermal conductivity.

11.2.4 Summary

As shown by the examples, inelastic neutron scattering is a powerful technique to study the nanoscale heat transfer. The measured phonon dispersion or density of states are strongly coupled to the thermal properties of the material. INS not only allows investigating the thermal transport mechanism in specific materials, but also refining the heat transfer theory. INS can be used to determine the phonon frequency and lifetime under external conditions, such as various temperatures and pressures, which may affect the lattice behavior. In addition, INS can be used to determine the coupling between the phonons and between phonons and other excitations and particles, such as magnons and electrons. In conclusion, the INS technique provides the entire picture of the phonons and has many useful applications, in particular for understanding phonon nanoscale heat transfer.

References

[1] Shirane G, Shapiro S M and Tranquada J M 2002 *Neutron Scattering with a Triple-Axis Spectrometer: Basic Techniques* (Cambridge: Cambridge University Press)
[2] Squires G L 2012 *Introduction to the Theory of Thermal Neutron Scattering* (Cambridge: Cambridge University Press)
[3] Maradudin A A and Fein A E 1962 *Phys. Rev.* **128** 2589
[4] Abernathy D L, Stone M B, Loguillo M J, Lucas M S, Delaire O, Tang X, Lin J Y Y and Fultz B 2012 *Rev. Sci. Instrum.* **83** 015114
[5] Stone M B, Niedziela J L and Abernathy D L 2014 *Rev. Sci. Instrum.* **85** 045113
[6] Willendrup P, Farhi E, Knudsen E, Filges U and Lefmann K 2014 *J. Neutron Res.* **17** 35
[7] Zheludev A 2009 Reslib https://neutron.ethz.ch/research/resources/reslib.html (ResLib v3.4c)
[8] Li C W, Hong J, May A F, Bansal D, Chi S, Hong T, Ehlers G and Delaire O 2015 *Nat. Phys.* **11** 1063
[9] Cowley E R, Darby J K and Pawley G S 1969 *J. Phys. C: Solid State Phys.* **2** 1916
[10] Fultz B, Robertson J L, Stephens T A, Nagel L J and Spooner S 1998 *J. Appl. Phys.* **79** 8318
[11] Li C W, Tang X, Munoz J A, Keith J B, Tracy S J, Abernathy D L and Fultz B 2011 *Phys. Rev. Lett.* **107** 195504
[12] Kresch M, Lucas M, Delaire O, Lin J Y Y and Fultz B 2008 *Phys. Rev.* B **77** 024301

[13] May A F, Delaire O, Niedziela J L, Lara-Curzio E, Susner M A, Abernathy D L, Kirkham M and McGuire M A 2016 *Phys. Rev.* B **93** 064104

[14] Pang J W L, Buyers W J L, Chernatynskiy A, Lumsden M D, Larson B C and Phillpot S R 2013 *Phys. Rev. Lett.* **110** 157401

[15] Delaire O *et al* 2011 *Nat. Mater.* **10** 614

Chapter 12

Thermal transport measurements of nanostructures using suspended micro-devices

Sunmi Shin and Renkun Chen

Thermal transport in nanostructures, such as nanowires, nanotubes, nanofibers and two-dimensional materials, has been an active field of research due to the unique thermophysical properties of these nanostructures and their promise as technologically relevant building blocks. One of the major experimental tools used to study their thermal transport properties is the suspended micro-device platform, which was first developed in the early 2000s and has been continuously evolving and improving over the past two decades. The platform has been employed to measure the thermal conductivity of a large variety of one- and two-dimensional nanostructures and consequently has revealed rich thermal transport physics. In this chapter, we will review the basic characteristics and recent developments of the suspended micro-developed platform. We will discuss the resolution limit and contact resistance of the platform. We will then summarize recent developments over the past decade, including methods to improve the measurement sensitivity, quantification and reduction of contact resistance, discuss various heat loss mechanisms, and the utilization of these devices to study new phenomena such as diffuson and photon mediated thermal transport.

12.1 Introduction

Since the turn of this century, various nanostructures, including nanotubes, nanowires and nanofibers, have gained increasing attention due to both their interesting fundamental properties and promising applications in a broad range of areas, such as computing [1–3], photovoltaics [4], thermoelectrics [5, 6], piezoelectrics [7], photoelectrochemistry [8], energy storage [9], bio- and chemical sensing [10] and so on. For many of these applications, such as thermoelectrics, thermal transport properties play an important role. On the other hand, nanostructures also serve as an important platform to investigate and understand fundamental thermal transport

doi:10.1088/978-0-7503-1738-2ch12

phenomena at the nanoscale, a field that has seen tremendous growth in the past two decades [11, 12]. As in any other field, new understanding is often gained through innovations in instrumentation and measurement [13]. In the field of nanoscale thermal transport, there have been a number of important measurement techniques, such as the 3ω method [14], scanning thermal microscopy [15], ultrafast optical thermoreflectance (time domain [16] and frequency domain [17]), and the suspended micro-device [18].

Since its invention in 2001 by Majumdar *et al* [19], the suspended micro-device technique has been used widely to measure the thermal conductivity of a large variety of one-dimensional-like nanostructures, such as nanotubes, nanowires and nanofibers. The measured nanostructures include carbon and boron nitride nanotubes [19–23], smooth vapor liquid solid (VLS) Si nanowires [24], rough electrolessly etched Si nanowires [5] or roughened VLS Si nanowires [25], ZnO nanobelts [26], Bi nanowires [27], silicide nanowires [28], Bi_2Te_3 and related alloy nanowires [29, 30], Ge nanowires [31], Si nanotubes [32], Si nanoribbons [33], SiGe based nanowires [34, 35], boron carbide nanowires [36], and many other materials and structures. These measurements have revealed new insights into nanoscale thermal transport phenomena, such as the high thermal conductivity (over 3000 W m^{-1} K^{-1} at room temperature) in multi-wall carbon nanotubes (MWCNTs) [19], ballistic phonon transport over 2 μm length in single-wall carbon nanotubes (SWCNTs) [22], significant reduction in thermal conductivity of Si nanowires due to comparable diameter and phonon mean free path (tens to hundreds of nanometers) [37], the effect of roughness on nanowire thermal conductivity (e.g. in electrolessly etched Si nanowires [5], roughened VLS Si nanowires [25] and electron-beam lithography defined Si nanowires [38]), two-dimensional phonon transport in supported single layer graphene [39], the phonon softening effect in thin Si nanotubes [32] and nanoribbons [33].

Due to the rich phonon physics this technique has revealed, it is worthwhile to discuss the instrumentational aspects of the technique in detail, which is the focus of this chapter. We will first briefly describe the working principles of the technique, and then discuss at length the uncertainties and contact resistance issue of the technique. We will then summarize recent advances made on this technique, including the improvement of the measurement resolution, quantification of the background conductance, new instrumentation to quantify and minimize the contact resistance, and several versions of the micro-devices with integrated nanostructures.

12.2 Suspended micro-device platform

12.2.1 Basic principles and configuration

The basic suspended micro-device platform was first conceived and developed by Shi, Kim, Li and Majumdar at Berkeley in early 2000s [19, 37, 40]. Figures 12.1(a) and (b) show an SEM image and schematic drawing of the measurement set-up, respectively. As shown in figure 12.1(a), the device consists of two suspended silicon nitride (SiN_x) membranes that are supported by five long SiN_x beams. The thickness of the SiN_x layer is typically on the order of 100 microns. The long (>400 μm) and

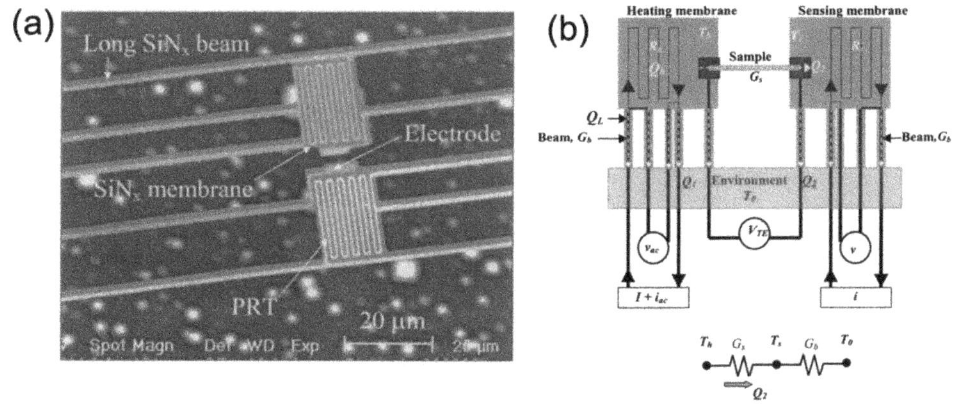

Figure 12.1. (a) Scanning electron microscopy (SEM) image of a suspended micro-device consisting of two SiN_x membranes, each of which is supported by five long SiN_x beams. A serpentine Pt line is integrated on each membrane for heating and thermometry. (b) Electrical and thermal diagram of the device. On the left heating membrane, a dc heating current (I) is applied to the Pt heating resistor. Thermometry is performed by coupling a small ac current (i_{ac}) with the dc current to measure the resistance of the Pt resistor (R_h) using a lock-in amplifier. On the sensing membrane, a small ac current (i) is applied to measure the resistance and temperature of the sensing Pt resistor (R_S) using another lock-in amplifier. Reproduced with permission from [18]. Copyright 2003 the American Society of Mechanical Engineering.

narrow (<3 μm) SiN_x beams ensure good thermal insulation between the membrane and the substrate, characterized by a low thermal conductance of the beams (G_b in figure 12.1(b)) which is on the order of 100 nW K^{-1}. Each of the SiN_x membranes is also integrated with narrow serpentine Pt lines, serving as the heater on one membrane and the thermometry on both membranes. The resistance of the serpentine heater lines on the heating and sensing membranes is R_h and R_s, respectively, and that of each of the long Pt legs on the SiN_x beams is R_L. During the measurement, a nanostructure sample (nanowire, nanotube, nanofiber, etc) is placed between the two membranes, as denoted by 'Sample' in figure 12.1(b). A small current is applied on the Pt heater line to raise the temperature of the heating membrane by ΔT_h, which is defined as $T_h - T_o$, the difference between the membrane temperature and the substrate temperature. A portion of the generated heat is conducted along the sample and reaches the sensing membrane; this amount of heat is denoted as Q_2 in figure 12.1(b), raising the temperature of the sensing side by ΔT_s (defined as $T_s - T_o$). Through simple energy balance and heat transfer analysis, one can find that

$$Q_1 = G_b \Delta T_h \tag{12.1}$$

$$Q_2 = G_s(\Delta T_h - \Delta T_s) = G_b \Delta T_s \tag{12.2}$$

$$Q_{tot} = Q_1 + Q_2 = I_h^2(R_h + R_L), \tag{12.3}$$

where R_L, rather than $2R_L$, is used in equation (12.3), because only half of the heat generated on the two Pt legs is dissipated to the heating membrane, while the other half is conducted to the substrate. Combining equations (12.1)–(12.3) results in the following equation:

$$Q_{\rm tot} = \Delta T_{\rm h}\left(G_{\rm b} + \frac{1}{G_{\rm b}^{-1} + G_{\rm s}^{-1}}\right). \tag{12.4}$$

The second term on the right-hand side represents the effective thermal conductance of the specimen and the sensing beams (two thermal resistors in series). Combining equations (12.2) and (12.4) yields

$$G_{\rm b} = \frac{Q_{\rm tot}}{\Delta T_{\rm h} + \Delta T_{\rm s}} \tag{12.5}$$

and

$$G_{\rm s} = \frac{Q_{\rm tot}\Delta T_{\rm s}}{\Delta T_{\rm h}^2 - \Delta T_{\rm s}^2}. \tag{12.6}$$

The analysis shown above means that by measuring $\Delta T_{\rm h}$, $\Delta T_{\rm s}$ and $Q_{\rm tot}$, one can obtain both $G_{\rm s}$ and $G_{\rm b}$. In a typical measurement, the changes in the heating and sensing Pt resistors ($\Delta R_{\rm h}$ and $\Delta R_{\rm s}$) are measured as a function of the total heating power $Q_{\rm tot}$ (figure 12.2(a)). In addition, the temperature coefficient of resistance (TCR) of the same Pt resistive thermometer (PRT) on each membrane is separately calibrated (figure 12.2(b)). From these two quantities, one can determine the temperature increases on the heating and sensing membranes ($\Delta T_{\rm h}$ and $\Delta T_{\rm s}$) as a function of $Q_{\rm tot}$, which can be used to obtain $G_{\rm s}$ and $G_{\rm b}$. As a sanity check, one needs to ensure that the measured $G_{\rm b}$ values are consistent between different devices with the same design. The $G_{\rm b}$ value is on the order of 100 nW K^{-1} for a common design of the SiN_x beams (e.g. 2 μm wide, 400 μm long and 200 nm thick).

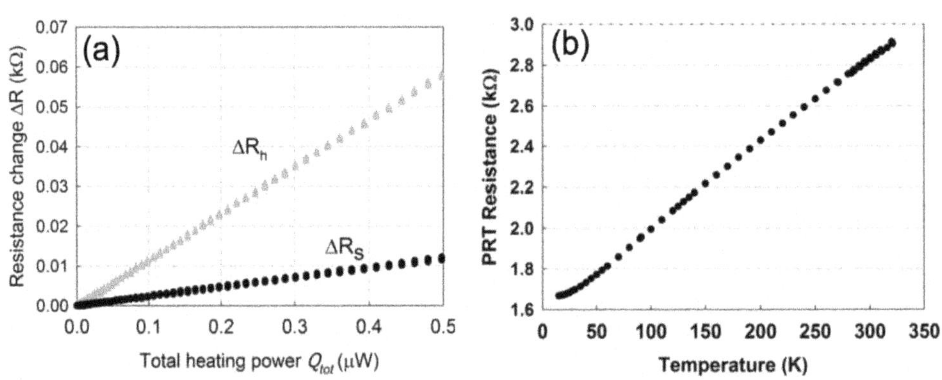

Figure 12.2. (a) Measured changes in the heating and sensing Pt resistors ($\Delta R_{\rm h}$ and $\Delta R_{\rm s}$) as a function of the total heating power $Q_{\rm tot}$ in a typical measurement. (b) Calibration of the temperature coefficient of resistance (TCR) of a representative Pt resistive thermometer (PRT) on the membrane. Reproduced with permission from [41]. Copyright 2008 Chen.

There are several advantages to using this fully suspended device platform. First, the method eliminates potential influence from the substrates that could either complicate the analysis of the heat flow (i.e. into the substrates and along the nanostructures) or affect the phonon transport in the nanostructures. Some earlier work attempted to measure the thermal conductivity of one-dimensional type nanostructures when they are placed on a substrate. It was also well known that phonon in the nanostructures could interact with the substrate, i.e. remote interfacial phonon (RIP) for phonons in carbon nanotubes or graphene. Second, the fully suspended design allows the precise quantification of minute heat flow through the nanostructures, by virtue of the temperature increase measurement on the sensing membrane, as shown in equation (12.2). Since G_b, the thermal conductance of the supporting beams to the sensing membranes, can be small, usually on the order of tens to 100 nW K^{-1}, the platform enables the measurement of a very small heat flow going through the nanostructured specimen, thus allowing the measurement of small thermal conductance of the specimen. Compared to the T-shape method, where the 'sensing' side is the substrate and the heat flow through the nanostructure is extracted by comparing the difference in the temperature rise on a heating beam with and without the nanostructures, the fully suspended device has a much higher sensitivity. Third, the temperature difference at the two ends of the nanostructures is measured by a resistive thermometer (e.g. the serpentine Pt lines shown in figure 12.1(a)), which represents the true temperature when the ends of the nanostructures are in thermal equilibrium with the two membranes.

12.2.2 Sensitivity and uncertainties

Based on (12.6), the sensitivity of the G_s measurement depends on the measurements of Q_{tot}, ΔT_h and ΔT_s. However, both Q_{tot} and ΔT_h have very small relative noise due to the high accuracy of the electrical measurements and the large ΔT_h (usually about 5–10 K). Therefore, the sensitivity of the measurement is primarily dictated by how accurate one can measure the sensing side temperature rise ΔT_s, as shown in (12.6). The sensitivity in ΔT_s measurement can be further divided into two categories: one is the instrumentation sensitivity and the other is the temperature fluctuation of the platform.

This first one can be estimated from the sensitivities of the instrumentation: a small ac current, i_s, at frequency f_s is applied to the sensing membrane, and then a lock-in amplifier (e.g. Stanford Research SR-830 or SR-850) is used to measure the voltage across the serpentine Pt lines using a four-point probe configuration (see figure 12.1(b)). As such, the noise equivalent temperature of the sensing side ($\mathrm{NET_s}$) can be expressed as

$$\mathrm{NET_s} = \frac{\mathrm{NEV_s}}{i_s}\frac{1}{\alpha_s R_s}, \tag{12.7}$$

where $\mathrm{NEV_s}$ is the noise equivalent voltage, as measured by the lock-in amplifier at the frequency f_s, α_s is the temperature coefficient of the resistance (TCR) of the sensing resistor, defined as $\alpha_s = \frac{dR_s}{dT}\frac{1}{R_s}$, and R_s is the sensing side resistance.

NEV_s could arise from two main noise sources: Johnson noise and instrumentation noise (shot noise is nonexistent, as expected for the metallic Pt heater as well as experimentally demonstrated on this type of device [42]), namely:

$$\mathrm{NEV_s} = \sqrt{\mathrm{NEV_{s,J}} + \mathrm{NEV_{s,i}}}, \tag{12.8}$$

where the Johnson noise is determined from

$$\mathrm{NEV_{s,J}} = \sqrt{4k_BTR_s\Delta f} \tag{12.9}$$

and is estimated to be about 5.86 nV/$\sqrt{\text{Hz}}$, which is similar to the input noise of an SR-830 lock-in amplifier ($\mathrm{NEV_{s,i}}$ = 6 nV/ $\sqrt{\text{Hz}}$). Therefore, NEV_s is estimated to be about 8 nV/$\sqrt{\text{Hz}}$. In typical measurements, the time constant setting (τ) of the lock-in amplifier is set to be 300 ms, 1 s or 3 s. If we use τ = 1 s, then the equivalent noise bandwidth (ENBW, $\Delta f = \frac{5}{64\tau}$) is 78 mHz and the NEV_s is 2.23 nV. With other typical values in this type of experiment (e.g. $i_s = 0.5$–$1\ \mu$A, $\alpha_s \sim 2000$ ppm $\mathrm{K^{-1}}$, $R_s \sim 2\ \mathrm{k\Omega}$), then the estimated NET_s is about 1 mK, which is a small uncertainty in the temperature measurement.

One could further reduce the NET_s by applying a higher i_s, or having a larger α_s or R_s, as shown by (12.7). α_s of typical metallic thermometry materials are usually of the order of 2000 ppm $\mathrm{K^{-1}}$ and there is not much room for improvement if a metal is used (semiconducting films could possess much higher α_s, such as NbN [43]). R_s is limited by the total area of the membrane and the smallest feature size one can attain on the serpentine heater lines. Additionally, i_s cannot be too high to avoid excess temperature excursion on the sensing side. If one wants to limit the temperature increase on the sensing side induced by the applied i_s to be within 5 K, then i_s has to be less than 500 μA ($\frac{i_s^2R_s}{G_s} < 5$ K). With these numbers, NET_s is estimated to be about 1 μK.

The analysis shown above suggests that the instrumentation noise is perhaps not the dominant noise for the thermometry on the sensing side. Rather, in a typical measurement where a dc heating current is used, the main noise comes from the temperature fluctuation on the sensing membrane, which is caused by the ambient temperature drift in the vacuum chamber. As the data collection at each temperature point usually takes a long time (about 1 h or longer), the temperature fluctuation can be up to 100 mK for a typical set-up using the traditional four-point probe resistance measurement technique, even when the sample stage temperature is precisely controlled by a PID control circuit [44–46]. This relatively large temperature fluctuation on the sensing membrane is partially caused by the excellent thermal insulation (low G_b) between the membrane and the substrate (or the sample stage), which, on the other hand, is desirable to attain high measurement sensitivity on G_s. Therefore, there is an inherent thermal design trade-off that limits the minimum measurable G_s.

Equation (12.2) relates the NET_s to the minimum measurable G_s (or noise equivalent thermal conductance, NEG_s):

$$\mathrm{NEG_s} = \frac{\mathrm{NET_s}}{\Delta T_h - \Delta T_s} G_b. \tag{12.10}$$

If one uses 50 mK as the $\mathrm{NET_s}$, then the minimum measurable G_s (or noise equivalent thermal conductance, $\mathrm{NEG_s}$) is about 0.5 nW $\mathrm{K^{-1}}$, ($G_b \sim 100$ nW $\mathrm{K^{-1}}$ and assuming ΔT_h is up to 10 K). This $\mathrm{NEG_s}$ suggests that the platform is applicable to measuring specimens with relatively high conductance (e.g. Si nanowires with a diameter greater than 20 nm [37], or nanostructures with high thermal conductivity, such as carbon nanotubes [19, 22]). However, as the conductance of the specimens is high, thermal contact resistance between the specimens and the devices could become important and great care must be taken to either minimize or quantify its impact, as discussed in detail in the next section.

12.2.3 Thermal contact resistance

It should be noted that the device only measures the total thermal resistance (R_{th}), which includes both the intrinsic conduction thermal resistance of the specimen ($R_{th,s}$) and the contact between the nanostructures and the suspended pads ($R_{th,c}$): $R_{th} = R_{th,s} + R_{th,c}$. Therefore, it is important to either quantify the $R_{th,c}$ or ensure it is negligible compared to $R_{th,s}$. The latter assumption is usually valid for a specimen with large $R_{th,s}$, i.e. low thermal conductivity, small cross-sectional area, or long length. However, for samples with relatively high thermal conductivity (e.g. carbon nanotubes and large-diameter nanowires), the influence of thermal contact resistance needs to be carefully examined. In this section, we analyzed the thermal contact resistance and summarized earlier work to quantify or minimize it on several materials. In sections 12.3.5 and 12.3.6, we will also discuss more recent work related to quantifying the contact resistance.

The thermal contact resistance can be analyzed as a cylinder in contact with a flat substrate with van der Waals (vdW) interaction. As one could perceive, the actual contact area between these two objects, which is the contact width ($2b$) multiplied by the contact length (l), will play a critical role in dictating the contact resistance. In reality, it is a rather complicated issue to precisely calculate the contact width, because of the surface roughness of the substrate and the nanostructures. If we assume perfectly smooth surfaces, we can estimate the contact width using the contact mechanics [47]:

$$2b = \sqrt{\frac{16 F_{vdW} E_m D}{\pi}}, \tag{12.11}$$

where D is the diameter of the nanostructure (nanowire, nanotube or nanofiber), E_m is the effective modulus of the nanostructure and the pad, and F_{vdW} is the vdW attractive force. Expressions for E_m and F_{vdW} can be found in the literature (e.g. [48–50]). The estimated contact width is generally on the order of ~10 nm for nanostructures with a diameter of ~100 nm. For example, Yu *et al* [49] estimated a contact width of ~10 nm for a carbon nanofiber 152 nm in diameter; Zhong *et al* [50] obtained contact widths of ~80 nm and ~20 nm for Nylon-11 nanofibers of 400 and 70 nm in diameter, respectively. Yang *et al* [21] estimated the contact width between a multi-wall carbon nanotube (MWCNT) and a planar surface to be 3.8 nm and 0.92 nm for an MWCNT of 66 nm and 10 nm in diameter, respectively.

With the contact width, $R_{th,c}$ between the nanostructure and the SiN_x substrate can then be estimated using [47, 50]

$$R_{th,c} = \frac{1}{\pi l k_n}\ln\left(\frac{2D}{b}\right) - \frac{1}{2lk_n} + \frac{1}{\pi l k_P}\ln\left(\frac{D}{\pi b}\right), \tag{12.12}$$

where k_n and k_P are the thermal conductivity of the nanostructure and the SiN_x pad, respectively, ($k_P \approx 4$ W m^{-1} K^{-1} for SiN_x). The first two terms describe the constriction resistance imposed on the surface of the cylindrical nanostructure with a contact width of $2b$, and the third term shows the constriction resistance imposed on the planar SiN_x pad. When dealing with anisotropic materials such as graphitic materials, both the in-plane and cross-plane thermal conductivities need to be considered in the analysis [20, 21, 49]. Note that equation (12.12) is an approximate model with several assumptions. For instance, since the contact area is very small, ballistic transport of the phonon through the constriction could impose additional resistance [51].

For typical material properties and geometries (k_n = 1 – 100 W m^{-1} K^{-1}, $D \sim 100$ nm and $b \sim 10$ nm), $R_{th,c}$ is on the order of 0.1–1 × 10^7 K W^{-1}. This can also be understood from the typical contact areas ($2bl$, where l is about 1 μm) and the areal interfacial thermal resistance of vdW interfaces ($R'_{th,c} < 3 \times 10^{-8}$ m^2 KW^{-1}), and hence $R_{th,c} = 2\frac{R'_{th,c}}{2bl} < 3 \times 10^6$ KW^{-1} for two nanostructure/substrate junctions. Wingert *et al* [50] estimated the $R_{th,c}$ to be 2.7 ×10^6 KW^{-1} and 3.8 × 10^6 K W^{-1} for polymer nanofibers of 400 and 70 nm in diameter, respectively, while $R_{th,s}$ of the fiber specimens was estimated to be between 5 × 10^8 KW^{-1} and 3 × 10^9 KW^{-1} (with thermal conductivity values less than 2 W m^{-1} K^{-1} for the fibers). Therefore, the contact resistance is not expected to be important for low thermal conductivity materials, as long as an intimate contact between the specimen and the substrate can be ensured.

It should be noted that (12.11) for the contact width is for atomically smooth and clean surfaces. In reality, surface roughness and contamination could reduce the contact width. In order to enhance the contact area, a Pt–C composite or a similar material has usually been deposited at the nanostructure/membrane junctions, typically by electron beam induced deposition (EBID) in an SEM or using the e-beam in a dual-beam focused ion beam (FIB) chamber (the focused Ga ions would damage the nanostructures and greatly reduce their thermal conductivity [52, 53]). The idea is to attain a total contact width of at least 10 nm to reduce the $R_{th,c}$ to below 3 × 10^6 KW^{-1}. Yu *et al* [49] analyzed and measured $R_{th,c}$ between a 152 nm diameter carbon nanofiber and the Pt heater on the suspended membrane (left inset in figure 12.3(a)). They used an EBID Pt–C composite at the fiber/pad junction to increase the contact area. By measuring the total thermal resistances before and after the EBID, they could attribute the difference to $R_{th,c}$ prior to the EBID, which was found to be $R_{th,c}$ = 3 –5 × 10^6 K W^{-1} (right inset in figure 12.3(a)). This result is

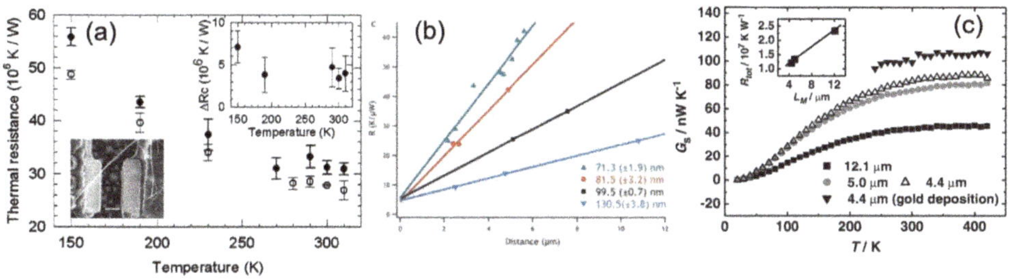

Figure 12.3. Thermal contact resistance between nanostructures and the thermal devices. (a) Measured total thermal resistance (R_{th}) of a 152 nm diameter carbon nanofiber, before and after Pt–C deposition (black and empty circles, respectively). Inset: contact resistance ($R_{th,c}$) between the fiber and the device prior to the EBID, which was extracted from the reduction in R_{th} after the EBID. Reproduced with permission from [49]. Copyright 2006 the American Society of Mechanical Engineering. (b) Measured total thermal resistance of VLS Si nanowires as a function of wire length, for four different diameter groups. The intercept of the fitted lines at the *y*-axis represents the thermal contact resistance. Reproduced with permission from [54]. Copyright 2012 the American Chemical Society. (c) Measured thermal conductance of an MWNT specimen (G_s) with different suspend lengths (L = 4.4 to 12.1 μm) before EBID of Au and a 4.4 μm long specimen with Au. Reproduced with permission from [21]. Copyright 2011 Wiley.

consistent with the theoretical expectation for $R_{th,c}$ due to vdW interactions, as we discussed earlier. It was also shown that EBID is an effective strategy to reduce the contact resistance to a negligible level on carbon nanofibers, which has a thermal conductivity of about 10 W m^{-1} K^{-1} at room temperature.

A similar conclusion was also reached in Si nanowires. Li found that for Si nanowires grown from a vapor–liquid–solid (VLS) process, the EBID treatment was effective in eliminating the influence of the contact resistance to yield intrinsic nanowire thermal conductivity, which ranges from ~10 to ~40 W m^{-1} K^{-1} for nanowire diameters ranging from 22 to 115 nm [37]. Lim *et al* [54] further quantified the $R_{th,c}$ between VLS Si nanowires and the suspended SiN_x membranes when the junctions were enhanced with the Pt–C composite by EBID. They varied the nanowire length and measured the corresponding total thermal resistance R_{th} as a function of length for nanowires with diameters ranging from ~70 to 130 nm, shown in figure 12.3(b). For each diameter group, the measured R_{th} showed a linear increase with the length, indicating a constant $R_{th,c}$ and a constant nanowire thermal conductivity (κ_{NW}): $R_{th} = R_{th,c} + L/(A_{NW}\kappa_{NW})$. The intercept on the *y*-axis for each line in figure 12.3(b) represents the $R_{th,c}$, which is about 4.5×10^6 K W^{-1}. This value is similar to what we analyzed earlier. This resistance represents less than 10% of the conduction resistance of the nanowire with 71 nm diameter and 5 μm length, and the measured results are in good agreement with the VLS Si nanowires measured by Li *et al* [37]. For the nanowire of 130 nm diameter and 5 μm length, however, $R_{th,c}$ could be significant. Therefore, with the Pt–C deposition, the thermal contact resistance is negligible for sufficiently thin and long Si nanowires, but could be significant for thick nanowires. Great care needs to be taken when measuring nanowires with diameters over 100 nm. For instance, sufficiently long nanowires (>5 μm) need to be chosen to ensure a much larger $R_{th,s}$ than $R_{th,c}$. The EBID Pt–C or a similar material must be applied to increase the contact area and reduce the contact resistance.

For materials that are prone to damage by e-beams such as polymers, alternative approaches need to be taken to ensure a sufficiently large contact area between the specimen and the membranes. For example, Shen *et al* developed a capillary-assisted adhesion method to enhance the contact area of ultra-drawn polyethylene (PE) nanofibers [55]. In this method, they placed a droplet of isopropanol onto the thermal devices. The surface tension induced by the evaporation of the solvent pulls the PE nanofibers into intimate contact with the SiN_x membranes. They have shown the effectiveness of this method by obtaining high thermal conductivity of over 60 W m^{-1} K^{-1} in the PE nanofibers. A similar method was used by Lee *et al* to measure electrospun poly(acrylic acid) nanofibers [56, 57], but with a nonpolar solvent of cyclohexane. For other polymer nanofibers with lower thermal conductivity, no contact enhancement treatment was used and the results did not seem to be influenced by the contact resistance, for example, on nanofibers of nylon-11 (highest $\kappa = 2$ W m^{-1} K^{-1} [50]), electrospun PE nanofibers (highest $\kappa = 9.3$ W m^{-1} K^{-1} [58]), solution drawn PE nanofibers (highest $\kappa = 8.8$ W m^{-1} K^{-1} [59]), vinyl polymer nanofibers with various molecular weights and side groups (highest $\kappa \sim 5$ W m^{-1} K^{-1} [60]) and polystyrene nanowires (highest $\kappa \sim 14.4$ W m^{-1} K^{-1} measured using a dual-cantilever platform [61]). The relatively less important contact resistance can be attributed to the low thermal conductivity of the specimens (generally less than 20 W m^{-1} K^{-1}) and inherently better contact between the soft polymer fibers and the SiN_x membranes. As mentioned earlier, Wingert *et al* estimated $R_{th,c}$ on the order of 10^6 KW^{-1} for Nylon-11 fibers 70–400 nm in diameter [50], and work by Ma *et al* [58] showed that the PE nanofibers can be anchored to the membranes steadily without additional contact treatment. However, measurements by Zeng *et al* suggested a very high $R_{th,c}$, about 1×10^8 KW^{-1} on a 395 nm diameter epoxy resin fiber [62]. The exact reason for this particularly high $R_{th,c}$ [62] is unknown, and warrants further investigation. A possible reason is that in Ma *et al*'s work, they found residual solvent trapped in the PE nanofibers as evidenced from Raman signals (either solution drawn [59] or electrospinning at high voltage [58]). In Zhong *et al*'s work [50], the Nylon-11 nanofibers were directly collected onto the thermal measurement micro-devices during the electrospinning process, and it is very likely that there was also residual solvent in the fibers. In both cases, the residual solvent could be evaporated when the fibers were placed in the measurement vacuum chambers, which may have improved the contacts, as in the capillary-assisted adhesion method. It is possible that this unintentional solvent evaporation induced effect was absent in Zeng's work, either due to the lack of the residual solvent or insufficient evaporation.

The presence of the contact resistance makes it more challenging to measure nanostructures with high thermal conductivity, such as carbon nanotubes and graphene. For instance, Yang *et al* [21] systematically studied the $R_{th,c}$ of a 66 nm diameter MWCNT with varying nanotube lengths and also with EBID of Au. Figure 12.3(c) shows the measured results for this nanotube sample. The measured thermal conductance of the specimen (G_s) decreases with increasing nanotube length from 4.4 to 12.1 μm (all prior to EBID). This length-dependent thermal resistance (shown in the inset of figure 12.3(c)) can be used to extract the $R_{th,c}$, which was found to be about 5–6$\times10^6$ K W^{-1} at room temperature, similar to that of the Si nanowires on the suspended devices. With EBID of Au, the thermal conductance of the sample with 4.4 μm length is significantly increased, e.g. by ~100% at room temperature. This indicates that the $R_{th,c}$ could contribute to ~50% of the total thermal resistance prior to EBID. The intrinsic thermal conductivity of the MWCNT was found to be slightly higher than 200 W m^{-1} K^{-1} by using the length-dependent data (inset of figure 12.3(c)). By subtracting the intrinsic $R_{th,s}$ of the CNT from the measured total resistance of the 4.4 μm long sample after EBID, the $R_{th,c}$ after the EBID of Au was estimated to be 3–4$\times10^6$ K W^{-1} in the temperature range of 240–420 K, which is still significant compared to $R_{th,s}$ due to the high thermal conductivity of the CNT and thus cannot be neglected. Therefore, additional care must be taken for the contact resistance when measuring high thermal conductivity materials.

The discussion in this section highlights the importance of properly addressing the impact of contact resistance on the measurement using the suspended device platform. To obtain the intrinsic thermal conductivity of the nanostructures, it is necessary to reduce the contact resistance to below the conduction resistance of the specimens. Good practice would be to choose sufficiently long nanowires for higher sample resistance, and to enhance the contact area between the specimens and the suspended membranes, for example by using EBID to deposit a Pt–C composite. This protocol has been successfully used to measure the thermal conductivity of a variety of nanostructures, including carbon nanotubes [19–22], smooth and rough Si nanowires [5, 24, 37, 54], and many other materials. In the case of polymer nanofibers that are sensitive to irradiation damage induced by an electron beam, the adhesion and contact between the fibers and the membranes can be enhanced by the evaporation of a solvent [55–57]. For low thermal conductivity polymer nanofibers, the contact resistance may be negligible and additional contact treatment is not always necessary [50, 58, 59]. However, great care still needs to be taken as there is a report of unusually high thermal contact resistance [62].

12.3 Recent developments

12.3.1 The differential bridge method

The previous section shows that the dominant noise source in the early version of the suspended platform originated from the extrinsic ambient temperature fluctuation and was far away from the intrinsic instrumentation noise level. The relatively large temperature drift (NET_s ~ 50 − 100 mK) limits the minimum thermal conductance of the specimen (G_s) to be around 0.5–1 nW K^{-1}. However, there are a large number of nanostructures that could possess much lower G_s. For example, there had been prediction of the phonon confinement effect, which modifies the phonon dispersion when the nanowire diameter is below 10 nm [63]. The experimental study of thermal transport in nanowires with such small diameters would be valuable but was limited by the insufficient sensitivity of the suspended device platform. Furthermore, it was noticed that the suspended device might have an inherent background conductance, that is, the conductance between the two membranes when there was no specimen bridging them. This background conductance needs to be accounted for if the specimen conductance is also small. However, a precise determination of this background conductance had not been feasible, although it could cause a relatively large error in measuring specimens with thermal conductance on par with the background conductance.

One of the strategies to reduce the influence of the ambient temperature fluctuation is to employ a differential method, such as a Wheatstone bridge, to measure the sensing side resistance R_s [44]. As shown in figures 12.4(a) and (b), R_s is not measured by the four-point probe method, but rather by a Wheatstone bridge, with respect to three other resistors with known resistances: R_1, R_2 and R_3. In particular, R_1 is on the same chip as R_s and they have the same nominal resistance (figure 12.4(c)), and R_2 and R_3 are a precision resistor and a variable resistor, respectively, placed outside the chamber (at room temperature). During the measurement, a small ac source voltage (v_s) is applied to the bridge and the voltage difference between points A and B (v_g) is measured using a lock-in amplifier at the first harmonic (figure 12.4(b)). R_s can be determined from [44]

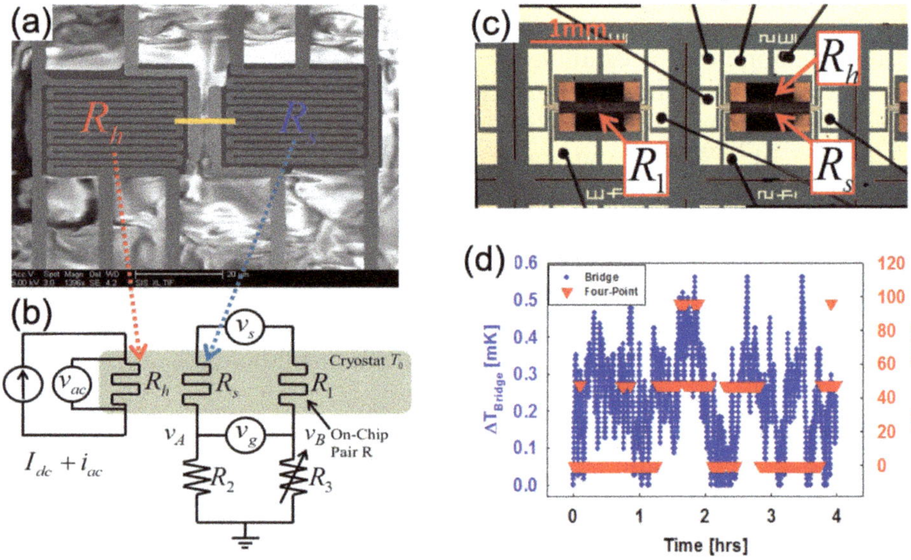

Figure 12.4. A differential bridge scheme to measure the changes in sensing side resistance (ΔR_s) and temperature (ΔT_s). (a) An SEM image of the suspended micro-device used for the measurement. (b) Circuit diagram for measurement of the heating side resistance (R_h) which is the same as in the traditional method (see figure 12.1(b)), but the sensing side resistance is measured in the Wheatstone bridge configuration, with the on-chip pairing resistance being another Pt serpentine resistor of the identical design fabricated on the same Si substrate. (c) Photograph of the heating, sensing and pair resistors on the Si substrate. (d) Measured temperature fluctuation over the course of four hours, as measured by the bridge method (blue, left axis) and the four-point method (red, right axis). Reproduced with permission from [44]. Copyright 2012 the American Institute of Physics.

$$R_s = \frac{R_2}{\frac{v_g}{v_s} + \frac{R_3}{R_1 + R_3}} - R_2. \tag{12.13}$$

Since both R_s and R_l are on the same chip with a high thermal conductivity Si substrate, they would experience similar temperature fluctuations. Therefore, the noise on R_s measurement due to the temperature fluctuation is very small, and can be reduced to ~1 mK (figure 12.4(d)) when the experimental conditions are optimized, such as the setting of the lock-in amplifier (e.g. a long time constant of 3 to 10 s) and a larger applied v_s.

It should be noted that the bridge method is useful to substantially reduce the influence of the temperature fluctuation on R_s measurement by using the principle of common mode noise rejection. One important issue to clarify is whether or not the temperature fluctuations on R_l and R_s are correlated, as only the correlated noise between the two resistors can be canceled in the bridge configuration. One can hypothesize that the temperature fluctuations on the two membranes originating from the temperature drift of the sample stage are largely correlated and hence can be canceled using the bridge scheme, due to the high thermal conductivity of the common Si substrate and the small thermal time constant of the supporting SiN_x beams and the membranes (~100 ms, much smaller than the time constant of the measurements). If there were additional temperature fluctuations experienced by the membranes that do not originate from the common Si substrate, the noise would have been uncorrelated and cannot be canceled in the present bridge scheme. As analyzed by Wingert *et al* [44], the best NET_s attained was about 0.6 mK when the instrument conditions were properly set (such as the time constant of the measurement, the driving current on Wheatstone bridge, etc) (figure 12.4(d)), which is comparable to the intrinsic noise of the instrumentation (Johnson and amplifier noises). As the NET_s is now reduced to below 1 mK using the bridge method, NEG_s is similarly reduced to below 10 pW K^{-1} (again, assuming $G_b \sim 100$ nW K^{-1} and $\Delta T_h = 10$ K). Therefore, the bridge method represents a simple and yet powerful tool to greatly improve the sensitivity of the measurements.

12.3.2 Modulated heating

The differential bridge scheme shown in section 12.3.1 can improve the measurement sensitivity of the G_s (or NEG_s) to around 10 pW K^{-1}. Further improvement of NEG_s along the same lines may not be feasible. Equation (12.9) suggests that Johnson-noise limited NEV_s can be reduced by using a smaller NEBW (Δf), or a longer averaging time in the measurement (i.e. the time constant setting in the lock-in amplifier), for either the four-point or the bridge method. However, the bridge scheme is still sensitive to the uncorrelated temperature fluctuations as well as the $1/f$ electronic noise experienced by the sensing and pairing resistors. One can recognize that these two types of the noise are more pronounced at low frequency, due to the system thermal mass for the temperature fluctuation and the $1/f$ nature of the electronic noise. Since the Wheatstone bridge scheme by Wingert *et al* [44] was operating with the dc heating condition ($f_h = 0$ Hz), these low frequency noises are expected to be pronounced.

In order to further reduce the noise, one can move the frequency of the measured signal (temperature in this case) to a higher frequency, which can be achieved using a modulated (ac) heating current on the heating membrane. Earlier work by Reddy *et al* developed a resistive thermometry technique operated with modulated heating to largely suppress the $1/f$ noise and achieved a low NET_s of 30–50 μK, and consequently demonstrated ~1 pW calorimetry on a single suspended membrane [64]. We further utilized this concept and applied it to probe thermal transport in a nanostructure bridging the two suspended membranes [65]. Figure 12.5 shows the circuit diagram of the thermometry on both the heating and sensing membranes using the modulated heating scheme. Instead of a dc current as shown in figures 12.1(b) and 12.4(b), the heating side resistor (R_h) is now heated by a modulated current I_h with a modulation frequency of f_h. This current will generate a heat flux Q_h modulated at $2f_h$, which will lead to a heating side temperature increase ΔT_h also oscillated at $2f_h$. By measuring the third harmonic voltage of the heating beam $V_h(3f_h)$, one can extract $\Delta T_h(2f_h)$, similar to a typical 3ω measurement on bulk and thin film samples [14]. In practice, $V_h(3f_h)$ and $\Delta T_h(2f_h)$ of the heating beam is measured with respect to that of a reference resistor with the same resistance as R_h but placed on a constant temperature thermal reservoir. This half-bridge configuration on the heating beam cancels the higher harmonic noise inherently contained in the heating current I_h.

Similar to that of the dc heating scheme shown in figure 12.1(b), a portion of the heating flux Q_h, now modulated at $2f_h$, will now conduct along the nanostructure specimen and reach the sensing beam, raising the sensing beam temperature to ΔT_s that is also oscillating at $2f_h$. The associated resistance change due to the temperature increase is now measured by a full Wheatstone bridge circuit, similar to the one shown in figure 12.4(b), with the exception that the bridge driving current is now a dc

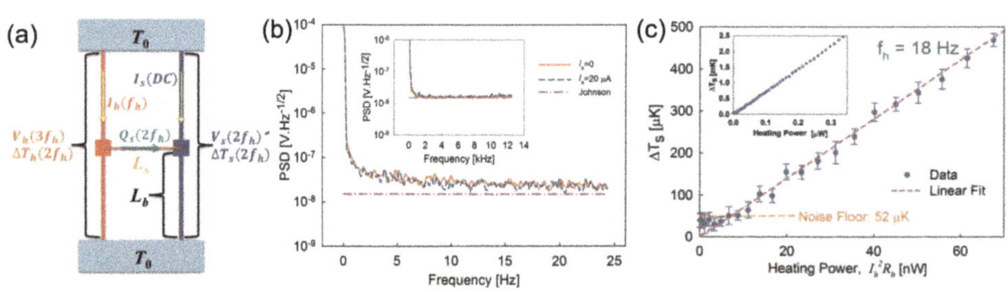

Figure 12.5. (a) Circuit diagram of the thermometry on the heating and sensing sides using the modulated heating scheme. (b) Power spectral density (PSD) of the noise measured from the Wheatstone bridge circuit on the sensing side under two different sensing side dc currents: $I_s = 0$ and 20 μA. The inset shows the PSD of the noise up to 12 kHz. (c) Measured sensing side temperature increase (ΔT_S) versus the heating power on the heating membrane up to 70 nW (up to 0.3 μW in the inset) at f_h = 18 Hz, showing a noise floor of 52 μK. Reproduced with permission from [65]. Copyright 2013 the American Institute of Physics.

current (I_s(dc)) and the voltage difference between A and B (v_g) is now measured at $2f_h$ using a lock-in amplifier with the reference signal from $I_h(f_h)$.

Figure 12.5(b) shows the power spectrum density (PSD) of the noise of v_g in the ac bridge circuit. It clearly shows that the PSD is greatly reduced at higher frequencies and is approaching that of the Johnson noise of the four resistors in the bridge, when the frequency is higher than about 10 Hz. The inset in figure 12.5(b) further shows the closeness to the Johnson-noise limit when the frequency is up to 12 kHz. This result confirms the hypothesis that when the heating current is modulated at a sufficiently high frequency, the $1/f$ noises, including the ambient temperature fluctuation, can be eliminated, and one can reach the Johnson-noise limit. For instance, at $f_h = 15$ Hz, the measured PSD is about 20 nV/$\sqrt{\text{Hz}}$, which is only slightly higher than the calculated combined noise (15.2 nV/$\sqrt{\text{Hz}}$) of the Johnson noise (14.3 nV/$\sqrt{\text{Hz}}$) of the four resistors in the bridge and the input noise of the spectrum analyzer (5 nV/$\sqrt{\text{Hz}}$). Figure 12.5(c) shows a measurement within this operation regime with $f_h = 18$ Hz, and with an NEBW (Δf) of 7.8 mHz in the lock-in amplifier. At a very small heating power on the heating side, ΔT_s on the sensing side is indistinguishable from the noise. From this we can determine the noise floor of ΔT_s (NET_s) to be 52 μK, which is about ten times better than the lowest NET_s attained in the dc Wheatstone bridge scheme. This noise floor is consistent with the PSD of the noise measured at this frequency in figure 12.5(b). This work [65] also shows that the noise floors were slightly higher at lower frequency, namely at 65 and 84 μK with f_h of 8 and 2 Hz, respectively, due to the small increase in the $1/f$ noise. These noise floors are still much lower than what could be achieved in the dc Wheatstone bridge scheme. Therefore, one can conclude that the modulated heating scheme can be used to achieve an ultrahigh sensitivity. With the NET_s as low as about 50 μK, it can potentially resolve nanostructure specimens with thermal conductance as low as 0.5 pW K^{-1} based on (12.10) ($\Delta T_h = 10$ K and $G_b = 100$ nW K^{-1}).

There is an important complication one needs to keep in mind when implementing the modulated heating scheme: ac heating could lead to the attenuation of the temperature response when the time scale of the modulation ($1/(2f_h)$) is shorter than the thermal time constant of the suspended beams, or when the thermal penetration depth (L_p, scales as $\sqrt{\alpha/f_h}$, where α is the thermal diffusivity of the beams) associated with the modulated heat current is shorter than or comparable to the length of the beams. Therefore, if one wants to attain the same temperature response with the ac heat current as in the case of the dc heating, the applied frequency f_h has to be lower than the cut-off value, which is called the roll-off frequency [66]. Analysis by Zheng *et al* [65] shows that the roll-off frequencies were 3.78, 10.20, 39.33 and 156.83 Hz when the beam lengths (L_b, see figure 12.5(a)) were 400, 200, 100 and 50 μm, respectively. These numbers were consistent with the results from the thermal time constant of the beams (64.6, 16.2, 4.0 and 1.0 ms for $L_b = 400$, 200, 100 and 50 μm, respectively) [65]. Note that the thermal time constant of the beams (τ_b) scales as C_b/G_b, or the heat capacity divided by the thermal conductance of the beams, so it scales as L_b^2. Therefore, for a given beam length f_h needs to be carefully chosen such that it is below the roll-off frequency but also sufficiently high to largely suppress the $1/f$ noise (figure 12.5(b)). This operation window can be enlarged if one uses a shorter L_b, but it could increase G_b which would increase the NEG_s for a given NET_s. Therefore, one has to carefully evaluate the trade-off associated with the design of the device geometry if an ultrahigh conductance measurement sensitivity (low NEG_s) is needed.

The frequency-dependent attenuation of the thermal response is not necessarily a drawback. For example, it can be utilized to measure the specific heat of the bridging nanostructure, as shown by Zheng *et al* [67]. This is based on the fact that the thermal response is dependent on the thermal penetration depth of the heating current with respect to the length of the specimen. For the set-up shown in figure 12.5(a), the heat conduction equation in the frequency domain can be shown as

$$\frac{d^2 \Delta T(x, \omega)}{dx^2} = a^2 \Delta T(x, \omega), \tag{12.14}$$

where ω is the thermal angular frequency ($\omega = 4\pi f_h$). Under the condition that the nanostructure conductance is much smaller than the beam conductance (or the heat flux conducted through the nanostructure is negligible compared to joule heating on the heating beam), this equation yields the frequency-dependent temperature increase ratio between the sensing and heating beams [67]:

$$\frac{\Delta T_s(\omega)}{\Delta T_h(\omega)} = \frac{1}{\cosh(a_2 L_2) + \frac{\kappa_1 A_1 a_1 \sinh(a_2 L_2)}{\kappa_2 A_2 a_2 \tanh(a_1 L_1)}}, \tag{12.15}$$

where $a = \sqrt{\frac{j\omega\rho c}{\kappa}}$ (the subscripts 1 and 2 represent the sensing beam and the nanostructure, respectively), j is the imaginary unit, and ρ, c, κ, L and A represent the density, specific heat, thermal conductivity, length and cross-sectional area of the nanostructure or the beam, respectively.

Equation (12.15) is reduced to equation (12.2) in the low frequency limit, i.e. when $a_2 L_2 \ll 1$ and $a_1 L_1 \ll 1$. This means the measurement at the low frequency is not frequency-dependent and is only sensitive to thermal conductance and thermal conductivity of the nanostructure ($G_s = \kappa_2 A_2 / L_2$) (figure 12.6(a)), the same as in the dc heating case, as expected. At the higher frequency ($a_2 L_2 > 1$), measurement of $\Delta T_s(\omega)/\Delta T_h(\omega)$ is dependent on the frequency and can be used to obtain a_2 (figure 12.6(a)), which is related to the thermal diffusivity of the nanostructure specimen, provided that a_1 for the beam is known, which can be separately characterized on the heating beam alone. Therefore, from this frequency-dependent measurement with the modulated heating, one can obtain both thermal conductivity and specific heat values. Figure 12.6(b) shows an example of the measured specific heat of individual Nylon-11 nanofibers with diameters of 615 and 693 nm [67]. The measured specific heat is very close to those of the bulk values and the theoretical modeling, as expected for such thick fibers. The technique can potentially be extended for simultaneous thermal conductivity and specific heat measurements of individual nanostructures with interesting phonon spectra, such as one-dimensional nanotubes and two-dimensional materials.

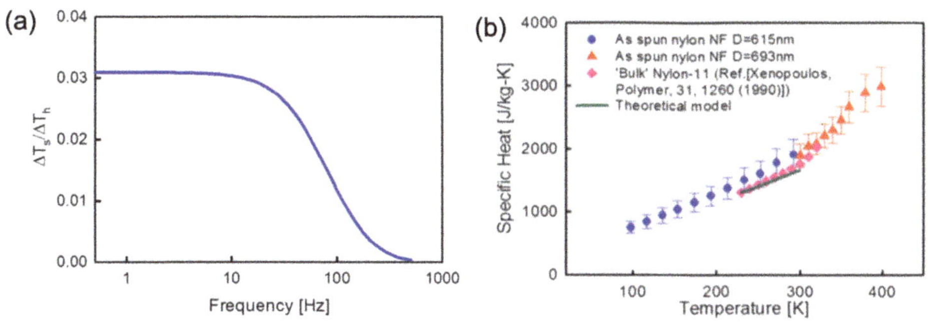

Figure 12.6. Specific heat measurement of individual nanostructures using the modulation heating scheme. (a) Modeled $\Delta T_s/\Delta T_h$ ratio as a function of the heating frequency. (b) Measured specific heat of two large Nylon-11 nanofibers. Reproduced with permission from [67]. Copyright 2016 the Institute of Physics.

12.3.3 Background conductance

The improved sensitivity enabled by the dc and ac bridge methods allows us to measure the small heat transfer signal from the heating to sensing beams. This provides an opportunity to precisely probe the background conductance (G_{BG}) of the devices, that is, the conductance between the heating and sensing membranes when there is no nanostructure specimen bridging them. This background conductance could be significant if it is comparable to the intrinsic conductance of the specimen (G_s) and needs to be subtracted from the measured total thermal conductance. This issue has been largely resolved by efforts by multiple groups [42, 44, 68].

G_{BG} can be subtracted using both indirect and direct methods. In the indirect method [44], the total thermal conductance of the nanowire and the background ($G_{NW} + G_{BG}$) was measured when there is a nanowire bridging the device (figure 12.7(a)). Then, the nanowire was removed, for instance, by cutting with a focused ion beam (FIB). The blank device with the nanowire removed was then measured again to yield G_{BG} (red triangles in figure 12.7(b)). From these two measurements, one can obtain G_s ($=(G_{NW} + G_{BG}) - G_{BG}$). As shown in figure 12.7(b), G_{BG} is about 200 pW K^{-1} at 300 K, and it increases with temperature. This G_{BG} is comparable to that of small-diameter specimens with low thermal conductivity, such as the 15 nm diameter Ge nanowire shown in figure 12.7(a), which has G_{NW} of about 250 pW K^{-1} at room temperature (figure 12.7(b)). Therefore, one needs to carefully quantify the G_{BG} when dealing with such types of specimens.

Alternatively, and perhaps more conveniently, G_{NW} can also be directly obtained by using the same differential bridge scheme shown earlier (figure 12.4). This method was first reported by Weathers *et al* using a dc heating-differential scheme [68] and was also implemented by Zheng *et al* [42] with the modulated heating differential scheme shown in section 12.3.2. Here the results from Zheng *et al* were shown for the sake of consistency (i.e. the direct method in Zheng *et al* [42] and the indirect method in Wingert *et al* [44] can be directly compared, as they were on devices with the same design). As shown in figure 12.7(c), the R_s of the working device with a nanowire

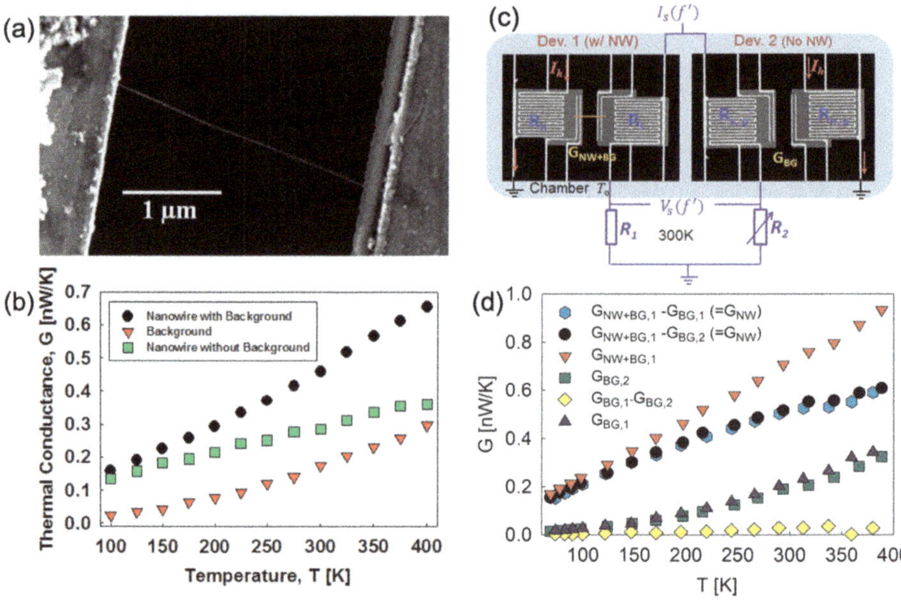

Figure 12.7. Measurement of background conductance (G_{BG}) using the indirect and direct methods. (a) An SEM image of a thin (15 nm diameter) Ge nanowire bridging two suspended membranes in a micro-device. (b) Measured total conductance of the nanowire with background, background conductance of the same device (after the nanowire was cut) and the extracted nanowire conductance. (c) Electrical circuit diagram of the differential scheme to directly obtain the nanowire conductance, by measuring the sensing side temperature increase of the device with a nanostructure specimen (Dev. 1) with respect to that of an identical blank device (Dev. 2). (d) Measured conductance showing the same G_{NW} results obtained from the direct and indirect methods. See the text for a detailed explanation. (a) and (b) reproduced with permission from [44]. Copyright 2012 the American Institute of Physics. (c) and (d) reproduced with permission from [42]. Copyright 2013 the American Institute of Physics.

(Dev. 1) is measured with respect to that of the pairing blank device with no nanowire (Dev. 2) using the same Wheatstone bridge configuration. The only difference is that the heating membrane of Dev. 2 ($R_{h,p}$) is also heated up by the same heating current as R_h in Dev. 1. Since the heating power between the two devices is identical, we can assume the same amount of heat conducted via G_{BG} to the sensing sides in both devices. As a result, the temperature increases and resistance changes due to G_{BG} in R_s and $R_{s,p}$ are also the same. The resistance increase on R_s relative to that of $R_{s,p}$, which are directly measured using the bridge circuit, is then solely due to the nanowire conductance. The validity of this direct differential measurement scheme, therefore, hinges on the similarities in G_{BG} between identical devices under the same ambient conditions (i.e. $G_{BG,1} = G_{BG,2}$). This has been confirmed in various control experiments. For instance, figure 12.7(d) shows that the background conductance of blank Dev. 1 and Dev. 2 are the same ($G_{BG,1} = G_{BG,2}$). Therefore, the nanowire conductance (G_{NW}) obtained from the direct differential scheme (black circles, $G_{NW} = G_{NW+BG,1} - G_{BG,2}$) is the same as that from the indirect scheme, i.e. subtracting $G_{BG,1}$ from the total conductance of

Dev. 1 with the nanowire ($G_{NW} = G_{NW+BG,1} - G_{BG,1}$, blue hexagons). The similarity between $G_{BG,1}$ and $G_{BG,2}$ is further demonstrated with zero (within the measurement sensitivity) conductance difference between the two blank devices directly measured using the bridge (yellow diamonds, $G_{BG,1} - G_{BG,2}$). This series of measurements shows that G_{BG} across different devices with identical designs are the same and the contribution from the G_{BG} can be directly subtracted in the bridge circuit by applying the same heating power on the heating membranes between the working and pairing devices. This result, namely, the consistent background conductance measured by both direct and indirect methods, was also shown in a recent work by Xu *et al* [69]. Xu's work also extended the measurement to high temperature (up to 740 K), where the background conductance is higher and its subtraction is necessary to obtain accurate results.

The relatively large background conductance naturally prompts one to wonder about its origin. Possible causes of the G_{BG} could be the heat conduction due to residual air molecules in the vacuum chamber and the radiation heat transfer. However, the former cause is not very likely based on the analysis of the mean free path of molecular air at the relevant pressure (e.g. MFP ~ 1 m at 0.1 mTorr). Figure 12.8(a) also experimentally shows the near constant room-temperature G_{BG} when the chamber pressure was below about 10 mTorr. G_{BG} was also measured over a long period on separate days after removing and replacing samples (first and second runs). At pressure around ~10^{-4} Torr with the turbo pump running, G_{BG} was measured to be around 215 ± 7 pW K^{-1}. This was verified over multiple runs and with multiple devices with the same gap distance between the heating and sending membranes (~4 μm) and the same beam length ($L_b = 400$ μm). At a pressure between 10 and 100 mTorr, G_{BG} started to increased slightly, by up to 20%. This increase is likely to have originated from the heat conduction by air molecules, as their MFP is around 1–10 mm within this pressure range, which is comparable to the device dimensions.

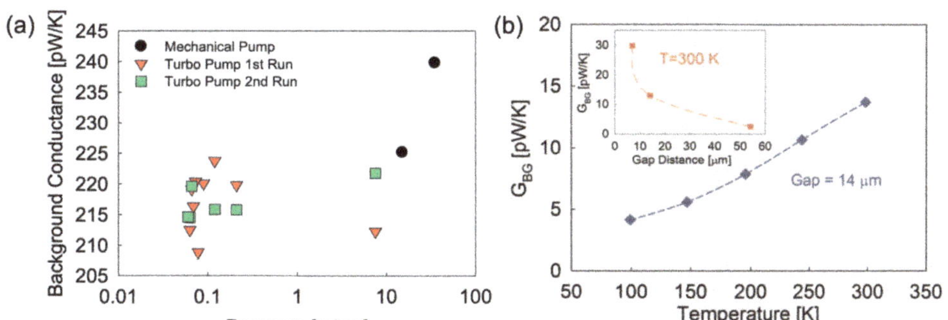

Figure 12.8. (1) Measured room-temperature background conductance (G_{BG}) versus chamber pressure for a suspended device. The data were collected from multiple devices with the same nominal gap distance between the heating and sending membranes (~4 μm) and the same beam length ($L_b = 400$ μm). (b) Measured G_{BG} as a function of temperature for a suspended device with a shorter beam length ($L_b = 50$ μm) and a longer gap distance between the two beams (14 μm). The inset shows the room-temperature G_{BG} as a function of gap distances. Reproduced with permission from [42]. Copyright 2013 the American Institute of Physics.

The other possible cause for G_{BG} is the radiation between the suspended membranes and/or beams. This seems to be consistent with several observations on the dependence of G_{BG} on the device geometry and temperature. When the beam was shortened to $L_b = 50\ \mu m$ and the gap between the two beams was increased to $14\ \mu m$, G_{BG} was greatly reduced. Figure 12.8(b) shows that the G_{BG} of this device is about 14 pW K^{-1} at 300 K, more than one order of magnitude lower than that of the device with $L_p = 400\ \mu m$, and the gap distance is $4\ \mu m$ (shown in figure 12.8(a)). Figure 12.8(b) and its inset further show that G_{BG} increases with temperature and decreases with increasing gap distance. These observations are consistent with the behaviors of the radiation heat transfer between the two suspended beams, with longer beams giving larger emission/receiving surfaces and the shorter gaps leading to a larger view factor. However, a quantitative analysis of the radiative heat transfer between two gray body beams (with emissivity of 0.88, see section 12.3.4) with the exact view factors shows that the measured G_{BG} is much larger than the expected values (e.g. measured $G_{BG} = 29.82$, 13.70 and 2.45 pW K^{-1} at 300 K for gap distances of 7, 14 and 54 μm, respectively, while the calculated values were 2.6, 1.2 and 0.24 pW K^{-1}, respectively) [42]. Clearly, the calculation using the view factor between the two suspended beams cannot explain the measured G_{BG}. Recently, theoretical and experimental studies [70, 71] discovered over 100-fold enhancement in radiative heat transfer compared to the blackbody limit calculated from the view factor between two dielectric pads. Fluctuational electrodynamics was able to explain the giant radiative conductance, which was attributed to the directional emission and absorption of the radiative energy between the two thin membranes. This directionality is caused by the propagating surface phonon polariton of the polar dielectric SiN_x pads.

12.3.4 Characterization of heat loss from suspended beams

In the previous section, we discussed the background conductance between the heating and sensing beams in a blank device which can be attributed to radiation heat transfer. The existence of the background conductance indicates that there is emissive heat loss from the heating and sensing beams, which prompts the question of the validity of the assumptions used in the heat transfer model, namely the parabolic and linear temperature distributions in the heating and sensing beams, respectively. This issue has been carefully examined by Weathers *et al* [68] and was also evaluated by Zheng *et al* [42]. The radiative heat loss from the beam can be measured as the effective heat transfer coefficient h. We used a simple fin model to extract both h and the thermal conductivity of the SiN_x beams (κ_b) with varying lengths (100, 200, 400 μm), as shown in figure 12.9(a). The top surface of the beams was coated with Pt for self-heating. We measured the average temperature rise ($\overline{\Delta T}$) of the self-heated beams (figure 12.9(b)), and obtained the following equation:

$$\frac{Q}{\overline{\Delta T}} = \frac{hPL}{\left[1 - \frac{2(\cosh(mL) - 1)}{mL\sinh(mL)}\right]}, \tag{12.16}$$

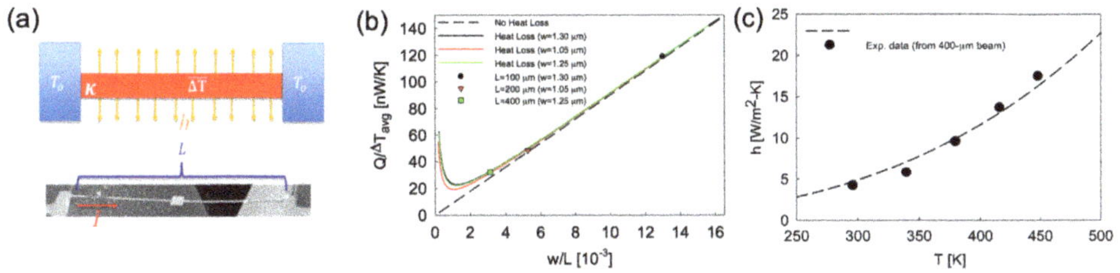

Figure 12.9. Thermal conductivity (κ_b) and heat transfer coefficient (h) determination of the SiN_x beams. (a) Schematic of the thermal fin model applied to a suspended beam of total length L self-heated by an electrical current I. The average temperature rise of the beam ($\overline{\Delta T}$) was measured using the coated Pt lines. (b) Measured (symbols) and fitted (lines) of the $\frac{Q}{\overline{\Delta T}}$ ratios as a function of w/L ratio. (c) Measured (dots) and calculated (dashed line, (12.17)) heat transfer coefficient (h) for the beam with $L = 400\ \mu\text{m}$. Reproduced with permission from [42]. Copyright 2013 the American Institute of Physics.

where Q is the total electrical power dissipated in the beam, A is the cross-sectional area of the beam, L is the total length of the beam (note that $L = 2L_b$), P is the perimeter of the SiN_x surfaces on the beam and m is the fin parameter, $m = \sqrt{\frac{hP}{k_b A}}$.

Figure 12.9(b) shows $Q/\overline{\Delta T}$ for beams with varying width/length ratios (w/L). For short beam lengths ($L < 200\ \mu\text{m}$), there is no difference in $Q/\overline{\Delta T}$ between the no heat loss case ($h = 0$) and the heat loss case ($h > 0$), which means the heat loss can be completely neglected for shorter beams. In this regime, one can extract the κ_b from equation (12.16) in the limit of $mL \to 0$. For longer beams, however, heat loss becomes increasingly significant, such that neglecting the heat loss can lead to an overestimated temperature increase (or a smaller $Q/\overline{\Delta T}$ ratio). With the knowledge of κ_b, we can calculate h from the $Q/\overline{\Delta T}$ data for longer beams (as shown in figure 12.9(b)) by using equation (12.16). Figure 12.9(c) shows the extracted h from 300 to 450 K, using the 400 μm long beam. In the limit of small temperature increase due to the self-heating (< 10 K), one can also estimate the heat transfer coefficient due to radiation heat transfer from the SiN_x beam to the ambient from

$$h = 4\varepsilon\sigma T^3, \tag{12.17}$$

where ε and σ are the emissivity (0.88 for SiN_x [72]) and Stefan–Boltzmann constant, respectively. The calculated h is also plotted as the dashed line in figure 12.9(c). The measured and calculated h values are in good agreement with each other, confirming that the heat loss from the suspended beams is primarily due to radiative thermal exchange with the sample surroundings.

12.3.5 Electron-beam heating

Thong *et al* developed a novel 'electron-beam heating' scheme to quantify the contact resistance of the nanostructure specimen and the suspended membranes [53, 73]. The basic principle of this scheme is to introduce a movable heating source with a focused electron beam (e-beam), and essentially measure the thermal resistance as a functional of length. As shown in figure 12.10(a), the focused e-beam can be moved

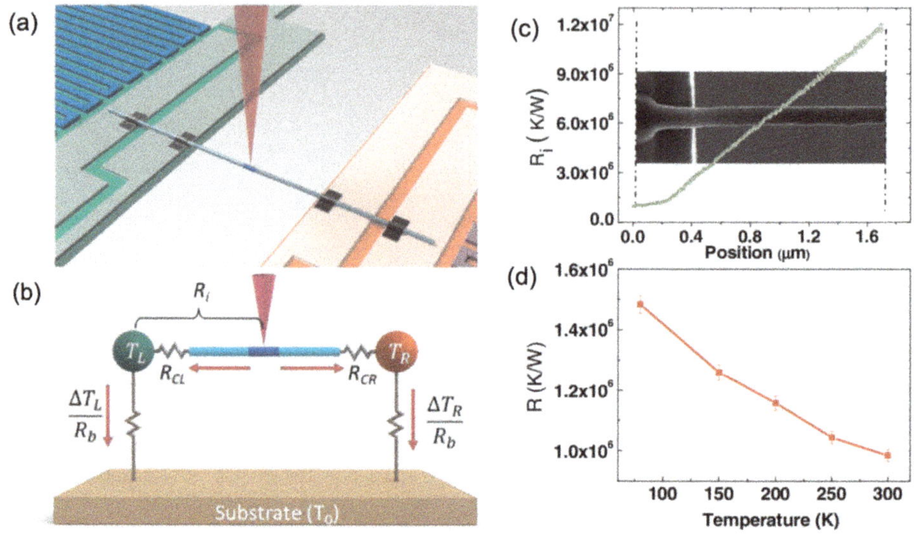

Figure 12.10. E-beam heating method to resolve the contact resistance. (a) Schematics of the experimental set-up for the e-beam heating technique. A focused e-beam (purple cone) inside an SEM chamber is incident on the suspended nanostructure and is moving along the nanostructure as well as the two suspended membranes. The e-beam can then be modeled as a moving point heat source. (b) The equivalent thermal resistance circuit. (c) Measured cumulative thermal resistance (R_i) from the left contact area between the nanowire/membrane (see inset) to the suspended section of a ZnO nanowire. (d) Temperature dependence of R_i (a) and (b) reproduced with permission from [73]. Copyright 2014 the American Chemical Society. (c) and (d) reproduced with permission from [53]. Copyright 2012 Wiley.

continuously on the suspended membrane and along the nanowire, with nanometer resolution. Figure 12.10(b) shows the equivalent thermal resistance circuit. R_{CL} and R_{CR} are the thermal contact resistances of the left and right membrane/nano-structure contacts, respectively. R_i is the cumulative thermal resistance from the left membrane to the e-beam, R_b is the thermal resistance of the suspending beams between the membranes and the substrates ($R_b = 1/G_b$), and ΔT_L and ΔT_R are the temperature increases measured on the left and right membranes, respectively. Here, R_i includes the thermal resistance of the membrane itself due to the temperature non-uniformity (heat spreading) in the membrane, left side contact resistance (R_{CL}) and the conduction resistance of the segment of the nanowire from the left junction to the e-beam heating spot. As the e-beam moves from left to right, the measured R_i would gradually increase, as shown in figure 12.10(c) for a ZnO nanowire. The R_i is shown from the membrane to a bridging ZnO nanowire, overlaid on the SEM image of the pad and the nanowire shown in the inset. The linear portion of the R_i versus position (from approximately 0.2 c) indicates the linear increase in the resistance of the nanowire as a function of length, as one would expect for diffusive transport. This slope can be used to obtain the intrinsic thermal conductivity of the nanowire ($\kappa = (A\frac{dR_i}{dL})^{-1}$). The left portion of the R_i versus position plot before it becomes linear (i.e. position <0.2 μm in figure 12.10(c)), which is about 1×10^6 K W^{-1} at 300 K, can be attributed to the sum of the internal resistance of the membrane and the contact resistance. Bui *et al* [53] show that the contact resistance of the nanowire/membrane junction, enhanced with the EBID Pt–C composite, is negligible compared to the internal resistance of the membrane. Figure 12.10(d) further shows the measured cumulative resistance (contact resistance plus the internal resistance of the membrane) as a function of temperature. The cumulative resistance decreases with temperature, presumably due to the increasing thermal conductivity of the membrane and also the lower contact resistance with temperature. This cumulative resistance is in the range of 1–2 $\times 10^6$ K W^{-1}, suggesting that it has a negligible impact on intrinsic nanowire thermal conductivity measurement as long as the wires are sufficiently thin and long and a proper contact enhancement technique is used (e.g. Pt–C by EBID), a conclusion that was similarly derived from earlier nanowire measurements [37, 54].

The measured cumulative resistance is on the same order of magnitude, but a factor of 3–5 smaller than what was measured by Lim *et al* [54] and Yang [21]. This difference may be caused by the different designs in the SiN_x membranes. The membranes in Lim *et al* [54] are about 2–3 times larger compared to Bui *et al* [53] and their measured contact resistance should also include the internal spreading resistance of the membrane. Also, the nanowire/membrane contact resistance is dependent on the exact contact areas, which could vary depending on many factors, such as how the EBID process was performed. Nevertheless, the work from both Bui *et al* [53] and Lim *et al* [54] showed that the contact resistance (including the internal resistance of the SiN_x membrane) is negligible if the EBID Pt–C is used and if the thermal resistance of the nanostructure is much larger than 10^6 K W^{-1}.

The e-beam heating method has the advantage of being able to measure the contact resistance with only one sample, as opposed to multiple samples or multiple lengths of a sample in Lim *et al* [54] and Yang *et al* [21]. Moreover, since the movement of the e-beam can be controlled with nanometer resolution, the technique is also capable of probing thermal resistance with the same spatial resolution. As a demonstration, Liu *et al* [73] showed the nanoscale mapping of thermal resistance across Si/silicide interfaces in a nanowire. However, the widespread application of this technique could be hindered by the need to use the e-beam, for example, in an SEM. Also, the method is not compatible with materials that are sensitive to e-beam damage, such as polymers. Care also needs to be taken to avoid damage or amorphous carbon coating on the nanostructures during the e-beam exposure. Finally, the e-beam heating is a dc heating technique, so the resolution of the thermometry and calorimetry could also be limited by the ambient temperature stability in the SEM chamber. This limitation could be eliminated if modulated e-beam heating can be introduced, for instance, via a beam blanker. The principle demonstrated by the technique is not limited to the e-beam; one could conceivably use a focused laser beam, which is much more accessible and can easily be modulated, and still possess a desirable spatial resolution (diffraction limited spot size, around hundreds of nanometers).

12.3.6 Four-point thermal measurement

Another approach to address the contact resistance issue is to use a four-point thermal measurement, in analogy to the four-probe current–voltage method in electrical measurements. This method was invented by Shi *et al* [74]. As shown in figure 12.11(a), four suspended straight Pt/SiN_x beams were used as heaters and thermometers. A nanostructure was placed around the middle of the four beams. The specimen shown in figure 12.11(a) was a fabricated Si nanoribbon with a rectangular cross-section 240 nm wide and 220 nm thick. Figure 12.11(b) shows the equivalent thermal circuit of the measurement set-up. $R_{b,i}$ is the thermal resistance of the ith beam ($i = 1, 2, 3, 4$). $R_{c,i}$ is the contact resistance between the nanowire and the ith beam. R_1, R_2, R_3 are the thermal resistances of the left, middle and right segments of the nanowire, respectively. This method is capable of obtaining all the $R_{c,i}$ and the thermal resistance of the middle segment of the nanowire (R_2) (as well as $R_1 + R_{c,1}$ and $R_2 + R_{c,2}$), as shown in figure 12.11(c). This was achieved by performing a total of four sets of measurements, each of which used one of the four beams as the heater while measuring the average temperature increases of all

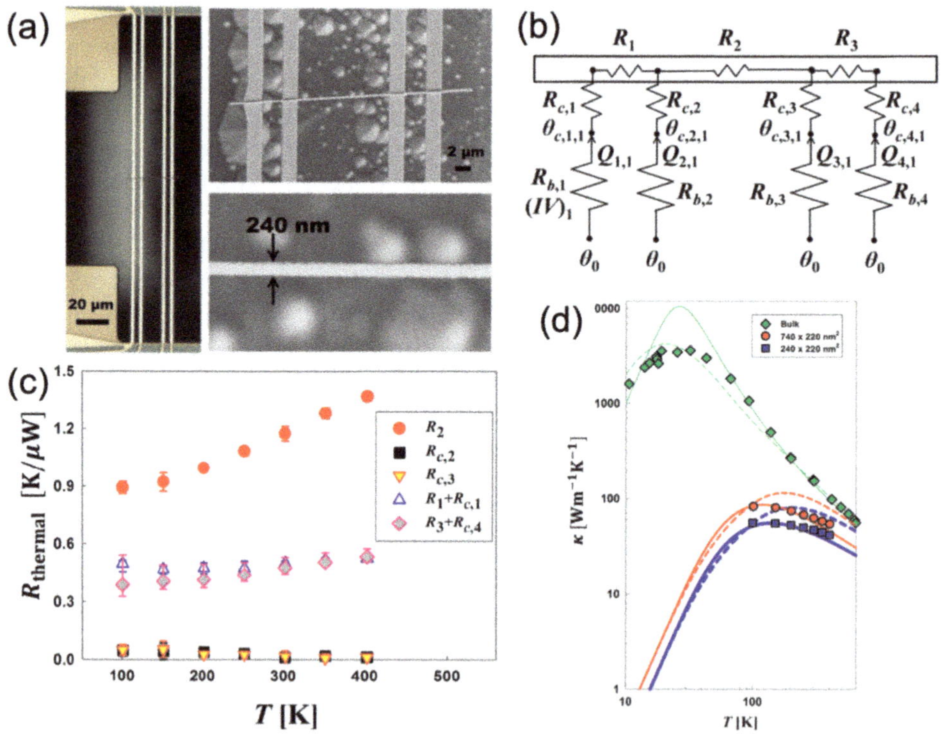

Figure 12.11. Four-probe thermal measurement method developed by Shi *et al* [74]. (a) Photograph and SEM images of the device with four straight Pt/SiN_x beams bridged by a Si nanoribbon. (b) Equivalent circuit of the device. (c) Measured thermal resistances as a function of the temperature. (d) Extracted thermal conductivity of the two Si nanoribbons (740 and 220 nm wide, and 220 nm thick). Reproduced with permission from [74]. Copyright 2015 the American Institute of Physics.

the beams. Therefore, there were a total of 16 measurement data sets, which were able to yield the thermal resistance mentioned above. Thus, this method is capable of measuring both the contact and intrinsic thermal resistance of a nanowire interfacing with the Pt/SiN_x beams using the electrothermal heating method, which is generally more accessible than the e-beam heating method. With this method, the intrinsic thermal conductivities of two relatively large Si nanoribbons (740 and 220 nm wide, and 220 nm thick) were obtained, which were in good agreement with the theoretical model results (figure 12.11(d)), showing the effectiveness of the four-probe thermal method.

Figure 12.11(c) also shows that the thermal contact resistances ($R_{c,1}$ and $R_{c,2}$)are negligible compared to the resistance of the middle suspended segment of the nanowire (R_2), due to the large contact area associated with the flat surface of the rectangular nanoribbon (as opposed to the common cylindrical nanowires, e.g. in [54]). This was also similarly observed in Si nanoribbons with smaller width, as studied by Li *et al* [33]. The interfacial thermal conductance was extracted to be from 38 to 230 × 10^6 W m^{-2} K^{-1} using the apparent contact area [74]. This value is fairly high, presumably due to the effect of the high temperature (350 °C) annealing on the Si/Pt interfaces [74], but it is also within the expected range for a vdW interface, as in the case of the cylindrical nanowire/Pt contacts discussed in section 12.2.3. Therefore, the contact resistance values obtained from the rectangular Si nanoribbons [33, 74] and from the cylindrical nanowires [54] in general agree with each other, if one takes into account the difference in the actual contact area.

12.3.7 Integrated devices

So far, we have discussed the experimental platform of placing a nanostructure on micro-devices to bridge the two suspended membranes. This method relies on a somewhat random and sometimes low-yield process of manipulating the nanostructures onto the suspended devices. For certain materials and structures, it could be desirable to have a higher device yield or to precisely control the dimension of the nanostructures. For some other measurements, such as the ones looking for non-diffusive thermal transport phenomena, it is desirable to completely eliminate the contact resistance, which becomes more significant at low temperature (figure 12.10(d) and figure 12.11(c)). In recent years, there have been several studies involving the 'integrated devices' version of the suspended micro-devices to attain these desirable features.

Figure 12.12(a) shows a fabricated micro-device with the bridging Si nanowires connected to the Si layers underneath the suspended SiN_x membranes [38]. Serpentine Pt resistors similar to the ones in the regular micro-devices were also fabricated on the membranes. The Si nanowires and the Si layers were fabricated from the same device layer of an SOI wafer, such that the two were monolithically connected to eliminate the thermal contact resistance. The nanowire width (40–150 nm) and length (4–100 μm) were defined by e-beam lithography (EBL), and the thickness was controlled by that of the device layer in the SOI wafer. Measurements on nanowires with variable lengths showed negligible thermal

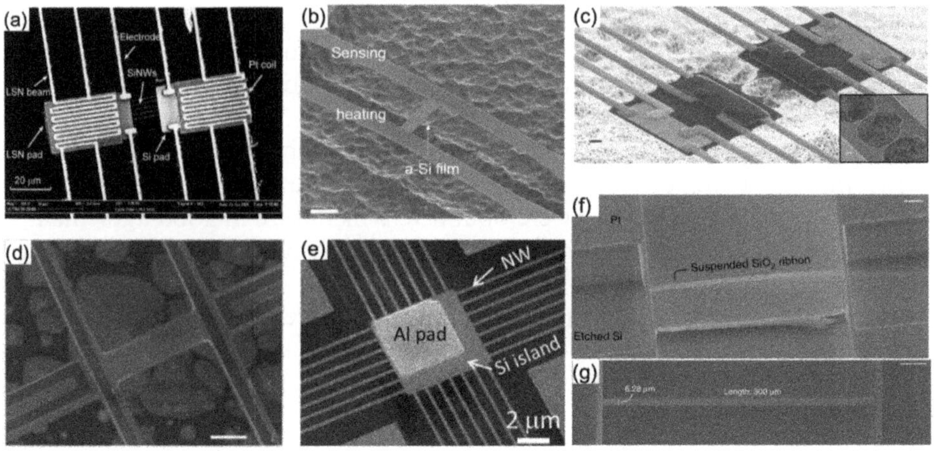

Figure 12.12. Examples of integrated devices. (a) Single-crystal Si nanowires fabricated by EBL and monolithically connected to the Si layers under the SiN_x membranes. Reproduced with permission from [38]. Copyright 2010 the American Chemical Society. (b) Amorphous Si ribbons bridging the heating and sensing beams for in-plane thermal conductivity measurement of amorphous Si films. The scale bar is 5 μm. Reproduced with permission from [75]. Copyright 2017 the American Chemical Society. (c) Low-temperature SiN_x phonon waveguides connected to the devices with catenoidal junctions. The scale bar is 2 μm. Inset: magnified image on the waveguides showing the catenoidal junctions. The scale bar is 1 μm. Reproduced with permission from [82]. Copyright 2018 Nature Publishing Group. (d) An integrated SiO_2 ribbon as a support for measured thermal conductivity of single layer graphene (SLG). The scale bar is 3 μm. Reproduced with permission from [39]. Copyright 2010 the American Association for the Advancement of Science. (e) An Al pad serving as a laser absorber and thermoreflectance transducer to measure thermal transport in nanostructures (shown here are Si nanowires) that bridge the pad and the substrate. Reproduced with permission from [85]. Copyright 2015 AIP Publishing. (f) A device with integrated SiO_2 nanoribbons with variable lengths to extract the thermal emissivity of the ribbons. The scale bar is 30 μm. Reproduced with permission from [89]. Copyright 2019 Nature Publishing Group.

contact resistance [38]. The measured intrinsic thermal conductivity of the EBL nanowires was found to be considerably smaller than that of the smooth VLS Si nanowires [37] but larger than that of the rougher electrolessly etched Si nanowires at the same critical diameters [5], which correlates well with the surface roughness [38]. Figure 12.11(b) shows a similar device platform for measuring in-plane thermal conductivity of amorphous Si thin films [75]. This approach was effective in eliminating the contact resistance experienced in cross-plane thermal conductivity measurement of amorphous Si thin films. The measured thermal conductivity showed a strong size dependence, indicating the prominent role of the propagon on the thermal conductivity and its broad mean free path distribution [76–81]. The thickest amorphous Si film (1.7 μm) exhibited room-temperature thermal conductivity over 5 W m^{-1} K^{-1}, with about 80% from the propagons and the rest from the diffusons. Figure 12.11(c) shows an integrated device to measure ballistic phonon transport in SiN_x waveguides at very low temperature (~0.1 K to 5 K) [82]. The inset of figure 12.11(c) shows the catenoid-shaped junctions between the waveguides and the membranes to enhance the phonon transmission, as first studied by Schwab [83].

To enable the low-temperature thermometry with high sensitivity, niobium nitride (NbN_x) was used, which has been shown to possess a much higher temperature coefficient of resistance compared to Pt at low temperature due to its semiconducting nature [43, 84]. Shi *et al* developed a device with an integrated SiO_2 ribbon, which served as a supporting substrate for single layer graphene (SLG) [39], as shown in figure 12.11(d). By measuring the thermal conductance of the ribbon with and without the SLG, the thermal conductivity of the graphene was extracted. The measured thermal conductivity of SLG was as high as ~600 W m^{-1} K^{-1} due to the two-dimensional phonon transport. In the past few years, Nomura *et al* developed an integrated platform with the nanostructures bridging a middle suspended pad and the substrate (figure 12.11(e)) [85–88]. The middle pad was covered with an Al layer to serve as the heater and thermometer (transducer) in a laser pump–probe scheme. They have used this platform to probe thermal transport in a variety of structures, mostly based on Si, including Si nanowire and phononic crystals. Lastly, Shin *et al* developed an integrated device platform to measure the thermal emissivity of nanoribbons [89]. Figures 12.11(f) and (g) show a device with an integrated suspended SiO_2 nanoribbon 100 nm thick and about 6 μm wide. By using a thermal fin model and measuring thermal transport along the ribbons with the same cross-sectional areas but with variable lengths (50–600 μm), both the intrinsic thermal conductivity and thermal emissivity of the ribbons could be extracted. The study showed an enhanced thermal emissivity of narrow ribbons compared to that of thin films with the same thickness, due to the coherent phonon polariton resonance effect [89]. This platform can be utilized to far-field thermal emission measurements on individual nanostructures.

12.4 Summary and outlook

In this chapter, we have reviewed the basic characteristics and recent developments of the suspended micro-developed platform. We discussed the noise sources and thermal contact resistance associated with this platform. Earlier work revealed that the dominant thermometry noise originated from the ambient temperature fluctuation. This noise has been greatly suppressed to approach the instrumentation noise limit (Johnson noise), by using the differential bridge scheme and the modulated heating method. By doing so, specimens with low thermal conductance could be measured. The noise equivalent conductance was found to be ~10 pW K^{-1} with the dc heating–differential bridge scheme, and down to <1 pW K^{-1} with the modulated heating scheme. Recent work has also quantified and found ways to subtract the background thermal conductance, which could be important for specimens with low thermal conductance. Both the noise reduction strategy and background conductance measurement would enable the measurement of nanostructures with low thermal conductance below 1 pW K^{-1}.

For nanostructures with high thermal conductance, the impact of the thermal contact resistance always needs to be carefully considered and mitigated. The contact resistance was found to be generally lower than 3–5 10^6 K W^{-1} if a proper contact enhancement method was used, such as EBID of Pt and capillary-assisted adhesion promotion. Therefore, for nanostructures with conduction thermal resistance much greater than 10^6 K W^{-1}, the impact of the thermal contact resistance is minimal. However, for structures with potentially low thermal resistance, such as high thermal conductivity materials or large cross-sectional area, the measurements could be influenced by the contact resistance. Therefore, great care still needs to be taken when dealing with the contacts for this measurement. Recent developments in the e-beam heating method and four-probe thermal method have both successfully quantified the thermal contact resistance and extracted the intrinsic thermal conductivity of the nanostructures.

Finally, several versions of the integrated devices, that is, suspended micro-devices with integrated nanostructures, have been developed recently and utilized to explore interesting transport phenomena, such as low-temperature (0.1–5 K) one-dimensional ballistic phonon wave guiding, two-dimensional phonon transport in graphene and enhanced far-field thermal emission from polar dielectric nanoribbons.

Due to its capability of measuring thermal transport in individual nanostructures, the suspended micro-device platform will continue to be used. Recent developments have made the platform more accurate and sensitive. However, innovation in instrumentations should not stop. New features can be added onto the platform for studying new thermal transport phenomena. For example, the platform can be integrated with the ultrafast pump–probe technique, to eliminate the influence of the contact resistance, as in the case of e-beam heating but with more accessible and easily modulated lasers. The optical method can also introduce a new knob in the time domain or frequency domain to probe non-diffusive thermal transport, which has only been measured in several nanostructures based on length-dependent steady-state thermal conductance that could be complicated by contact resistance [90–92]. The integrated devices have also shown great promise for exploring interesting thermal transport physics. It is likely that more work will be carried out along this direction, for example, to study wave guiding effects and thermal transport phenomena with coupled phonon and other energy carriers (e.g. photon, electrons, magnons).

Acknowledgments

The authors acknowledge financial support from the National Science Foundation (CBET-1140 121, CBET-1336 428, DMR-1508 420 and CMMI-1762 560), the Solar Energy Technology Office, US Department of Energy (Award Nos. DE-EE0008529, DE-EE0008379, DE-EE0007113 and DE-EE0005802), a Hellman Faculty Award and an ACS-PRF grant (54 109-ND10).

References

[1] Tans S J, Verschueren A R M and Dekker C 1998 Room-temperature transistor based on a single carbon nanotube *Nature* **393** 49–52

[2] Shulaker M M, Hills G, Patil N, Wei H, Chen H Y, PhilipWong H S and Mitra S 2013 Carbon nanotube computer *Nature* **501** 526

[3] Huang Y, Duan X F, Cui Y, Lauhon L J, Kim K H and Lieber C M 2001 Logic gates and computation from assembled nanowire building blocks *Science* **294** 1313–7
[4] Tian B Z, Zheng X L, Kempa T J, Fang Y, Yu N F, Yu G H, Huang J L and Lieber C M 2007 Coaxial silicon nanowires as solar cells and nanoelectronic power sources *Nature* **449** 885
[5] Hochbaum A I, Chen R K, Delgado R D, Liang W J, Garnett E C, Najarian M, Majumdar A and Yang P D 2008 Enhanced thermoelectric performance of rough silicon nanowires *Nature* **451** 163
[6] Boukai A I, Bunimovich Y, Tahir-Kheli J, Yu J K, Goddard W A and Heath J R 2008 Silicon nanowires as efficient thermoelectric materials *Nature* **451** 168–71
[7] Wang Z L and Song J H 2006 Piezoelectric nanogenerators based on zinc oxide nanowire arrays *Science* **312** 242–6
[8] Su Y D, Liu C, Brittman S, Tang J Y, Fu A, Kornienko N, Kong Q and Yang P D 2016 Single-nanowire photoelectrochemistry *Nat. Nanotechnol.* **11** 609
[9] Chan C K, Peng H L, Liu G, McIlwrath K, Zhang X F, Huggins R A and Cui Y 2008 High-performance lithium battery anodes using silicon nanowires *Nat. Nanotechnol.* **3** 31–5
[10] Cui Y, Wei Q Q, Park H K and Lieber C M 2001 Nanowire nanosensors for highly sensitive and selective detection of biological and chemical species *Science* **293** 1289–92
[11] Cahill D G, Ford W K, Goodson K E, Mahan G D, Majumdar A, Maris H J, Merlin R and Sr P 2003 Nanoscale thermal transport *J. Appl. Phys.* **93** 793–818
[12] Cahill D G *et al* 2014 Nanoscale thermal transport. II. 2003–2012 *Appl. Phys. Rev.* **1** 011305
[13] Cahill D G, Goodson K E and Majumdar A 2002 Thermometry and thermal transport in micro/nanoscale solid-state devices and structures *J. Heat Trans.* **124** 223–41
[14] Cahill D G 2002 Erratum: "Thermal conductivity measurement from 30 to 750 K: the 3ω method" [*Rev. Sci. Instrum.* **61**, 802 (1990)] *Rev. Sci. Instrum.* **73** 3701
[15] Majumdar A 1999 Scanning thermal microscopy *Annu. Rev. Mater. Sci.* **29** 505–85
[16] Cahill D G 2004 Analysis of heat flow in layered structures for time-domain thermoreflectance *Rev. Sci. Instrum.* **75** 5119–22
[17] Schmidt A J, Cheaito R and Chiesa M 2009 A frequency-domain thermoreflectance method for the characterization of thermal properties *Rev. Sci. Instrum.* **80** 094901
[18] Shi L, Li D Y, Yu C H, Jang W Y, Kim D, Yao Z, Kim P and Majumdar A 2003 Measuring thermal and thermoelectric properties of one-dimensional nanostructures using a microfabricated device *J. Heat Trans.* **125** 881–8
[19] Kim P, Shi L, Majumdar A and McEuen P L 2001 Thermal transport measurements of individual multiwalled nanotubes *Phys. Rev. Lett.* **87** 215502
[20] Yang J K *et al* 2014 Phonon transport through point contacts between graphitic nanomaterials *Phys. Rev. Lett.* **112** 205901
[21] Yang J K, Yang Y, Waltermire S W, Gutu T, Zinn A A, Xu T T, Chen Y F and Li D Y 2011 Measurement of the intrinsic thermal conductivity of a multiwalled carbon nanotube and its contact thermal resistance with the substrate *Small* **7** 2334–40
[22] Yu C H, Shi L, Yao Z, Li D Y and Majumdar A 2005 Thermal conductance and thermopower of an individual single-wall carbon nanotube *Nano Lett.* **5** 1842–6
[23] Chang C W, Okawa D, Majumdar A and Zettl A 2006 Solid-state thermal rectifier *Science* **314** 1121–4
[24] Chen R, Hochbaum A I, Murphy P, Moore J, Yang P D and Majumdar A 2008 Thermal conductance of thin silicon nanowires *Phys. Rev. Lett.* **101** 105501

[25] Zhang Q *et al* 2018 Thermal transport in quasi-1D van der Waals crystal $Ta_2Pd_3Se_8$ nanowires: size and length dependence *ACS Nano* **12** 2634–42
[26] Shi L, Hao Q, Yu C H, Mingo N, Kong X Y and Wang Z L 2004 Thermal conductivities of individual tin dioxide nanobelts *Appl. Phys. Lett.* **84** 2638–40
[27] Roh J W, Hippalgaonkar K, Ham J H, Chen R K, Li M Z, Ercius P, Majumdar A, Kim W and Lee W 2011 Observation of anisotropy in thermal conductivity of individual single-crystalline bismuth nanowires *ACS Nano* **5** 3954–60
[28] Zhou F, Szczech J, Pettes M T, Moore A L, Jin S and Shi L 2007 Determination of transport properties in chromium disilicide nanowires via combined thermoelectric and structural characterizations *Nano Lett.* **7** 1649–54
[29] Zhou J H, Jin C G, Seol J H, Li X G and Shi L 2005 Thermoelectric properties of individual electrodeposited bismuth telluride nanowires *Appl. Phys. Lett.* **87** 133109
[30] Mavrokefalos A, Moore A L, Pettes M T, Shi L, Wang W and Li X G 2009 Thermoelectric and structural characterizations of individual electrodeposited bismuth telluride nanowires *J. Appl. Phys.* **105** 104318
[31] Wingert M C, Chen Z C Y, Dechaumphai E, Moon J, Kim J H, Xiang J and Chen R K 2011 Thermal conductivity of Ge and Ge–Si core–shell nanowires in the phonon confinement regime *Nano Lett.* **11** 5507–13
[32] Wingert M C, Kwon S, Hu M, Poulikakos D, Xiang J and Chen R K 2015 Sub-amorphous thermal conductivity in ultrathin crystalline silicon nanotubes *Nano Lett.* **15** 2605–11
[33] Yang L *et al* 2016 Thermal conductivity of individual silicon nanoribbons *Nanoscale* **8** 17895–901
[34] Li D Y, Wu Y, Fan R, Yang P D and Majumdar A 2003 Thermal conductivity of Si/SiGe superlattice nanowires *Appl. Phys. Lett.* **83** 3186–8
[35] Lee E K *et al* 2012 Large thermoelectric figure-of-merits from sige nanowires by simultaneously measuring electrical and thermal transport properties *Nano Lett.* **12** 2918–23
[36] Zhang Q *et al* 2017 Defect facilitated phonon transport through kinks in boron carbide nanowires *Nano Lett.* **17** 3550–5
[37] Li D Y, Wu Y Y, Kim P, Shi L, Yang P D and Majumdar A 2003 Thermal conductivity of individual silicon nanowires *App. Phys. Lett.* **83** 2934–6
[38] Hippalgaonkar K, Huang B L, Chen R K, Sawyer K, Ercius P and Majumdar A 2010 Fabrication of microdevices with integrated nanowires for investigating low-dimensional phonon transport *Nano Lett.* **10** 4341–8
[39] Seol J H *et al* 2010 Two-dimensional phonon transport in supported graphene *Science* **328** 213–6
[40] Shi L, Li D Y, Yu C H, Jang W Y, Kim D Y, Yao Z, Kim P and Majumdar A 2003 Erratum: "Measuring thermal and thermoelectric properties of one-dimensional nanostructures using a microfabricated device" [*Journal of Heat Transfer*, 2003, **125**(5), pp. 881–888] *J. Heat Trans.* **125** 1209
[41] Chen R 2008 Nanowires for thermal energy conversion and management *Mechanical Engineering.* (Berkeley, CA: University of California)
[42] Zheng J L, Wingert M C, Dechaumphai E and Chen R K 2013 Sub-picowatt/kelvin resistive thermometry for probing nanoscale thermal transport *Rev. Sci. Instrum.* **84** 114901
[43] Dechaumphai E and Chen R K 2014 Sub-picowatt resolution calorimetry with niobium nitride thin-film thermometer *Rev. Sci. Instrum.* **85** 094903

[44] Wingert M C, Chen Z C Y, Kwon S, Xiang J and Chen R K 2012 Ultra-sensitive thermal conductance measurement of one-dimensional nanostructures enhanced by differential bridge *Rev. Sci. Instrum.* **83** 024901
[45] Hippalgaonkar K, Seol J H, Xu D Y and Li D Y 2017 Experimental studies of thermal transport in nanostructures *Thermal Transport in Carbon-Based Nanomaterials.* ed G Zhang (Amsterdam: Elsevier), pp 319–57
[46] Yang L 2019 *Phonon Transport in Nanowires—Beyond Classical Size Effects, in Mechanical Engineering.* (Nashville, TN: Vanderbilt University)
[47] Bahadur V, Xu J, Liu Y and Fisher T S 2005 Thermal resistance of nanowire–plane interfaces *J. Heat Trans.* **127** 664–8
[48] Bahadur V, Xu J, Liu Y and Fisher T S 2006 Erratum: "Thermal resistance of nanowire-plane interfaces" [*Journal of Heat Transfer*, 2005, **127**(6), pp. 664–668] *J. Heat Trans.* **128** 858
[49] Yu C H, Saha S, Zhou J H, Shi L, Cassell A M, Cruden B A, Ngo Q and Li J 2006 Thermal contact resistance and thermal conductivity of a carbon nanofiber *J. Heat Trans.* **128** 234–9
[50] Zhong Z X, Wingert M C, Strzalka J, Wang H H, Sun T, Wang J, Chen R K and Jiang Z 2014 Structure-induced enhancement of thermal conductivities in electrospun polymer nanofibers *Nanoscale* **6** 8283–91
[51] Prasher R 2005 Predicting the thermal resistance of nanosized constrictions *Nano Lett.* **5** 2155–9
[52] Xia M G, Cheng Z F, Han J Y, Zheng M R, Sow C H, Thong J T L, Zhang S L and Li B W 2014 Gallium ion implantation greatly reduces thermal conductivity and enhances electronic one of ZnO nanowires *AIP Adv.* **4** 057128
[53] Bui C T, Xie R G, Zheng M R, Zhang Q X, Sow C H, Li B W and Thong J T L 2012 Diameter-dependent thermal transport in individual Zno nanowires and its correlation with surface coating and defects *Small* **8** 738–45
[54] Lim J W, Hippalgaonkar K, Andrews S C, Majumdar A and Yang P D 2012 Quantifying surface roughness effects on phonon transport in silicon nanowires *Nano Lett.* **12** 2475–82
[55] Shrestha R *et al* 2018 Crystalline polymer nanofibers with ultra-high strength and thermal conductivity *Nat. Commun.* **9** 1664
[56] Park Y, Lee S, Ha S S, Alunda B, Noh D Y, Lee Y J, Kim S and Seol J H 2019 Crosslinking effect on thermal conductivity of electrospun poly(acrylic acid) nanofibers *Polymers* **11** 858
[57] Park Y, You M, Shin J, Ha S, Kim D, Heo M H, Nah J, Kim Y A and Seol J H 2019 Thermal conductivity enhancement in electrospun poly(vinyl alcohol) and poly(vinyl alcohol)/cellulose nanocrystal composite nanofibers *Sci. Rep.* **9** 3026
[58] Ma J, Zhang Q, Mayo A, Ni Z H, Yi H, Chen Y F, Mu R, Bellan L M and Li D Y 2015 Thermal conductivity of electrospun polyethylene nanofibers *Nanoscale* **7** 16899–908
[59] Ma J, Zhang Q, Zhang Y, Zhou L, Yang J K and Ni Z H 2016 A rapid and simple method to draw polyethylene nanofibers with enhanced thermal conductivity *Appl. Phys. Lett.* **109** 033101
[60] Zhang Y, Zhang X, Yang L, Zhang Q, Fitzgerald M L, Ueda A, Chen Y F, Mu R, Li D Y and Bellan L M 2018 Thermal transport in electrospun vinyl polymer nanofibers: effects of molecular weight and side groups *Soft Matt.* **14** 9534–41
[61] Canetta C, Guo S and Narayanaswamy A 2014 Measuring thermal conductivity of polystyrene nanowires using the dual-cantilever technique *Rev. Sci. Instrum.* **85** 104901
[62] Zeng X L, Xiong Y C, Fu Q, Sun R, Xu J B, Xu D Y and Wong C P 2017 Structure-induced variation of thermal conductivity in epoxy resin fibers *Nanoscale* **9** 10585–9

[63] Balandin A and Wang K L 1998 Significant decrease of the lattice thermal conductivity due to phonon confinement in a free-standing semiconductor quantum well *Phys. Rev.* B **58** 1544–9
[64] Sadat S, Meyhofer E and Reddy P 2012 High resolution resistive thermometry for micro/nanoscale measurements *Rev. Sci. Instrum.* **83** 084902
[65] Zheng J L, Wingert M C, Dechaumphai E and Chen R K 2013 Sub-picowatt/kelvin resistive thermometry for probing nanoscale thermal transport *Rev. Sci. Instrum.* **84** 114901
[66] Sadat S, Meyhofer E and Reddy P 2013 Resistance thermometry-based picowatt-resolution heat-flow calorimeter *Appl. Phys. Lett.* **102** 163110
[67] Zheng J L, Wingert M C, moon J and Chen R K 2016 Simultaneous specific heat and thermal conductivity measurement of individual nanostructures *Semicond. Sci. Technol.* **31** 084005
[68] Weathers A, Bi K D, Pettes M T and Shi L 2013 Re-examination of thermal transport measurements of a low-thermal conductance nanowire with a suspended micro-device *Rev. Sci. Instrum.* **84** 084903
[69] Wang X M, Yang J K, Xiong Y C, Huang B L, Xu T T, Li D Y and Xu D Y 2018 Measuring nanowire thermal conductivity at high temperatures *Meas. Sci. Technol.* **29** 025001
[70] Fernandez-Hurtado V, Fernandez-Dominguez A I, Feist J, Garcia-Vidal F J and Cuevas J C 2018 Super-Planckian far-field radiative heat transfer *Phys. Rev.* B **97** 045408
[71] Thompson D, Zhu L X, Mittapally R, Sadat S, Xing Z, McArdle P, Qazilbash M M, Reddy P and Meyhofer E 2018 Hundred-fold enhancement in far-field radiative heat transfer over the blackbody limit *Nature* **561** 216
[72] Ravindra N M, Abedrabbo S, Chen W, Tong F M, Nanda A K and Speranza A C 1998 Temperature-dependent emissivity of silicon-related materials and structures *IEEE Trans. Semicond. Manuf.* **11** 30–9
[73] Liu D, Xie R G, Yang N, Li B W and Thong J T L 2014 Correction to Profiling nanowire thermal resistance with a spatial resolution of nanometers *Nano Lett.* **14** 4195
[74] Kim J, Ou E, Sellan D P and Shi L 2015 A four-probe thermal transport measurement method for nanostructures *Rev. Sci. Instrum.* **86** 044901
[75] Kwon S, Zheng J L, Wingert M C, Cui S and Chen R K 2017 Unusually high and anisotropic thermal conductivity in amorphous silicon nanostructures *ACS Nano* **11** 2470–6
[76] Regner K T, Sellan D P, Su Z H, Amon C H, McGaughey A J H and Malen J A 2013 Broadband phonon mean free path contributions to thermal conductivity measured using frequency domain thermoreflectance *Nat. Commun.* **4** 1640
[77] Larkin J M and McGaughey A J H 2014 Thermal conductivity accumulation in amorphous silica and amorphous silicon *Phys. Rev.* B **89** 144303
[78] DeAngelis F, Muraleedharan M G, moon J, Seyf H R, Minnich A J, McGaughey A J H and Henry A 2019 Thermal transport in disordered materials *Nanoscale Microscale Thermophys. Eng.* **23** 81–116
[79] moon J, Latour B and Minnich A J 2018 Propagating elastic vibrations dominate thermal conduction in amorphous silicon *Phys. Rev.* B **97** 024201
[80] Braun J L, Baker C H, Giri A, Elahi M, Artyushkova K, Beechem T E, Norris P M, Leseman Z C, Gaskins J T and Hopkins P E 2016 Size effects on the thermal conductivity of amorphous silicon thin films *Phys. Rev.* B **93** 140201

[81] Wingert M C, Zheng J L, Kwon S and Chen R K 2016 Thermal transport in amorphous materials: a review *Semicond. Sci. Technol.* **31** 113003

[82] Tavakoli A, Lulla K, Crozes T, Mingo N, Coilin E and Bourgeois O 2018 Heat conduction measurements in ballistic 1D phonon waveguides indicate breakdown of the thermal conductance quantization *Nat. Commun.* **9** 4287

[83] Schwab K, Henriksen E A, Worlock J M and Roukes M L 2000 Measurement of the quantum of thermal conductance *Nature* **404** 974–7

[84] Bourgeois O, Andre E, Macovei C and Chaussy J 2006 Liquid nitrogen to room-temperature thermometry using niobium nitride thin films *Rev. Sci. Instrum.* **77** 126108

[85] Nomura M, Nakagawa J, Kage Y, Maire J, Moser D and Paul O 2015 Thermal phonon transport in silicon nanowires and two-dimensional phononic crystal nanostructures *Appl. Phys. Lett.* **106** 143102

[86] Anufriev R, Ramiere A, Maire J and Nomura M 2017 Heat guiding and focusing using ballistic phonon transport in phononic nanostructures *Nat. Commun.* **8** 15505

[87] Maire J, Anufriev R and Nomura M 2017 Ballistic thermal transport in silicon nanowires *Sci. Rep.* **7** 41794

[88] Maire J, Anufriev R, Yanagisawa R, Ramiere A, Volz S and Nomura M 2017 Heat conduction tuning by wave nature of phonons *Sci. Adv.* **3** e1700027

[89] Shin S M, Elzouka M, Prasher R and Chen R K 2019 Far-field coherent thermal emission from polaritonic resonance in individual anisotropic nanoribbons *Nat. Commun.* **10** 1377

[90] Hsiao T K, Chang H K, Liou S C, Chu M W, Lee S C and Chang C W 2013 Observation of room-temperature ballistic thermal conduction persisting over 8.3 μm SiGe nanowires *Nat. Nanotechnol.* **8** 534–8

[91] Xu X F *et al* 2014 Length-dependent thermal conductivity in suspended single-layer graphene *Nat. Commun.* **5** 3689

[92] Lee V, Wu C H, Lou Z X, Lee W L and Chang C W 2017 Divergent and ultrahigh thermal conductivity in millimeter-long nanotubes *Phys. Rev. Lett.* **118** 135901

Chapter 13

Recent advances in structured surface enhanced condensation heat transfer

Hyeongyun Cha, Soumyadip Sett, Patrick Birbarah, Tarek Gebrael, Junho Oh and Nenad Miljkovic

Condensation is a ubiquitous process in nature and industry. Since the 1930s, researchers have attempted to improve the efficiency of condensation due to its particular importance in energy production and efficiency. Recent developments in promoter deposition techniques have offered new opportunities to satisfy the need for phase change heat transfer performance, low cost and integration penalty, and high durability, which have been bottlenecks in industrial implementation for the past eight decades. Furthermore, rationally designed micro/nanoengineered surfaces have opened up the potential to create stable dropwise condensation of refrigerants as well as enable coalescence-induced droplet shedding. In this review, we present a broad overview of promising surface functionalization techniques for next-generation condensing surfaces. We also discuss recent insights gained for tailoring the dropwise condensation of low-surface-tension fluids and using electric fields to enhance condensation heat transfer.

13.1 Introduction

Condensation heat transfer is an essential process found in a plethora of energy–water nexus applications such as power generation [1, 2], heating, ventilation, air conditioning and refrigeration (HVAC&R) systems [3, 4], electronics thermal management [5], energy harvesting [6], water desalination [7, 8] and water harvesting [9, 10]. The majority of condenser surfaces for existing applications are made of metals such as stainless steel, copper, or aluminum. On such metallic surfaces, the condensing vapor forms a film of liquid condensate, called filmwise condensation (FWC). The metallic substrates and their oxides have higher surface energy (>1 J m^{-2}), leading to a lower condensate advancing contact angle, higher contact angle hysteresis and reduced activation energy for heterogeneous nucleation. The condensate liquid film can grow

doi:10.1088/978-0-7503-1738-2ch13

up to ~100 μm in thickness, deteriorating the heat transfer rate by acting as a thermal resistance due to the low thermal conductivity of typical condensates ($k_{H_2O} = 0.6$ W m^{-1} · K^{-1}) [11]. In contrast, condensing water vapor forms distinct droplets on lower surface energy substrates (<20 mJ m^{-1}). The larger difference in affinity between polar water molecules (a surface tension of $\gamma = 72.4$ mN m^{-1}) and the low-surface-energy substrates results in higher droplet contact angles and a higher thermodynamic energy barrier for heterogeneous nucleation. Termed dropwise condensation (DWC), continuous removal of condensate occurs via droplet shedding and re-nucleation. On suitably promoted hydrophobic surfaces, DWC yields a 5–10 times higher heat transfer coefficient (HTC) than FWC [12].

In recent decades, scalable superhydrophobic surfaces achieving DWC were developed based on advancements in micro/nanoscale fabrication techniques and low-surface-energy coating methods. In the late 2000s, self-propelled jumping-droplet condensate with water was discovered on suitably designed superhydrophobic surfaces [13]. Upon coalescence, droplets leap from the surface via the conversion of excess surface energy to kinetic energy [14]. Jumping-droplet condensation promotes removal of condensate resulting in further enhancements in HTC compared to DWC [15–24]. However, the lifetime of current hydrophobic and structured superhydrophobic coatings during DWC and jumping-droplet condensation (~months), respectively, is still limited compared to the lifetime of heat transfer surfaces (~years) in practice.

In this chapter, we will review recent advances in the development of structured surfaces and low-surface-energy coating materials to enhance condensation heat transfer with water and other low-surface-tension fluids. We broadly define here the term 'structured surface' as a surface having micro, and nanostructures, or nano-engineered features. Note, this review only comments briefly on industrially relevant macroscale structured surfaces sometimes termed 'enhanced surfaces' or 'tech tubes' [25–27], and is not meant to be a thorough review of currently implemented industrial solutions. Furthermore, this chapter is not intended to cover every topic in condensation heat transfer or superhydrophobicity, and mainly focuses on scientific challenges to the condensation heat transfer community with an emphasis on DWC. Specifically, we focus on the durability of the low-surface-energy coatings and the surface structures, difficulty in achieving DWC of low-surface-tension fluids, the recent discovery of the importance of ambient conditions and hydrocarbon adsorption on condensation dynamics, and device scale-up focusing on electric field enhanced (EFE) condensation heat transfer. Moreover, this review focuses on condensation on external surfaces, which represents the vast majority of the structured surface scientific literature conducted in the past decade. Lastly, with the publication of many great reviews in the past decade on structured surface enhanced phase change heat transfer [12, 28–38], we differentiate our work here by focusing on what we deem is needed with a critical and selective view of promising solutions and methods.

13.2 Advancements in coating materials and the durability of coatings

Dropwise condensation on industrial condenser materials (e.g. copper, aluminum, steel and stainless steel) can be achieved by coating with a low-energy non-wetting hydrophobic promoter material. Since its discovery in 1930, no satisfactory approaches for promoting DWC under industrial conditions have been demonstrated on the timescales needed to warrant their use. Although research activity into DWC has waned over the years, recent advances in successful promoter deposition techniques have revived interest. New developments and potential hydrophobic promoter materials are discussed in the following subsections.

13.2.1 Self-assembled monolayers

Self-assembled monolayers (SAMs) are organic molecules which spontaneously chemisorb to select substrates. SAMs have received considerable attention for their hydrophobicity and capability to promote DWC. A secondary significant advantage of SAMs is their negligible thermal resistance due to their molecular-level thickness (~1 nm) [39–44]. In DWC of steam, SAMs have been used to functionalize a variety of surfaces using fatty acids [43, 45, 46], thiols [47–52] and silanes [15, 44, 53–57]. Yet, long-term durability during steam condensation has not been demonstrated regardless of the types of SAMs used [43, 58–60]. Although carbine based formulations have recently been developed demonstrating significant robustness and thermal stability [61], their durability to steam condensation remains to be demonstrated.

13.2.2 Polymers

Polymer coatings have been used to promote DWC for many decades [12, 14, 62–67]. Although polymer coatings show sufficient durability to steam DWC when thick (>10 μm), they also suffer from two to four orders of magnitude lower thermal conductivity (~0.1 W m^{-1}·K^{-1}) compared to the metallic substrates they reside on during DWC. The low thermal conductivity results in significantly lower condensation heat transfer performance at the polymer thicknesses needed to achieve suitable longevity [68, 69]. Recent advances in polymer deposition techniques, such as plasma-enhanced chemical vapor deposition (PECVD) and initiated chemical vapor deposition (iCVD), have enabled the super-conformal growth of ultrathin polymer coatings.

Parylene is a polymer of para-xylylene which can be vapor-deposited using PECVD to achieve ultrathin (10 nm) and super-conformal films. The films are pinhole free, cross linked, chemically inert and physically stable with high dielectric breakdown [70]. Parylene shows great thermal stability at continuous temperatures as high as 130 °C in air or remains stable at 220 °C in the absence of oxygen. Parylene has very good mechanical durability in a wide range of temperatures from 200 °C to 275 °C. Furthermore, Parylene films have low permeability to moisture and other corrosive gases. Parylene has been used as a protective coating for more than three decades and its applications have widened in several industries. Parylene coatings have high potential as a more durable solution in phase change applications. From the automobile to medical and electronic industries, Parylene has enjoyed continued success [71–73]. However, a limited number of condensation experiments have been conducted with this promising polymer. The Parylene coatings (Parylene N and D) produced stable dropwise condensation with good durability but only showed heat transfer enhancement factors of approximately 2 to 3 when compared to filmwise condensation due to the larger coating thicknesses ($>1\ \mu$m) utilized in past studies [74–76]. Parylene C, which is modified Parylene N by substitution of a chlorine atom for one of the aromatic hydrogens, is one of the most promising candidates among the Parylene series for phase change systems since it provides excellent moisture barrier properties. Water vapor transmission rates for Parylene C ($\approx$ 0.08 g·mm/(m^2·day)) are an order of magnitude lower than most polymeric materials (e.g. $\approx$ 0.94 g·mm/(m^2·day) for epoxy) [77, 78]. Hydrophobic functionalization using a 10 nm layer of Parylene C has been demonstrated to induce dropwise condensation with limited success when compared to self-assembled monolayers [79]. The durability and heat transfer performance of intermediate Parylene coating thicknesses (between 100 nm and 1 μm) during condensation needs to be addressed in the future with an emphasis on longevity.

Polymers deposited by initiated chemical vapor desposition (iCVD) represent a second category of promising polymers for sustained DWC. The iCVD is a convenient single-step vapor-phase method to deposit polymeric films [80–84]. Volatile monomers and initiator flow into a reactor, and initiators are cleaved to form radicals by plasma, UV radiation, or thermal decomposition. The resulting radicals initiate polymerization with absorbed monomers on the temperature-controlled substrate. The iCVD process can provide super-conformal coatings and enhanced durability due to grafting of the polymer to the substrate [85]. Furthermore, iCVD is able to deposit polymeric film without the use of a solvent, providing environmental benefits by removing the need for manufacturers to expose themselves to chemicals.

The iCVD process has been used to synthesize copolymer from fluorinated monomer, 1H,1H,2H,2H-perfluorodecyl acrylate (PFDA) with the hydrocarbon monomer, di-vinyl benzene (DVB) [86]. The perfluorinated side chains of the PFDA units segregate to the interface under dry conditions in order to minimize surface energy. Recently, researchers have demonstrated that grafted fluoropolymer surfaces undergo stable dropwise condensation for prolonged exposure (>48 h) to steam at high temperature ($\approx$100 °C) [58]. Further studies including longer duration longevity tests and the mechanisms of water permeation, swelling and delamination on iCVD surfaces are needed during water vapor condensation in steam environments.

Lastly, significant effort was placed on demonstrating enhanced mechanical robustness of polymeric surfaces by a number of researchers utilizing composite organic polymers [87–90] and modified structures [91–95]. However, heat transfer measurements during condensation on these surfaces are currently not available.

13.2.3 Diamond-like carbon (DLC)

Diamond-like carbon (DLC) is a metastable form of amorphous carbon which has the beneficial properties of diamond, such as a high mechanical hardness, chemical and electrochemical inertness, elastic modulus and optical transparency [96]. The great versatility of DLC arises from crystalline and disordered structures, with their physical properties depending on the ratio of different hybridization, sp^3 (diamond-like), sp^2 (graphite-like) and even sp^1 (acetylene-like) bonds. DLC is defined as amorphous carbon with a significant fraction of sp^3 configuration [97]. The mechanical properties of DLC coatings have been characterized in detail, with manufacturing methods for growth of DLC films maturing over the past four decades. While DLC is much cheaper than diamond, the similar properties to diamond have led to its widespread application as a protective film in a variety of fields such as optical windows, wear resistance, biomedical components and micro-electromechanical systems (MEMS) [98–100]. For example, the Mach 3 razor from Gillette is a DLC coated product. Furthermore, DLC coatings are heavily used in the automotive industry, with more than 100 million parts per year coated to improve their performance. In the automotive sector alone, DLC coatings form a $100 million dollar industry [101–103].

Despite its successful path to industrial implementation, DLC has not received much attention in the field of dropwise condensation. Past studies have shown that the surface energy of DLC coatings can be decreased to as low as that of PTFE ($\approx$ 19 mN m^{-1}, microhardness $\approx$ 0.3 GPa) by modifying the network structure with the addition of other elements without losing hardness (e.g. fluorine $\approx$ 20 mN m^{-1}, 2 GPa) [104–106]. Furthermore, previous studies have shown that DLC coatings can provide higher thermal conductivity than traditional hydrophobic promoters [107] and can promote dropwise condensation with enhanced phase change heat transfer [108, 109].

13.2.4 Rare earth oxides (REOs)

Initial studies on the wettability of rare earth oxide ceramics (REOs) reported intrinsic hydrophobicity due to their unique electronic structure which prevents hydrogen bonding with water molecules [65, 110–114]. A follow-up study suggested that freshly fabricated REOs are intrinsically hydrophilic and gradually become hydrophobic when the oxygen-to-metal ratio decreases due to surface relaxation under ultra-high vacuum environments [111]. In contrast to this explanation, several other studies have reported that REOs are intrinsically hydrophilic, obtaining hydrophobicity by adsorption of volatile organic compounds from the surrounding atmosphere [115–120].

Although the mechanism of REO hydrophobicity is under debate, REOs have the potential for practical DWC implementation due to their mechanical and thermal stability [110, 121], and their relatively low cost approaching 1% of noble metals [115]. Moreover, REOs have an order of magnitude higher thermal conductivity ($\approx$ 1–10 W $m^{-1} \cdot K^{-1}$) when compared to traditional hydrophobic promoters such as SAMs and polymers ($\approx$0.1 W $m^{-1} \cdot K^{-1}$) [115, 122]. Using droplet size distribution analysis, DWC heat transfer performance of REOs after hydrocarbon adsorption has been recently estimated to result in 2–5 times enhanced heat transfer performance when compared to FWC [119]. However, durability and direct heat transfer measurements during condensation are not available at this time.

13.2.5 Hydrocarbon adsorption

The aforementioned mechanism of spontaneous hydrocarbon adsorption on clean metal oxide surfaces has been demonstrated on noble metals. Until the 1970s, the wettability of noble metals was a contentious topic, with some researchers showing hydrophobicity [123–131] and others showing hydrophilicity [132–137]. Research eventually showed that hydrophobicity was obtained from airborne hydrophobic contamination (i.e. organic materials). After decades of debate, the first direct measurement of carbonaceous contamination using Auger electron spectroscopy showed evidence of monolayer formation of carbon containing contamination on gold [138]. A number of follow-up studies have concluded that a significant amount of organic compound adsorption is responsible for stable DWC on noble metals [139–141]. Although excellent durability ($\approx$5 years) during steam DWC had been demonstrated on a thick layer of gold ($\approx$50 μm) [142], noble metals have not received much attention due to their high cost and questionable durability.

Recently, hydrocarbon-induced hydrophobicity/superhydrophobicity has been demonstrated on a variety of surfaces including noble metals [143, 144], metal alloys [145], metal oxides [146–149], boron nitride [150, 151] and graphene [59, 152]. Hydrocarbon adsorption has the potential to enable durable superhydrophobicity due to its self-healing properties stemming from atmospheric adsorption, in particular for systems that operate in ambient conditions such as air-side heat exchangers and atmospheric water harvesters.

Previous studies of hydrocarbon detection via x-ray photoelectron spectroscopy (XPS) and time of flight secondary ion mass spectroscopy (ToF-SIMS) have shown that non-polar hydrocarbon groups (e.g. C–C or C–H) were responsible for enhancing hydrophobicity [146, 147], while the adsorption dynamics of airborne hydrocarbons remains debated, as to whether physisorption (multilayer of molecules) or chemisorption (covalently boned monolayer) takes place [147, 153, 154].

13.2.6 Slippery omniphobic covalently attached liquids (SOCALs)

Recently, researchers have introduced covalently attached flexible groups onto silicon wafer substrates, termed slippery omniphobic covalently attached liquids (SOCALs) [155]. The SOCAL surfaces are fabricated by acid-catalyzed graft polycondensation of dimethyldimethoxysilane ($Me_2Si(OMe)_2$). The SOCAL coating provides extremely low contact angle hysteresis ($\leqslant 1°$) for a broad range of liquids having surface tensions from 18.4 (hexane) to 78.2 mN m^{-1} (water), allowing them to potentially achieve DWC of low-surface-tension fluids such as refrigerants. However, the $Me_2Si(OMe)_2$ chemistry requires a silicon (Si) group to be grafted to the surface and does not work with regular metallic oxides. To achieve DWC on the majority of industrial condenser materials (e.g. copper, steel, aluminum),

SOCALs necessitate the deposition of thin and conformal seed layers of SiO_2 using scalable and cost-effective means such as electrodeposition or PVD [156]. Heat transfer results during DWC on SOCAL surfaces are currently not available.

13.2.7 Degradation of coatings

While a variety of hydrophobic promoters have been developed in the past century and have been shown to degrade over time with exposure to steam condensation, the fundamental mechanism of degradation remains poorly understood. In the following section, we present several different theories of degradation.

It has been recently reported that the evaporation of sessile droplets on promoted hydrophobic substrates results in hydrophobic molecule detachment at the contact line [60]. The authors hypothesized that the degraded state was energetically favorable considering the driving force to remove the hydrophobic promoter. During DWC, the hydrophobic promoter will eventually experience contact line forces due to continuous condensation and shedding of water droplets, which indicate that stronger adhesion or bonding between the promoter coating and substrate is required for greater durability. Moreover, current theory points to the presence of nanoscale defect sites on hydrophobic promoted surfaces as a source of nucleation and long-term degradation of the coating [48, 50, 157–159]. Although qualitatively correct and able to explain observed trends during heterogeneous nucleation at low supersaturation, the experimental observation of SAM defects has been difficult to achieve. Future work utilizing techniques with nanoscopic resolution, such as environmental atomic force microscopy or scanning probe microscopy, are required to study the formation and dynamic growth of defects. Furthermore, the study of the effect of evolving contact line motion on defect site formation is necessary to test the aforementioned degradation hypothesis.

On polymeric promoters, water permeation, swelling and delamination have been qualitatively shown to lead to long-term degradation [76, 160–163]. The lack of good adhesion to the substrate via weak van der Waals interactions at the polymer/substrate interface allows for relatively easy water permeation through the film and accumulation of condensate at the high surface energy interface. The accumulation of water at the interface can lead to droplet growth beneath polymer films and delamination from the substrate [76]. Surprisingly, the degradation mechanism stemming from water permeation during water vapor condensation has not been studied thoroughly in the context of DWC. Although blister tests have been developed and used to study interfacial adhesion between coatings and substrates [164–166], condensation induced failures during DWC have not. Better fundamental understanding of blister formation and nucleation beneath hydrophobic promoter coatings has the potential to provide rational design guidelines for next-generation durable promoter coatings.

In addition to degradation of the promoter materials themselves, a recent study found that sulfur-containing volatile organic compounds (VOCs) and secondary organic aerosols (SOAs) can accumulate on substrates during steady and un-steady DWC which has the potential to create degradation and condensate-side fouling

[159]. The VOCs were identified to be organic compounds with relatively high vapor pressure, typically containing carbon, oxygen and sulfur. The oxidation of VOCs is known to form SOAs, with a major source of VOCs stemming from fossil fuel combustion [167]. Organosulfur compounds (OSCs) are generated by agricultural practices and biological processes [168–172]. For example, ocean, livestock and farming practices are significant sources of dimethyl sulfide (CH_3SCH_3) [171], methanethiol (CH_3SH), dimethyl disulfide (CH_3SSCH_3) and dimethyl trisulfide (CH_3SSSCH_3) airborne hydrocarbons [173–177]. It is important to note that although VOC emissions from the transportation sector have decreased markedly over the last few decades, the use of volatile chemical products (VCPs) including pesticides, coatings, print inks, cleaning agents, adhesives and personal care products, has led to the expansion of alternative VOC sources of emission to the atmosphere [178, 179]. Thus, water soluble solid-phase SOAs generated through oxidation of vapor-phase VOCs in the atmosphere [180] accumulate inside the bulk condensate fluid during DWC [181, 182].

Accumulation of VOCs and SOAs has important implications toward the potential development of durable functional coatings. The agglomerate deposition mechanism shows that the surrounding environment plays an important role in phase change processes. Indeed, even for systems which utilize condensation and do not operate in ambient conditions, such as heat pipes [183], industrial condensers [184] and vapor chambers [51], ultra-clean and ultra-high vacuum conditions may contain VOCs [185]. The VOC and SOA absorption process results in the accumulation of contaminants in the condensate stream, posing a threat to working fluid purity and water-side fouling.

The difficulty of achieving long-lasting DWC has led some researchers to propose creative techniques to enhance the condensation heat transfer rate on intrinsically hydrophilic substrates, including thin film condensation [186, 187], and gravitational-wicking enhanced superhydrophilic condensation [188, 189].

13.3 Structured surfaces for low-surface-tension fluids

Although significant effort has been placed on developing structured surfaces promoting DWC of steam, achieving DWC for low-surface-tension fluids remains a distant reality. From a fundamental perspective, the non-polarity of low-surface-tension fluids (i.e. fluorinated refrigerants) results in wetting of promoter surfaces composed of non-polar hydrophobic molecules, resulting in FWC. Dropwise condensation of low-surface-tension fluids can greatly benefit several industrial processes, in particular those that use refrigerants as the working fluid [30, 190]. Current state-of-the-art strategies [191] involve passive techniques such as fabrication of enhanced macroscale structured or enhanced surfaces composed of fins and channels on the surface for enhancement of FWC heat transfer. However, enhanced surfaces remain limited by the fundamental constraint of filmwise wetting behavior. In addition to refrigeration, systems that use non-refrigerant low-surface-tension process fluids such as chemical refineries [192], natural gas production facilities [193], biomass combustion [194] and food industry processes [195], stand to significantly benefit from achieving DWC via condenser size reduction (capital cost reduction) and energy cost savings (operating cost reduction).

13.3.1 Re-entrant structured surfaces

For the majority of state-of-the-art superhydrophobic surfaces, the promoter material has surface energy comparable to the surface tension of the condensing organic fluids, resulting in wetting and FWC. The unique challenge posed by low-surface-tension fluids necessitates additional considerations for developing surfaces capable of promoting DWC. Over the past few decades, superoleophobic surfaces have received considerable attention for potential applications in oil–water separation [196–198], oil capture [199], self-cleaning [200, 201], anti-oil coatings [202, 203] and bioadhesion [204, 205]. In addition to surface microstructures and low-surface-energy functionalization, re-entrant structured surfaces achieving Cassie wetting states have been shown to play a crucial role in designing omniphobic surfaces [206–212]. Re-entrant surfaces utilize surface geometry to create local energy barriers for liquid to impale the surface microstructures by maintaining air pockets beneath the liquid droplets on the surface. Researchers have developed modified electrospun fiber mats with a series of fluorodecyl polyhedral oligomeric silsesquioxane (POSS) molecules which led to the fiber nets showing re-entrant surface curvature displaying superoleophobicity [208, 209]. Similarly, re-entrant surface structures with silicon nanonail arrays were developed by reactive ion etching [212], displaying superoleophobicity for a wide range of liquids having surface tension ranging from 21.8 mN m^{-1} (ethanol) to 72 mN m^{-1} (water). Others have developed micro-pillar array structures on a silicon wafer substrates using the Bosch process resulting in advancing contact angles approaching 158° with contact angle hysteresis of only 10° for deposited hexadecane droplets [210]. Inspired by the success of re-entrant surfaces, recent works have focused on doubly re-entrant structures which are capable of repelling even lower surface energy fluids [213].

Although re-entrant structured surfaces have been proven to repel low-surface-tension liquid droplets, they fail during condensation and other phase change applications [214, 215]. When vapor condenses on re-entrant structures designed to repel deposited liquid droplets, the benefits of the modified geometry is lost due to the spatially random heterogeneous nucleation process within the microstructures [215]. The trapped air in the re-entrant microstructures is displaced by condensate, and the surface loses its repellent properties, transitioning to condensate droplet morphologies having the Wenzel wetting state and FWC [214, 216].

In an attempt to overcome flooding and FWC of re-entrant surface structures during condensation, a recent study [214] utilized doubly re-entrant surfaces with disconnected pores (pitch $\approx$ 100 nm). The lack of communication between air pockets resulted in nonwicking nanoscale re-entrant features maintaining fluid repellency during condensation without any low-surface-energy coatings. The re-entrant cavity surface prevents nucleating droplets from spreading between structures, thereby preserving the trapped air in the cavity and retaining wetting properties during condensation [214].

13.3.2 Slippery liquid-infused porous surfaces (SLIPSs) and lubricant-infused surfaces (LISs)

To avoid nucleation of condensate droplets within the surface structures, a novel category of surfaces infused with lubricating liquids has recently been developed. Inspired by the *Nepenthes* pitcher plant which uses microstructures to lock-in an intermediate liquid [201], slippery liquid-infused porous surfaces (SLIPSs) or lubricant-infused surfaces (LISs) are porous or structured solid surfaces stabilized by a liquid lubricant [201, 217, 218]. The liquid lubricant is stabilized within the structured surface by capillary forces [201, 217, 219, 220]. The infused lubricant, which is immiscible with the condensate, creates a chemically homogeneous and atomically smooth liquid–liquid interface, drastically reducing structural and chemical inhomogeneity, resulting in minimal contact line pinning, ultra-low contact angle hysteresis and easy droplet removal [220, 221]. LISs/SLIPSs infused with perfluorinated lubricant oils show very low adhesion to a variety of liquids with surface tensions as low as 17.2 mN m^{-1} (pentane) [201, 222].

Condensation of water vapor on LISs/SLIPSs exhibits DWC, demonstrating rapid condensate droplet shedding and re-nucleation [216, 221, 223]. Compared to conventional hydrophobic surfaces, where the critical diameter of shedding droplets is on the order of a few millimeters for water, droplets as small as 20 μm in diameter become mobile on LISs/SLIPSs [223]. Furthermore, the heterogeneous nucleation rate LISs/SLIPs during DWC is higher than on solid hydrophobic surfaces due to a lower energy barrier for nucleation [223, 224]. Condensation heat transfer of LISs/SLIPs was experimentally measured to be twice that of conventional hydrophobic and superhydrophobic surfaces in conditions comparable to those encountered in industrial condenser operations [224]. A follow-up study performed water vapor condensation in the absence of any non-condensable gases (NCGs), and found the heat transfer performance of LISs/SLIPSs to exceed that of dropwise condensation on a hydrophobic surface by $\approx$30% and filmwise condensation by $\approx$400% (figure 13.1(a)) [216].

Inspired by the success of achieving DWC of steam, recent studies have focused on studying the condensation behavior of low-surface-tension fluids on LISs/SLIPSs [215, 216]. Rykaczewski *et al* tested the condensation of a number of low-surface-tension liquids with surface tension ranging from 12 to 28 mN m^{-1} on LISs with Krytox as the lubricant and compared the resulting condensate morphologies with smooth hydrophobic and re-entrant structured surfaces [215]. Although toluene ($\gamma = 27.9$ mN m^{-1}), pentane ($\gamma = 15.1$ mN m^{-1}), hexane ($\gamma = 18$ mN m^{-1}) and octane ($\gamma = 21.1$ mN m^{-1}) exhibited DWC, polar liquids such as ethanol ($\gamma = 24.8$ mN m^{-1}), isopropanol ($\gamma = 20.9$ mN m^{-1}) and perfluorohexane ($\gamma = 12$ mN m^{-1}) displayed FWC on the Krytox infused nanotextured surface [215] (figure 13.1(b)). Furthermore, the study demonstrated that the choice of lubricant and underlying surface texture plays a crucial role in stabilizing the lubricant and reducing pinning of the condensate. By experimentally measuring droplet departure diameters and average contact angles, in combination with theoretical models to predict the heat transfer coefficients [225], their study demonstrated a four- to eight-fold enhancement

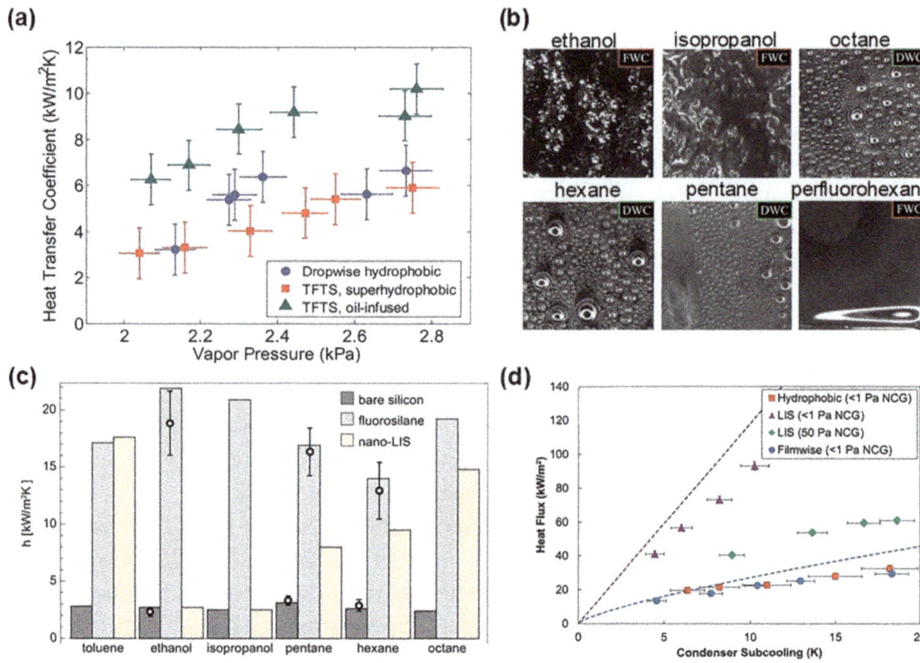

Figure 13.1. (a) Comparison of overall heat transfer coefficient during condensation of steam on a smooth hydrophobic surface, an SAM-coated superhydrophobic surface, and an LIS/SLIPS surface. (b) Condensation of various low-surface-tension fluids on a Krytox-impregnated nanotextured surface. Both DWC and FWC condensation was observed. (c) Theoretical predictions of heat transfer coefficients (bars) for the condensation of low-surface-tension liquids on a smooth silicon surface, a smooth hydrophobic surface and a Krytox-impregnated surface. Experimental measurements are indicated with circular points. (d) Heat flux as a function of condenser subcooling for toluene condensation on a smooth hydrophobic copper tube (FWC) and a Krytox infused CuO tube (DWC). (a) and (b) Reproduced with permission from [215]. Copyright 2014 Nature Publishing. (c) Reproduced with permission from [224]. Copyright 2013 Nature Publishing. (d) Reproduced with permission from [216]. Copyright 2018 Nature Publishing.

in heat transfer coefficient compared to FWC for the four low-surface-tension fluids promoting DWC on LISs (figure 13.1(c)). A follow-up study confirmed the effectiveness of LISs in promoting DWC by experimentally measuring heat transfer performance during toluene condensation (figure 13.1(d)) [216]. Using a superhydrophobic CuO nanostructured surface infused with Krytox (GPL 101) fluorinated oil, they showed a 450% enhancement in heat transfer when compared to an uncoated surface exhibiting FWC. However, within 1 h of DWC of toluene, the surface began to transition to FWC and the heat transfer coefficient degraded by ≈78%.

13.3.3 LIS/SLIPS stability

Even though initial results showed the promise of LISs/SLIPs as plausible solutions for promoting DWC of low-surface-tension fluids, these surfaces come with a unique set of challenges not encountered with classical solid promoter coatings. The primary and most important factor in developing durable, stable and robust LISs/SLIPSs is the choice of lubricant for the condensate working fluid. Usually, liquids having low surface energy (<20 mJ m^{-1}) and vapor pressure (<0.1 mPa at STP) form the most stable lubricants for LISs/SLIPSs. These include fluorinated Krytox oils [201, 215, 221, 223, 226], silicone oils [220, 226–228], mineral oils [221] and ionic liquids [220, 223, 229]. Although these lubricants are immiscible with water due to the strong intramolecular interactions (hydrogen bonding) of water molecules, the similar surface tension and molecular structure of the low-surface-tension liquids and lubricants along with the polarity of the molecules, limit the choices. Hence, proper care must be taken when choosing the lubricant for low-surface-tension condensates. Using the van Oss, Chaudhury and Good (vOCG) method [230], the miscibility/immiscibility of low-surface-tension liquids with fluorinated Krytox oils (VPF 1506 and GPL 101) was predicted, and concluded both to be fairly immiscible [233]. In parallel, researchers studied the miscibility of alcohols and hydrocarbons ($\gamma = 12$–73 mN m^{-1}) with lubricants having a wide range of interfacial parameters, vapor pressures (5×10^{-8}–0.7 kPa) and viscosities (4–5300 mPa·s) including fluorinated Krytox oils (VPF 1506, 1525 and 16256), silicone oils, Carnation mineral oil and ionic liquid [232]. By mechanically mixing the working fluids and lubricants and allowing the mixture solution to settle and form distinct liquid–liquid interfaces, they showed most of the non-fluorinated lubricants to be miscible with the working fluids. The fluorinated Krytox oils were the only lubricants immiscible with the tested condensates.

For immiscible condensate–lubricant pairs, the lubricant creates a defect-free liquid–liquid interface, thereby facilitating easy gravitational removal of condensate during DWC. Although the additional lubricant layer prevents contact line pinning, it has the potential to lead to lubricant cloaking. Cloaking is defined as the encapsulation of the condensate droplets by the surface lubricant, resulting in coating of the condensate liquid with a thin (<100 nm) layer of lubricant. During condensation, cloaking of the condensate creates a vapor diffusion barrier, reducing direct condensation and inhibiting droplet growth, coalescence [221, 229] and density [223]. The lubricant cloak forms almost immediately after condensate droplets nucleate at the lubricant–air interface [227], leading to non-coalescence of the droplets on the LIS/SLIPS surface [226]. Depending on the properties of the condensate liquid, impregnating lubricant and surface structures, the condensate droplet can take on a variety of thermodynamically favorable states [217, 220]. A lubricant cloak will form if the spreading coefficient of the lubricant oil on the condensate liquid, S_{ol} is greater than zero [233]:

$$S_{ol} = \gamma_l - \gamma_o - \gamma_{ol}, \tag{13.1}$$

where γ_l, γ_o and γ_{ol} are the liquid–vapor surface tension of the condensate, the liquid–vapor surface tension of the lubricant and the interfacial tension between the lubricant and condensate, respectively. A spreading coefficient $S_{ol} > 0$ implies that the lubricants on the LIS/SLIPS surface will cloak the condensate. Hence, for effective selection of a lubricant in the design of LISs and SLIPSs, $S_{ol} < 0$ is desired.

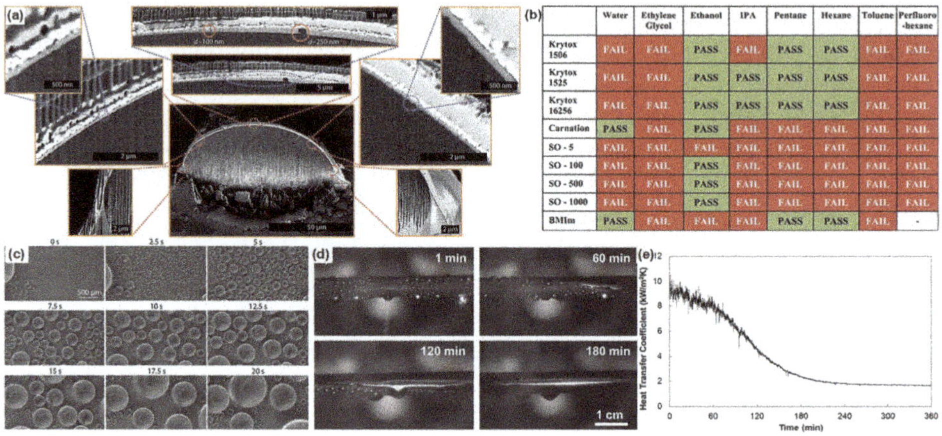

	Water	Ethylene Glycol	Ethanol	IPA	Pentane	Hexane	Toluene	Perfluoro-hexane
Krytox 1506	FAIL	FAIL	PASS	FAIL	PASS	PASS	FAIL	FAIL
Krytox 1525	FAIL	FAIL	PASS	PASS	PASS	PASS	FAIL	FAIL
Krytox 16256	FAIL	FAIL	PASS	PASS	PASS	PASS	FAIL	FAIL
Carnation	PASS	FAIL	PASS	FAIL	FAIL	FAIL	FAIL	FAIL
SO - 5	FAIL	FAIL	FAIL	FAIL	FAIL	FAIL	FAIL	FAIL
SO - 100	FAIL	FAIL	PASS	FAIL	FAIL	FAIL	FAIL	FAIL
SO - 500	FAIL	FAIL	PASS	FAIL	FAIL	FAIL	FAIL	FAIL
SO - 1000	FAIL	FAIL	PASS	FAIL	FAIL	FAIL	FAIL	FAIL
BMIm	PASS	FAIL	FAIL	FAIL	PASS	PASS	FAIL	-

Figure 13.2. (a) Silicone oil cloak around a condensed water droplet suspended on an LIS obtained by cryo-FIB-SEM. The light gray sandwiched layer shows the thin lubricant layer. (b) Combined miscibility and cloaking results for a variety of low-surface-tension liquids with wide range of lubricants. Green cells labeled PASS correspond to immiscible and non-cloaking lubricant/condensate pairs. (c) Time evolution of water vapor condensate droplets on a fluorinated Krytox (VPF 1514) oil infused aluminum nanostructured surface. (d) Toluene condensation on a Krytox (GPL 101) infused LIS transitioning from dropwise to filmwise within 1 h of steady DWC, resulting in (e) degradation of heat transfer coefficient by ≈ 78%. (a) Reproduced with permission from [227]. Copyright 2015 the Royal Society of Chemistry. (b) Reproduced with permission from [232]. Copyright 2017 American Chemical Society. (c) Reproduced with permission from [221]. Copyright 2017 Elsevier. (d) and (e) Reproduced with permission from [216]. Copyright 2018 Nature Publishing.

By balancing the disjoining pressure due to the interaction between air and condensate liquid with the thin lubricant film, and the Laplace pressure due to droplet curvature, the thickness of the lubricant (FC-70) cloak on a 1 mm water droplet was estimated to be 20 nm [229]. Experimental methods such as scanning electron microscopy [227, 234] (figure 13.2(a)), confocal microscopy [229], high resolution x-ray tomography [235] and imaging [236] have shown additional evidence of lubricant cloaking. Recent studies have determined the interfacial tension between the lubricant and the condensate for different lubricant/condensate pairs using the vOCG theoretical model [230, 237]. Using the calculated interfacial, the spreading coefficient tensions S_{ol}, and hence the propensity for cloaking, was determined. To verify the theoretical vOCG results, the pendant drop method was used to experimentally measure the lubricant/condensate interfacial tensions for the immiscible lubricant/condensate pairs [232]. Considering miscibility and cloaking, the study showed that only a few lubricant options exist for designing stable LISs/SLIPSs for low-surface-tension fluid applications (figure 13.2(b)).

13.3.4 Durability of LISs/SLIPs

Miscibility and cloaking represent the most important criteria for designing stable LISs/SLIPs. However, additional failure modes exist, which needs to be investigated in detail prior to acceptance in DWC applications. For applications covering a wide range of low-surface-tension fluids, the liquid–liquid interface between the lubricant and the condensate fluid is of utmost importance, and can be the limiting factor determining the performance of LISs/SLIPSs. For instance, even for apparent immiscible lubricant–condensate pairs [231, 232], partial miscibility may be present, a phenomena common with organic fluids [238, 239]. Even minute levels of partial miscibility lead to the depletion of lubricant. Further studies are required for utilizing analytical techniques such as gas chromatography–mass spectrometry (GC-MS), thermogravimetric analysis (TGA), Fourier transform infrared spectroscopy (FTIR) and nuclear magnetic resonance (NMR) to determine the definitive extent of miscibility of the chosen lubricant with the desired condensate fluid.

For robust and long-term performance, the drainage of lubricants from LISs/SLIPs is a major concern [217, 220, 223]. For sustained DWC, condensate droplets are continuously removed from the condenser surface by gravity. Cloaking can thus lead to significant depletion of the lubricant due to shedding of the condensate droplets. Furthermore, cloaking may not be the only mechanism of lubricant loss. Condensate shedding from LISs/SLIPs leads to shear-induced drainage [240], depleting the lubricant even for immiscible and non-cloaking condensate/lubricant pairs. For water vapor condensation on Krytox-based LIS (figure 13.2(c)), a previous study showed that surfaces infused with low-viscosity lubricants (12 cSt) degraded faster due to lubricant drainage, whereas those infused with higher viscosity lubricants (140 cSt) remained stable for more than 10 h of steady operation [221]. The use of an immiscible and non-cloaking lubricant (Carnation mineral oil) with water also showed lubricant drainage, indicating the presence of shear-based lubricant loss. Condensation of toluene ($\gamma \approx 28$ mN m^{-1}) on CuO nanostructures infused with fluorinated Krytox (GPL 101) oil (17.4 cSt) initially exhibiting dropwise condensation, transitioned to filmwise condensation within 1 h of steady condensation conditions (figure 13.2(d)) [216]. The heat transfer coefficient in this case degraded by $\approx$78% (figure 13.2(e)). The reason for heat transfer degradation was most likely due to the low lubricant viscosity and rapid drainage.

In addition to shear, erosion based damage to the lubricant and surface structures due to elevated velocities of the incoming vapor towards the condensing surface present a degradation issue. Furthermore, the evaporation of lubricant due to condensate and vapor cycling and mass transfer gradients remains to be studied. Although achieving the low equilibrium vapor pressure at operating conditions results in minute lubricant evaporation, condensation in industrial conditions is not an equilibrium process. The removal of condensate and post-condensation of vapor would result in continual evaporation and removal of lubricant. Much future work is required to conduct rigorous experimental studies to quantify and classify these fundamental degradation mechanisms on LIS/SLIPS DWC performance.

13.4 Electric field enhanced (EFE) condensation

The advent of structured surfaces has opened up the possibility to remove condensate during DWC at length scales much smaller than possible with classical gravitational shedding. Specifically, coalescence-induced droplet shedding has been used in conjunction with external electric fields to control condensation dynamics. Charge separation at the solid–liquid interface during the jumping process has been shown to result in positive electrostatic charge build-up in departing droplets [52, 241, 242]. Prior to examining jumping, we first focus here on the forces that are exerted on a fluid exposed to an external electric field ($\overrightarrow{E_{\text{ext}}}$) [243]. From the electrodynamics (EHD) perspective, two categories of fluids exist: dielectric fluids and conducting fluids. The force per unit volume ($\overrightarrow{f_{\text{e,c}}}$) acting on a conductor is

$$\overrightarrow{f_{\text{e,c}}} = q_{\text{ex}} \overrightarrow{E_{\text{ext}}} + \frac{1}{2} \nabla(\alpha \overrightarrow{E_{\text{ext}}} \cdot \overrightarrow{E_{\text{ext}}}), \tag{13.2}$$

where q_{ex} is the volumetric charge density, ε_0 is the permittivity of free space, α is the polarizability tensor which depends solely on the shape of the conductor and the symbol ∇ is the gradient operator. The first term is applicable if the conductor has a net charge while the second term is non-zero for non-uniform electric fields. Since charges are free in a conductor, the external applied field will cause the charges to move to the surface causing an electric dipole moment $\vec{\mu}$ over the whole body of volume V, $\vec{\mu} = V\alpha \overrightarrow{E_{\text{ext}}}$. The torque on the body can be determined by $\vec{\mu} \times \overrightarrow{E_{\text{ext}}}$. If both the force and the torque on a conductor are zero, then the conductor experiences a pressure causing it to deform (expand), $\Delta p = (1/2)(\alpha \overrightarrow{E_{\text{ext}}} \cdot \overrightarrow{E_{\text{ext}}})$, which is termed electrostriction.

As for an isothermal, isotropic dielectric fluid, that deforms homogeneously under stress, the electric-induced force per unit volume $\overrightarrow{f_{\text{e,d}}}$ is given by

$$\overrightarrow{f_{\text{e,d}}} = q_{\text{ex}}\vec{E} + \frac{1}{2} \nabla\left(E^2\rho\left(\frac{\partial \varepsilon}{\partial \rho}\right)_T\right) - \frac{E^2}{2}\nabla\varepsilon, \tag{13.3}$$

where $\vec{E}$ is the local electric field, ρ is the density and ε is the permittivity of the dielectric. The first term describes the electrophoretic force similar to the conducting fluid, replacing the external electric field by the local electric field in the dielectric, whereas the remaining terms arise mainly from the non-uniformity of the electric field, permittivity and density.

13.4.1 Electrohydrodynamic (EHD) enhancement of condensation heat transfer

Since its discovery in 1965 [244], EHD-enhanced condensation heat transfer has seen increased interest as an active enhancement technique for condensation. The EHD approach is based on applying a high voltage (ac or dc) that acts on a dielectric fluid and its vapor to exert a force on the interface as determined by equation (13.3), mainly due to the gradient in permittivity and non-uniformities in the density, electric field or temperature. The force on the interface acts to enhance condensation heat transfer through disturbing the condensate film. (i) It would cause the thickness of the film to be reduced through liquid removal and hence decrease the thermal resistance of the film [245, 246]. (ii) If the film is thinned enough, transition from filmwise condensation to pseudo-dropwise condensation, a more efficient condensation process, can occur [247]. (iii) The movement of the liquid would disrupt the non-condensable gas layer at the liquid–vapor interface, which acts as a diffusion barrier for mass transfer [248–250]. The formation of interfacial instabilities and waviness in the condensate film can also accelerate some of these processes and lead to an effective heat transfer enhancement [251].

In order for EHD to result in a substantive force on the fluid, by disturbing the condensate layer, high external electric fields (~1 kV cm^{-1}) are needed. Although these fields are not high enough to cause dielectric breakdown in air (a dielectric strength of ~30 kV cm^{-1}), high voltages may lead to currents flowing in the fluid if fluid is conductive and not well insulated. Because of this electrical hazard, EHD condensation has been limited to use with non-polar dielectric working fluids such as refrigerants and synthetic heat transfer fluids. This eliminates the possibility of using water, a superior heat transfer fluid, for EHD purposes; although water as a pure substance has good electrical insulation properties, it can easily become more conductive due to contamination with salts that ionize in solution. EHD can in addition provide other practical limitations. Adding electrodes and electronics to the heat exchanger will increase the volume, complexity and potential failure modes of the heat exchanger. These factors contribute to increased fixed costs and maintenance costs for the heat exchanger that may not outweigh the benefits, or not satisfy the constraints of the application. Lastly, EHD has not been studied as a long-term durable (>5 years of operation) technique to enhance heat transfer. Potential breakdown of the fluids from the extensive use of high voltages is yet to be investigated.

13.4.2 Electric field induced condensation (EIC)

The vapor pressure of a liquid is known to be altered by the curvature of the surface, according to the Kelvin equation:

$$\ln\frac{P_{\text{v}}}{P_{\text{v},0}} = C\frac{\gamma V_{\text{m}}}{RT}, \tag{13.4}$$

where P_{v} and $P_{\text{v},0}$ are the vapor pressures for a curved and flat interface, respectively, and γ, V_{m}, R and T represent the surface tension, molar volume of the liquid, gas constant and temperature, respectively. The geometric factor $C = 1/r_1 + 1/r_2$ takes into account the radii of curvature r_1 and r_2 of the interface, with a positive value counted for a concave shape of the condensed phase (liquid droplet), and a negative value for a convex shape (bubble in a liquid). The Gibbs free energy is altered further by the presence of the electric field and this can be accounted for in a modified Kelvin equation giving the vapor pressure $P_{\text{v},E}$ with the field E applied [252]:

$$\ln\frac{P_{\text{v},E}}{P_{\text{v},0}} = C\frac{\gamma V_m}{RT} + (1 - \varepsilon)\varepsilon_0\frac{E^2}{2}, \tag{13.5}$$

where ε represents the permittivity of the liquid. Equation (13.5) enables the initiation of condensation without a change in temperature, and may be used as an active mechanism for condensation enhancement. One barrier to the EFI approach is the need for very high electric fields (~10^8 V m^{-1} for water).

13.4.3 Electric field enhanced (EFE) jumping-droplet condensation

Following the discovery of jumping-droplet condensation on superhydrophobic surfaces [14, 15], studies have revealed that these droplets contain a net positive charge (~10 fC), due to electric double layer (EDL) separation at the solid interface, with adsorption of hydroxide ions (OH^-) at the solid surface [52]. The charge separation enables the use of external fields to act on the droplets via the Coulomb force (electrophoretic force) expressed by the first term in equation (13.2). The droplet removal mechanism via external fields can reduce condensate build-up on the condensing surface hence delaying flooding, and decreasing average droplet size that tends to enhance the heat transfer coefficient by decreasing the conduction thermal resistance through the droplets [241]. Models have been developed to design adequate heat exchangers, condensers in particular, that enable collection of water droplets (figure 13.3(a)) [253]. The very low currents generated by the low charge densities of the droplets provide low power input for increased heat transfer coefficients of 20% or more, as defined by an incremental coefficient of performance $COP_{inc} = \Delta Q/W$ describing the enhancement in heat transfer per unit work supplied (figure 13.3(b)).

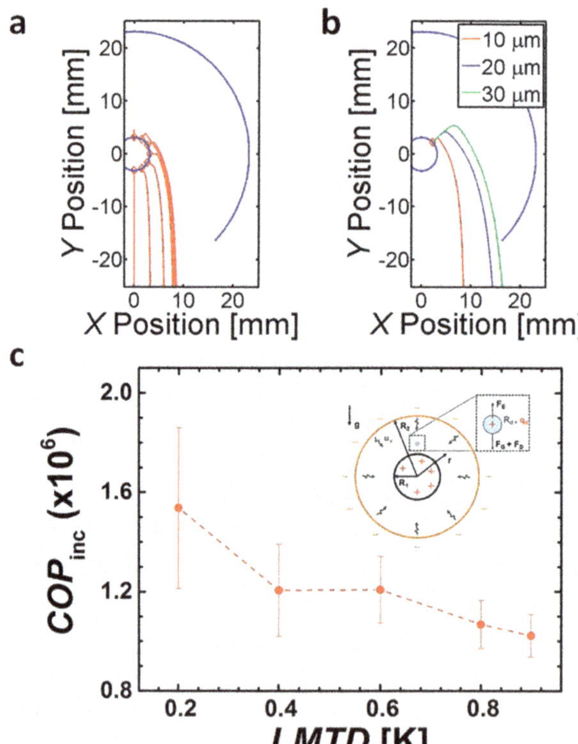

Figure 13.3. (a) and (b) Electric field based collection of jumping droplets. (c) Incremental coefficient of performance ($COP_{inc} = \Delta Q/W$) for electric field enhanced jumping condensation droplet condensation. Reproduced with permission from [253]. Copyright 2015 American Chemical Society.

EFE jumping-droplet condensation requires fields on the order of 100 V cm^{-1} or greater, which is slightly lower than EHD or EIC. However, issues remain about adding electrodes to the system and failure potentials. Another current research focus is on the durability of the superhydrophobic surfaces, which is still not well understood and tested.

One potential application arises in thermal management applications, such as electronics cooling where the droplets can be attracted towards the hot component for localized evaporative cooling [5].

13.4.4 Potential research avenues for EFE condensation

The previous works on EFE condensation present good opportunities for further development for practical applications. From the EHD standpoint, work remains in durability studies and design integration with actual heat exchangers. Designs need to consider electrically insulating the fluid as a potential solution to current leaks, to extrapolate the usage of EHD for conducting fluids. This can be achieved, for example, by coating the heat exchanger with an electrically insulating film such as Parylene C. Furthermore, fluid mixtures that might optimize the parameters for EHD forces still need to be investigated. With respect to EFE jumping-droplet condensation, opportunities exist for integration with wickless vapor chambers where jumping droplets return to the evaporator with the help of an external electric fields. The vapor chambers can be used in applications where other forces need to be overcome, such as gravitational forces from accelerating vehicles (aircraft).

The electric (and magnetic) manipulation of non-coalescing droplets deserves more investigation as a way to improve DWC performance. The electrical properties of the condensing droplets permit their removal from the surface without the need for coalescence-induced jumping and associated net charge gain. For instance, the H^+ and OH^- ions in deionized water act as free charge carriers that feel a force when subjected to an electric field. These ions appear because water molecules dissociate at thermodynamic equilibrium [254]. A problem that is widely studied in the literature is that of an electrically conducting droplet sitting on the lower electrode of a parallel plate capacitor [255, 256]. The charges inside the droplet migrate to the surface in such a way as to cancel the interior electric field which creates a net electric force that acts on the droplet. The competition between the Maxwell electric stresses, buoyancy and surface tension leads to the deformation of the droplet and the decrease of its contact angle until de-wetting occurs when the contact angle reaches below the receding contact angle. The droplet lifts off if the electric force is larger than the buoyancy force [257]. For an undeformed spherical droplet, the estimate for the net electrostatic force is

$$F_{\mathrm{el}} = 4\pi\epsilon E^2 r^2 \left(\zeta(3) + \frac{1}{6}\right), \tag{13.6}$$

where ϵ is the permittivity of the surrounding medium, E is the electric field, r is the radius of the sphere and ζ is the Riemann zeta function [258]. Equation (13.6) predicts the critical electric field that leads to lift-off when the electric force is equal to the buoyancy force. The term E_{cr} is approximately equal to 5 KV cm^{-1} for a 1 mm radius water droplet, a value that is close to the order of electric field needed for EHD. Dielectric droplets are known to exhibit a similar behavior [259], but more work is needed to achieve their lift-off.

Preventing droplets from coalescing is beneficial when electric fields are used to remove droplets directly from a surface, in particular since larger droplets need higher electric fields for lift-off. A recent study showed that oppositely charged water droplets exposed to an electric field are repelled from one another instead of coalescing after contact, due to a fast charge transfer between the droplets [260]. The observed behavior occurs when the electric field strength exceeds a critical value on the order of kV cm^{-1}. Alternatively, the charge redistribution or dielectrophoretic effects triggered by the application of an electric field create inter-particle forces that can potentially be leveraged to control the spacing and relative motion between interacting droplets. This concept has already been used successfully in a variety of processes such as de-emulsification [261] and two-dimensional assembly of monolayers [262]. It was shown that the force acting between two dielectric particles is attractive when the line joining them and the electric field applied are parallel, and repulsive when they are perpendicular [263, 264]. Hence, exposing a surface on which droplets are forming due to DWC in the presence of a normal electric field creates a repulsive force that can possibly be useful in controlling their clustering, preventing coalescence and achieving spatial control of droplet position.

References

[1] Beér J M 2007 High efficiency electric power generation: the environmental role *Prog. Energy Combust. Sci.* **33** 107–34

[2] Rosen M A 2001 Energy- and exergy-based comparison of coal-fired and nuclear steam power plants *Exergy Inter. J.* **1** 180–92

[3] Mazzei P, Minichiello F and Palma D 2005 HVAC dehumidification systems for thermal comfort: a critical review *Appl. Therm. Eng.* **25** 677–707

[4] Pérez-Lombard L, Ortiz J and Pout C 2008 A review on buildings energy consumption information *Energy Build.* **40** 394–8

[5] Oh J, Birbarah P, Foulkes T, Yin S L, Rentauskas M, Neely J, Pilawa-Podgurski R C N and Miljkovic N 2017 Jumping-droplet electronics hot-spot cooling *Appl. Phys. Lett.* **110** 123107

[6] Miljkovic N, Preston D J, Enright R and Wang E N 2014 Jumping-droplet electrostatic energy harvesting *Appl. Phys. Lett.* **105** 013111

[7] Humplik T, Lee J, O'hern S, Fellman B, Baig M, Hassan S, Atieh M, Rahman F, Laoui T and Karnik R 2011 Nanostructured materials for water desalination *Nanotechnology* **22** 292001

[8] Vinoth Kumar K and Kasturi Bai R 2008 Performance study on solar still with enhanced condensation *Desalination* **230** 51–61

[9] Kim H, Yang S, Rao S R, Narayanan S, Kapustin E A, Furukawa H, Umans A S, Yaghi O M and Wang E N 2017 Water harvesting from air with metal–organic frameworks powered by natural sunlight *Science* **356** 430

[10] Park K-C, Kim P, Grinthal A, He N, Fox D, Weaver J C and Aizenberg J 2016 Condensation on slippery asymmetric bumps *Nature* **531** 78
[11] Nusselt W 1916 The surface condensation of water vapor *Z. Ver. Dtsch. Ing.* **60** 541–6
[12] Rose J W 2002 Dropwise condensation theory and experiment: a review *Proc. Inst. Mech. Eng.* A **216** 115–28
[13] Boreyko J B and Chen C-H 2010 Self-propelled jumping drops on superhydrophobic surfaces *Phys. Fluids* **22** 091110
[14] Enright R, Miljkovic N, Sprittles J, Nolan K, Mitchell R and Wang E N 2014 How coalescing droplets jump *ACS Nano* **8** 10352–62
[15] Miljkovic N, Enright R, Nam Y, Lopez K, Dou N, Sack J and Wang E N 2013 Jumping-droplet-enhanced condensation on scalable superhydrophobic nanostructured surfaces *Nano Lett.* **13** 179–87
[16] Birbarah P and Miljkovic N 2017 Internal convective jumping-droplet condensation in tubes *Int. J. Heat Mass Transfer* **114** 1025–36
[17] Birbarah P and Miljkovic N 2017 External convective jumping-droplet condensation on a flat plate *Int. J. Heat Mass Transfer* **107** 74–88
[18] Kim M-K, Cha H, Birbarah P, Chavan S, Zhong C, Xu Y and Miljkovic N 2015 Enhanced jumping-droplet departure *Langmuir* **31** 13452–66
[19] Enright R, Miljkovic N, Dou N, Nam Y and Wang E N 2013 Condensation on superhydrophobic copper oxide nanostructures *J. Heat Transfer* **135** 091304–091304-12
[20] Nenad M, Daniel John P, Ryan E, Solomon A, Youngsuk N and Evelyn N W 2013 Jumping droplet dynamics on scalable nanostructured superhydrophobic surfaces *J. Heat Transfer* **135** 080907–080907-1
[21] Miljkovic N, Enright R and Wang E N 2012 Effect of droplet morphology on growth dynamics and heat transfer during condensation on superhydrophobic nanostructured surfaces *ACS Nano* **6** 1776–85
[22] Wiedenheft K F, Guo H A, Qu X, Boreyko J B, Liu F, Zhang K, Eid F, Choudhury A, Li Z and Chen C-H 2017 Hotspot cooling with jumping-drop vapor chambers *Appl. Phys. Lett.* **110** 141601
[23] Qu X, Boreyko J B, Liu F, Agapov R L, Lavrik N V, Retterer S T, Feng J J, Collier C P and Chen C-H 2015 Self-propelled sweeping removal of dropwise condensate *Appl. Phys. Lett.* **106** 221601
[24] Liu F, Ghigliotti G, Feng J J and Chen C-H 2014 Numerical simulations of self-propelled jumping upon drop coalescence on non-wetting surfaces *J. Fluid Mech.* **752** 39–65
[25] Murase T, Wang H S and Rose J W 2006 Effect of inundation for condensation of steam on smooth and enhanced condenser tubes *Int. J. Heat Mass Transfer* **49** 3180–9
[26] Belghazi M, Bontemps A and Marvillet C 2002 Condensation heat transfer on enhanced surface tubes: experimental results and predictive theory *J. Heat Transfer* **124** 754–61
[27] Guo S-P, Wu Z, Li W, Kukulka D, Sundén B, Zhou X-P, Wei J-J and Simon T 2015 Condensation and evaporation heat transfer characteristics in horizontal smooth, herring-bone and enhanced surface EHT tubes *Int. J. Heat Mass Transfer* **85** 281–91
[28] Cho H J, Preston D J, Zhu Y and Wang E N 2016 Nanoengineered materials for liquid–vapour phase-change heat transfer *Nat. Rev. Mater.* **2** 16092
[29] Shahriari A, Birbarah P, Oh J, Miljkovic N and Bahadur V 2017 Electric field-based control and enhancement of boiling and condensation *Nanoscale Microscale Thermophys. Eng.* **21** 102–21

[30] Cavallini A, Censi G, Del Col D, Doretti L, Longo G A, Rossetto L and Zilio C 2003 Condensation inside and outside smooth and enhanced tubes—a review of recent research *Int. J. Refrig.* **26** 373–92
[31] Ma X, Rose J W, Xu D, Lin J and Wang B 2000 Advances in dropwise condensation heat transfer: Chinese research *Chem. Eng. J.* **78** 87–93
[32] Tanasawa I 1991 Advances in condensation heat transfer *Advances in Heat Transfer* ed J P Hartnett, T F Irvine and Y I Cho vol 21 (Amsterdam: Elsevier), pp 55–139
[33] Attinger D, Frankiewicz C, Betz A R, Schutzius T M, Ganguly R, Das A, Kim C-J and Megaridis C M 2014 Surface engineering for phase change heat transfer: a review *MRS Energy Sustainability* **1** E4
[34] Enright R, Miljkovic N, Alvarado J L, Kim K and Rose J W 2014 Dropwise condensation on micro- and nanostructured surfaces *Nanoscale Microscale Thermophys. Eng.* **18** 223–50
[35] Dalkilic A S and Wongwises S 2009 Intensive literature review of condensation inside smooth and enhanced tubes *Int. J. Heat Mass Transfer* **52** 3409–26
[36] Zhang P and Lv F Y 2015 A review of the recent advances in superhydrophobic surfaces and the emerging energy-related applications *Energy* **82** 1068–87
[37] Miljkovic N and Wang E N 2013 Condensation heat transfer on superhydrophobic surfaces *MRS Bull.* **38** 397–406
[38] Wen R, Xu S, Ma X, Lee Y-C and Yang R 2018 Three-dimensional superhydrophobic nanowire networks for enhancing condensation heat transfer *Joule* **2** 269–79
[39] Rose J 1964 Dropwise condensation of steam on vertical planes *PhD Dissertation* University of London (Queen Mary College)
[40] Utaka Y, Saito A, Ishikawa H and Yanagida H 1986 Study on dropwise condensation curves: dropwise to filmwise transition of propylene glycol, ethylene glycol and glycerol vapors on a copper surface using a monolayer type promoter (Part 1) *Bull. JSME* **29** 4228–34
[41] Utaka Y, Saito A, Ishikawawa H and Yanagida H 1988 Study on dropwise condensation curves (dropwise to filmwise transition of propylene glycol, ethylene glycol and glycerol vapors on a copper surface using a monolayer type promoter—part 2) *JSME Inter. J.* **31** 73–80
[42] Das A K, Kilty H P, Marto P J, Andeen G B and Kumar A 1999 The use of an organic self-assembled monolayer coating to promote dropwise condensation of steam on horizontal tubes *J. Heat Transfer* **122** 278–86
[43] Vemuri S, Kim K J, Wood B D, Govindaraju S and Bell T W 2006 Long term testing for dropwise condensation using self-assembled monolayer coatings of *n*-octadecyl mercaptan *Appl. Therm. Eng.* **26** 421–9
[44] Vemuri S and Kim K J 2006 An experimental and theoretical study on the concept of dropwise condensation *Int. J. Heat Mass Transfer* **49** 649–57
[45] Pauporté T, Bataille G, Joulaud L and Vermersch F J 2010 Well-aligned ZnO nanowire arrays prepared by seed-layer-free electrodeposition and their Cassie–Wenzel transition after hydrophobization *J. Phys. Chem.* C **114** 194–202
[46] Razavi S M R *et al* 2017 Superhydrophobic surfaces made from naturally derived hydrophobic materials *ACS Sustain. Chem. Eng.* **5** 11362–70
[47] Love J C, Wolfe D B, Haasch R, Chabinyc M L, Paul K E, Whitesides G M and Nuzzo R G 2003 Formation and structure of self-assembled monolayers of alkanethiolates on palladium *JACS* **125** 2597–609
[48] Love J C, Estroff L A, Kriebel J K, Nuzzo R G and Whitesides G M 2005 Self-assembled monolayers of thiolates on metals as a form of nanotechnology *Chem. Rev.* **105** 1103–70

[49] Chen L, Liang S, Yan R, Cheng Y, Huai X and Chen S 2009 *n*-octadecanethiol self-assembled monolayer coating with microscopic roughness for dropwise condensation of steam *J. Therm. Sci.* **18** 160–5

[50] Enright R, Miljkovic N, Al-Obeidi A, Thompson C V and Wang E N 2012 Condensation on superhydrophobic surfaces: the role of local energy barriers and structure length scale *Langmuir* **28** 14424–32

[51] Boreyko J B and Chen C H 2013 Vapor chambers with jumping-drop liquid return from superhydrophobic condensers *Int. J. Heat Mass Transfer* **61** 409–18

[52] Miljkovic N, Preston D J, Enright R and Wang E N 2013 Electrostatic charging of jumping droplets *Nat. Commun.* **4** 2517

[53] Varanasi K K, Hsu M, Bhate N, Yang W S and Deng T 2009 Spatial control in the heterogeneous nucleation of water *Appl. Phys. Lett.* **95** 094101

[54] Chen X, Wu J, Ma R, Hua M, Koratkar N, Yao S and Wang Z 2011 Nanograssed micropyramidal architectures for continuous dropwise condensation *Adv. Funct. Mater.* **21** 4617–23

[55] Cha H, Chun J M, Sotelo J and Miljkovic N 2016 Focal plane shift imaging for the analysis of dynamic wetting processes *ACS Nano* **10** 8223–32

[56] Chavan S *et al* 2016 Heat transfer through a condensate droplet on hydrophobic and nanostructured superhydrophobic surfaces *Langmuir* **32** 7774–87

[57] Rykaczewski K, Scott J H J, Rajauria S, Chinn J, Chinn A M and Jones W 2011 Three dimensional aspects of droplet coalescence during dropwise condensation on superhydrophobic surfaces *Soft Matter* **7** 8749–52

[58] Paxson A T, Yagüe J L, Gleason K K and Varanasi K K 2014 Stable dropwise condensation for enhancing heat transfer via the initiated chemical vapor deposition (iCVD) of grafted polymer films *Adv. Mater.* **26** 418–23

[59] Preston D J, Mafra D L, Miljkovic N, Kong J and Wang E N 2015 Scalable graphene coatings for enhanced condensation heat transfer *Nano Lett.* **15** 2902–9

[60] Luo H, Liu T, Ma J, Wang P, Wang Y, Leprince-Wang Y and Jing G 2016 Evaporation-induced failure of hydrophobicity *Phys. Rev. Fluids* **1** 053901

[61] Crudden C M *et al* 2014 Ultra stable self-assembled monolayers of *N*-heterocyclic carbenes on gold *Nat. Chem.* **6** 409

[62] Erb R and Thelen E 1965 Promoting permanent dropwise condensation *Ind. Eng. Chem.* **57** 49–52

[63] Brown A R and Thomas M A 1966 Filmwise and dropwise condensation of steam at low pressures (International Heat Transfer Conference Digital Library) (Danbury, CT: Begell House)

[64] Ma X, Xu D and Lin J 1994 A study of dropwise condensation on the ultra-thin polymer surfaces. *International Heat Transfer Conference Digital Library* (Begell House Inc.)

[65] Oh I-K *et al* 2015 Hydrophobicity of rare earth oxides grown by atomic layer deposition *Chem. Mater.* **27** 148–56

[66] Ölçeroğlu E and McCarthy M 2016 Self-organization of microscale condensate for delayed flooding of nanostructured superhydrophobic surfaces *ACS Appl. Mater. Interfaces* **8** 5729–36

[67] Cha H, Xu C, Sotelo J, Chun J M, Yokoyama Y, Enright R and Miljkovic N 2016 Coalescence-induced nanodroplet jumping *Phys. Rev. Fluids* **1** 064102

[68] Hannemann R J and Mikic B B 1976 An experimental investigation into the effect of surface thermal conductivity on the rate of heat transfer in dropwise condensation *Int. J. Heat Mass Transfer* **19** 1309–17

[69] Rose J W 1978 The effect of surface thermal conductivity on dropwise condensation heat transfer *Int. J. Heat Mass Transfer* **21** 80–1

[70] Kuppusami S and Oskouei R H 2015 Parylene coatings in medical devices and implants: a review *Univ. J. Biomed. Eng.* **3** 9–14

[71] Chen T-N, Wuu D-S, Wu C-C, Chiang C-C, Chen Y-P and Horng R-H 2007 Improvements of permeation barrier coatings using encapsulated Parylene interlayers for flexible electronic applications *Plasma Processes Polym.* **4** 180–5

[72] Young Shik S, Keunchang C, Sun Hee L, Seok C, Sung-Jin P, Chanil C, Dong-Chul H and Jun Keun C 2003 PDMS-based micro PCR chip with Parylene coating *J. Micromech. Microeng.* **13** 768

[73] Noar J H, Wahab A, Evans R D and Wojcik A G 1999 The durability of parylene coatings on neodymium-iron-boron magnets *Eur. J. Orthod.* **21** 685–93

[74] 1965 Parylene coating promotes dropwise condensation on condenser tubes *Chem. Eng. News Archive* **43** 45

[75] Marto P J, Looney D J, Rose J W and Wanniarachchi A S 1986 Evaluation of organic coatings for the promotion of dropwise condensation of steam *Int. J. Heat Mass Transfer* **29** 1109–17

[76] Holden K M, Wanniarachchi A S, Marto P J, Boone D H and Rose J W 1987 The use of organic coatings to promote dropwise condensation of steam *J. Heat Trans.* **109** 768–74

[77] Spivack M A and Ferrante G 1969 Determination of the water vapor permeability and continuity of ultrathin Parylene membranes *J. Electrochem. Soc.* **116** 1592–4

[78] SCS Parylene Properties 2016 Specialty Coating Systems

[79] Chen C-H, Cai Q, Tsai C, Chen C-L, Xiong G, Yu Y and Ren Z 2007 Dropwise condensation on superhydrophobic surfaces with two-tier roughness *Appl. Phys. Lett.* **90** 173108

[80] Tenhaeff W E and Gleason K K 2008 Initiated and oxidative chemical vapor deposition of polymeric thin films: iCVD and oCVD *Adv. Funct. Mater.* **18** 979–92

[81] Lau K K S and Gleason K K 2006 Initiated chemical vapor deposition (iCVD) of poly (alkyl acrylates): an experimental study *Macromolecules* **39** 3688–94

[82] Chan K and Gleason K K 2005 Initiated chemical vapor deposition of linear and cross-linked poly(2-hydroxyethyl methacrylate) for use as thin-film hydrogels *Langmuir* **21** 8930–9

[83] Soto D, Ugur A, Farnham T A, Gleason K K and Varanasi K K 2018 Short-fluorinated iCVD coatings for nonwetting fabrics *Adv. Funct. Mater.* **28** 1707355

[84] Ölçeroğlu E, Hsieh C-Y, Rahman M M, Lau K K S and McCarthy M 2014 Full-field dynamic characterization of superhydrophobic condensation on biotemplated nanostructured surfaces *Langmuir* **30** 7556–66

[85] Alf M E *et al* 2010 Chemical vapor deposition of conformal, functional, and responsive polymer films *Adv. Mater.* **22** 1993–2027

[86] Yagüe J L and Gleason K K 2013 Enhanced cross-linked density by annealing on fluorinated polymers synthesized via initiated chemical vapor deposition to prevent surface reconstruction *Macromolecules* **46** 6548–54

[87] Peng C, Chen Z and Tiwari M K 2018 All-organic superhydrophobic coatings with mechanochemical robustness and liquid impalement resistance *Nat. Mater.* **17** 355–60

[88] Wang Y and Gong X 2017 Superhydrophobic coatings with periodic ring structured patterns for self-cleaning and oil–water separation *Adv. Mater.* **4** 1700190
[89] Tiwari M K, Bayer I S, Jursich G M, Schutzius T M and Megaridis C M 2010 Highly liquid-repellent, large-area, nanostructured poly(vinylidene fluoride)/poly(ethyl 2-cyanoacrylate) composite coatings: particle filler effects *ACS Appl. Mater. Interfaces* **2** 1114–9
[90] Bakir M, Henderson C N, Meyer J L, Oh J, Miljkovic N, Kumosa M, Economy J and Jasiuk I 2018 Effects of environmental aging on physical properties of aromatic thermosetting copolyester matrix neat and nanocomposite foams *Polym. Degrad. Stab.* **147** 49–56
[91] Kondrashov V and Rühe J 2014 Microcones and nanograss: toward mechanically robust superhydrophobic surfaces *Langmuir* **30** 4342–50
[92] Wu Y, Zhou S and Wu L 2016 Fabrication of robust hydrophobic and super-hydrophobic polymer films with onefold or dual inverse opal structures *Macromol. Mater. Eng.* **301** 1430–6
[93] Scarratt L R J, Hoatson B S, Wood E S, Hawkett B S and Neto C 2016 Durable superhydrophobic surfaces via spontaneous wrinkling of Teflon AF *ACS Appl. Mater. Interfaces* **8** 6743–50
[94] Hönes R, Kondrashov V, Huai H and Rühe J 2017 Wetting transitions in polymer nanograss generated by nanoimprinting *Macromol. Chem. Phys.* **218** 1700056
[95] Hönes R, Kondrashov V and Rühe J 2017 Molting materials: restoring superhydrophobicity after severe damage via snakeskin-like shedding *Langmuir* **33** 4833–9
[96] Robertson J 2002 Diamond-like amorphous carbon *Mater. Sci. Eng. Rep.* **37** 129–281
[97] Ferrari A C and Robertson J 2000 Interpretation of Raman spectra of disordered and amorphous carbon *Phys. Rev.* B **61** 14095–107
[98] Dearnaley G and Arps J H 2005 Biomedical applications of diamond-like carbon (DLC) coatings: a review *Surf. Coat. Technol.* **200** 2518–24
[99] Roy R K and Lee K-R 2007 Biomedical applications of diamond-like carbon coatings: a review *J. Biomed. Mater. Res.* B **83B** 72–84
[100] Robertson J 2008 Comparison of diamond-like carbon to diamond for applications *Phys. Status Solidi* a **205** 2233–44
[101] Treutler C P O 2005 Industrial use of plasma-deposited coatings for components of automotive fuel injection systems *Surf. Coat. Technol.* **200** 1969–75
[102] van der Kolk G J 2008 Wear resistance of amorphous DLC and metal containing DLC in industrial applications *Tribology of Diamond-Like Carbon Films: Fundamentals and Applications* ed C Donnet and A Erdemir (Boston, MA: Springer), pp 484–93
[103] Matthews A and Eskildsen S S 1994 Engineering applications for diamond-like carbon *Diamond Relat. Mater.* **3** 902–11
[104] Trojan K, Grischke M and Dimigen H 1994 Network modification of DLC coatings to adjust a defined surface energy *Phys. Status Solidi* a **145** 575–85
[105] Grischke M, Bewilogua K, Trojan K and Dimigen H 1995 Application-oriented modifications of deposition processes for diamond-like-carbon-based coatings *Surf. Coat. Technol.* **7475** 739–45
[106] Grischke M, Hieke A, Morgenweck F and Dimigen H 1998 Variation of the wettability of DLC-coatings by network modification using silicon and oxygen *Diamond Relat. Mater.* **7** 454–58
[107] Shamsa M, Liu W L, Balandin A A, Casiraghi C, Milne W I and Ferrari A C 2006 Thermal conductivity of diamond-like carbon films *Appl. Phys. Lett.* **89** 161921

[108] Koch G, Zhang D C and Leipertz A 1997 Condensation of steam on the surface of hard coated copper discs *Heat Mass Transfer* **32** 149–56

[109] Koch G, Zhang D C, Leipertz A, Grischke M, Trojan K and Dimigen H 1998 Study on plasma enhanced CVD coated material to promote dropwise condensation of steam *Int. J. Heat Mass Transfer* **41** 1899–906

[110] Azimi G, Dhiman R, Kwon H-M, Paxson A T and Varanasi K K 2013 Hydrophobicity of rare-earth oxide ceramics *Nat. Mater.* **12** 315–20

[111] Khan S, Azimi G, Yildiz B and Varanasi K K 2015 Role of surface oxygen-to-metal ratio on the wettability of rare-earth oxides *Appl. Phys. Lett.* **106** 061601

[112] Pedraza F, Mahadik S A and Bouchaud B 2015 Synthesis of ceria based superhydrophobic coating on $Ni_{20}Cr$ substrate via cathodic electrodeposition *Phys. Chem. Chem. Phys.* **17** 31750–7

[113] Carchini G, García-Melchor M, Łodziana Z and López N 2016 Understanding and tuning the intrinsic hydrophobicity of rare-earth oxides: a DFT+U study *ACS Appl. Mater. Interfaces* **8** 152–60

[114] Tam J, Palumbo G, Erb U and Azimi G 2017 Robust hydrophobic rare earth oxide composite electrodeposits *Adv. Mater. Interfaces* **4** 1700850

[115] Preston D J, Miljkovic N, Sack J, Enright R, Queeney J and Wang E N 2014 Effect of hydrocarbon adsorption on the wettability of rare earth oxide ceramics *Appl. Phys. Lett.* **105** 011601

[116] Lundy R, Byrne C, Bogan J, Nolan K, Collins M N, Dalton E and Enright R 2017 Exploring the role of adsorption and surface state on the hydrophobicity of rare earth oxides *ACS Appl. Mater. Interfaces* **9** 13751–60

[117] Külah E, Marot L, Steiner R, Romanyuk A, Jung T A, Wäckerlin A and Meyer E 2017 Surface chemistry of rare-earth oxide surfaces at ambient conditions: reactions with water and hydrocarbons *Sci. Rep.* **7** 43369

[118] Fu S-P, Rossero J, Chen C, Li D, Takoudis C G and Abiade J T 2017 On the wetting behavior of ceria thin films grown by pulsed laser deposition *Appl. Phys. Lett.* **110** 081601

[119] Shim J, Seo D, Oh S, Lee J and Nam Y 2018 Condensation heat-transfer performance of thermally stable superhydrophobic cerium-oxide surfaces *ACS Appl. Mater. Interfaces* **10** 31765–76

[120] Prakash S, Ghosh S, Patra A, Annamalai M, Motapothula M R, Sarkar S, Tan S J R, Zhunan J, Loh K P and Venkatesan T 2018 Intrinsic hydrophilic nature of epitaxial thin-film of rare-earth oxide grown by pulsed laser deposition *Nanoscale* **10** 3356–61

[121] Xu P, Coyle T W, Pershin L and Mostaghimi J 2018 Superhydrophobic ceramic coating: fabrication by solution precursor plasma spray and investigation of wetting behavior *J. Colloid Interface Sci.* **523** 35–44

[122] Mogensen M, Sammes N M and Tompsett G A 2000 Physical, chemical and electrochemical properties of pure and doped ceria *Solid State Ionics* **129** 63–94

[123] White M L 1964 The wetting of gold surfaces by water *J. Phys. Chem.* **68** 3083–85

[124] Erb R A 1965 Wettability of metals under continuous condensing conditions *J. Phys. Chem.* **69** 1306–9

[125] White M L and Drobek J 1966 The effect of residual abrasives on the wettability of polished gold surfaces *J. Phys. Chem.* **70** 3432–36

[126] Erb R A 1968 Wettability of gold *J. Phys. Chem.* **72** 2412–7

[127] Erb R A, Haigh T I and Downing T M 1970 Permanent systems for dropwise condensation for distillation plants *Symp. on Enhanced Tubes for Desalination Plants (US Dept. of Interior Washington, DC)* pp 177–201

[128] Morcos I 1971 Double layer structure and the phenomena of wetting *J. Colloid Interface Sci.* **37** 410–21

[129] Trasatti S 1971 Work function, electronegativity, and electrochemical behaviour of metals: II. Potentials of zero charge and 'electrochemical' work functions *J. Electroanal. Chem. Interfacial Electrochem.* **33** 351–78

[130] Trasatti S 1974 Operative (electrochemical) work function of gold *J. Electroanal. Chem. Interfacial Electrochem.* **54** 19–24

[131] Morcos I 1975 The electrocapillary phenomena of solid electrodes *J. Electroanal. Chem. Interfacial Electrochem.* **62** 313–40

[132] Bewig K W and Zisman W A 1965 The wetting of gold and platinum by water *J. Phys. Chem.* **69** 4238–42

[133] Bernett M K and Zisman W A 1970 Confirmation of spontaneous spreading by water on pure gold *J. Phys. Chem.* **74** 2309–12

[134] Schrader M E 1970 Ultrahigh-vacuum techniques in the measurement of contact angles II. Water on gold *J. Phys. Chem.* **74** 2313–7

[135] Wilkins D G, Bromley L A and Read S M 1973 Dropwise and filmwise condensation of water vapor on gold *AlChE J.* **19** 119–23

[136] Gardner J R and Woods R 1977 The hydrophilic nature of gold and platinum *J. Electroanal. Chem. Interfacial Electrochem.* **81** 285–90

[137] Parsegian V A, Weiss G H and Schrader M E 1977 Macroscopic continuum model of influence of hydrocarbon contaminant on forces causing wetting of gold by water *J. Colloid Interface Sci.* **61** 356–60

[138] Smith T 1980 The hydrophilic nature of a clean gold surface *J. Colloid Interface Sci.* **75** 51–5

[139] Woodruff D W and Westwater J W 1981 Steam condensation on various gold surfaces *J. Heat Transf.* **103** 685–92

[140] Westwater J W 1981 Gold surfaces for condensation heat transfer *Gold Bull.* **14** 95–101

[141] O'Neill G A and Westwater J W 1984 Dropwise condensation of steam on electroplated silver surfaces *Int. J. Heat Mass Transfer* **27** 1539–49

[142] Erb R A 1973 Dropwise condensation on gold *Gold Bull.* **6** 2–6

[143] Vorobyev A Y and Guo C 2015 Multifunctional surfaces produced by femtosecond laser pulses *J. Appl. Phys.* **117** 033103

[144] Kietzig A-M, Hatzikiriakos S G and Englezos P 2009 Patterned superhydrophobic metallic surfaces *Langmuir* **25** 4821–7

[145] Khorsand S, Raeissi K, Ashrafizadeh F and Arenas M A 2015 Super-hydrophobic nickel–cobalt alloy coating with micro-nano flower-like structure *Chem. Eng. J.* **273** 638–46

[146] Balajka J, Hines M A, DeBenedetti W J I, Komora M, Pavelec J, Schmid M and Diebold U 2018 High-affinity adsorption leads to molecularly ordered interfaces on TiO_2 in air and solution *Science* **361** 786–9

[147] Yang Z, Liu X and Tian Y 2019 Insights into the wettability transition of nanosecond laser ablated surface under ambient air exposure *J. Colloid Interface Sci.* **533** 268–77

[148] Zhang Y, Zou G, Liu L, Zhao Y, Liang Q, Wu A and Zhou Y N 2016 Time-dependent wettability of nano-patterned surfaces fabricated by femtosecond laser with high efficiency *Appl. Surf. Sci.* **389** 554–9

[149] Chang F-M, Cheng S-L, Hong S-J, Sheng Y-J and Tsao H-K 2010 Superhydrophilicity to superhydrophobicity transition of CuO nanowire films *Appl. Phys. Lett.* **96** 114101
[150] Boinovich L B, Emelyanenko A M, Pashinin A S, Lee C H, Drelich J and Yap Y K 2012 Origins of thermodynamically stable superhydrophobicity of boron nitride nanotubes coatings *Langmuir* **28** 1206–16
[151] Li L H and Chen Y 2010 Superhydrophobic properties of nonaligned boron nitride nanotube films *Langmuir* **26** 5135–40
[152] Kim G-T, Gim S-J, Cho S-M, Koratkar N and Oh I-K 2014 Wetting-transparent graphene films for hydrophobic water-harvesting surfaces *Adv. Mater.* **26** 5166–72
[153] Long J, Zhong M, Zhang H and Fan P 2015 Superhydrophilicity to superhydrophobicity transition of picosecond laser microstructured aluminum in ambient air *J. Colloid Interface Sci.* **441** 1–9
[154] Wang G and Zhang T-Y 2012 Oxygen adsorption induced superhydrophilic-to-superhydrophobic transition on hierarchical nanostructured CuO surface *J. Colloid Interface Sci.* **377** 438–41
[155] Wang L and McCarthy T J 2016 Covalently attached liquids: instant omniphobic surfaces with unprecedented repellency *Angew. Chem. Int. Ed.* **55** 244–8
[156] Castro Y, Ferrari B, Moreno R and Durán A 2005 Corrosion behaviour of silica hybrid coatings produced from basic catalysed particulate sols by dipping and EPD *Surf. Coat. Technol.* **191** 228–35
[157] Lopez G, Biebuyck H, Frisbie C and Whitesides G 1993 Imaging of features on surfaces by condensation figures *Science* **260** 647–49
[158] Cao P, Xu K, Varghese J O and Heath J R 2011 The microscopic structure of adsorbed water on hydrophobic surfaces under ambient conditions *Nano Lett.* **11** 5581–6
[159] Cha H, Wu A, Kim M-K, Saigusa K, Liu A and Miljkovic N 2017 Nanoscale-agglomerate-mediated heterogeneous nucleation *Nano Lett.* **17** 7544–51
[160] Cognard J 1994 Blistering of glass–epoxy amine adhesive joints in water vapour at high pressure an indication of interfacial crumpling *J. Adhes.* **47** 83–93
[161] Yang X F, Vang C, Tallman D E, Bierwagen G P, Croll S G and Rohlik S 2001 Weathering degradation of a polyurethane coating *Polym. Degrad. Stab.* **74** 341–51
[162] Yang X F, Tallman D E, Bierwagen G P, Croll S G and Rohlik S 2002 Blistering and degradation of polyurethane coatings under different accelerated weathering tests *Polym. Degrad. Stab.* **77** 103–9
[163] Tucker W C and Brown R 1989 Blister formation on graphite/polymer composites galvanically coupled with steel in seawater *J. Compos. Mater.* **23** 389–95
[164] Hinkley J A 1983 A blister test for adhesion of polymer films to SiO_2 *J. Adhes.* **16** 115–25
[165] Jensen H M 1991 The blister test for interface toughness measurement *Eng. Fract. Mech.* **40** 475–86
[166] Cao Z, Wang P, Gao W, Tao L, Suk J W, Ruoff R S, Akinwande D, Huang R and Liechti K M 2014 A blister test for interfacial adhesion of large-scale transferred graphene *Carbon* **69** 390–400
[167] Perraud V *et al* 2015 The future of airborne sulfur-containing particles in the absence of fossil fuel sulfur dioxide emissions *PNAS* **112** 13514–9
[168] Aneja V P 1990 Natural sulfur emissions into the atmosphere *J. Air Waste Manage. Assoc.* **40** 469–76

[169] Bates T S, Lamb B K, Guenther A, Dignon J and Stoiber R E 1992 Sulfur emissions to the atmosphere from natural sourees *J. Atmos. Chem.* **14** 315–37
[170] Barnes I, Hjorth J and Mihalopoulos N 2006 Dimethyl sulfide and dimethyl sulfoxide and their oxidation in the atmosphere *Chem. Rev.* **106** 940–75
[171] Lana A *et al* 2011 An updated climatology of surface dimethlysulfide concentrations and emission fluxes in the global ocean *Global Biogeochem. Cycles* **25** GB1004
[172] Jardine K *et al* 2015 Dimethyl sulfide in the Amazon rain forest *Global Biogeochem. Cycles* **29** 19–32
[173] Filipy J, Rumburg B, Mount G, Westberg H and Lamb B 2006 Identification and quantification of volatile organic compounds from a dairy *Atmos. Environ.* **40** 1480–94
[174] Trabue S, Scoggin K, Mitloehner F, Li H, Burns R and Xin H 2008 Field sampling method for quantifying volatile sulfur compounds from animal feeding operations *Atmos. Environ.* **42** 3332–41
[175] Feilberg A, Liu D, Adamsen A P S, Hansen M J and Jonassen K E N 2010 Odorant emissions from intensive pig production measured by online proton-transfer-reaction mass spectrometry *Environ. Sci. Technol.* **44** 5894–900
[176] Hansen M J, Toda K, Obata T, Adamsen A P S and Feilberg A 2012 Evaluation of single column trapping/separation and chemiluminescence detection for measurement of methanethiol and dimethyl sulfide from pig production *J. Anal. Methods Chem.* **7** 489239
[177] Rumsey I C, Aneja V P and Lonneman W A 2014 Characterizing reduced sulfur compounds emissions from a swine concentrated animal feeding operation *Atmos. Environ.* **94** 458–66
[178] Gligorovski S and Abbatt J P D 2018 An indoor chemical cocktail *Science* **359** 632–3
[179] McDonald B C *et al* 2018 Volatile chemical products emerging as largest petrochemical source of urban organic emissions *Science* **359** 760–4
[180] Kürten A *et al* 2014 Neutral molecular cluster formation of sulfuric acid–dimethylamine observed in real time under atmospheric conditions *PNAS* **111** 15019–24
[181] Karl M, Gross A, Leck C and Pirjola L 2007 Intercomparison of dimethylsulfide oxidation mechanisms for the marine boundary layer: gaseous and particulate sulfur constituents *J. Geophys. Res. Atmos.* **112** D15304
[182] Dawson M L, Varner M E, Perraud V, Ezell M J, Gerber R B and Finlayson-Pitts B J 2012 Simplified mechanism for new particle formation from methanesulfonic acid, amines, and water via experiments and *ab initio* calculations *PNAS* **109** 18719–24
[183] Faghri A 1995 *Heat Pipe Science and Technology* 2nd ed (Columbia, MO: Global Digital Press)
[184] Bejan A 2002 Fundamentals of exergy analysis, entropy generation minimization, and the generation of flow architecture *Int. J. Energy Res.* **26** 0–43
[185] Landoulsi J, Genet M J, Fleith S, Touré Y, Liascukiene I, Méthivier C and Rouxhet P G 2016 Organic adlayer on inorganic materials: XPS analysis selectivity to cope with adventitious contamination *Appl. Surf. Sci.* **383** 71–83
[186] Oh J, Zhang R, Shetty P P, Krogstad J A, Braun P V and Miljkovic N 2018 Thin film condensation on nanostructured surfaces *Adv. Funct. Mater.* **28** 1707000
[187] Ölçeroğlu E, Hsieh C-Y, Lau K K S and McCarthy M 2017 Thin film condensation supported on ambiphilic microstructures *J. Heat Transfer* **139** 020910
[188] Wang R and Antao D S 2018 Capillary-enhanced filmwise condensation in porous media *Langmuir* **34** 13855–63

[189] Preston D J, Wilke K L, Lu Z, Cruz S S, Zhao Y, Becerra L L and Wang E N 2018 Gravitationally driven wicking for enhanced condensation heat transfer *Langmuir* **34** 4658–64
[190] Park K-J, Seo T and Jung D 2007 Performance of alternative refrigerants for residential air-conditioning applications *Appl. Energy* **84** 985–91
[191] Dincer I 2017 *Refrigeration Systems and Applications* (New York: Wiley)
[192] Allen D T and Shonnard D R 2001 *Green Engineering: Environmentally Conscious Design of Chemical Processes* (Upper Saddle River, NJ: Prentice Hall)
[193] Lim W, Choi K and Moon I 2013 Current status and perspectives of liquefied natural gas (LNG) plant design *Ind. Eng. Chem. Res.* **52** 3065–88
[194] Koppejan J and Van Loo S 2012 *The Handbook of Biomass Combustion and Co-firing* (London: Earthscan)
[195] Sun D-W and Zheng L 2006 Vacuum cooling technology for the agri-food industry: past, present and future *J. Food Eng.* **77** 203–14
[196] Xue Z, Cao Y, Liu N, Feng L and Jiang L 2014 Special wettable materials for oil/water separation *J. Mater. Chem.* A **2** 2445–60
[197] Gao S J, Shi Z, Zhang W B, Zhang F and Jin J 2014 Photoinduced superwetting single-walled carbon nanotube/TiO_2 ultrathin network films for ultrafast separation of oil-in-water emulsions *ACS Nano* **8** 6344–52
[198] Xue Z, Wang S, Lin L, Chen L, Liu M, Feng L and Jiang L 2011 A novel superhydrophilic and underwater superoleophobic hydrogel-coated mesh for oil/water separation *Adv. Mater.* **23** 4270–3
[199] Jin M, Wang J, Yao X, Liao M, Zhao Y and Jiang L 2011 Underwater oil capture by a three-dimensional network architectured organosilane surface *Adv. Mater.* **23** 2861–4
[200] Nishimoto S and Bhushan B 2013 Bioinspired self-cleaning surfaces with superhydrophobicity, superoleophobicity, and superhydrophilicity *RSC Adv.* **3** 671–90
[201] Wong T-S, Kang S H, Tang S K, Smythe E J, Hatton B D, Grinthal A and Aizenberg J 2011 Bioinspired self-repairing slippery surfaces with pressure-stable omniphobicity *Nature* **477** 443
[202] Pan S, Kota A K, Mabry J M and Tuteja A 2012 Superomniphobic surfaces for effective chemical shielding *JACS* **135** 578–81
[203] Kota A K, Kwon G and Tuteja A 2014 The design and applications of superomniphobic surfaces *NPG Asia Mater.* **6** e109
[204] Chen L, Liu M, Bai H, Chen P, Xia F, Han D and Jiang L 2009 Antiplatelet and thermally responsive poly (N-isopropylacrylamide) surface with nanoscale topography *JACS* **131** 10467–72
[205] Yong J, Chen F, Yang Q, Huo J and Hou X 2017 Superoleophobic surfaces *Chem. Soc. Rev.* **46** 4168–217
[206] Bellanger H, Darmanin T, Taffin de Givenchy E and Guittard F D R 2014 Chemical and physical pathways for the preparation of superoleophobic surfaces and related wetting theories *Chem. Rev.* **114** 2694–716
[207] Chu Z and Seeger S 2014 Superamphiphobic surfaces *Chem. Soc. Rev.* **43** 2784–98
[208] Tuteja A, Choi W, Ma M, Mabry J M, Mazzella S A, Rutledge G C, McKinley G H and Cohen R E 2007 Designing superoleophobic surfaces *Science* **318** 1618–22
[209] Tuteja A, Choi W, Mabry J M, McKinley G H and Cohen R E 2008 Robust omniphobic surfaces *Proc. Natl Acad. Sci.* **105** 18200–5

[210] Zhao H, Law K-Y and Sambhy V 2011 Fabrication, surface properties, and origin of superoleophobicity for a model textured surface *Langmuir* **27** 5927–35
[211] Kota A K, Li Y, Mabry J M and Tuteja A 2012 Hierarchically structured superoleophobic surfaces with ultralow contact angle hysteresis *Adv. Mater.* **24** 5838–43
[212] Ahuja A, Taylor J, Lifton V, Sidorenko A, Salamon T, Lobaton E, Kolodner P and Krupenkin T 2008 Nanonails: a simple geometrical approach to electrically tunable superlyophobic surfaces *Langmuir* **24** 9–14
[213] Liu T and Kim C-J 2014 Turning a surface superrepellent even to completely wetting liquids *Science* **346** 1096–100
[214] Wilke K L, Preston D J, Lu Z and Wang E N 2018 Towards condensation-resistant omniphobic surfaces *ACS Nano* **12** 11013–21
[215] Rykaczewski K, Paxson A T, Staymates M, Walker M L, Sun X, Anand S, Srinivasan S, McKinley G H, Chinn J and Scott J H J 2014 Dropwise condensation of low surface tension fluids on omniphobic surfaces *Sci. Rep.* **4** 4158
[216] Preston D J, Lu Z, Song Y, Zhao Y, Wilke K L, Antao D S, Louis M and Wang E N 2018 Heat transfer enhancement during water and hydrocarbon condensation on lubricant infused surfaces *Sci. Rep.* **8** 540
[217] Solomon B R, Subramanyam S B, Farnham T A, Khalil K S, Anand S and Varanasi K K 2016 Lubricant-impregnated surfaces *Non-wettable Surfaces* (Cambridge: Royal Society of Chemistry), pp 285–318
[218] Manna U and Lynn D M 2015 Fabrication of liquid-infused surfaces using reactive polymer multilayers: principles for manipulating the behaviors and mobilities of aqueous fluids on slippery liquid interfaces *Adv. Mater.* **27** 3007–12
[219] Lafuma A and Quéré D 2011 Slippery pre-suffused surfaces *Europhys Lett.* **96** 56001
[220] Smith J D, Dhiman R, Anand S, Reza-Garduno E, Cohen R E, McKinley G H and Varanasi K K 2013 Droplet mobility on lubricant-impregnated surfaces *Soft Matter* **9** 1772–80
[221] Weisensee P B, Wang Y, Hongliang Q, Schultz D, King W P and Miljkovic N 2017 Condensate droplet size distribution on lubricant-infused surfaces *Int. J. Heat Mass Transfer* **09** 187–99
[222] Vogel N, Belisle R A, Hatton B, Wong T-S and Aizenberg J 2013 Transparency and damage tolerance of patternable omniphobic lubricated surfaces based on inverse colloidal monolayers *Nat. Commun.* **4** 2176
[223] Anand S, Paxson A T, Dhiman R, Smith J D and Varanasi K K 2012 Enhanced condensation on lubricant-impregnated nanotextured surfaces *ACS Nano* **6** 10122–9
[224] Xiao R, Miljkovic N, Enright R and Wang E N 2013 Immersion condensation on oil-infused heterogeneous surfaces for enhanced heat transfer *Sci. Rep.* **3** 1988
[225] Kim S and Kim K J 2011 Dropwise condensation modeling suitable for superhydrophobic surfaces *J. Heat Transfer* **133** 081502
[226] Boreyko J B, Polizos G, Datskos P G, Sarles S A and Collier C P 2014 Air-stable droplet interface bilayers on oil-infused surfaces *Proc. Natl Acad. Sci.* 201400381
[227] Anand S, Rykaczewski K, Subramanyam S B, Beysens D and Varanasi K K 2015 How droplets nucleate and grow on liquids and liquid impregnated surfaces *Soft Matter* **11** 69–80
[228] Subramanyam S B, Rykaczewski K and Varanasi K K 2013 Ice adhesion on lubricant-impregnated textured surfaces *Langmuir* **29** 13414–8

[229] Schellenberger F, Xie J, Encinas N, Hardy A, Klapper M, Papadopoulos P, Butt H-J and Vollmer D 2015 Direct observation of drops on slippery lubricant-infused surfaces *Soft Matter* **11** 7617–26
[230] Van Oss C J, Chaudhury M K and Good R J 1988 Interfacial Lifshitz–van der Waals and polar interactions in macroscopic systems *Chem. Rev.* **88** 927–41
[231] Preston D J, Song Y, Lu Z, Antao D S and Wang E N 2017 Design of lubricant infused surfaces *ACS Appl. Mater. Interfaces* **9** 42383–92
[232] Sett S, Yan X, Barac G, Bolton L W and Miljkovic N 2017 Lubricant-infused surfaces for low-surface-tension fluids: promise versus reality *ACS Appl. Mater. Interfaces* **9** 36400–408
[233] de Gennes P-G, Brochard-Wyart F and Queré D 2013 Capillarity and wetting phenomena: drops *Bubbles, Pearls, Waves* (Berlin: Springer), pp 291
[234] Rykaczewski K, Landin T, Walker M L, Scott J H J and Varanasi K K 2012 Direct imaging of complex nano- to microscale interfaces involving solid, liquid, and gas phases *ACS Nano* **6** 9326–34
[235] Cheng Y, Suhonen H, Helfen L, Li J, Xu F, Grunze M, Levkin P A and Baumbach T 2014 Direct three-dimensional imaging of polymer–water interfaces by nanoscale hard x-ray phase tomography *Soft Matter* **10** 2982–90
[236] Wu A and Miljkovic N 2018 Droplet cloaking imaging and characterization *J. Heat Transfer* **140** 030902
[237] Etzler F M 2013 Determination of the surface free energy of solids *Rev. Adhes. Adhesiv.* **1** 3–45
[238] Donahue D J and Bartell F 1952 The boundary tension at water–organic liquid interfaces *J. Phys. Chem.* **56** 480–4
[239] Vorobev A 2014 Dissolution dynamics of miscible liquid/liquid interfaces *Curr. Opin. Colloid Interface Sci.* **19** 300–08
[240] Wexler J S, Jacobi I and Stone H A 2015 Shear-driven failure of liquid-infused surfaces *Phys. Rev. Lett.* **114** 168301
[241] Miljkovic N, Preston D J, Enright R and Wang E N 2013 Electric-field-enhanced condensation on superhydrophobic nanostructured surfaces *ACS Nano* **7** 11043–54
[242] Preston D J, Miljkovic N, Wang E N and Enright R 2014 Jumping droplet electrostatic charging and effect on vapor drag *J. Heat Transfer* **136** 080909
[243] Landau L D, Lifshitz E M and Pitaevskiĭ L P 1984 *Electrodynamics of Continuous Media* 2nd edn (Oxford: Pergamon)
[244] Velkoff H R and Miller J H 1965 Condensation of vapor on a vertical plate with a transverse electrostatic field *J. Heat Transfer* **87** 197–201
[245] Yabe A 1991 *Active Heat Transfer Enhancement by Applying Electric Fields* (New York: American Society of Mechanical Engineers)
[246] Yamashita K, Kumagai M, Sekita S, Yabe A, Taketani T and Kikuchi K 1991 *Heat Transfer Characteristics on an EHD Condenser* (New York: American Society of Mechanical Engineers)
[247] Yabe A, Taketani T, Yoshizawa Y and Sunada K 1991 *Experimental Study of EHD Pseudo-Dropwise Condensation* (New York: American Society of Mechanical Engineers)
[248] Seth A K and Lee L 1974 The effect of an electric field in the presence of noncondensable gas on film condensation heat transfer *J. Heat Transfer* **96** 257–8
[249] Bologa M K, Savin I K and Didkovsky A B 1987 Electric-field-induced enhancement of vapour condensation heat transfer in the presence of a non-condensable gas *Int. J. Heat Mass Transfer* **30** 1577–85

[250] Omidvarborna H, Mehrabani-Zeinabad A and Esfahany M N 2009 Effect of electro-hydrodynamic (EHD) on condensation of R-134a in presence of non-condensable gas *Int. Commun. Heat Mass Transfer* **36** 286–91
[251] Budov V M, Kir'yanov V A and Shemagin I A 1987 Heat transfer in the laminar-wave section of condensation of a stationary vapor *J. Eng. Phys.* **52** 647–9
[252] Butt H-J, Untch M B, Golriz A, Pihan S A and Berger R 2011 Electric-field-induced condensation: an extension of the Kelvin equation *Phys. Rev.* E **83** 061604
[253] Birbarah P, Li Z E, Pauls A and Miljkovic N 2015 A comprehensive model of electric-field-enhanced jumping-droplet condensation on superhydrophobic surfaces *Langmuir* **31** 7885–96
[254] McCarty L S and Whitesides G M 2008 Electrostatic charging due to separation of ions at interfaces: contact electrification of ionic electrets *Angew. Chem. Int. Ed.* **47** 2188–207
[255] Corson L T, Tsakonas C, Duffy B R, Mottram N J, Sage I C, Brown C V and Wilson S K 2014 Deformation of a nearly hemispherical conducting drop due to an electric field: theory and experiment *Phys. Fluids* **26** 122106
[256] Roux J M, Achard J L and Fouillet Y 2008 Forces and charges on an undeformable droplet in the dc field of a plate condenser *J. Electrostat.* **66** 283–93
[257] Glière A, Roux J-M and Achard J-L 2013 Lift-off of a conducting sessile drop in an electric field *Microfluid. Nanofluid.* **15** 207–18
[258] Lebedev N and Skalskaya I 1962 Force acting on conducting ball placed in plane capacitor field *J. Tech. Phys.* **32** 375–8
[259] Wohlhuter F K and Basaran O A 1992 Shapes and stability of pendant and sessile dielectric drops in an electric field *J. Fluid Mech.* **235** 481–510
[260] Ristenpart W D, Bird J C, Belmonte A, Dollar F and Stone H A 2009 Non-coalescence of oppositely charged drops *Nature* **461** 377
[261] Ren B and Kang Y 2018 Demulsification of oil-in-water (O/W) emulsion in bidirectional pulsed electric field *Langmuir* **34** 8923–31
[262] Aubry N, Singh P, Janjua M and Nudurupati S 2008 Micro- and nanoparticles self-assembly for virtually defect-free, adjustable monolayers *Proc. Natl Acad. Sci.* **105** 3711
[263] Pohl H A 1978 *Dielectrophoresis: The Behavior of Neutral Matter in Nonuniform Electric Fields* (Cambridge: Cambridge University Press)
[264] Aubry N and Singh P 2006 Control of electrostatic particle–particle interactions in dielectrophoresis *Europhys. Lett.* **74** 623–9

Chapter 14

Thermionic energy conversion

Md Golam Rosul, Md Sabbir Akhanda and Mona Zebarjadi

Thermionic modules can convert heat directly into electricity. Despite their long history, they have had only limited applications to date. The advancement of vacuum thermionic energy converters has been limited by factors such as the unavailability of stable, low work function materials, the space charge effect and the inaccessibility of electrodes for efficient heat management. Solid-state thermionic converters were proposed in the late 1990s. They have also shown limited efficiency, mainly due to their inherently large heat leaks. In recent years, interest in thermionic energy conversion has been renewed as a result of increasing demands for clean energy and progress in nanotechnology. In this chapter, the history, basic performance, design strategy, and applications of vacuum-state and solid-state thermionic energy converters are discussed.

14.1 Introduction

Thermionic energy conversion is the process of converting heat directly into electricity using heat as the source of energy and electrons as the working fluid. A thermionic converter is essentially a heat engine converting thermal energy directly to electricity. Similar to most other heat engines, thermionic devices can operate in different operational modes. In its power generation mode, which is the focus of this chapter, heat is used to boil off electrons in the cathode (emitter). These electrons are then collected by a colder anode (collector). Part of the thermal energy is thus converted directly to electricity and the rest is rejected as heat to the cold side. The very same devices can also operate in the refrigeration mode, where current is passed through the device to pump heat from the cold side to the hot side. The most common operational mode of these devices is, however, their rectifying mode. Vacuum thermionic diodes were once the heart of the electronic industry. They have now been mostly replaced by solid-state devices, but still have a reasonable market and are the basis for cathode-ray tubes, radio tubes and broadcast transmitters. Thermionic devices are simple in construction and, having no moving parts, are

doi:10.1088/978-0-7503-1738-2ch14

quiet in operation. They are environmentally friendly as no emissions of greenhouse gases are involved and they are highly reliable. Because of these excellent features thermionic devices have attracted great interest for several decades.

There are two main types of thermionic converters: vacuum-state thermionic converters (VSTIC) and solid-state thermionic converters (SSTIC). VSTICs operate at very high temperatures (above 1500 K), produce high power, yet occupy a very small volume. They are suitable for applications such as solar concentrated power generators, waste heat conversion from nuclear reactors and fossil fuel combustion. SSTICs work at much lower temperatures and under smaller temperature differences and are more suitable for power generation applications involving smaller amounts of heat, such as residential and industrial waste heat recovery (less than 600 K). Both VSTICs and SSTICs can be used in the power generation as well as refrigeration mode as the parameters affecting their performance in both modes are similar. Hence, if a VSTIC or an SSTIC exhibits good performance in the power generation mode, a similar level of performance can be expected in its refrigeration mode. It is also possible to use TICs as thermal switches [1, 2] and/or active coolers [3] that actively pump heat from the hot side to the cold side. However, in these last two modes of operation, the design parameters are different.

Finally, the recently proposed photon-enhanced thermionic converter (PETIC) is essentially a hybrid version of the VSTIC. It utilizes both light and heat as its energy source and could be viewed as a hybrid photoelectric–thermionic device. PETIC has been demonstrated in the power generation mode [4].

14.2 History of thermionic converters

The discovery of thermionic emission by Edison in 1885 paved the way for developing a new method of energy conversion using the thermal emission of electrons. Thermionic diodes (also called thermionic valves or thermionic tubes) were invented by Fleming in 1904. They have been widely used in radio and telephone communications. A thermionic converter is very similar in operation to a thermionic diode. Even though the idea of thermionic energy conversion was first suggested by Schlicter [5] in 1915, extensive theoretical and experimental investigations for practical level power generation using this concept were not carried out until the 1950s. In 1956, Murphy and Good published a rigorous study of thermionic and field-emission theory [6]. Hatsopoulos studied vacuum thermionic and vapor thermionic converters during his PhD work on thermo-electron engines. There, he discussed the single and multiple emitter–collector configurations for power generation [7]. In 1957, Ioffe briefly discussed vacuum thermo-elements in his book [8] on thermoelectric conversion. Moss [9] evaluated the importance of many of the device parameters in her calculations on thermionic devices. Hernqvist *et al* demonstrated a practical thermionic converter with an efficiency of several percent in 1957 [10]. In 1958, Webster evaluated the performance of a high-vacuum thermionic converter using Langmuir's work on thermionic diodes [11]. Hatsopoulos *et al* used a cylindrical emitter and collector with a 0.125″ diameter in their preliminary experiment on thermionic diode model. With an inter-electrode separation of 10 μm to minimize the so-called space charge effect, and an emitter temperature of 1540 K, a 13% energy conversion efficiency was obtained [12]. Wilson used positively charged cesium ion gas (plasma diode) to cancel the negative inter-electrode space charge effect and have demonstrated 9.2% conversion efficiency and an output power of 3.1 W cm^{-2} [13].

In 1959, the first thermionic converter exhibiting promising results was installed into the water moderated core of the Omega West Reactor. A short circuit current of 35 A and an open circuit voltage of 3.5 V was produced by this converter, although it had a fairly large inter-electrode space of about 6 mm [14]. The initial development of a thermionic converter in the United States took place using solar energy and radioisotopes as the thermal source. In the 1960s, the Jet Propulsion Laboratory Solar Energy Technology Thermionic Program started a solar thermionic converter evaluation and generation program, where they developed and tested several converters. The majority of the converters were operated at an emitter temperature of about 1900 K with a lifetime of 11 000 h, including one particular converter with a lifetime of 20 000 h. The converters generated 150 W power exhibiting an energy conversion efficiency of 7% to 11%. However, the program was discontinued in the 1970s as the thermionic converters could no longer compete with the evolving solid-state based photovoltaic and thermoelectric energy conversion techniques [14, 15].

Later the focus shifted to the space power generation systems using nuclear heat sources. NASA and the Atomic Energy Commission (AEC) started to fund the development of in-core and out-of-core thermionic fuel elements. Experimental testing of a prototype of an in-core thermionic converter advanced to multi-cell TI fuel element design using tungsten-clad UO_2 emitters began in 1970, which operated at an emitter temperature of 1900 K with a lifetime of 10 000 h. In 1970, General Atomics (GA) developed a Mark III reactor that could operate at 1900 K with a lifetime of 12 500 h. The thermionic space reactor development program was canceled in 1973 as Congress and the Executive Branch shifted funding out of the space power field. In the 1970s, there were no major developments in thermionic research, except for the USSR's TOPAZ (a Russian acronym for thermionic experiment with conversion in active zone) project. The TOPAZ reactor system generated 5 to 10 kW of power with a lifetime of 3000 to 5000 h. TOPAZ-II, a 6 kW converter with an energy conversion efficiency of 10% and an inter-electrode gap of 100 μm, was flown by the Soviet space program in 1987. This project was then terminated due to budget restrictions [14–16].

In 1973, thermionic applications in the area of fossil fuel terrestrial power systems attracted the attention of researchers. Significant developments at this time include a new type of converter design applying advanced electrode and plasma technologies to improve the thermionic cell lifetime and performance. During this time, a flame fired thermionic conversion unit was developed to generate electricity using the waste heat produced from an environment where a high temperature is attained by combustion [17].

In 1979, a rebirth of the space program was brought about by the joint initiative of NASA, the Department of Energy (DOE) and the Department of Defense (DoD). Together they launched the space power advanced reactor (SPAR) program focusing exclusively on heat pipe technology. In 1982, the SPAR program was extended and renamed as the SP-100 program. The main focus of the SP-100 program was to design a 100 kW nuclear power system suitable for outer space applications. Research from 1984 to 1986 was primarily focused on a better understanding of the system design, but no major reactor was demonstrated [14, 15, 18].

A number of government programs as well as many individual researchers continued research in the field of thermionic converters over the next decade. In 2001, a report published by the National Research Council projected a negative perspective on the viability of thermionic energy conversion [15]. However, increasing demand for clean energy and advancements in nanotechnology helped create a renewed interest in thermionic conversion in recent years.

Difficulty in finding low work function electrodes restricts the performance of the VSTICs to very high temperatures. The need for a vacuum restricts direct access to the electrodes. These and many other challenges inspired the design of a solid-state replacement. In 1997, Shakouri and Bowers [19] proposed a single-layer solid-state thermionic diode in which the vacuum is replaced by a semiconducting material. In this structure, the semiconductor layer is the energy barrier that an electron experiences. In the following year, Mahan proposed the idea of using multilayer barriers in which each layer maintains a small temperature difference [20, 21]. These proposed structures created a sudden surge in SSTIC research. Over the next few years, extensive studies were carried out in search of suitable materials which would enhance the performance of SSTICs in both power generation and refrigeration modes. As these structures can be grown directly on a chip. Scientists soon realized the potential of SSTICs as integrated coolers for hotspots in electronic and optical devices [22–25]. A large number of thin-film coolers lattice matched to Si, GaAs or InP were fabricated and characterized. InGaAsP/InP [18, 26–28] and InGaAs/InP [29] lattice matched to InP and AlGaAs/GaAs [30] lattice matched to GaAs, all grown by metal organic chemical vapor deposition (MOCVD) and InGaAs/InAlAs [31], InGaAsSb/InGaAs [32], SiGe/Si [33–35] and SiGeC/Si [36] deposited by molecular beam epitaxy (MBE) were all reported around this period. In 2015, for the first time, Cronin used a 2D van der Waals heterostructure, a stack of atomic layers, to build an SSTIC [37]. The unstable nature of the van der Waals heterostructures and the difficulty in providing a proper metallic contact to these structures were identified as the main stumbling blocks preventing the experimental demonstration of highly efficient van der Waals heterostructure based SSTICs. Recent in-depth theoretical calculations indicate the potential these structures hold in thermionic energy research [38, 39].

14.3 Theory of thermionic converters

14.3.1 Basic working principle

The simplest form of a thermionic converter consists of two electrodes, an emitter (cathode) and a collector (anode). A general schematic diagram of a VSTIC is shown in figure 14.1(a) and that of an SSTIC is shown in figure 14.1(b). In the case of

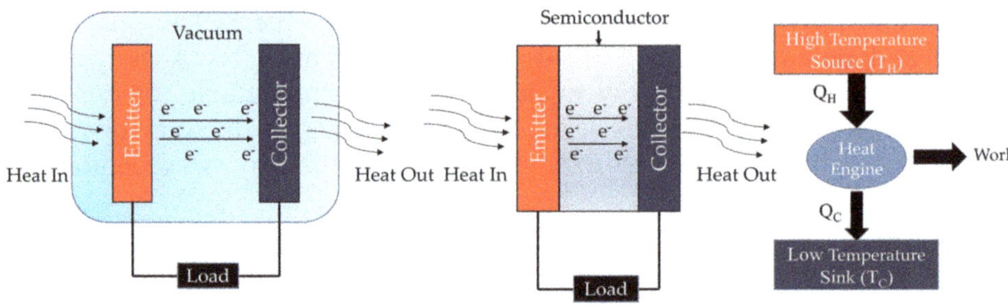

Figure 14.1. A general schematic diagram of (a) a VSTIC, (b) an SSTIC and (c) a heat engine.

VSTICs, the emitter and the collector electrodes are enclosed in a vacuum container. The electrodes are separated from one another by an inter-electrode vacuum gap. The emitter is in thermal contact with a heat source and the collector is in thermal contact with a cold heat sink. The heat source supplies thermal energy to the emitter raising the emitter temperature. Thus, the high energy electrons in the tail of the Fermi–Dirac distribution function acquire a sufficient amount of energy to overcome the work function energy barrier and escape the emitter. These electrons enter the vacuum gap and are then absorbed by the collector. Once in the collector, they reject their extra energy to the heat sink. Finally, the electrons flow back from the collector to the emitter through an external load. This flow of electrons delivers useful electrical work to the external load.

SSTICs operate in a similar manner. The vacuum gap is replaced by a semiconductor giving rise to several consequences. First, the energy barrier is smaller, so the SSTICs can operate at lower temperatures. The energy barrier in the case of VSTICs is the work function of the emitter (φ_E), but for the SSTICs, it is the difference between the emitter work function and the electron affinity of the semiconducting layer ($\varphi_E - \chi_S$). Second, the vacuum gap is micron to millimeter size while the semiconducting layer thickness is much smaller (<100 nm) to ensure ballistic transport. Third, radiation (the only heat transfer mechanism in a vacuum) in VSTICs is replaced by conduction in SSTICs. As a consequence, SSTICs suffer from large heat leaks compared to VSTICs and in comparison, they can only operate at much smaller temperature differences. Fourth, a lack of background positive charges creates a space charge effect in VSTICs which to an extent is absent in SSTICs when a proper level of doping is provided in the semiconducting layer. The absence of the space charge effect is an advantage for SSTICs as the presence of space charges would otherwise lower the performance of the thermionic converters. We will discuss the space charge effect in detail in section 14.4.1.2. Finally, connection to the heat source and heat sink is much easier in SSTICs due to the absence of a vacuum.

A thermionic converter can be viewed as a heat engine (figure 14.1(c)) in which the emitter receives heat from a high-temperature source, the collector rejects heat to a cold sink, and some part of the input thermal energy converts to electrical energy as useful work. The energy conversion efficiency of a heat engine is limited by the Carnot efficiency, which is defined as

$$\eta_{\mathrm{C}} = 1 - \frac{T_{\mathrm{C}}}{T_{\mathrm{H}}},$$

where η_{C} is the Carnot efficiency, T_{C} is the temperature of the cold side (the collector in this case) and T_{H} is the temperature of the hot side (emitter). The upper bound of the emitter temperature is limited by the melting point and the chemical and mechanical stability of the emitter material. The emitter and the collector, in a VSTIC, are separated by a vacuum gap. This gap allows only a very small amount of radiative heat to be transferred through it. Consequently, it is possible to maintain a very high temperature difference between the emitter and the collector and achieve a higher Carnot efficiency compared to other heat engines. For example, a typical VSTIC operating with a collector temperature of 1100 K and an emitter temperature of 2000 K will reach to a Carnot efficiency of 45%. On the other hand, SSTICs operate at lower temperatures and smaller temperature differences. As a result, they are more suitable for applications such as waste heat recovery wherein the available temperature is in the 400–600 K range. As the temperature difference is smaller the Carnot efficiency of SSTICs is also relatively smaller compared to VSTICs.

14.3.2 Ideal output current, voltage and power

Consider the simplest type of thermionic converter, a single-barrier structure. Figure 14.2 shows its energy diagram. The energy barrier is represented by Φ_{E} which is equal to φ_{E} for VSTICs and is equal to $\varphi_{\mathrm{E}} - \chi_{\mathrm{S}}$ for SSTICs. Similarly, we named the barrier height (the energy offset) at the collector side Φ_{C}, which is equal to φ_{c} for VSTICs and $\varphi_{\mathrm{c}} - \chi_{\mathrm{S}}$ for SSTICs. In figure 14.2, it is assumed that the emitter energy barrier Φ_{E} is larger than the collector energy barrier, Φ_{C}, plus the total voltage V ($V = IR + IR_{\mathrm{l}}$, R_{l} is the load resistance and R is the internal resistance of the thermionic converter).

In order to overcome the potential energy barrier, the electrons inside the emitter must be energized above the energy barrier. When thermal energy from the heat source is supplied to the electrons inside the emitter, they obtain enough energy to escape the emitter surface. The net current density of the electron flux from the

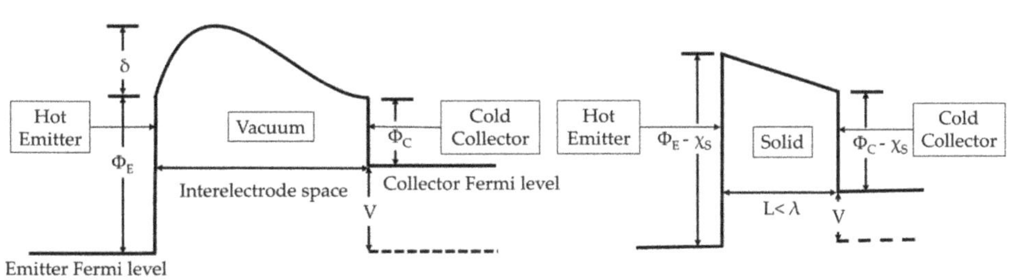

Figure 14.2. Potential energy diagram of (a) a VSTIC and (b) an SSTIC.

emitter to the collector is given by the Richardson–Dushman equation [40, 41], which can be written as

$$J = A_R T_H^2 \exp\left(-\frac{q\phi_E}{k_B T_H}\right) - A_R T_C^2 \exp\left(-\frac{q(\phi_E - V)}{k_B T_C}\right), \tag{14.1}$$

where T_H is the emitter temperature, T_C is the collector temperature, Φ_E is the work function of the emitter material, k_B is the Boltzmann constant, q is the electron charge constant, V is the total voltage and A_R is the Richardson constant. The theoretical value of $A_R = \frac{4\pi q m^* K_B^2}{h^3}$ is calculated assuming a parabolic band structure. Assuming the effective mass to be the same as the mass of a free electron, $A_R = 120\frac{A}{\mathrm{cm}^2\,\mathrm{K}^2}$. However, the experimental values of A_R vary with materials, arising from the complex Fermi-surface of the metals. As current depends on the energy barrier exponentially (equation (14.1)), a small change in the value of the energy barrier results in a significant change in the current density. The first term in equation (14.1) is the flux of electrons from the emitter to the collector and the second term is the leak current flux going backward from the collector to the emitter. The power density delivered to the load is

$$p_{\mathrm{load}} = (V - IR)J. \tag{14.2}$$

Similarly, the thermal current of the thermionic converters can be written as

$$\begin{aligned} J_{Qc} = {} & A_R T_H^2 \exp\left(-\frac{q\phi_E}{k_B T_H}\right)(\phi_E + 2k_B T_H/q) \\ & - A_R T_C^2 \exp\left(-\frac{q(\phi_E - V)}{k_B T_C}\right)\left(\phi_E + \frac{2k_B T_C}{q}\right) + J_{Q\text{-leak}} \end{aligned} \tag{14.3}$$

$$J_{Q\text{-leak}} = \begin{cases} \dfrac{T_H - T_C}{R_t} \text{ for SSTICs} \\ \sigma\epsilon\left(T_H^4, -, T_C^4\right) \text{ for VSTICs} \end{cases}.$$

Here, $\phi_E + \frac{2k_B T_H}{q}$ is the average energy of the electrons passing above the energy barrier. The $2k_B T$ factor comes from Fermi–Dirac statistics and is considered as the excess energy factor that lowers the efficiency. $J_{Q\text{-leak}}$ is the thermal leak current which is mainly due to radiation in VSTICs and conduction in SSTICs. In the case of VSTICs, it is proportional to the effective emissivity of the cathode and anode, ϵ. To minimize the radiation, one can use cathode and anode materials with low emissivity at the operating temperature of VSTICs. At the end, radiation is much weaker compared to other channels of heat transport. In this context, we will see later that in most VSTICs plasma gas is used to minimize the space charge effect. When used, plasma gas creates a convective channel of heat transport which increases the heat leak significantly. In the case of SSTICs, the leak problem is even more serious.

$J_{Q\text{-leak}}$ in this case is due to conduction and is inversely proportional to the thermal resistance, R_t. To minimize $J_{Q\text{-leak}}$, we need to maximize R_t which is a very difficult task considering the small required size of the semiconducting layer (<100 nm).

Finally, the conversion efficiency of a thermionic converter is defined as the ratio of the output electrical power to the heating power supplied to the emitter and can be calculated as

$$\eta = \frac{p_{\text{load}}}{J_{\text{Qc}}}, \tag{14.4}$$

where p_{load} is the power density delivered to the load and J_{Qc} is the thermal current. To optimize the efficiency, instead of matching resistance conditions, the resistance of the load (R_l) should be adjusted according to the internal resistance of the thermionic diode (R) and the Richardson current to satisfy [42]

$$R_l = \frac{k_B}{A_R T_C A q} \exp\left(q, \frac{\phi_E - V}{k_B T_a}\right) + R. \tag{14.5}$$

The barrier height can also be optimized, and it is shown that the optimum barrier height to maximize the efficiency is on the order of 2–5 $k_B T$. Theoretical analysis demonstrated that the total energy conversion efficiency of a VSTIC can exceed 30% but cannot be greater than 90% of the Carnot efficiency [43]. Mahan theoretically showed that the efficiency of a thermionic refrigerator can be greater than 80% of the Carnot value [44].

14.4 Design of thermionic converters

14.4.1 Vacuum-state thermionic converters

There are several non-ideal effects that lower the efficiency and need to be considered in the design of highly efficient VSTICs. Radiation leaks from the hot cathode to the anode lower the efficiency as shown in the theory section (equations (14.3) and (14.4)). We note that radiation is weak at low temperatures, however, the VSTICs work at high temperatures and as a result the radiation leak plays a role in lowering the efficiency. This leak could be minimized if the cathode and the anode are made out of materials with low emissivity. Tungsten, for example, is a material with low emissivity (spectral radiant emissivity of 2 μm thick tungsten is 0.184 at 1244 K and 0.229 at 2002 K [45]) and is suitable for highly efficient VSTICs.

Any internal electrical resistance including the resistances of the leads further lowers the efficiency as the RI^2 generated as a result of internal resistance is subtracted from the total power generated (see equation (14.2)). Another problem is the inaccessibility of the cathode and anode for the purpose of cooling and heating. The electrodes are inside a vacuum, and it is not possible to put their surface in direct contact with the external heat source and the heat sink. The space charge effect also has a significant effect on the performance of a VSTIC and will be discussed later in section 14.4.1.2. Considering that VSTICs are only operating at very high temperatures, it is desirable to extend their operation to lower temperatures where the heat sources are more abundant. The biggest challenge in lowering the operating temperature comes from the high energy barrier developed due to the large work function of the electrodes which the electrons cannot overcome at lower temperatures. There are several ways to resolve the problem, which will be discussed in the following section.

14.4.1.1 Work function

The work function, one of the most fundamental properties of a surface, is defined as the amount of energy required for an electron to be elevated from the Fermi level to the vacuum level. A lower work function material can emit electrons more easily. Therefore, the work function is important in determining a material's applicability as an electron emitter in a VSTIC. The output power density, which is the product of the output voltage and the thermionic current, can be increased by employing a low work function emitter and collector. At the same time a large work function difference between the electrodes needs to be maintained. In this section we summarize several methods to produce low work function emitter and collector materials for VSTIC.

It has been known since the 1930s that it is possible to create low work function surfaces using coating [46, 47]. For example, the adsorption of alkali metals on metal or semiconductor substrates reduces the work function. This reduction mechanism is explained by Langmuir [48] as follows. If the electronegativity value of the alkali metal is lower than the substrate surface, the adsorption of alkali metal on the substrate surface leads to the transfer of valence electrons from the alkali metal to the substrate. This transfer of electrons creates a dipole moment between the positive adatom ion core and the surface negative image charges. This dipole moment, being antiparallel to the substrate surface dipole layer, causes a reduction in the work function. A systematic study of work function reduction caused by alkali metal adsorption both on a metal substrate and a semiconductor surface was carried out by Sinsarp *et al* [49]. Platinum was used as the metal surface and silicon was used as the semiconductor surface. A clean surface of Pt (111) and Si (111)-(7 $\times$ 7) was prepared and Cs was deposited on both of them. As the amount of adsorbed cesium increased, the work functions of both platinum and silicon surface decreased rapidly and finally reached a minimum value. Maximum change in the work function for the platinum and silicon were 4.8 eV and 3.3 eV, respectively. Significant reduction in the work function of clean tungsten (110) from 5.25 eV to 1.45 eV under cesium exposure has also been observed [50]. Similarly, a cesium covered molybdenum emitter and collector exhibited work functions of 2.2 eV at 1200 °C and 1.7 eV at 500 °C, respectively [51]. It was also reported that the adsorption of cesium on a platinum substrate reduces the work function from 5.6 eV to 1.4 eV [52]. The co-adsorption of oxygen and alkali metal on the semiconductor substrate further reduces the work function. The additional work function reduction is attributed to the co-adsorption of oxygen on a semiconductor substrate covered by alkali metal. The work function of a hydrogen passivated (100) p-type silicon is reduced from 4.7 eV to 1.35 eV by coating its surface with potassium in an oxidant atmosphere at room temperature [53]. A work function value of 0.85 eV was also observed in a

Cs-covered Si (100) system due to the oxygen co-adsorption [54]. In the same way, an extremely low work function value of 0.3 eV is reported on a K_2O_2/Si (100) surface [55]. Oxide film coatings, normally consisting of one of the three alkaline earth oxides BaO, SrO and CaO, or a mixture of these, can also be used to significantly lower the work function. Lee *et al* designed a SiC emitter suspended on a Si collector substrate with a nominal 1.7 μm electrode gap. For high-temperature operation micro-thermionic converters with both barium and barium oxide coatings on thin films of tungsten were tested. The coatings reduced the work function of the SiC emitter to ~2.14 eV and increased the thermionic current by 5–6 orders of magnitude [56]. Feng *et al* created a carbon nanotube thermionic cathode on a thin tungsten ribbon surface. The carbon nanotubes were then coated with barium strontium oxide (BaO/SrO). The final structure exhibited a 2.1 eV reduction in the work function from 4.2 eV resulting in 50 times higher thermionic current density [57]. Some of the emitter materials with low work function values along with their Richardson constants are listed in table 14.1 [58].

The low work function characteristics observed in diamond thin films and carbon nanomaterials such as graphene and carbon nanotubes have led to numerous studies of thermionic emission in these materials in recent years. Koeck *et al* prepared phosphorus-doped polycrystalline diamond films on metallic substrates using plasma assisted chemical vapor deposition, which exhibited a work function of 0.9 eV and sustained temperatures of up to 765 °C [59]. Koeck *et al* also reported a work function value of 1.5–1.9 eV for nitrogen-doped diamond films grown on molybdenum substrates [60]. Sherehiy *et al* reported reduced work function values of phosphorus-doped diamond nanocrystals grown on conical carbon nanotubes and phosphorus-doped diamond film on silicon substrate. The carbon nanotubes exhibited a work function of 2.23 eV, which was attributed to the presence of a mid-gap state. The phosphorus-doped diamond film showed a work function value of 1.8 eV [61]. Nitrogen incorporated ridged nano-diamond films on silicon substrate exhibited a work function of 1.39 eV, and were thermally stable up to a maximum testing temperature of 900 °C [62]. Nitrogen-doped polycrystalline diamond films and single crystal diamond surfaces demonstrated work functions of about 1.4 eV and 2 eV. The band bending effect was reported to be responsible for the higher work function of the single crystal surface [59, 63]. The atomic structure of carbon nanotubes, graphite and few-layer graphene offers the opportunity for intercalation of the lattice with alkali metals, such as potassium, which greatly reduces their work functions. Intercalation occurs when guest molecules or ions are inserted into narrow spaces between host layers. A single monolayer of potassium absorbed on graphite decreased the work function from 4.7 eV to 2.3 eV [64]. Michel *et al* reported a reduced work function of graphitic carbon nanofiber intercalated by molten potassium through a stoichiometric reaction. Stage-1 K/herringbone graphitic carbon nanofiber intercalate exhibited a work function of 2.2 eV and remained thermally stable up to 1000 °C [65]. Westover *et al* experimentally demonstrated that the intercalation of potassium into single-walled and multi-walled carbon nanotubes reduced the work function from 4.5 eV to 2 eV [66].

Table 14.1. The work function values and the Richardson constants of different metals for thermionic emission, extracted from [58].

Metal–film	Work function (eV)	Richardson constant (A cm^{-2} K^{-2})
Alkali metal adsorption		
Titanium–cesium (Ti–Cs)	1.07	0.10
Nickel–cesium (Ni–Cs)	1.65	2.09
Tantalum–cesium (Ta–Cs)	1.11	0.15
Molybdenum–cesium (Mo–Cs)	1.22	0.45
Niobium–cesium (Nb–Cs)	1.02	0.05
Carbon–cesium (C–Cs)	1.37	10.0
Tungsten–cesium (W–Cs)	1.36	3.20
	1.38	3.26
	1.41	3.55
Tungsten–sodium (W–Na)	1.76	—
Tungsten–potassium (W–K)	1.64	—
Tungsten–lithium (W–Li)	1.83	—
Platinum–cesium (Pt–Cs)	1.38	—
Stainless steel–cesium	1.41	4.68
Co-adsorption of alkali metals and oxygen		
Tungsten–oxygen–cesium (W–O–Cs)	0.72	0.003
	1.44	—
Tungsten–oxygen–sodium (W–O–Na)	1.72	—
Tungsten–oxygen–potassium (W–O–K)	1.76	—
Silver–oxygen–cesium (Ag–O–Cs)	1.00	—
Oxide film coating		
Tungsten–calcium oxide (W–CaO)	2.1	—
Tungsten–strontium oxide (W–SrO)	1.1	—
Tungsten–barium oxide (W–BaO)	1.1, 1.34, 1.36	—
Nickel alloy–barium oxide (Ni alloy–BaO)	1.503–1.830	0.087–2.18
Nickel–barium oxide (Ni–BaO)	1.27, 1.32, 1.35	—
Nickel–strontium oxide (Ni–SrO)	2.0	—
Platinum–barium oxide (Pt–BaO)	1.68	2.88
Platinum–strontium oxide (Pt–SrO)	1.86	4.07
Nickel–barium oxide, strontium oxide (Ni–BaO/SrO)	1.201.15	0.960.25
Platinum–barium oxide, strontium oxide (Pt–BaO/SrO)	1.51–1.89	—
Platinum–iridium alloy–calcium oxide (Pt/Ir –CaO)	1.77	—
Platinum–iridium alloy–strontium oxide (Pt/Ir–SrO)	1.27	10^{-4}–10^{-2}
Platinum–iridium alloy–barium oxide, strontium oxide (Pt/Ir–BaO/SrO)	1.03	10^{-3}–10^{-2}
Platinum–nickel alloy–barium oxide, strontium oxide (Pt/Ni–BaO/SrO)	1.37	2.45

In addition to exploring low work function emitter materials, the enhancement of electron emission through other mechanisms has been studied as well. One discovery in this field is the photon-enhanced thermionic emission (PETE) [4]. PETE converts solar thermal energy to electrical energy by combining the photovoltaic effect, and the thermionic emission. Instead of lowering the work function, in PETE, electrons are pushed to higher energies by absorbing photons. When excited, they encounter a smaller energy barrier. Smestad first experimentally demonstrated that the photo-enhanced thermionic effect can produce higher electrical output than the sum of the outputs of photoelectric current and the thermionic current measured separately [67]. Figure 14.3(a) shows an energy diagram of a PETE process and one possible implementation of a parallel-plate PETE converter. The structure of a PETE is similar to the vacuum thermionic converter in which two different electrodes in parallel-plate formation are separated by a vacuum gap and connected by an external load resistance. One major difference is that the typical metallic emitter is replaced with a p-type semiconductor. In PETE, electrons transfer from the emitter to the collector in three steps. First, the thermal energy irradiated from the Sun excites electrons into the conduction band. Second, the electrons rapidly thermalize in the conduction band based on the equilibrium thermal distribution at the emitter's temperature and diffuse throughout the emitter. Finally, electrons with energies greater than the emitter's electron affinity will be emitted directly into the vacuum and collected by the collector. Therefore, with a combined boost of photon energy and thermal energy each electron overcomes the material bandgap and electron affinity.

14.4.1.2 Space charge effect

Another major problem that significantly affects the performance of a vacuum thermionic converter is the space charge effect caused by a build-up of negative charges in the inter-electrode gap. The electrons emitted from the emitter sense a negative space charge and are forced to go back to the emitter, consequently

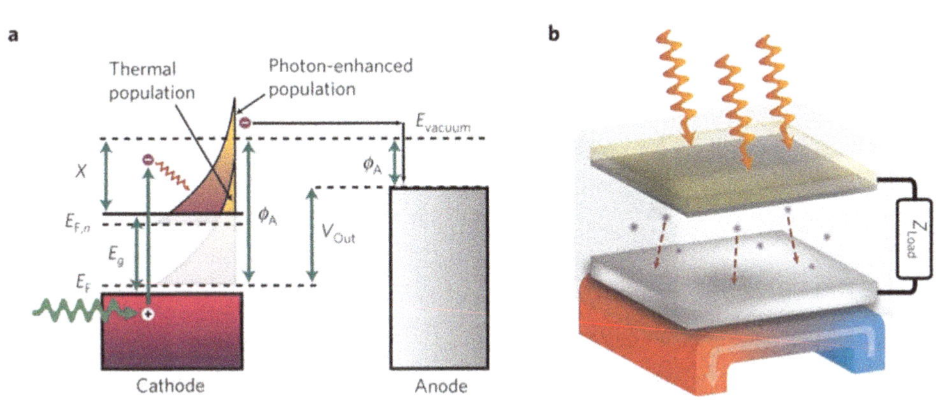

Figure 14.3. (a) Energy diagram of a PETE process. (b) One possible implementation of a parallel-plate PETE converter. Reproduced with permission from [4]. Copyright 2010 MacMillan Publishers.

decreasing the number of electrons transferred to the collector. In other words, the space charge effect creates an additional energy barrier in between the emitter and collector which adds up to the work function barrier. Only those electrons which have enough kinetic energy to surmount the total potential barrier created by the emitter work function and space charges can reach the collector. Therefore, the space charge effect limits the current in the vacuum thermionic converter and degrades the energy conversion efficiency. Over the years several approaches have been adopted to eliminate the space charge effect. Some of the strategies are summarized below and in figure 14.4.

The space charge problem can be solved by decreasing the gap between the emitter and collector (figure 14.4(a)). If the gap between the emitter and the collector is very small, a sufficient number of electrons cannot accumulate in the emitter–collector gap to create an additional energy barrier. The electrons are collected by the collectors before they can collide with each other. An inter-electrode gap greater than 100 μm significantly decreases the conversion efficiency of a VSTIC. The efficiency of the VSTICs can be greatly increased if the inter-electrode gap is kept in the 5 to 10 μm range [15]. A thermionic converter with an inter-electrode gap of 6 μm was built in Russia, but was found to be mechanically unstable [15]. Modern semiconductor fabrication technologies are used to fabricate thermionic converters with a few micron inter-electrode gap. Belbachir *et al* fabricated a thermionic converter with an inter-electrode space of 10 μm using a SiC emitter, but thermal

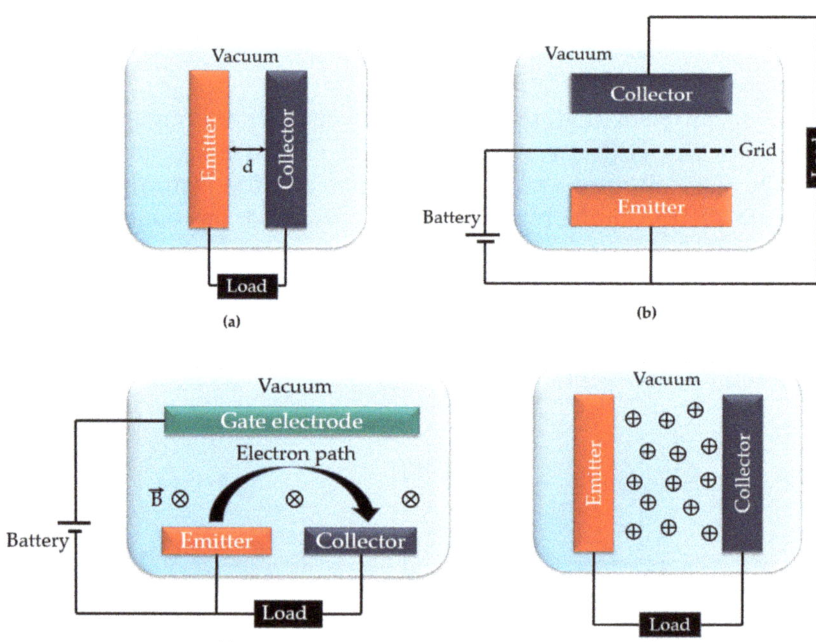

Figure 14.4. Proposed methods to reduce the space charge effect: (a) reducing the gap between the emitter and collector, (b) using a triode configuration, (c) applying an electric field and a magnetic field perpendicular to each other, and (d) inserting positive charge (cesium ion) in the inter-electrode space.

loss due to heat conduction was predominant in that device [68]. Littau *et al* reported construction of a thermionic converter using barium dispenser cathodes and thin-film tungsten anodes having a 5 μm gap [69]. However, in a micro-gap thermionic converter, as the distance between the emitter and collector becomes smaller, it becomes harder to maintain a large temperature difference between the electrodes due to the near field radiative heat transfer. If the distance between the electrodes is smaller than a certain value, the near field radiative heat transfer increases by many orders of magnitude leading to significant reduction in the power conversion efficiency of a VSTIC. Therefore, there is an optimal gap between the electrodes which maximizes the energy conversion efficiency. Lee *et al* theoretically calculated the optimal inter-electrode gap range of cesiated tungsten electrode for maximizing the efficiency of the VSTICs [70]. They showed that for an inter-electrode gap between 900 nm to 3 μm the energy conversion efficiency is nearly optimal.

The space charge problem can also be solved using positively biased gate electrodes or grids to accelerate the electrons toward the collector (figure 14.4(b)). In this approach the gate electrode creates a potential trough which accelerates electrons away from the emitter. This acceleration of electrons removes the space charge from the inter-electrode gap near the surface of the emitter. However, the negatively charged electrons also tend to be pulled by the positive gate electrode. Consequently, the electrons accumulate in the gate electrode which increases the gate leakage current, decreasing the energy conversion efficiency significantly. Moreover, the gate current scales with the gate voltage. The gate current problem can be solved by applying a perpendicular magnetic field that forces the electrons to flow in a cycloidal path to keep the electrons from hitting the gate (figure 14.4(c)) [71]. This process prevents the formation of a space charge cloud and provides a smooth transfer path for emitted electrons towards the collector. This process of thermionic power conversion is characterized as thermoelectronic generation as only the electrons, but no ions, are involved. When a magnetic field of order 0.5 T is applied, the electrons are guided through the gate opening and reach the collector, reducing the gate current. The output current is proportional to the geometrical transparency of the of the gate electrode. However, the generation of magnetic fields by permanent magnets or high-TC superconducting coils add complexity to the fabrication process. Also, the magnet increases the mass of the thermoelectronic generator. An electrically conducting electron transparent material such as graphene can be used to eliminate the necessity for a magnetic field. Graphene has an electron transparency of 60% but for a practical thermoelectronic generator an electron transparency of above 90% is required. It has been shown that the transparency of the graphene gate electrode can be increased up to 82% by patterning nanoscale holes in the graphene sheet [72].

Another approach to solve the space charge effect is the introduction of positive charges in the inter-electrode space for neutralizing the accumulated negative electrons (figure 14.4(d)). Low ionization gas such as cesium is commonly used for this purpose [73]. The function of the cesium vapor in the inter-electrode space is twofold. First, a layer of cesium vapor is formed in the emitter and collector reducing the work function of both electrodes, thus increasing the energy conversion efficiency. Second, part of the cesium vapor is positively ionized in the inter-electrode space which attracts the negative space charges and neutralizes them. However, there are some disadvantages to this approach. The ionization process of cesium takes some energy and consequently decreases the overall energy conversion efficiency. Moreover, a cesium reservoir is required for a continuous supply of cesium, which reduces the lifetime of the thermionic converter. Finally, adding cesium plasma increases the heat leak of the device as heat can transfer via convection from the cathode to anode.

The space charge effect can also be mitigated using materials which exhibit negative electron affinity (NEA). NEA lowers the vacuum level below the conduction band in a semiconductor. As a result, the electron emission originates from the conduction band minimum. Therefore, the electrons can escape the emitter surface if enough energy is supplied. Moreover, in a VSTIC, low energy electrons are mainly responsible for the formation of the space charge. The NEA property of the material acts as a low energy electron filter, which prevents slower electrons from escaping the emitter surface, thus further reducing the space charge effect. Smith [74] developed a model considering the space charge effect in a thermionic converter where the emitter features the NEA property. This model showed that the NEA property can alleviate and, in some cases, remove the space charge effect. Hydrogen terminated diamond material shows the NEA property. The hydrogen passivation layer creates a strong dipole moment, which lowers the vacuum level below the conduction band minimum [75]. Furthermore, the wide bandgap semiconductors, such as aluminum nitride (AlN) and boron nitride (BN), have shown the NEA property and are being considered for thermionic emission applications [76, 77].

14.4.2 Solid-state thermionic converters

14.4.2.1 Design considerations

SSTICs were introduced to resolve many of the challenges of VSTICs but they also posed some new difficulties. We discussed the differences briefly above. In particular, the high work function of the metals and the space charge effect are not relevant to SSTICs. Instead, SSTICs suffer from conduction heat leakage due to the small thickness of the semiconducting layer. Thermionic conditions impose the requirement of ballistic transport in the semiconducting layer. Note that if transport is diffusive in the semiconducting layer, then the device should be described as a thermoelectric device working based on the Seebeck effect. In that case, within the bulk semiconducting layer, electrons will lose their memory and reach near equilibrium with the lattice. To maintain ballistic transport of the hot electrons, the semiconducting layer thickness, L, should be equal to or less than the mean free path, λ, of the electrons inside the barrier: $L \leqslant \lambda$. On the other hand, too thin a barrier results in tunneling of electrons, which is undesirable since low energy electrons (with energies smaller than the chemical potential, μ) will act as holes and lower the efficiency of the electron transmitting device (similar to the bipolar effect in the case of thermoelectric transport). If the height of a square shaped barrier of length L_t is $e\phi$, the probability of an electron tunneling through is proportional to

$\exp(2L_t\sqrt{\frac{2m^*e\phi}{\hbar^2}})$ while the probability of thermionic emission is proportional to $\exp(\frac{e\phi}{k_BT})$ [21]. Thus, the minimum thickness L_t which makes the thermionic emission dominant while suppressing the tunneling part, is $L_t = \frac{\hbar}{2k_BT}\sqrt{\frac{e\phi}{2m^*}}$ [21]. So, the semiconducting layer thickness should satisfy the condition $L_t < L \leqslant \lambda$.

It is known that for an ideal SSTIC, internal electrical resistance, R, is zero. Non-zero values of R lower the performance [9]. Therefore, having a small thickness and ballistic transport is beneficial since it will result in smaller values of R. The effect of a non-zero value of R could be considered negligible for $R \ll 0.5(k_BT_C/eJA)$, where J is the Richardson current, A is the electrode area and T_C is the temperature of the cold side [42]. For example, the output power of an SSTIC operating with a barrier height of $5k_BT$, an anode and cathode area of 1 cm^2, T_H at 400 K and T_C at 360 K, would be close to that of the ideal TICs if the value of is R smaller than 10^{-7} Ω. Therefore, any internal resistance below the limit could be neglected and the SSTIC, in that case, could be approximated by an ideal diode. In SSTICs, R represents the total electrical resistance that has a contribution from the semiconductor layer, the electrodes and the semiconductor–electrode interfaces. Electrical resistance of the semiconductor layer is very small as its thickness is less than 100 nm. Typical electrodes with 1 mm thickness and 1 cm^2 area have very small electrical resistances on the order of 10^{-7} Ω. Therefore, the interfacial electrical resistance plays the most important role among the three components of R. To minimize the internal resistance, the work function of the barrier and the cathode should be aligned to prevent the formation of a Schottky barrier and to form Ohmic contacts with low interfacial resistance [42].

Small thickness is inherently useful for maintaining low electrical resistances but is also the cause of a decrease in the thermal resistance which increases the heat leak of SSTICs. The low thermal resistance prevents a large temperature difference between the electrodes to develop and sustain. While there are no optimum thermal resistance values, it is desirable to have thermal resistance, R_t, larger than (e/k_BJ). This criterion essentially reduces the conduction loss to a minimum value that arises due to the use of a solid barrier. If R_t is much larger than (e/k_BJ), then the thermal leak is negligible. If it is only larger, the thermal leak is not negligible but low and the device can still work with high efficiency. To be highly efficient, an SSTIC operating with the same parameter as described above would need $R_t > 9 \times 10^{-6}$ m^2 KW^{-1} or $G_t < 0.1$ MW m^{-2} K^{-1}, which is a very small number to attain practically considering the size of the device. R_t represents the total thermal resistance contribution from the semiconductor layer and the semiconductor–electrode interface. In the diffusive limit, a way of increasing R_t is by increasing the thickness of the barrier. However, thermionic devices are working in the ballistic limit where resistance is independent of length. Therefore, a good design should address the issue of making the semiconductor–electrode interfacial thermal resistance be as large as possible and find a suitable semiconductor layer which will have very high thermal resistance within a thickness less than the electron mean free path. An SSTIC with all the optimization discussed above is estimated to achieve efficiency higher than that of the state-of-the-art thermoelectric modules [42].

14.4.2.2 Single versus multi-barrier structures

Both single-layer and multilayer SSTICs have been studied thus far. In 1997, it was predicted that an optimized single-layer SSTIC, operating in refrigeration mode, will result 20–30 K of cooling with a cooling power density exceeding kW cm^{-2} [19]. A high barrier height at the anode side was suggested as a means to reduce the backflow of the reverse current in this configuration [19]. Shakouri *et al* introduced the idea of using high barriers in superlattices and theoretically demonstrated an order of magnitude improvement in efficiency with respect to the bulk materials [78]. An increased Seebeck coefficient and reduced thermal conductivity in the superlattice compared to the bulk were cited as the main reasons behind this improvement. Although ballistic transport was not taken into account during the calculation, it was suggested that addition of this transport will improve the efficiency further. Later it was found that non-ideal effects such as contact resistance, finite thermal resistance of the substrate and the heat sink limit the actual cooling to 1° to 4° experimentally [79]. Monte Carlo simulation of a single-barrier InGaAs/GaAs/InGaAs thermionic cooler indicated that most of the heating and the cooling happens at the contact region [80]. At low temperatures and low carrier concentrations, the linear Peltier coefficient reduces remarkably but the nonlinear part of the Peltier coefficient (that is nonlinear with respect to current) survives and dominates the transport. This dominant behavior can be achieved at a current on the order of 10^5 A cm^{-2}. A single-barrier device operated in this condition could achieve a seven-order increase in maximum cooling efficiency [81].

Mahan and Woods, in 1998, argued that the optimum temperature drop which gives the maximum efficiency for an SSTIC is only about 20 K. To obtain larger temperature differences they proposed a multilayer structure [21]. The idea is to maintain a small temperature difference in each layer, the sum of which results in a large temperature difference and high efficiency. This device was estimated to have a performance twice that of the conventional thermoelectric devices. A suitable superlattice which facilitates the flow of hot electrons but blocks that of the cold electrons was used for this purpose. Later, the reduction in thermal conductivity was identified as the only benefit of using a multilayer geometry [82, 83]. Superlattices, including metallic ones, with very low thermal conductivity were studied extensively. Regular metals have a low Seebeck coefficient because of the symmetric distribution of conduction electrons around the Fermi energy. The symmetry can break when tall energy barriers are inserted, filtering only high energy electrons (electrons with energies above the Fermi energy) to enhance the Seebeck coefficient [84]. HfN/ScN and ZrN/ScN metal/semiconductor superlattices were studied theoretically and found to exhibit low thermal conductivity in the cross-plane direction [22, 23]. The thermoelectric properties of mercury cadmium telluride (HgCdTe) based superlattices have also been studied, which is the primary material for high performance infrared imaging systems. Bulk HgCdTe has a low Seebeck coefficient because of its low effective mass and non-degenerate single conduction band. Tall barriers were found to create asymmetric differential conductivity near the Fermi energy and increase the Seebeck coefficient. $Hg_{1-x}Cd_xTe$ in different compositions has also been studied for thermoelectric applications [21]. An increase in the Seebeck coefficient was predicted to be observed by increasing the carrier concentration. However, doping HgCdTe superlattices to as high as 10^{19} cm^{-3} has proven to be a difficult challenge to overcome [85].

Vining and Mahan carried out a comparative study between a thermoelectric and an SSTIC module using a linearized Richardson's equation and showed that the efficiency of the thermoelectric module is always dominant if they both have similar parameters [83]. Their analysis is valid and reliable when transport is linear and most importantly when thermal interfacial resistances are small compared to the thermal conductance of the semiconducting layer. The latter can easily break in nanoscale devices. Later, in a detailed study of SSTICs operating in a nonlinear regime, it was found that thermionic power generators can achieve efficiencies higher than those of the state-of-the-art thermoelectric modules [42].

14.4.2.3 Integrated systems

A large number of thin-film coolers lattice matched to Si, GaAs or InP have been fabricated and characterized thus far. A schematic of a thin-film 60 μm × 60 μm micro-cooler using $Si_{0.75}Ge_{0.25}$/Si superlattices that exhibited 4.2 K cooling while operating at room temperature is shown in figure 14.5(a) [35]. This 3 μm thick microrefrigerator was made up of 200 SiGe/Si superlattices (3 nm Si/12 nm Si.Ge.) grown on top of a boron doped Si (001) substrate. A 1 μm thin buffer layer was deposited between the substrate and the superlattice to reduce the strain that would otherwise develop due to lattice mismatch. Both the superlattice and the buffer layer were doped to a concentration of 5×10^{19} cm^{-3}. To make a good ohmic contact on the top, a highly doped 0.3 μm $Si_{0.8}Ge_{0.2}$ layer was deposited between the metal contact and the superlattice. Many identical coolers from 30×30 μm^2 to 150×150 μm^2 in size were fabricated on a wafer at the same time using standard silicon IC processing technology. A maximum cooling power density of 600 W m^{-2} was

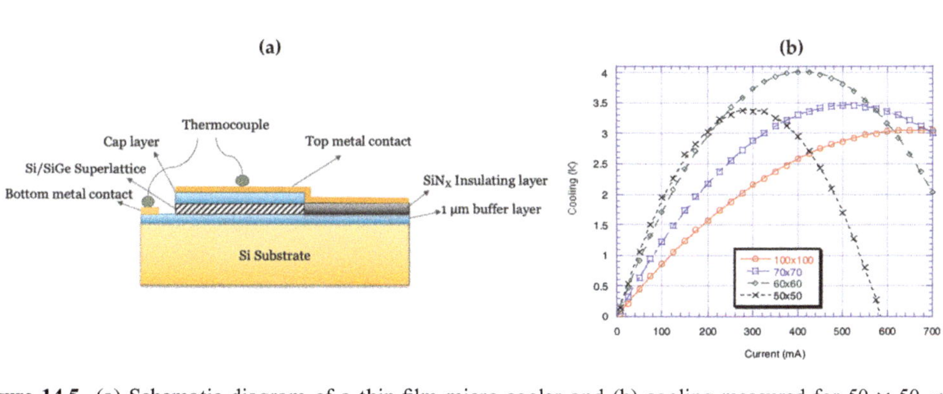

Figure 14.5. (a) Schematic diagram of a thin-film micro-cooler and (b) cooling measured for 50×50 μm^2, 60×60 μm^2, 70×70 μm^2 and 100×100 μm^2 SiGe/Si coolers. Replotted with permission from [84]. Copyright 2004 AIP Publishing.

reported. It was shown that a bulk silicon based micro-cooler with the same dimensions has a cooling of one fourth of the above. Non-ideal effects such as contact resistance, non-ideal heat sink, heat conduction from the top contact metal, poor TE property of the silicon substrate, etc, restricted the cooling to 4.2 K. Cooling in the device is proportional to the current and was found to depend on the device and substrate resistance. While the joule heating, proportional to the square of the current, was found to only be dependent on the substrate resistance. The buffer layer was identified as the major source of heating due to the area dependence of cooling and heating and an optimized device size was adopted to achieve a better cooling performance [79, 86, 87].

A very attractive feature of these micro-coolers is their 20–30 ms thermal transient response time, which is four orders of improvement to that found in a conventional Bi_2Te_3 cooler [88–90]. The maximum cooling power, defined as the heat load power that makes the devices' maximum cooling temperature equal to zero, as a function of device dimension was also studied for devices fabricated following the same procedure as described above. Smaller devices were found to exhibit greater cooling power density, as shown in figure 14.5(b). In this paper, the smallest of the samples fabricated were 40×40 μm^2 in area and showed a598 W cm^{-2} cooling power density [91]. Although 10 °C cooling was expected from a InGaAsP/InP cooler, a poor heat sink at the anode plus the wire connected at the cold junction limited the actual cooling to 0.5 °C and 1 °C over a 1 μm thick barrier at 20 °C and 80 °C, respectively [27]. A 3600 μm^2 AlGaAs/GaAs cooler showed maximum cooling of 2 °C at 100 °C [30]. InGaAs/InP with improved packaging resulted in cooling of 2.3 °C at 90 °C temperature over a 1 μm thick barrier [29]. Bulk $Si_{1-x}Ge_x$, with a ZT of 0.1 at $x = 0.3$, is predicted to exhibit 10° of cooling and as a result, has also been studied for monolithic integration with Si-based micro-electronic devices [92].

14.4.2.4 2D van der Waals heterostructures

2D materials and, in particular, stacks of 2D materials are the latest class of low-dimensional materials studied for solid-state thermionic converters. We refer to such stacks as 2D van der Waals heterostructures (2DvdWH). In these structures, the in-plane atoms are covalently bonded to each other while the cross-plane atoms are weakly bonded by van der Waals forces. Due to weak interlayer bonding, it is possible to stack different 2D materials on top of each other without any strain that would otherwise develop because of lattice mismatch in the presence of a strong bonding [93]. The bandgap of 2D materials can be tuned by applying strain, the electric field and also by changing the number of stacked layers. For example, silicene and germanene are semimetals, but their bandgaps open up when a vertical field is applied and the gap size increases linearly with the electric field [94]. The $TiSe_2$ gap opens up under biaxial strain [95].

Another important property of 2D van der Waals heterostructures is their low value of thermal conductance in the cross-plane direction developed due to the presence of the weak van der Waals bonding between each layer. Earlier we discussed that the thermal conductance should be as small as possible and extremely small values of 0.1 MW m^{-2} K^{-1} are desirable for achieving high efficiency in SSTICs [42]. The super low thermal conductance together with the other beneficial properties of the 2D van der Waals heterostructure mentioned above have thus grabbed the attention of researchers in this field. Several theoretical studies of SSTICs based on this structure have been reported predicting high ZT [38, 39, 96]. Chen *et al* experimentally measured a very low thermal conductance of 4.25×10^{-7} W K^{-1} for a graphene/h-BN/graphene heterostructure [37]. Although this structure exhibited superior thermal properties suitable for thermionic devices, its electronic properties were poor. Insulating the nature of h-BN means too large a barrier for the electrons to overcome and consequently they found a very small ZT of 1.05×10^{-6}. By using seven layers of MoS_2, Yuan *et al* experimentally obtained thermal conductance smaller than 1 MW m^{-2} K^{-1} [97]. Afterwards, a cross-plane ZT of 2.8 has been calculated theoretically for a graphene/MoS_2/graphene heterostructure [98]. Massicotte *et al* experimentally reported a cross-plane thermal conductance of 0.5 MW m^{-2} K^{-1} in a graphene/WSe_2/graphene heterostructure that was tested for photo-thermionic emission [99]. The cross-plane phonon thermal conductance of the Au/G/P/G/Au heterostructure was found to be as small as 4.1 MW m^{-2} K^{-1} from first-principles calculation and an equivalent ZT of 0.13 was predicted for a thermionic device based on this structure [38]. Recently, a ZT of 1.2 at room temperature and 3 at 600 K has been calculated theoretically using first-principles calculations and Green's function formalism for a Sc/WSe_2/$MoSe_2$/WSe_2/Sc van der Waals heterostructure [39]. In 2DvdWH structures, it is crucial to have enough layers to block the tunneling current. Transmission function in Au/G/P/G/Au heterostructure with a varying number of phosphorene layers is shown in figure 14.6 [38]. The averaged electron transmission of 2P and 5P structures was theoretically studied by Wang *et al* and is shown in figure 14.6(a). In the case of 1P and 2P, phosphorene layers are still not thick enough to eliminate the quantum tunneling effect, which results in non-zero transmission within the bandgap, as seen in figure 14.6(d). They observed no quantum tunneling for 5P, as indicated by the zero transmission right above the Fermi level shown in figure 14.6(a) and the white region in figure 14.6(e) [38].

Achieving a good degree of chemical and thermal stability in the 2D materials is very difficult [100, 101]. Stacking them is the next challenging task. The interfaces between the layers need to be clean to obtain large electrical conductance. An ohmic contact is required between the metallic contact and the 2DvdWHs. At the same time, the thermal conductance needs to be extremely small. For these reasons, there has not been any experimental demonstration of SSTICs with a large equivalent figure of merit based on 2DvdWHs thus far.

14.5 Application of thermionic converters

VSTICs are considered as a candidate for both terrestrial and extraterrestrial applications. Cost and longtime reliability issues are ruling factors for terrestrial applications whereas mass, compactness and efficiency are the decisive factors for space power applications. For space power applications, a few watts to megawatts

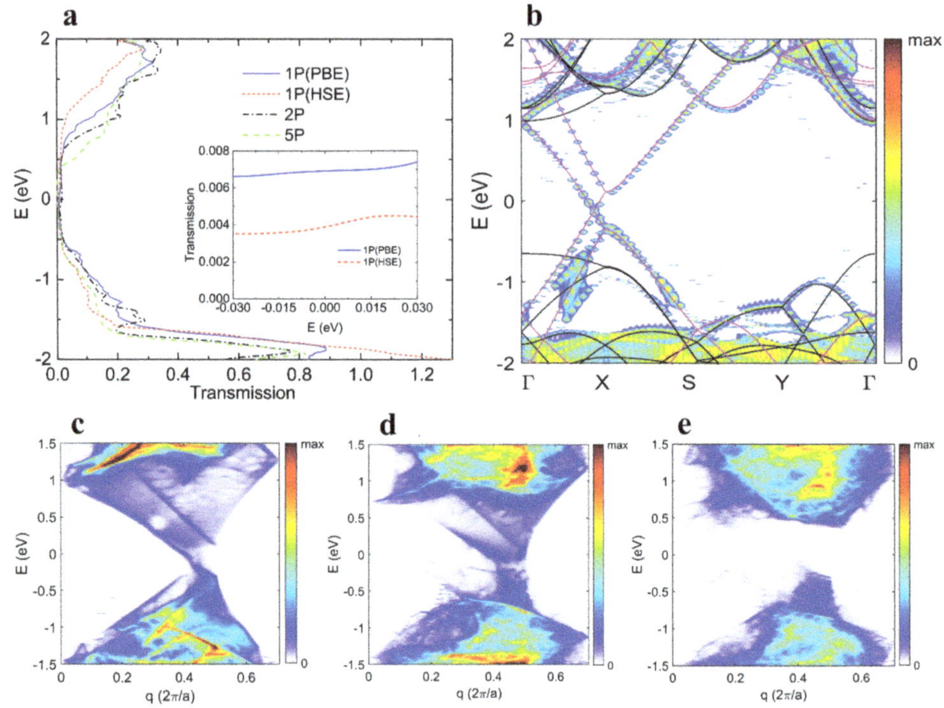

Figure 14.6. (a) The averaged electron transmission of the heterostructures. 1P, 2P and 5P are short for the heterostructures with monolayer, bilayer and quintuple layer phosphorene, respectively. The inset shows the zoom-in of the 1P transmission around the Fermi level. (b) The band-resolved transmission of 1P by HSE. The black and magenta curves are the local band structures of phosphorene and graphene, respectively. The 1D q-resolved transmission of (c) 1P, (d) 2P and (e) 5P. The data of 2P and 5P are calculated by PBE.

range of power is required depending on the mission type and the duration of time that peak power is required. Thermionic power systems are the best choice for missions which require a high electric power output. Converters heated by a nuclear reactor and radioisotopes are considered for space power applications. USA and the former USSR funded several projects for space applications. Under the USSR's TOPAZ program, two nuclear reactors equipped with a 6 kW converter with an energy conversion efficiency of 10% successfully orbited in space in 1987 [15]. A cascaded, high efficiency, static power conversion concept for space power applications using a thermionic and alkali metal thermal to electric converter (AMTEC) cell was proposed and investigated by Van Hagan *et al* [102]. As the heat source temperature of the AMTEC cell is the same as the heat rejected by the thermionic cell, a high energy conversion efficiency of 35%–40% was predicted for this type of cascaded converter. The TI fuel element reactor has been extensively studied for space power applications due to their compactness and ability to produce high power output. The high heat source available from the nuclear fission leads to high power output and high conversion efficiency, and the high-power density available from the nuclear fuel make the TI fuel element compact. The overall size of the system can be further decreased by integrating the power conversion package into the core, which makes TI fuel elements more attractive for space power applications. High-power advanced low mass (HPALM) solar space power systems based on thermionic conversion are being studied and developed for high-power space applications. Energy conversion efficiency as high as 40% could be achieved by cascading this cylindrical inverted multi-cell converter with another power conversion cycle such as AMTEC [103–105]. Overall, the capability of producing very high power makes thermionic power conversion systems ideal candidates for future space missions.

The VSTICs can be used for remote or small power systems that require low voltage supply and low current for a longer period. Due to low voltage and low current requirements, the conversion system can operate at low temperatures, which exert low stress on the power conversion package thus increasing the lifetime. Considering the pollution created by internal combustion engines, Dennis Gabor suggested VSTICs for automotive applications [106]. The dc current produced by the VSTIC can be used to run a dc motor that can directly drive the wheels. Another advantage of a dc motor driven by VSTICs over an internal combustion engine is that it produces maximum torque on starting and generates electricity during braking. Many industrial and power production systems produce heat at higher temperatures than their actual need. For example, conventional steam cycle central power plant combustion of fossil fuels usually produces heat at over 2000 K, but due to material limitations only low temperatures below 1400 K are used to power up a steam turbine [106]. Since VSTICs are able to use heat at the highest temperatures available, they can maximize the system conversion efficiency of a power plant when employed as a topping cycle. In a topping cycle, a VSTIC can utilize the higher temperature from combustion of fossil fuel to heat up the emitter, convert part of the heat to electricity and then deliver rejected heat from the collector at a high enough temperature to power a steam turbine. VSTICs can also be used for cogeneration applications [107]. For example, the same heat source can be used to heat up the emitter of a VSTIC and the boiler of a steam power plant to increase the energy conversion efficiency of the whole system.

In general, VSTICs are not appropriate for low temperature applications, as has been discussed throughout this chapter. SSTICs are more suitable for such applications. In recent years, wearable electronics have attracted much interest due to the availability of a large number of low-power electronic devices. One possibility to power wearable electronics is to generate power harvested from human body heat. SSTICs have the potential to be used as wearable power generators for applications such as powering wireless sensors designed for tracking vital body signs or monitoring chronic diseases in a patient, or powering electronic devices integrated into the clothes of an athlete to monitor his/her movement. SSTICs can replace batteries and hence have the potential to reduce the average weight that military personnel have to carry on various missions.

Another potential application of SSTICs is in electronic cooling. As the size of the electronic chips decreases, their power densities increase, hence they generate higher temperature and localized hotspots under operation. The path to smaller electronic chips is currently blocked by our incapability to maintain a low temperature in small scale electronics. In particular, a localized hotspot several hundred in diameter can be generated in a typical IC if there is a loss of synchronicity among its different functional units. These hotspots are generally 10 °C–40 °C hotter than the comparatively cooler adjacent areas. Most microelectronic devices suffer from localized overheating which ultimately reduces the lifetime and reliability of the device. These hotspots are very small in size compared to that of the whole chip. A conventional passive cooling method such as natural convection or forced air-cooling cannot reach these embedded hotspots. Thermoelectric coolers based on Bi_2Te_3 are in millimeter to centimeter dimensions, which is too large to cool these localized hotspots. Thin-film micro-coolers, micrometers in dimension, on the other hand, are small enough to be integrated with common semiconductor devices based on Si, GaAs and InP. This also removes the issue of packaging. Thus, *in situ* cooling by an integrated micro-cooler is an attractive option to achieve increased control over the temperature management of the electronic and optoelectronic devices for boosting their performance. For example, integrating the micro-cooler in a temperature sensitive laser diode, e.g. a quantum cascade diode, based on GaAs [108] could enhance the spectral stability and power output while reducing the threshold current level. GaAs based integrated circuits would exhibit lower noise and higher gain if effective measures are taken to remove the heat produced by its transistors [109]. Thermionic coolers made using a superlattice can facilitate this monolithic integration with infrared sensors [110].

14.6 Summary and future directions

Vacuum-state thermionic diodes have been replaced extensively with solid state ones. The vacuum-state thermionic converters that generate electricity out of heat have been studied for more than 70 years now. While there is some renewed interest in this technology, there are many challenges to implementing it in a practical commercial setting. Perhaps their best application is in space where a vacuum exists. SSTICs are the modern version of VSTICs. They are a bridge in between thermoelectric modules and VSTICs. They were proposed and studied about 20 years ago. However, the interest in these structures has shifted more toward thermoelectric materials. Superlattices for example, which were originally proposed for SSTICs, could be viewed either as multi-barrier thermionic structures or as a bulk thermoelectric material wherein the dispersions of the electrons and the phonons are those of the superlattice. Superlattices are expensive to build compared to nanostructured bulk samples and they have not shown superior performance compared to them. Therefore, over the years, research shifted toward nanostructured bulk thermoelectric ingots. Among the possibilities for SSTICs, 2DvdWHs are the most promising ones thus far. They are relatively new and unexplored. There are many challenges in fabricating practical packaged devices out of these structures. However, they seem to be one of the only options if we are to move toward nanoscale solid-state energy generation and heat management.

Integrated electronic cooling is the most optimistic application for SSTICs. We emphasized in this chapter that the biggest challenge in the design of SSTICs is to maintain a large thermal resistance at small scales. For electronic cooling, heat is pumped from the hotspots to the ambient. Therefore, there is no need for a large thermal resistance. In contrast, a small thermal resistance is needed in this case to combine passive and active cooling and to pump the heat flux effectively [3, 111].

References

[1] Kim K and Kaviany M 2016 Thermal conductivity switch: optimal semiconductor/metal melting transition *Phys. Rev.* B **94** 155203

[2] Adams M J, Verosky M, Zebarjadi M and Heremans J P 2019 High switching ratio variable-temperature solid-state thermal switch based on thermoelectric effects *Int. J. Heat Mass Transf.* **134** 114–8

[3] Zebarjadi M 2015 Electronic cooling using thermoelectric devices *Appl. Phys. Lett.* **106** 203506

[4] Schwede J W *et al* 2010 Photon-enhanced thermionic emission for solar concentrator systems *Nat. Mater.* **9** 762–7

[5] Schlichter W 1915 Die spontane Elektronenemission glühender Metalle und das glühelektrische Element *Ann. Phys.* **352** 573–640

[6] Murphy E L and Good R H 1956 Thermionic emission, field emission, and the transition region *Phys. Rev.* **102** 1464–73

[7] Hatsopoulos G N 1965 *The Thermo-Electron Engine* (Cambridge, MA: Massachusetts Institute of Technology)

[8] Ioffe A F 1957 *Semiconductor Thermoelements and Thermoelectric Cooling* (London: Infosearch Ltd)

[9] Moss H 1957 Thermionic diodes as energy converters *J. Electron. Control* **2** 305–22

[10] Hernqvist K G, Kanefsky M and Norman F H 1958 Thermionic energy converter *RCA Rev.* **19** 244–58

[11] Webster H F 1959 Calculation of the performance of a high-vacuum thermionic energy converter *J. Appl. Phys.* **30** 488–92

[12] Hatsopoulos G N and Kaye J 1958 Measured thermal efficiencies of a diode configuration of a thermo electron engine *J. Appl. Phys.* **29** 1124–5

[13] Wilson V C 1959 Conversion of heat to electricity by thermionic emission *J. Appl. Phys.* **30** 475

[14] White M C 1996 *An Overview of Thermionic Power Conversion Technology* (Gainesville, FL: University of Florida)

[15] National Research Council 2001 *Termionics Quo Vadis? An Assessment of the DTRA's Advanced Termionics Research and Development Program* (Washington DC: The National Academies Press)

[16] Usov N N P-S V M T V A 2000 Russian space nuclear power and nuclear thermal propulsion systems *Nucl. News* **43** 33–46

[17] Merrill O S 1980 The changing emphasis of the DOE thermionic program *IEEE International Conference of Plasma Science* (Piscataway, NJ: IEEE), p 14

[18] Rasor N S 1991 Thermionic energy conversion plasmas *IEEE Trans. Plasma Sci.* **19** 1191–208

[19] Shakouri A and Bowers J E 1997 Heterostructure integrated thermionic coolers *Appl. Phys. Lett.* **71** 1234–6

[20] Mahan G D, Sofo J O and Bartkowiak M 1998 Multilayer thermionic refrigerator and generator *J. Appl. Phys.* **83** 4683

[21] Mahan G D and Woods L M 1998 Multilayer thermionic refrigeration *Phys. Rev. Lett.* **80** 4016–9

[22] Saha B, Sands T D and Waghmare U V 2012 Thermoelectric properties of HfN/ScN metal/semiconductor superlattices: a first-principles study *J. Phys. Condens. Matter.* **24** 415303

[23] Saha B, Sands T D and Waghmare U V 2011 First-principles analysis of ZrN/ScN metal/semiconductor superlattices for thermoelectric energy conversion *J. Appl. Phys.* **109** 083717

[24] Sofo J O, Mahan G D and Baars J 1994 Transport coefficients and thermoelectric figure of merit of n-$Hg_{1-x}Cd_xTe$ *J. Appl. Phys.* **76** 2249–54

[25] Velicu S *et al* 2008 Thermoelectric characteristics in MBE-grown HgCdTe-based superlattices *J. Electron. Mater.* **37** 1504–8

[26] LaBounty C, Shakouri A, Abraham P and Bowers J E 2000 Monolithic integration of thin-film coolers with optoelectronic devices *Opt. Eng.* **39** 2847

[27] Shakouri A, LaBounty C, Piprek J, Abraham P and Bowers J E 1999 Thermionic emission cooling in single barrier heterostructures *Appl. Phys. Lett.* **74** 88–9

[28] Singh R *et al* 2003 Experimental characterization and modeling of InP-based microcoolers *MRS Proc.* **793** S11.4

[29] Labounty C J *et al* 2001 Experimental investigation of thin film InGaAsP coolers *Mater. Res. Soc. Symp. Proc.* **626** Z14.4.1-6

[30] Zhang J, Anderson N G and Lau K M 2003 AlGaAs superlattice microcoolers *Appl. Phys. Lett.* **83** 374–6

[31] Zhang Y *et al* 2003 Influence of doping concentration and ambient temperature on the cross-plane Seebeck coefficient of InGaAs/InAlAs superlattices *MRS Proc.* **793** S2.4

[32] LaBounty C, Almuneau G, Shakouri A and Bowers J E 2000 Sb-based III–V cooler *19th Int. Conf. on Thermoelectrics*

[33] Zeng G *et al* 1999 SiGe micro-cooler *Electron. Lett.* **35** 2146

[34] Fan X *et al* 2001 High cooling power density SiGe/Si micro-coolers *Electron. Lett.* **37** 126

[35] Fan X *et al* 2001 Integrated cooling for Si-based microelectronics *Proc. 20th Int. Conf. on Thermoelectrics* p 405–8 doi: 10.1109/ICT.2001.979917

[36] Fan X *et al* 2001 SiGeC/Si superlattice microcoolers *Appl. Phys. Lett.* **78** 1580–2

[37] Chen C-C, Li Z, Shi L and Cronin S B 2015 Thermoelectric transport across graphene/hexagonal boron nitride/graphene heterostructures *Nano Res.* **8** 666–72

[38] Wang X, Zebarjadi M and Esfarjani K 2016 First principles calculations of solid-state thermionic transport in layered van der Waals heterostructures *Nanoscale* **8** 14695–704

[39] Wang X, Zebarjadi M and Esfarjani K 2018 High-performance solid-state thermionic energy conversion based on 2D van der Waals heterostructures: a first-principles study *Sci. Rep.* **8** 9303

[40] Richardson O W 1912 LI. Some applications of the electron theory of matter *Philos. Mag. J. Sci.* **23** 594–627

[41] Dushman S 1923 Electron emission from metals as a function of temperature *Phys. Rev.* **21** 623

[42] Zebarjadi M 2017 Solid-state thermionic power generators: an analytical analysis in the nonlinear regime *Phys. Rev. Appl.* **8** 014008
[43] Houston J M 1959 Theoretical efficiency of the thermionic energy converter *J. Appl. Phys.* **30** 481
[44] Mahan G D 1994 Thermionic refrigeration *J. Appl. Phys.* **76** 4362–6
[45] Dmitriev V D and Kholopo G K 1965 Radiant emissivity of tungsten in the infrared region of the spectrum *J. Appl. Spectr.* **2** 315–20
[46] Taylor J B and Langmuir I 1933 The evaporation of atoms, ions and electrons from caesium films on tungsten *Phys. Rev.* **44** 423–58
[47] Gurney R W 1935 Theory of electrical double layers in adsorbed films *Phys. Rev.* **47** 479–82
[48] Langmuir I 1932 Vapor pressures, evaporation, condensation and adsorption *J. Am. Chem. Soc.* **54** 2798–832
[49] Sinsarp A, Yamada Y, Sasaki M and Yamamoto S 2003 Microscopic study on the work function reduction induced by Cs-adsorption *Jpn J. Appl. Phys.* **42** 4882–6
[50] Desplat J-L and Papageorgopoulos C A 1980 Interaction of cesium and oxygen on W(110) *Surf. Sci.* **92** 97–118
[51] Hernqvist K G 1963 Analysis of the arc mode operation of the cesium vapor thermionic energy converter *Proc. IEEE* **51** 748–54
[52] Hishinuma Y, Geballe T H, Moyzhes B Y and Kenny T W 2001 Refrigeration by combined tunneling and thermionic emission in vacuum: use of nanometer scale design *Appl. Phys. Lett.* **78** 2572–4
[53] Morini F, Dubois E, Robillard J F, Monfray S and Skotnicki T 2014 Low work function thin film growth for high efficiency thermionic energy converter: coupled Kelvin probe and photoemission study of potassium oxide *Phys. Status Solidi.* **211** 1334–7
[54] Ortega J E, Oellig E M, Ferrón J and Miranda R 1987 Cs and O adsorption on Si(100) 2×1: a model system for promoted oxidation of semiconductors *Phys. Rev.* B **36** 6213–6
[55] Wu J X *et al* 1999 Photoemission study of the effect of annealing temperature on a K_2O_2/Si (100) surface *Phys. Rev.* B **60** 17102
[56] Lee J-H *et al* 2014 Microfabricated thermally isolated low work-function emitter *J. Microelectromechanical Syst.* **23** 1182–7
[57] Jin F, Liu Y and Day C M 2006 Thermionic emission from carbon nanotubes with a thin layer of low work function barium strontium oxide surface coating *Appl. Phys. Lett.* **88** 163116
[58] Fomenko V S and Samsonov G V (ed) 1966 *Handbook of Thermoionic Properties* (Boston, MA: Springer)
[59] Koeck F A M, Nemanich R J, Lazea A and Haenen K 2009 Thermionic electron emission from low work-function phosphorus doped diamond films *Diam. Relat. Mater.* **18** 789–91
[60] Koeck F A M and Nemanich R J 2006 Emission characterization from nitrogen-doped diamond with respect to energy conversion *Diam. Relat. Mater.* **217** 15
[61] Sherehiy A *et al* 2014 Thermionic emission from phosphorus (P) doped diamond nano-crystals supported by conical carbon nanotubes and ultraviolet photoelectron spectroscopy study of P-doped diamond films *Diam. Relat. Mater.* **50** 66–76
[62] Paxton W F, Wisitsoraat A, Raina S, Davidson J L and Kang W P 2010 P2.14: Characterization of the thermionic electron emission properties of nitrogen-incorporated

'ridged' nanodiamond for use in thermal energy conversion *International Vacuum Nanoelectronics Conference* pp 149–50 (Piscataway, NJ: IEEE)

[63] Nemanich R J *et al* 2010 Thermionic and field electron emission devices from diamond and carbon nanostructures *3rd International Nanoelectronics Conference (INEC)* pp 56–7 (Piscataway, NJ: IEEE)

[64] Österlund L, Chakarov D V and Kasemo B 1999 Potassium adsorption on graphite(0001) *Surf. Sci.* **420** 174–89

[65] Michel J a *et al* 2008 Synthesis and characterization of potassium metal/graphitic carbon nanofiber intercalates *J. Nanosci. Nanotechnol.* **8** 1942–50

[66] Westover T L, Franklin A D, Cola B A, Fisher T S and Reifenberger R G 2010 Photo- and thermionic emission from potassium-intercalated carbon nanotube arrays *J. Vac. Sci. Technol.* B **28** 423–34

[67] Smestad G 2004 Conversion of heat and light simultaneously using a vacuum photodiode and the thermionic and photoelectric effects *Sol. Energy Mater. Sol. Cells* **82** 227–40

[68] Belbachir R Y, An Z and Ono T 2014 Thermal investigation of a micro-gap thermionic power generator *J. Micromechanics Microengineering* **24** 085009

[69] Littau K A *et al* 2013 Microbead-separated thermionic energy converter with enhanced emission current *Phys. Chem. Chem. Phys.* **15** 14442–6

[70] Lee J-H, Bargatin I, Melosh N A and Howe R T 2012 Optimal emitter-collector gap for thermionic energy converters *Appl. Phys. Lett.* **100** 173904

[71] Meir S, Stephanos C, Geballe T H and Mannhart J 2013 Highly-efficient thermoelectronic conversion of solar energy and heat into electric power *J. Renew. Sustain. Energy* **5** 043127

[72] Wanke R *et al* 2016 Magnetic-field-free thermoelectronic power conversion based on graphene and related two-dimensional materials *J. Appl. Phys.* **119** 244507

[73] Rasor N S 1963 Emission physics of the thermionic energy converter *Proc. IEEE* **51** 733–47

[74] Ryan Smith J 2013 Increasing the efficiency of a thermionic engine using a negative electron affinity collector *J. Appl. Phys.* **114** 164514

[75] Takeuchi D *et al* 2005 Direct observation of negative electron affinity in hydrogen-terminated diamond surfaces *Appl. Phys. Lett.* **86** 152103

[76] Nemanich R J *et al* 1996 Negative electron affinity surfaces of aluminum nitride and diamond *Diam. Relat. Mater.* **5** 790–6

[77] Sugino T, Kimura C and Yamamoto T 2002 Electron field emission from boron–nitride nanofilms *Appl. Phys. Lett.* **80** 3602–4

[78] Shakouri A, LaBounty C, Abraham P, Piprek J and Bowers J E 1998 Enhanced thermionic emission cooling in high barrier superlattice heterostructures *MRS Proc.* **545** 449

[79] Vashaee D *et al* 2005 Modeling and optimization of single-element bulk SiGe thin-film coolers *Microscale Thermophys. Eng.* **9** 99–118

[80] Zebarjadi M, Shakouri A and Esfarjani K 2006 Thermoelectric transport perpendicular to thin-film heterostructures calculated using the Monte Carlo technique *Phys. Rev.* B **74** 195331

[81] Zebarjadi M, Esfarjani K and Shakouri A 2007 Nonlinear Peltier effect in semiconductors *Appl. Phys. Lett.* **91** 122104

[82] Radtke R J, Ehrenreich H and Grein C H 1999 Multilayer thermoelectric refrigeration in $Hg_{1-x}Cd_xTe$ superlattices *J. Appl. Phys.* **86** 3195–8

[83] Vining C B and Mahan G D 1999 The B factor in multilayer thermionic refrigeration *J. Appl. Phys.* **86** 6852–3

[84] Vashaee D and Shakouri A 2004 Electronic and thermoelectric transport in semiconductor and metallic superlattices *J. Appl. Phys.* **95** 1233–45
[85] Schmidt J, Ortner K, Jensen J E and Becker C R 2002 Molecular beam epitaxially grown n-type $Hg_{0.80}Cd_{0.20}Te(112)B$ using iodine *J. Appl. Phys.* **91** 451
[86] Dilhaire S *et al* 2003 Thermal and thermomechanical study of micro-refrigerators on a chip based on semiconductor heterostructures *Proceedings 22nd International Conference on Thermoelectrics (January)* pp 519–23 (Piscataway, NJ: IEEE)
[87] Zhang Y *et al* 2003 3D electrothermal simulation of heterostructure thin film micro-coolers *ASME Int. Mechanical Engineering Congress and Exposition* pp 39–48
[88] Fitting A *et al* 2001 Transient response of thin film SiGe micro coolers *Int. Mechanical Engineering Congress and Exhibition*
[89] Shakouri A and Zhang Y 2005 On-chip solid-state cooling for integrated circuits using thin-film microrefrigerators *IEEE Trans. Components Packag. Technol.* **28** 65–9
[90] Zhang Y *et al* 2003 High speed localized cooling using SiGe superlattice microrefrigerators *19th Annual IEEE Semiconductor Thermal Measurement and Management Symposium* pp 61–5 (Piscataway, NJ: IEEE)
[91] Zeng G *et al* 2003 Cooling power density of SiGe/Si superlattice micro refrigerators *MRS Proc.* **793** S2.2
[92] Rowe D M (ed) 1995 *CRC Handbook of Thermoelectrics* (Boca Raton, FL: CRC)
[93] Liu Y *et al* 2016 Van der Waals heterostructures and devices *Nat. Rev. Mater.* **1** 16042
[94] Ni Z *et al* 2012 Tunable bandgap in silicene and germanene *Nano Lett.* **12** 113–8
[95] Nayeb Sadeghi S, Zebarjadi M and Esfarjani K 2019 Non-linear enhancement of thermoelectric performance of a $TiSe_2$ monolayer due to tensile strain, from first-principles calculations *J. Mater. Chem.* C **7** 7308–17
[96] Liang S-J, Liu B, Hu W, Zhou K and Ang L K 2017 Thermionic energy conversion based on graphene van der Waals heterostructures *Sci. Rep.* **7** 46211
[97] Yuan P, Li C, Xu S, Liu J and Wang X 2017 Interfacial thermal conductance between few to tens of layered-MoS_2 and c-Si: effect of MoS_2 thickness *Acta Mater.* **122** 152–65
[98] Sadeghi H, Sangtarash S and Lambert C J 2016 Cross-plane enhanced thermoelectricity and phonon suppression in graphene/MoS_2 van der Waals heterostructures *2D Materials* **4** 015012
[99] Massicotte M *et al* 2016 Photo-thermionic effect in vertical graphene heterostructures *Nat. Commun.* **7** 12174
[100] Vogt P *et al* 2012 Silicene: compelling experimental evidence for graphenelike two-dimensional silicon *Phys. Rev. Lett.* **108** 155501
[101] Fleurence A *et al* 2012 Experimental evidence for epitaxial silicene on diboride thin films *Phys. Rev. Lett.* **108** 245501
[102] Van Hagan T H, Smith J N and Schuller M 1997 Thermionic/AMTEC cascade converter concept for high-efficiency space power *IEEE Aerosp. Electron. Syst. Mag.* **12** 10–5
[103] Begg L L *et al* 2004 Conceptual design of high power advanced low mass (HPALM) solar thermionic power system *Proc. 37th Intersoc. Energy Convers. Eng. Conf.* pp 7–11
[104] Clark P N, Desplat J L, Streckert H H, Adams S F and Smith J W 2006 Solar thermionic test in a thermal receiver *AIP Conf. Proc.* **813** 598
[105] Adams S F 2006 Solar thermionic space power technology testing: a historical perspective *AIP Conf. Proc.* **813** 590–7

[106] Fraser D A, Tanner P G and Irving A D 2005 Developments in thermionic energy converters *IEE Proc.—Sci. Meas. Technol.* **152** 1

[107] Fitzpatrick G O *et al* 1997 Updated perspective on the potential for thermionic conversion to meet 21st century energy needs *IECEC-97 Proceedings of the 32nd Intersociety Energy Conversion Engineering Conference* pp 1045–51 (Piscataway, NJ: IEEE)

[108] Page H *et al* 2001 300 K operation of a GaAs-based quantum-cascade laser at $\lambda \approx 9$ μm *Appl. Phys. Lett.* **78** 3529–31

[109] Das N K and Bertoni H L 1997 *Directions for the Next Generation of MMIC Devices and Systems* (New York: Springer)

[110] Vashaee D and Shakouri A 2006 HgCdTe superlattices for solid-state cryogenic refrigeration *Appl. Phys. Lett.* **88** 132110

[111] Adams M J, Verosky M, Zebarjadi M and Heremans J P 2019 Active Peltier coolers based on correlated and magnon-drag metals *Phys. Rev. Appl.* **11** 054008

Chapter 15

Recent advances in frosting for heat transfer applications

Shreyas Chavan, Kalyan Boyina, Longnan Li and Nenad Miljkovic

Ice accretion and frost formation on solid surfaces is common in HVAC&R (heating, ventilation, air conditioning and refrigeration), energy conversion and transportation. Specifically, frost formation on heat exchanger surfaces in energy systems reduces heat transfer efficiency and results in significant economic and energy losses. Recent advancements in functional surface engineering have offered new opportunities for delaying frost formation, minimizing frost/ice adhesion and enhancing defrosting/deicing. In this chapter, we review recent fundamental developments in classical frosting theory on hydrophobic/superhydrophobic surfaces, discuss various surface fabrication methods and identify their anti-frosting/icing mechanisms. We discuss scale-up strategies for superhydrophobic surfaces, test methodologies for superhydrophobic heat exchangers and durability testing strategies from an application standpoint. We end our review by discussing efficient defrosting/deicing techniques which can be applied to energy systems.

15.1 Introduction

Ice accretion and frost formation on solid surfaces is common in industrial applications, such as HVAC&R (heating, ventilation, air conditioning and refrigeration) [1–7], aircraft [8–15], energy conversion [16, 17] and transportation [18–21]. Specifically, frost formation on heat exchanger surfaces in HVAC&R systems reduces heat transfer efficiency and results in significant economic and energy losses [2, 22–25]. As an example, for moderate climate zones (Memphis, TN), 5% of the annual electric energy consumption can be attributed to the frost/defrost cycle, or 5860 GWh/year [26]. Furthermore, current heat exchangers in HVAC&R systems have designs in which heat transfer is compromised to accommodate frost formation. The development of durable anti-frosting coatings on optimally designed heat exchangers can decrease energy consumption in HVAC&R systems, leading to an additional 7330

doi:10.1088/978-0-7503-1738-2ch15

GWh/year of US primary energy savings for commercial refrigeration alone [2, 26]. Furthermore, heat exchangers in HVAC&R systems are defrosted periodically in order to remove frost and maintain optimal performance [70]. The reduction of defrosting cycles and input energy is another efficient way to save energy.

Understanding the mechanisms of frost formation on solid surfaces is essential to develop suitable anti-frosting coatings for different surfaces used in industry. The dominant mechanism governing condensation frosting on a solid surface is inter-droplet frost wave propagation, or ice bridging [27–33]. Water vapor first condenses on the supercooled solid surface, followed by supercooled droplet freezing through heterogeneous nucleation [27, 34–37]. Upon freezing, frozen droplets interact with neighboring droplets due to the vapor pressure difference above them, resulting in ice bridge growth and frost wave propagation across the entire surface [27, 38, 39].

In the last decade, much effort has been placed on the modification of surface wettability to improve anti-frosting performance in HVAC&R systems [3, 7, 23, 24, 33, 40–42]. Functional surfaces with altered wettability, e.g. hydrophobic or super-hydrophobic surfaces, showed not only higher condensation heat transfer efficiency [43–51], but also reasonable anti-frosting performance [4, 29, 35, 37, 39, 52–55]. Compared to uncoated hydrophilic surfaces, hydrophobic surfaces showed lower frost density and decreased condensation frost growth rate [56–59]. In recent years, the development of micro/nanotechnology has provided an opportunity to mimic natural superhydrophobic surfaces [4, 53–55, 60–63]. Suitably designed superhydrophobic surfaces can delay heterogeneous nucleation of ice and thus condensation frosting [27, 33, 35, 54, 55]. In addition to frost delay, minimization of the adhesion between frost and substrate has been investigated as a frost removal strategy [35, 37, 64–67].

From a utility perspective, the durability and scalability of the superhydrophobic surfaces are critical in HVAC&R applications [5, 33, 36, 39, 68]. To date, the performance of anti-frosting superhydrophobic surfaces has been investigated and improved in different ways. However, the mechanical and chemical durability of the surfaces mentioned above has received little attention. One key reason for the absence of durability quantification is the lack of an industrial test method designed to test functional coatings under a cyclic frost/defrost load. Therefore, an immediate need exists to develop and standardize a durability test method for mechanically and chemically modified surfaces used in industrial components such as heat exchangers and air ducts. Furthermore, the fabrication scalability, manufacturability, and economic feasibility of functional surfaces on large-scale and low-cost heat exchangers is also a challenging problem [37, 51, 69]. Unfortunately, few scale-up fabrication strategies for superhydrophobic heat exchangers exist.

This chapter will review recent developments on both fundamentals and applications of superhydrophobic anti-frosting surfaces on heat exchangers in HVAC&R systems. Although lubricant infused surfaces (LIS or SLIPS) have shown improved dropwise condensation heat transfer performance and anti-frosting ability [35, 71–77], we mainly focus here on dry, solid superhydrophobic anti-frosting surfaces. We first summarize key developments in classical frosting theory on hydrophobic/superhydrophobic surfaces. We then discuss the development of fabrication methods for superhydrophobic surfaces together with their anti-frosting mechanisms. We discuss recent developments in durability test methods for superhydrophobic surfaces. Several scale-up strategies for fabricating superhydrophobic surfaces are introduced followed by a discussion of the performance test methods that can be used to characterize superhydrophobic heat exchanger performance. We end our review by discussing efficient defrosting methods which can be applied to HVAC&R systems.

15.2 Classical condensation frosting theory

When water vapor in the ambient condenses on a chilled substrate in the form of liquid water and then freezes, it is known as condensation frosting [28, 29]. Condensation frosting is markedly different from 'ablimation' (or desublimation) where water vapor in the ambient converts directly to ice on a chilled surface [78, 79], or 'icing' wherein water droplets fall on a cold surface and then freeze upon contact [4, 80–84]. The dominant mechanism governing the spread of condensation frosting is inter-droplet ice bridge frost wave propagation as compared to ice nucleation in icing [28, 29]. Thus, engineered surfaces which show delayed ablimation or icing cannot be directly used to delay condensation frosting [75, 85–102].

In order to delay condensation frosting on a surface, we have to understand the various stages of frosting. When a dry hydrophobic surface is cooled below the freezing temperature of water, five stages of condensation frosting are observed [27]: (1) condensation on a supercooled surface, (2) onset of freezing, (3) frost halo formation, (4) frost wave propagation, and (5) frost densification and scaffolding. We describe every stage in detail and determine how surface wettability can be used to influence and delay frosting.

In the first stage of condensation frosting, water nucleates on the chilled surface. The supersaturation required for heterogeneous nucleation of water can be determined by equating the Gibbs free energy change ΔG of a nucleating embryo to the change in free energy associated with supersaturation. The critical degree of supersaturation required for a particular mode of nucleation can be expressed as

$$\mathrm{SSD} = \frac{p_{\mathrm{n,w}} - p_{\mathrm{s,w}}}{p_{\mathrm{s,w}}}, \tag{15.1}$$

where $p_{\mathrm{n,w}}$ is the critical supersaturation required for nucleation in that mode (equal to the vapor pressure of water vapor in the ambient) and $p_{\mathrm{s,w}}$ is the saturation vapor pressure with respect to water (for condensation) or ice (desublimation) corresponding to the wall temperature of the substrate [79] Using classical nucleation theory and embryo formation kinetics, the SSD required for condensation and desublimation can be determined [79, 103, 104]. Figures 15.1(a) and (b) show the SSD required for condensation and desublimation, respectively, with respect to wall temperature for surfaces with varying wettability (contact angle, θ) [79]. It is observed that the SSD required for embryo formation is significantly higher for hydrophobic surfaces as compared to hydrophilic surfaces. Thus, a hydrophobic surface will delay the nucleation of water and thereby delay condensation frosting. In the literature, the current approach to fabricate frost reducing surfaces focuses on the development of rough hydrophobic surfaces to increase the energy barrier for ice nucleation

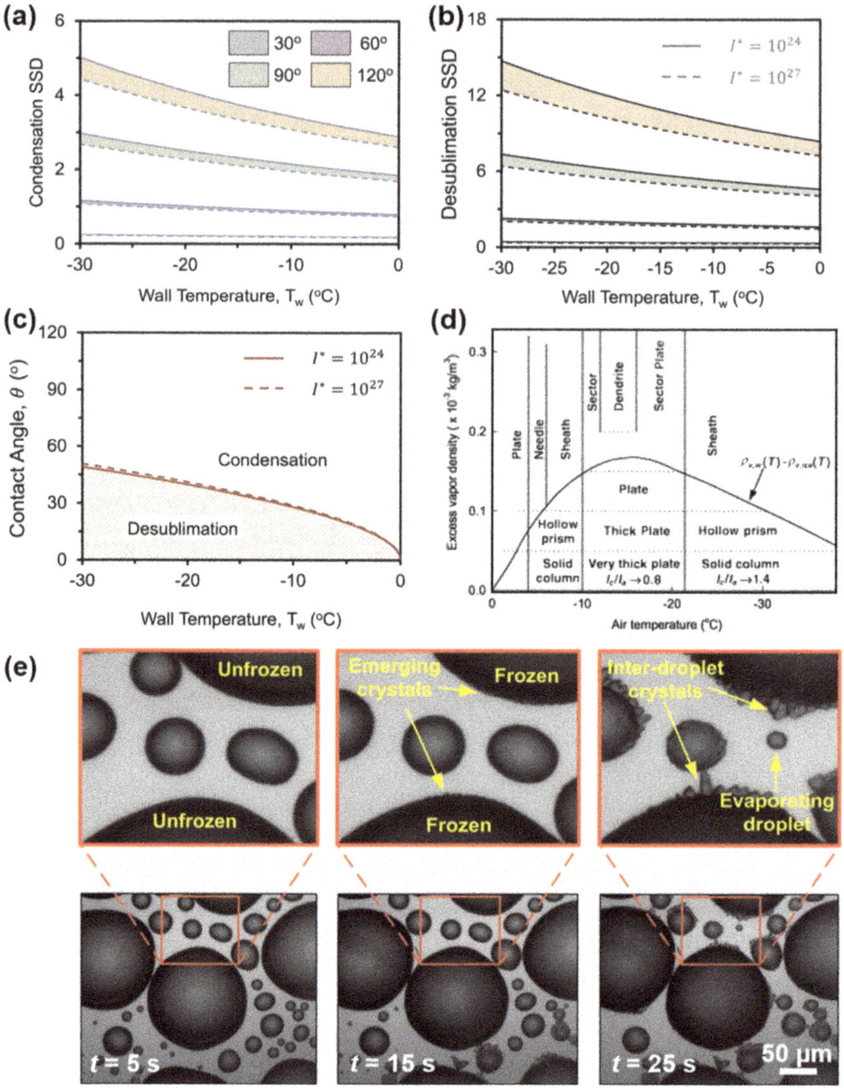

Figure 15.1. Supersaturation degree (SSD) required for (a) condensation or (b) desublimation embryos to nucleate on a substrate as a function of surface temperature (T_W) for different surface wettabilities ($\theta = 30°$, 60°, 90° and 120°) and embryo formation rates ($I^* = 10^{24}$ and 10^{27}). (c) Phase diagram for the preferred mode of nucleation for any surface temperature and wettability, where supercooled condensation is thermodynamically favorable in the phase space above the critical line and desublimation is favorable below. (d) Diagram showing the frost forms depending on excess vapor density and air temperature. (e) Frost wave propagation on a SiAl substrate at $T_S = -10$ °C. (a)–(c) Reproduced with permission from [79]. Copyright 2016 American Chemical Society. (d) Reproduced with permission from [117]. Copyright 2004 Elsevier. (e) Reproduced with permission from [28]. Copyright 2010 Texas A&M University.

[34, 86, 94, 105–112] and to further reduce both the contact angle hysteresis and the ice adhesion strength [64, 95, 113–116].

Recently, superhydrophobic (SHP) surfaces that enable a new mode of condensation, called jumping-droplet condensation, were employed to delay frost formation [48, 118]. Coalescence induced droplet jumping takes place when very small condensate droplets (~10–100 μm, i.e. much smaller than the water capillary length ~2 mm) coalesce on low solid fraction superhydrophobic surfaces, and the excess surface energy during coalescence is converted to kinetic energy [48, 118]. Jumping-droplet condensation is harnessed to remove the condensed water droplets prior to freezing [29, 119, 120]. Apart from delayed frosting, jumping-droplet condensation enhances condensation heat transfer [36, 50, 121–131]. Figure 15.1(c) shows the regime map that suggests the preferred mode of nucleation with respect to surface wettability and temperature. It is observed that for HVAC&R applications which use hydrophobic surfaces, condensation is preferred.

During the second stage of condensation frosting, ice nucleates inside a supercooled water droplet. The nucleation can be homogeneous (if it starts inside the water droplet) or heterogeneous (if it starts at the water–solid interface). The Gibbs free energy barrier for heterogeneous nucleation is generally lower than for homogeneous nucleation. Thus heterogeneous nucleation is preferred [78]. It has been shown in the literature that heterogeneous nucleation can be delayed by increasing the hydrophobicity of the surface [57, 82–84, 86, 88–90, 103, 132–134]. Once ice nucleates inside a water droplet, the droplet freezes in two stages: (i) recalescence, defined as the formation of a porous ice scaffold inside the supercooled liquid droplet [135–137] (which starts at the onset of freezing and ends when the droplet is at 0 °C) and (ii) isothermal (at 0 °C) freeze front propagation. During recalescence, only a fraction of the liquid volume freezes into an ice scaffold while the latent heat of freezing is almost entirely absorbed by the resulting ice–water mixture, resulting in an increase of its temperature from the supercooled state to 0 °C. When the ice–water droplet reaches 0 °C, freeze front propagation ensues isothermally (at 0 °C) within the droplet, converting the ice–water droplet to a homogeneous ice droplet and rejecting the latent heat released into the substrate. It is important to note that the ice bridge led frost wave propagation timescale (~1 s) observed on superhydrophobic surfaces is much slower than the individual droplet freezing timescale (~10 ms) [138]. Thus, frost wave propagation via ice bridge growth dominates surface condensation frosting dynamics, and a delay in individual microscale droplet freezing time has a negligible direct effect on frosting time on non-wetting surfaces [138].

The third stage of condensation frosting, the formation of frost halos [139], overlaps with the second stage. During the recalescence stage of freezing, if the supersaturated pressure required for heterogeneous nucleation of condensate on the substrate, p_n, is less than that of the pressure at the interface of the freezing ice–water droplet, $p_{i,0}$, water molecules will diffuse from the liquid–vapor interface of the large freezing droplet to the ambient surroundings [27, 138, 140]. These water vapor molecules re-condense as microdroplets in the form of an annular ring surrounding the frozen droplet that can rapidly freeze, forming a frost halo. The survival and extent of the frost halo formation has been shown to depend on the duration of the transient pressure conditions ($p_{i,0} > p_n$), which are governed by the time taken after recalescence, τ_f to complete freeze front propagation [140]. With an increase in the substrate thermal conductivity k_s, τ_f decreases [140].

The fourth stage of condensation frosting is the most dominant stage—ice bridge led frost wave propagation [28, 29, 32, 33, 112, 141]. When a supercooled condensate water droplet freezes on a hydrophobic or superhydrophobic surface, neighboring droplets still in the liquid phase immediately begin to evaporate [28, 29]. The evaporated water molecules deposit on the frozen droplet and initiate the growth of ice bridges directed toward the droplets being depleted. The neighboring liquid droplets freeze as soon as the ice bridge connects and provides a heterogeneous nucleation site. Coalescence induced droplet jumping on superhydrophobic surfaces decreases the droplet distribution density [72, 125, 131, 142], thereby reducing the frost wave propagation speeds, delaying frosting [29]. However, given enough time, frost formation still takes place due to the propagation of an inter-droplet frost wave from neighboring edge defects, which are present at the edges of the substrate where ice nucleation generally initiates [29, 143]. Figure 15.1(e) shows one of the first reported observations of inter-droplet ice bridging across supercooled droplets on an SiAl substrate at surface temperature $T_s = -10$ °C [28]. During inter-droplet ice bridging, two mechanisms govern the evaporation of neighboring water droplets: (i) the difference in saturation pressure of the water vapor surrounding the liquid droplet and the frozen droplet induces a vapor pressure gradient [28, 79, 105] and (ii) the latent heat released by freezing droplets locally heats the substrate, evaporating nearby droplets [29]. It was recently discovered that the latent heat released into the substrate does not affect frost wave propagation or frost halo formation [138]. Ice bridge formation during frost wave propagation is governed solely by the vapor pressure gradient caused by the difference in saturation pressure of the water vapor surrounding the liquid droplet and the frozen droplet [138]. The characteristic velocity of ice bridge growth can be determined by conservation of mass flux between an evaporating droplet and the growing ice bridge [79].

In the fifth stage of condensation frosting, the frost wave propagates throughout the surface via a network of inter-connected frozen droplets. On the top of the inter-connected frozen droplet network, out-of-plane frost growth takes place. The thermodynamics of frost densification has been studied extensively [110, 144–155]. Figure 15.1(d) shows the regime map for frost crystal formation depending on excess vapor density and air temperature [117]. Figure 15.1(c) suggests that desublimation takes place on a surface with zero contact angle irrespective of the wall temperature. Accordingly, desublimation is observed on the inter-connected frozen droplet network because the contact angle of water on ice is zero.

Although the aforementioned individual models of the five stages of condensation frosting are useful, work remains to develop an integrated, holistic model that is capable of simulating all of the stages within a computationally tractable framework.

15.3 Anti-frosting superhydrophobic surfaces

According to classical condensation frosting theory, an anti-frosting surface can fall into three different categories [25, 27, 33, 35, 37, 39, 54, 55, 68, 156]. First, the surface can prevent the formation of solid water (frost, glaze, rime, snow and ice) [37, 53]. Second, the surface can significantly decrease the frost accumulation rate [34, 84, 91]. Finally, the adhesion between the ice and substrate can be so low that any accumulated ice is removed easily [75, 95, 157]. Surface topography (smooth or rough) [35, 37], elasticity (hard or soft) [37, 158, 159] and liquid extent (dry or wet) [37, 75, 160] should be considered when designing an icephobic surface. Here, we direct our focus to recent developments of micro/nanotextured, hard and dry superhydrophobic surfaces and their application in anti-frosting HVAC&R systems.

As mentioned in the previous section, superhydrophobic surfaces that enable coalescence induced droplet jumping delay heterogeneous ice nucleation as well as ice bridge propagation [27, 35, 37, 68]. Boreyko and Collier [29] showed that the growth of inter-drop frost fronts on a superhydrophobic surface is up to three times slower than on a control hydrophobic surface due to the reduced droplet size distribution stemming from droplet jumping. To overcome limitations in supersaturation conditions, Zhang *et al* [52] showed a novel hierarchical structure which combined a nanostructure and a micropore array. The surface design had micro pitch spacing comparable to the diameter of coalescing microdroplets such that it could maximize the liquid/air interfacial area beneath the coalescing microdroplets. Frost front propagation velocity on the hierarchical structured superhydrophobic surface was up to an order of magnitude slower than on a control hydrophobic surface. Another merit of superhydrophobic surfaces is defrosting performance. Superhydrophobic surfaces can be engineered to promote frost growth in the suspended Cassie state, thus partially melted ice can be removed with small tilt angles [39, 161]. It could be a benefit to save energy during the defrosting process. The melting of a thin interfacial layer of frost at the solid surface is enough to result in the removal of the bulk frost layer residing on top due to hydrodynamic slip (lubrication).

15.4 Fabrication of superhydrophobic surfaces

Superhydrophobic surfaces were originally inspired by nature, by plants such as lotus leaves and insects such as water striders [53, 54, 61, 63, 160, 162–164]. The development of superhydrophobic surfaces has focused on using a combination of surface roughness [164, 165] and low surface energy coatings [36, 45, 50, 51, 123, 166–169]. Bottom-up (e.g. chemical oxidation, direct growth), top-down (e.g. lithography with wet and dry etching, laser irradiation) and hybrid (combination of bottom-up and top-down) approaches have been used to create a plethora of micro/nanotextures on surfaces.

Figure 15.2 shows superhydrophobic surfaces fabricated on both aluminum substrates (top row) and silicon substrates (bottom row). The structures created on the substrate could be nanoscale (figures 15.2(a) and (d)) [170, 171], microscale (figures 15.2(b) and (e)) [172, 173] or hierarchical (a hybrid of micro- and

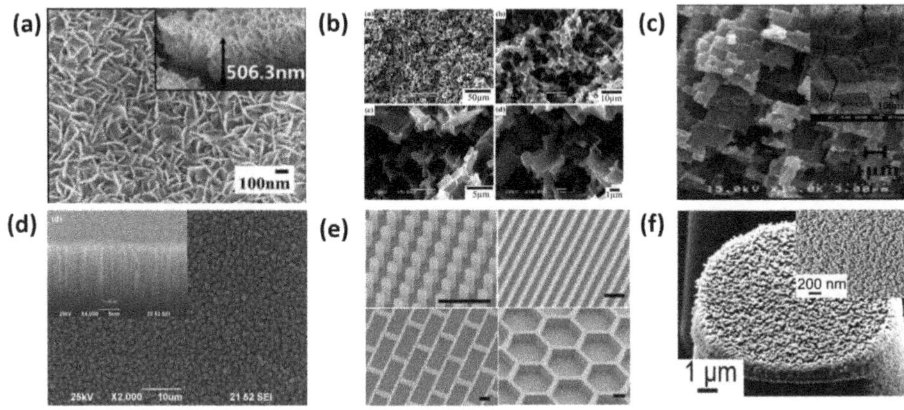

Figure 15.2. Superhydrophobic surfaces fabricated on an aluminum substrate (first row) and a silicon substrate (second row): (a) aluminum boehmite nanostructure (bottom-up); (b) etched aluminum microstructure (top-down); (c) hierarchical structure on aluminum substrate (hybrid); (d) nano-grass structure (top-down); (e) microstructures with various patterns (top-down); and (f) hierarchical structure on silicon substrate which combines micro-posts and nano-grass on top of it. Bottom-up fabrication is used to grow micro/nanostructures on the substrate through thermal, spray coating and self-assembling methods. Top-down fabrication forms roughness on the substrate through wet/dry etching and laser irradiation. (a) Reproduced with permission from [170]. Copyright 2013 Elsevier. (b) Reproduced with permission from [172]. Copyright 2010 American Chemical Society. (c) Reproduced with permission from [174]. Copyright 2013 American Chemical Society. (d) Reproduced with permission from [171]. Copyright 2016 Elsevier. (e) Reproduced with permission from [173]. Copyright 2014 Elsevier. (f) Reproduced with permission from [175]. Copyright 2012 Springer.

nanostructures, figures 15.2(c) and (f)) [174, 175] scale structures. The well-defined structures (pillars, wires, pits, cones, etc) on silicon substrates usually facilitate well controlled samples for conducting experiments to gain an understanding of the wetting and micro/nanoscale phase change phenomena. However, there is a lack of reliable and scalable methods of fabricating superhydrophobic surfaces on metal substrates that are used in real world applications, in particular on copper and aluminum.

Miljkovic *et al* [51, 123] studied scalable functionalized copper oxide (CuO) nanostructured surfaces. Copper is a typical heat exchanger material where chemical-oxidation based CuO nanostructuring allows for self-limiting growth behavior, resulting in low characteristic oxide structure heights ($h = 1\ \mu$m) and a low parasitic conduction thermal resistance. On aluminum, Miwa *et al* [170, 176] showed that boehmite nanostructures had a favorable flake-like nanoscale morphology (figure 15.2(a)), which was also able to achieve continuous droplet jumping [52, 143, 177]. Furthermore, in addition to copper and aluminum, opportunities remain to create scalable nanostructured superhydrophobic surfaces with zinc, stainless steel and titanium, etc.

Although many studies exist which analyze the jumping-droplet mechanism on a multitude of surfaces, there is still a lack of surface design guidelines for enhancing jumping-droplet performance. Miljkovic *et al* [169] proposed an optimum structure design model based on analyzing wetting energy barriers by connecting emergent droplet shape and structure geometry to develop a droplet condensation regime map. Boreyko *et al* [29] fabricated nanostructured superhydrophobic surfaces and tested their jumping performance while also proposing a model based on energetic considerations. Their model and experiments revealed that properly designed nanostructures should enable nanometric jumping droplets and subsequently enhanced heat transfer performance.

Emergent droplet type, contact angle behavior and surface adhesion need to be appropriately characterized and understood in order to create design guidelines for generating and optimizing enhanced micro-/nanostructured surfaces [45, 50, 51, 164, 165, 169]. Furthermore, in order to expand these surface designs for anti-frosting applications, future work should focus on a surface design that promotes consistent jumping droplets in specific environmental conditions. Environmental conditions characterized by the supersaturation and the presence of non-condensable gases (NCGs, mostly air) can significantly alter the heat transfer performance [45, 51, 178, 179]. NCGs can accumulate at the liquid–vapor interface and form a mass transfer boundary layer to hinder vapor diffusion towards the liquid or frost interface. The resulting diffusion boundary layer introduces additional heat and mass transfer resistances at the liquid–vapor interface and significantly degrades the total heat transfer performance [180, 181]. At high supersaturations, pinned Wenzel state water droplets lead to flooding of the surface which increases the adhesion at the frost–surface interface. Further investigation is required to understand the role of geometrical factors (e.g. surface roughness, solid fraction and hierarchy) on jumping droplets, surface wetting and frost propagation behavior.

15.5 Durability/robustness/fouling of superhydrophobic anti-frosting surfaces

The durability of anti-frosting superhydrophobic surfaces with micro/nanostructures can be defined as the ability of the surface to maintain its surface topography and low surface energy under practical working conditions for a long period of time [37, 39, 45, 54]. The anti-frosting superhydrophobic surface should ensure ice nucleation delay after thermal, mechanical and chemical durability tests, indicating that the surface can retain its wettability after many (thousands) frosting/defrosting cycles. In addition to withstanding cycling, the surface should be resistant to the failure of micro/nanostructures upon mechanical impact (e.g. scratches, abrasion or particulate impact) [37, 182, 183]. Furthermore, the low surface energy coating should be capable of resisting chemical, biological and solar degradation [184–187]. Lastly, the surface modification should not increase the corrosion rate of the metal substrate.

Figure 15.3 shows some previously developed durability quantification methods for superhydrophobic surfaces. Frosting/defrosting cycles (figure 15.3(a)) [35, 37, 39, 42, 188, 192] and ice adhesion tests (figure 15.3(d)) [35, 37, 188, 193–197] are some of the most common methods used for anti-frosting superhydrophobic surfaces because they are directly related to practical applications. Wang *et al* [188] conducted ice-breaking/ice-melting and icing/ice-melting tests on superhydrophobic aluminum surfaces. Their results showed that both 20 icing/ice-breaking samples and 40 icing/ice-melting samples lost part of their hydrophobicity with a decreased contact

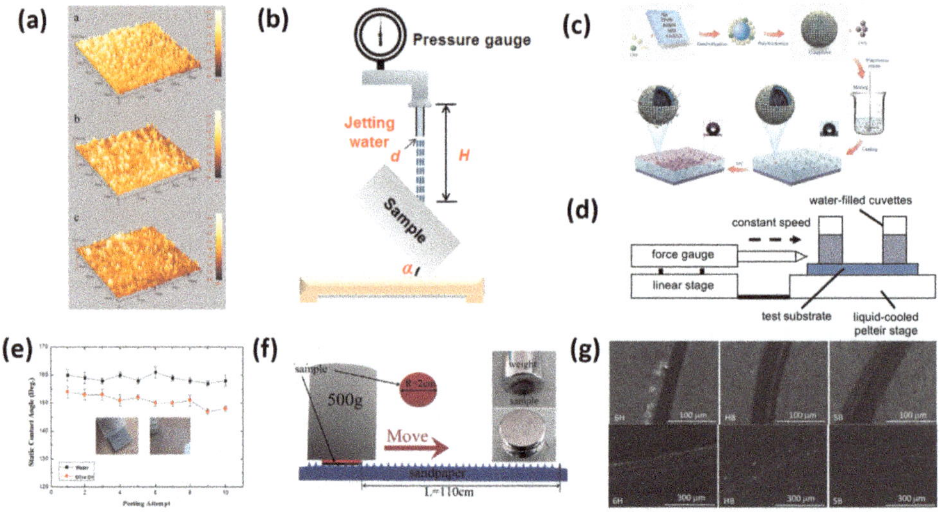

Figure 15.3. Superhydrophobic/icephobic surface durability quantifying methods: (a) frosting/defrosting cycle test; (b) water dripping/jetting test; (c) UV expose test (positive and negative effect); (d) ice adhesion test; (e) tape adhesion test; (f) abrasion test; and (g) pencil hardness test. Methods (a)–(d) are testing functional durability in operation conditions, e.g. frosting/icing, impacting and exposing conditions. Methods (f) and (g) mainly focus on the mechanical durability of superhydrophobic surfaces. (a) Reproduced with permission from [188]. Copyright 2013 American Chemical Society. (b) Reproduced with permission from [189]. Copyright 2014 Elsevier. (c) and (f) Reproduced with permission from [183]. Copyright 2015 American Chemical Society. (d) Reproduced with permission from [94]. Copyright 2010 AIP Publishing. (e) Reproduced with permission from [190]. Copyright 2013 American Chemical Society. (g) Reproduced with permission from [191]. Copyright 2012 Royal Society of Chemistry.

angle from 164.4° to 150.6° and 164.4° to 154.1°, respectively. Both icing/ice-breaking and icing/ice-melting have significant effects on reducing the surface roughness of the structures residing at the interface. Furthermore, water dripping (figure 15.3(b)) [189] and UV exposure tests (figure 15.3(c)) [183, 198–201] have been used to characterize the functional durability of anti-frosting superhydrophobic surfaces. Recently, UV responsive microcapsules synthesized by Pickering emulsion polymerization using titania (TiO_2) and silica (SiO_2) nanoparticles as the Pickering agents were used to fabricate all water based self-repairing UV-resistant super-hydrophobic surfaces [198]. The surface showed a higher water contact angle after exposure to UV. The tape adhesion test (figure 15.3(e)) [183, 190, 202], abrasion test (figure 15.3(f)) [183] and pencil hardness test (figure 15.3(g)) [191, 203, 204] are currently used to test the mechanical durability of superhydrophobic surfaces. Although some of these test methods are already standardized for commonly used substrates/surfaces, there is a need to develop a specific set of standards for anti-frosting superhydrophobic surfaces.

From an application standpoint, the durability/robustness of developed anti-icing surfaces should receive more attention. Current lab scale durability tests (functional and mechanical) can provide direction for the development of more rigorous tests that can mimic the conditions present in commercial applications. Lastly, durability testing poses a challenge of time, which can be addressed via the development of accelerated cycling tests for frost/defrost.

15.6 Anti-frosting coatings for HVAC&R heat exchangers

Durable and scalable anti-frosting coatings are prime candidates for manufacturing anti-icing heat exchangers. Frost forms on heat exchangers used as outdoor units of air source heat pumps during the heating cycle in winter months and cooling coils used in refrigerators [7]. Frost formation on the fins of the heat exchangers increases the thermal resistance by 75% and increases pumping costs by raising the pressure drop across the heat exchanger by blocking the gaps between the fins [23, 205]. Frost needs to be periodically removed from the heat exchanger to maintain optimum performance. Defrosting consumes energy to melt frost and ice from the surface [206, 207]. Heat exchangers with enhanced surfaces have the potential to increase heat transfer rates, delay the onset of frost, reduce defrosting time and improve the efficiency of HVAC&R systems [94, 106, 111, 208, 209].

15.6.1 Existing scalable coating methods

The outdoor evaporator coils used in the HVAC&R field are usually the lowest cost components in the system. Any coating applied to the heat exchangers needs to be economical. Easily scalable methods with low manufacturing and material costs that can be readily integrated into existing assembly lines are desirable. Coating methods that use complicated techniques such as laser etching [210–212] to create or grow nanostructures [76], or multi-step chemical or physical process that require high temperatures and pressures or expensive chemicals [212–214], may be too expensive to adapt in a large-scale manufacturing plant. Coatings that are durable will reduce the maintenance frequency and cost and will improve the appeal of the commercial product. Currently, few durable solutions (if any) exist. Furthermore, durability is a term relative to the application. For example, a coating used on an automotive heat exchanger needs to have excellent mechanical durability (impact of dust, debris and insects) and high temperature durability (proximity to the engine and radiator). However, a coating used on a residential heat exchanger does not require excellent mechanical durability but needs to be able to sustain constant cyclic loading. The coatings used in marine applications require significantly higher corrosion resistance compared to stationary applications. Durability can often be achieved by using coatings that are thick (>100 nm) [109, 163, 189, 215–218]; however, these coatings show poor thermal performance and can decrease the efficiency of the system [219]. Scalability can be achieved by using methods that use dip coating, spray coating, etching, chemical, electrochemical, or physical vapor deposition [189, 212, 220–223]. From the academic point of view, coating methods that can be applied to completely assembled heat exchangers are desirable because the coating does not have to withstand common HVAC&R manufacturing processes such as sheet metal straightening, cutting, tube expansion, bending, stamping and brazing [224], which may destroy or damage the coating. However, industrial manufacturers of heat exchangers prefer robust coatings that can be applied to finstock that are durable enough to withstand manufacturing processes for reasons of cost feasibility.

15.6.2 Performance quantification, testing methods and frost growth models

The performance of air-cooled heat exchangers is tested and certified according to Air-Conditioning Heating and Refrigeration Institute (AHRI) Standard 210/240 and AHRI standard 410. Heat exchangers are often tested in environmentally controlled chambers or wind tunnels that are designed according to the guidelines provided in AHRI standard 210/240. Inlet air temperature, humidity and flow rate are carefully regulated in order to replicate test conditions. The temperature, humidity and pressure change across the heat exchanger are measured to determine heat transfer rates, dehumidification performance as well as the increase in convection resistance stemming from frosting. The mass of the heat exchanger is usually measured to evaluate the amount of frost formed and the water retained on its surface [7, 225–231]. Researchers have also used various optical techniques to study the nature of frost growth on various samples and heat exchangers, including infrared imaging, endoscopy, high-speed photography and high-resolution photography. Much exciting work exists that explores the nature of frost growth on flat plates which have a minimal amount of imperfections [79, 94, 106, 111, 232, 233]. While past work has helped us to understand the fundamentals of frost growth on enhanced surfaces, it is essential for future work to consider how condensate and frost interact with surfaces of complex systems that have many defects and high-energy locations that can accelerate nucleation. Simulations that can model the interactions between condensate droplets and frost within the limited space available in the confines of the heat exchanger fins can provide important design tools for heat exchangers [234]. Numerical models [235–240] that can predict the frost growth rate in heat exchangers under varied inlet air conditions can also provide a design template for heat exchangers that are used in different climates [241].

15.6.3 Frosting, defrosting and re-frosting

As discussed earlier in this chapter, frost occurs through two means: (i) condensation frosting where water vapor condenses on fins, supercools and freezes, resulting in frost propagation between liquid droplets, and (ii) ablimation frosting where water vapor is directly deposited as ice on the surface, bypassing the condensate phase. Delay in frost growth can be achieved through the removal of condensed droplets before they start freezing or by decreasing the energy of the surface to minimize the heterogeneous nucleation rate. However, given enough time, frosting still occurs and the performance of the system deteriorates. Therefore, in addition to frost prevention, coatings need to be evaluated for their defrosting as well as meltwater retention performance.

Superhydrophobic heat exchanger coatings have the potential to increase the efficiency of HVAC&R systems by delaying frost growth rates and reducing defrosting frequency. Researchers have observed that superhydrophobic heat exchangers decrease the pressure drop during the frosting cycle due to the slower

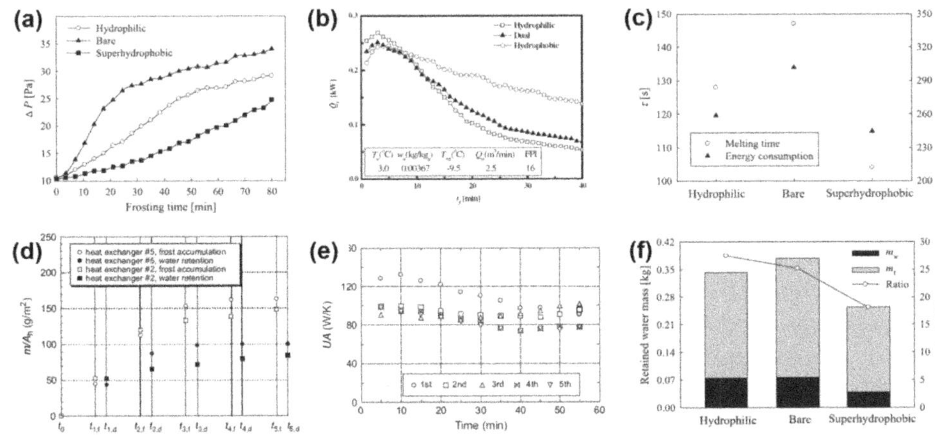

Figure 15.4. (a) Pressure drop across heat exchangers with varying surface wettabilities ($T_{f,in} = -10 \pm 0.5$ °C, $T_{a,in} = 5 \pm 0.5$ °C, $RH_{a,in} = 65 \pm 4\%$ and $u_a = 1.2$ m s^{-1}). (b) Heat transfer rates in modified plate louvered-fin heat exchangers ($T_{f,in} = -9.5$ °C, $T_{a,in} = 3$ °C, $w_{a,in} = 0.003\ 67$ kg/kg$_a$ and $Q_a = 2.5$ m^3 min^{-1}). (c) Melting time and energy consumption during the defrost cycle is effected by the surface wettability of the fins in a heat exchanger ($T_{defrost} = 60 \pm 0.5$ °C, $T_{a,in} = 5 \pm 0.5$ °C, $RH_{a,in} = 65 \pm 4\%$ and $u_a = 1.2$ m s^{-1}). (d) and (e) Water retained by the surface of a heat exchanger after the defrosting cycle can drastically affect the performance of consecutive frosting cycles. The retained water increases the mass of accumulated frost mass and decreases the heat transfer rate ($T_{f,in} = -11 \pm 0.2$ °C, $T_{a,in} = 0 \pm 0.2$ °C, $RH_{a,in} = 75 \pm 5\%$, $T_{defrost} = 10 \pm 0.2$ °C and $u_a = 1.0$ m s^{-1}). (f) A lower amount of water is retained on superhydrophobic fins when compared to hydrophilic and uncoated aluminum fins. (a) and (b) Reproduced with permission from [246]. Copyright 2013 Elsevier. (c) and (f) Reproduced with permission from [242]. Copyright 2015 Elsevier. (d) and (e) Reproduced with permission from [23]. Copyright 2006 Elsevier.

rate of frost growth [242] (figure 15.4(a)). Dropwise condensation with self-cleaning behavior leads to slower frost growth rates on superhydrophobic heat exchangers [208]. Additionally, it was also observed that filmwise condensation on hydrophilic heat exchangers leads to a flatter and denser frost layer which increases thermal resistance [58, 59, 243–245]. Consequently, slower frost growth and dropwise condensation lead to higher heat transfer rates in hydrophobic and superhydrophobic heat exchangers [246] as seen in figure 15.4(b). Previously researchers have investigated melting time and energy consumption (figure 15.4(c)) during the defrosting cycle and have concluded that a lower amount of energy input is required to remove frost from superhydrophobic heat exchangers when compared to uncoated heat exchangers. Any water that is retained on the fins of the heat exchanger post defrosting adversely affects its performance in the next frosting cycles. A variety of factors control the amount of water retained by a heat exchanger after defrosting, namely fin geometry, airflow rate, mode of defrosting and surface wettability. For example, closely spaced fins hinder drainage by providing an opportunity for increased water bridging that causes more water to be retained by the heat exchangers (figure 15.4(d)) [23, 247]. Retained water serves as frost nucleation sites and can decrease the heat transfer coefficient of the heat exchanger after the first frosting/defrosting cycle [23, 247] as detailed by figure 15.4(e).

The penalty imposed by water retention can be mitigated through the use of superhydrophobic coatings [242] (figure 15.4(f)). However, comprehensive data for combinations of superhydrophobic coatings with different fin types and geometries are not available. Therefore, heat transfer rates, heat transfer coefficients, air side pressure drop, defrosting time, meltwater retention and multi-cycle performance are important parameters that should be quantified for heat exchangers with various fin geometries and wettability.

Heat exchanger testing needs to include experiments that measure defrosting time, and meltwater retention in addition to the standard pressure drop, heat transfer rate and heat transfer coefficient measurements. A few new variables gain significance as we move towards implementing superhydrophobic heat exchangers. Water entrainment and moisture carryover may become relevant for heat exchangers that are used for comfort cooling applications. Melted water and ice chunk drainage may become crucial factors that need to be considered when designing enclosures for the heat exchangers. Tests performed in various ambient conditions with a wide range of heat exchanger geometries and coatings will prove to be invaluable. A comprehensive model of condensation and frost growth rate on heat exchangers may be hard to derive but models and simulations that can predict droplet coalescence, water bridging between fins and frost growth rate are invaluable and help us design heat exchangers that perform well in different climates and applications. Models that can predict the amount of water that is retained on the surface of a heat exchanger after defrosting can help in the production of heat exchangers with optimized geometry for condensate drainage and water retention.

15.7 Defrosting

Broadly, there are two methods of defrosting: passive and active [248]. Passive methods use engineered surfaces to delay or reduce frost formation without additional power consumption while active methods require additional power input for defrosting. We have discussed the passive methods in the previous section. There are three active defrosting methods: system defrosting, electro-hydrodynamics and oscillation/ultrasonic vibration [248].

System defrosting further consists of three types. The first type is reversed cycle defrosting (figure 15.5(a)). Here, the cycle direction is reversed to heat the working fluid and melt the frost [249, 250]. However, it has been shown that: (i) up to 27% of the total defrost heat input is used simply to heat up the evaporator due to specific heating [251] and (ii) up to 75% of the defrosting energy goes into heating the refrigeration system and not the ice [252]. The second type of system defrosting is via hot/cool gas defrosting (figure 15.5(b)). Here, cool vapor from the receiver or hot gas from the compressor is directed on the heat exchangers to speed up the defrosting process [253–256]. The method is intricate and comparatively expensive to install, however, the efficiency and the COP of defrosting are better than the other system defrosting methods [256]. The third type of system defrosting is via electric resistive heaters embedded in the heat exchanger to speed up defrosting [257–260]. Although easy to install and control, electric resistance defrosting has been reported to

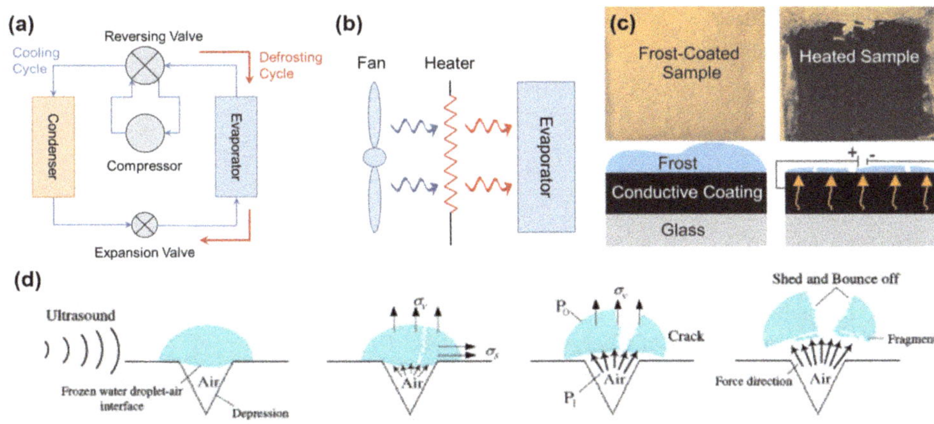

Figure 15.5. Defrosting strategies: (a) reversing the heat pump cycle to heat the evaporator and melt the frost; (b) flowing hot air over the evaporator to melt the frost; (c) Joule heating of the coating at the frost–surface interface; and (d) employing ultrasonic vibrations to remove the frost from the surface. (c) Reproduced with permission from [263]. Copyright 2016 American Chemical Society. (d) Reproduced with permission from [264]. Copyright 2014 Elsevier.

consume 24% more energy and have 8% more operational time as compared to reversed cycle defrosting [248, 261]. Recently, resistive heating at the frost–solid interface has been explored for instantaneous defrosting (figure 15.5(c)) [262, 263].

The second method of active defrosting involves charging the airflow with a high voltage electric field (electro-hydrodynamics, EHD). It has been shown in the literature that EHD has an effect on frost formation, making the frost very fragile in the presence of EHD [265]. EHD has also been shown to give rise to disordered ice growth and frost layers [266, 267] and enhanced defrosting (subject to natural convection) [268].

The third method of active defrosting utilizes vibrations to suppress frost formation. The effects of oscillations have been studied on frost formation [269, 270], and it has been reported that low-amplitude oscillation is not effective to combat frost formation [271]. Adopting high-frequency ultrasonic vibration, frost reduction of 60% has been reported [272, 273]. Ultrasonic vibrations (20 kHz) have also been reported to completely remove frozen droplets (2 mm to 30 mm diameter) from a supercooled vertical surface, as observed in the schematic of figure 15.5(d) [264]. However, the main limitation of utilizing vibrations to suppress frost formation is the high power consumption of ultrasonic sources.

Although active methods of defrosting enhance frost removal, one of the major unsolved setbacks is water retention. Water retention is a major problem during defrosting on heat exchangers as it results in a decrease in the heat transfer performance [23, 247]. Potentially, biphilic surfaces with surface patterns similar to banana leaves might help in combating water retention as they were shown to be very effective at removing water from the surface during condensation [274]. Moreover, the traditional methods of reversed cycle defrosting are highly inefficient, and new methods which employ ultrasonic vibrations are not economically feasible or scalable. The interfacial pulse defrosting method, which has been successfully implemented in the aviation industry, could be implemented in the HVAC&R sector [262, 275–277]. However, accurate theoretical models that simulate pulse defrosting physics are lacking.

References

[1] Radermacher R and Kim K 1996 Domestic refrigerators: recent developments *Int. J. Refrig.* **19** 61–9

[2] U.S. Department of Energy 2011 *Buildings Energy Data Book* (Washington DC: U.S. Dept. of Energy, Office of Energy Efficiency and Renewable Energy)

[3] Huang L, Liu Z, Liu Y, Gou Y and Wang J 2009 Experimental study on frost release on fin-and-tube heat exchangers by use of a novel anti-frosting paint *Exp. Therm. Fluid. Sci.* **33** 1049–54

[4] Lv J, Song Y, Jiang L and Wang J 2014 Bio-inspired strategies for anti-icing *ACS Nano* **8** 3152–69

[5] Zhang P and Lv F Y 2015 A review of the recent advances in superhydrophobic surfaces and the emerging energy-related applications *Energy* **82** 1068–87

[6] Machielsen C H M and Kerschbaumer H G 1989 Influence of frost formation and defrosting on the performance of air coolers: standards and dimensionless coefficients for the system designer *Int. J. Refrig.* **12** 283–90

[7] Moallem E, Cremaschi L, Fisher D E and Padhmanabhan S 2012 Experimental measurements of the surface coating and water retention effects on frosting performance of microchannel heat exchangers for heat pump systems *Exp. Therm. Fluid Sci.* **39** 176–88

[8] La Due J, Muller M and Swangler M 1996 Cratering phenomena on aircraft anti-icing films *J. Aircr.* **33** 131–8

[9] Broeren A P, Lee S and Clark C 2015 Aerodynamic effects of anti-icing fluids on a thin high-performance wing section *J. Aircr.* **53** 451–62

[10] Brown J, Raghunathan S, Watterson J, Linton A and Riordon D 2002 Heat transfer correlation for anti-icing systems *J. Aircr.* **39** 65–70

[11] Harireche O, Verdin P, Thompson C P and Hammond D W 2008 Explicit finite volume modeling of aircraft anti-icing and de-icing *J. Aircr.* **45** 1924–36

[12] Özgen S, Carbonaro M and Sarma G 2002 Experimental study of wave characteristics on a thin layer of de/anti-icing fluid *Phys. Fluids* **14** 3391–402

[13] Dong W, Zhu J, Zheng M and Chen Y 2015 Thermal analysis and testing of nonrotating cone with hot-air anti-icing system *J. Propul. Power* **31** 896–903

[14] Keith T G Jr, DEWITT K J, Nathman J K, Dietrich D A and Al-Khalil K M 1990 Thermal analysis of engine inlet anti-icing systems *J. Propul. Power* **6** 628–34

[15] Zilio C and Patricelli L 2014 Aircraft anti-ice system: evaluation of system performance with a new time dependent mathematical model *Appl. Therm. Eng.* **63** 40–51

[16] Giebel G, Brownsword R, Kariniotakis G, Denhard M and Draxl C 2011 The state-of-the-art in short-term prediction of wind power: a literature overview *ANEMOS Plus*

[17] Thomas D H 2010 *Energy Efficiency through Combined Heat and Power or Cogeneration* (New York: Nova Science Publishers)

[18] Andersson A K and Chapman L 2011 The impact of climate change on winter road maintenance and traffic accidents in West Midlands, UK *Accid. Anal. Prev.* **43** 284–9

[19] Aydın D, Kizilel R, Caniaz R O and Kizilel S 2015 Gelation-stabilized functional composite-modified bitumen for anti-icing purposes *Ind. Eng. Chem. Res.* **54** 12587–96

[20] Heymsfield E, Osweiler A, Selvam P and Kuss M 2014 Developing anti-icing airfield runways using conductive concrete with renewable energy *J. Cold Reg. Eng.* **28** 04014001

[21] Klein-Paste A and Wåhlin J 2013 Wet pavement anti-icing—a physical mechanism *Cold Reg. Sci. Technol.* **96** 1–7

[22] Li B and Yao R 2009 Urbanisation and its impact on building energy consumption and efficiency in China *Ren. Ener.* **34** 1994–8

[23] Xia Y, Zhong Y, Hrnjak P and Jacobi A 2006 Frost, defrost, and refrost and its impact on the air-side thermal-hydraulic performance of louvered-fin, flat-tube heat exchangers *Int. J. Refrig.* **29** 1066–79

[24] Chang Y-S 2011 Performance analysis of frostless heat exchanger by spreading antifreeze solution on heat exchanger surface *J. Therm. Sci. Technol.* **6** 123–31

[25] Nasr M R, Fauchoux M, Besant R W and Simonson C J 2014 A review of frosting in air-to-air energy exchangers *Renew. Sustain. Energy Rev.* **30** 538–54

[26] Goetzler W, Guernsey M, Foley K, Young J and Chung G 2016 *Energy Savings Potential and RD&D Opportunities for Commercial Building Appliances (2015 Update)* (Burlington, MA: Navigant Consulting)

[27] Nath S, Ahmadi S F and Boreyko J B 2017 A review of condensation frosting *Nanosc. Microsc. Thermophys. Eng.* **21** 81–101

[28] Dooley J B 2010 *Determination and characterization of ice propagation mechanisms on surfaces undergoing dropwise condensation* (College Station, TX: Texas A&M University)

[29] Boreyko J B and Collier C P 2013 Delayed frost growth on jumping-drop superhydrophobic surfaces *ACS Nano* **7** 1618–27

[30] Boreyko J B, Hansen R R, Murphy K R, Nath S, Retterer S T and Collier C P 2016 Controlling condensation and frost growth with chemical micropatterns *Sci. Rep.* **6** 19131

[31] Guadarrama-Cetina J, Mongruel A, González-Viñas W and Beysens D 2013 Percolation-induced frost formation *Europhys. Lett.* **101** 16009

[32] Zhang Y, Klittich M R, Gao M and Dhinojwala A 2017 Delaying frost formation by controlling surface chemistry of carbon nanotube-coated steel surfaces *ACS Appl. Mater. Interf.* **9** 6512–9

[33] Kim M-H, Kim H, Lee K-S and Kim D R 2017 Frosting characteristics on hydrophobic and superhydrophobic surfaces: a review *Energy Convers. Manage.* **138** 1–11

[34] Jung S, Tiwari M K, Doan N V and Poulikakos D 2012 Mechanism of supercooled droplet freezing on surfaces *Nat. Commun.* **3** 615

[35] Kreder M J, Alvarenga J, Kim P and Aizenberg J 2016 Design of anti-icing surfaces: smooth, textured or slippery? *Nat. Rev. Mater.* **1** 15003

[36] Miljkovic N and Wang E N 2013 Condensation heat transfer on superhydrophobic surfaces *MRS Bull.* **38** 397–406

[37] Sojoudi H, Wang M, Boscher N D, McKinley G H and Gleason K K 2016 Durable and scalable icephobic surfaces: similarities and distinctions from superhydrophobic surfaces *Soft Matter* **12** 1938–63

[38] Graeber G, Dolder V, Schutzius T M and Poulikakos D 2018 Cascade freezing of supercooled water droplet collectives *ACS Nano* **12** 11274–81

[39] Schutzius T M, Jung S, Maitra T, Eberle P, Antonini C, Stamatopoulos C and Poulikakos D 2015 Physics of icing and rational design of surfaces with extraordinary icephobicity *Langmuir* **31** 4807–21
[40] Roustan J, Kienlen J, Aubas P, Aubas S and Du Cailar J 1992 Comparison of hydrophobic heat and moisture exchangers with heated humidifier during prolonged mechanical ventilation *Inten. Care Med.* **18** 97–100
[41] Restuccia G, Freni A, Russo F and Vasta S 2005 Experimental investigation of a solid adsorption chiller based on a heat exchanger coated with hydrophobic zeolite *Appl. Therm. Eng.* **25** 1419–28
[42] Sommers A D and Jacobi A M 2006 Creating micro-scale surface topology to achieve anisotropic wettability on an aluminum surface *J. Micromech. Microeng.* **16** 1571
[43] Rykaczewski K, Scott J H J, Rajauria S, Chinn J, Chinn A M and Jones W 2011 Three dimensional aspects of droplet coalescence during dropwise condensation on superhydrophobic surfaces *Soft Matter* **7** 8749–52
[44] Chen X, Wu J, Ma R, Hua M, Koratkar N, Yao S and Wang Z 2011 Nanograssed micropyramidal architectures for continuous dropwise condensation *Adv. Funct. Mater.* **21** 4617–23
[45] Enright R, Miljkovic N, Alvarado J L, Kim K and Rose J W 2014 Dropwise condensation on micro- and nanostructured surfaces *Nanoscale Microscale Thermophys. Eng.* **18** 223–50
[46] Rose J 2002 Dropwise condensation theory and experiment: a review *Proc. Inst. Mech. Eng.* A **216** 115–28
[47] Marto P, Looney D, Rose J and Wanniarachchi A 1986 Evaluation of organic coatings for the promotion of dropwise condensation of steam *Int. J. Heat Mass Transfer* **29** 1109–17
[48] Boreyko J B and Chen C-H 2009 Self-propelled dropwise condensate on superhydrophobic surfaces *Phys. Rev. Lett.* **103** 184501
[49] Chen C-H, Cai Q, Tsai C, Chen C-L, Xiong G, Yu Y and Ren Z 2007 Dropwise condensation on superhydrophobic surfaces with two-tier roughness *Appl. Phys. Lett.* **90** 173108
[50] Miljkovic N, Enright R and Wang E N 2012 Effect of droplet morphology on growth dynamics and heat transfer during condensation on superhydrophobic nanostructured surfaces *ACS Nano* **6** 1776–85
[51] Miljkovic N, Enright R, Nam Y, Lopez K, Dou N, Sack J and Wang E N 2012 Jumping-droplet-enhanced condensation on scalable superhydrophobic nanostructured surfaces *Nano Lett.* **13** 179–87
[52] Zhang Q, He M, Chen J, Wang J, Song Y and Jiang L 2013 Anti-icing surfaces based on enhanced self-propelled jumping of condensed water microdroplets *Chem. Commun.* **49** 4516–8
[53] Sun X, Damle V G, Liu S and Rykaczewski K 2015 Bioinspired stimuli-responsive and antifreeze-secreting anti-icing coatings *Adv. Mater. Interf.* **2** 1400479
[54] Zhang S, Huang J, Cheng Y, Yang H, Chen Z and Lai Y 2017 Bioinspired surfaces with superwettability for anti-icing and ice-phobic application: concept, mechanism, and design *Small* **13** 1701867
[55] Li Q and Guo Z 2018 Fundamentals of icing and common strategies for designing biomimetic anti-icing surfaces *J. Mater. Chem.* A **6** 13549–81
[56] Hoke J, Georgiadis J and Jacobi A 2004 Effect of substrate wettability on frost properties *J. Thermophys. Heat Transfer* **18** 228–35

[57] Chavan S, Carpenter J, Nallapaneni M, Chen J and Miljkovic N 2017 Bulk water freezing dynamics on superhydrophobic surfaces *Appl. Phys. Lett.* **110** 041604
[58] Lee H, Shin J, Ha S, Choi B and Lee J 2004 Frost formation on a plate with different surface hydrophilicity *Int. J. Heat Mass Transfer* **47** 4881–93
[59] Kim K and Lee K-S 2011 Frosting and defrosting characteristics of a fin according to surface contact angle *Int. J. Heat Mass Transfer* **54** 2758–64
[60] Liu K, Yao X and Jiang L 2010 Recent developments in bio-inspired special wettability *Chem. Soc. Rev.* **39** 3240–55
[61] Yao X, Song Y and Jiang L 2011 Applications of bio-inspired special wettable surfaces *Adv. Mater.* **23** 719–34
[62] Liu K and Jiang L 2011 Bio-inspired design of multiscale structures for function integration *Nano Today* **6** 155–75
[63] Nishimoto S and Bhushan B 2013 Bioinspired self-cleaning surfaces with superhydrophobicity, superoleophobicity, and superhydrophilicity *RSC Adv.* **3** 671–90
[64] Kulinich S and Farzaneh M 2009 How wetting hysteresis influences ice adhesion strength on superhydrophobic surfaces *Langmuir* **25** 8854–6
[65] Kulinich S and Farzaneh M 2009 Ice adhesion on super-hydrophobic surfaces *Appl. Surf. Sci.* **255** 8153–7
[66] Zou M, Beckford S, Wei R, Ellis C, Hatton G and Miller M 2011 Effects of surface roughness and energy on ice adhesion strength *Appl. Surf. Sci.* **257** 3786–92
[67] Bharathidasan T, Kumar S V, Bobji M, Chakradhar R and Basu B J 2014 Effect of wettability and surface roughness on ice-adhesion strength of hydrophilic, hydrophobic and superhydrophobic surfaces *Appl. Surf. Sci.* **314** 241–50
[68] Attinger D, Frankiewicz C, Betz A R, Schutzius T M, Ganguly R, Das A, Kim C-J and Megaridis C M 2014 Surface engineering for phase change heat transfer: a review *MRS Energy Sustainability* **1** E4
[69] Tian X, Shaw S, Lind K R and Cademartiri L 2016 Thermal processing of silicones for green, scalable, and healable superhydrophobic coatings *Adv. Mater.* **28** 3677–82
[70] Jhee S, Lee K-S and Kim W-S 2002 Effect of surface treatments on the frosting/defrosting behavior of a fin-tube heat exchanger *Int. J. Refrig.* **25** 1047–53
[71] Kim P, Kreder M J, Alvarenga J and Aizenberg J 2013 Hierarchical or not? Effect of the length scale and hierarchy of the surface roughness on omniphobicity of lubricant-infused substrates *Nano Lett.* **13** 1793–9
[72] Weisensee P B, Wang Y, Hongliang Q, Schultz D, King W P and Miljkovic N 2017 Condensate droplet size distribution on lubricant-infused surfaces *Int. J. Heat Mass Transfer* **109** 187–99
[73] Cao M, Guo D, Yu C, Li K, Liu M and Jiang L 2015 Water-repellent properties of superhydrophobic and lubricant-infused 'slippery' surfaces: a brief study on the functions and applications *ACS Appl. Mater. Interf.* **8** 3615–23
[74] Anand S, Paxson A T, Dhiman R, Smith J D and Varanasi K K 2012 Enhanced condensation on lubricant-impregnated nanotextured surfaces *ACS Nano* **6** 10122–9
[75] Kim P, Wong T-S, Alvarenga J, Kreder M J, Adorno-Martinez W E and Aizenberg J 2012 Liquid-infused nanostructured surfaces with extreme anti-ice and anti-frost performance *ACS Nano* **6** 6569–77
[76] Rykaczewski K, Anand S, Subramanyam S B and Varanasi K K 2013 Mechanism of frost formation on lubricant-impregnated surfaces *Langmuir* **29** 5230–8

[77] Xiao R, Miljkovic N, Enright R and Wang E N 2013 Immersion condensation on oil-infused heterogeneous surfaces for enhanced heat transfer *Sci. Rep.* **3** 1988
[78] Carey V P 2008 *Liquid-Vapor Phase-Change Phenomena* 2nd edn (Boca Raton, FL: CRC Press)
[79] Nath S and Boreyko J B 2016 On localized vapor pressure gradients governing condensation and frost phenomena *Langmuir* **32** 8350–65
[80] Parent O and Ilinca A 2011 Anti-icing and de-icing techniques for wind turbines: critical review *Cold Reg. Sci. Technol.* **65** 88–96
[81] Laforte J, Allaire M and Laflamme J 1998 State-of-the-art on power line de-icing *Atmos. Res.* **46** 143–58
[82] Schutzius T M, Jung S, Maitra T, Eberle P, Antonini C, Stamatopoulos C and Poulikakos D 2014 Physics of icing and rational design of surfaces with extraordinary icephobicity *Langmuir* **31** 4807–21
[83] Kreder M J, Alvarenga J, Kim P and Aizenberg J 2016 Design of anti-icing surfaces: smooth, textured or slippery? *Nat. Rev. Mater.* **1** 15003
[84] Jung S, Dorrestijn M, Raps D, Das A, Megaridis C M and Poulikakos D 2011 Are superhydrophobic surfaces best for icephobicity? *Langmuir* **27** 3059–66
[85] Wen M, Wang L, Zhang M, Jiang L and Zheng Y 2014 Antifogging and icing-delay properties of composite micro-and nanostructured surfaces *ACS Appl. Mater. Interfaces* **6** 3963–8
[86] Cao L, Jones A K, Sikka V K, Wu J and Gao D 2009 Anti-icing superhydrophobic coatings *Langmuir* **25** 12444–8
[87] Tourkine P, Le Merrer M and Quéré D 2009 Delayed freezing on water repellent materials *Langmuir* **25** 7214–6
[88] Alizadeh A, Yamada M, Li R, Shang W, Otta S, Zhong S, Ge L, Dhinojwala A, Conway K R and Bahadur V 2012 Dynamics of ice nucleation on water repellent surfaces *Langmuir* **28** 3180–6
[89] Boinovich L, Emelyanenko A M, Korolev V V and Pashinin A S 2014 Effect of wettability on sessile drop freezing: when superhydrophobicity stimulates an extreme freezing delay *Langmuir* **30** 1659–68
[90] Farhadi S, Farzaneh M and Kulinich S 2011 Anti-icing performance of superhydrophobic surfaces *Appl. Surf. Sci.* **257** 6264–9
[91] Guo P, Zheng Y, Wen M, Song C, Lin Y and Jiang L 2012 Icephobic/anti-icing properties of micro/nanostructured surfaces *Adv. Mater.* **24** 2642–8
[92] Boinovich L B and Emelyanenko A M 2013 Anti-icing potential of superhydrophobic coatings *Mendeleev Commun.* **1** 3–10
[93] Kulinich S, Farhadi S, Nose K and Du X 2010 Superhydrophobic surfaces: are they really ice-repellent? *Langmuir* **27** 25–9
[94] Varanasi K K, Deng T, Smith J D, Hsu M and Bhate N 2010 Frost formation and ice adhesion on superhydrophobic surfaces *Appl. Phys. Lett.* **97** 234102
[95] Meuler A J, Smith J D, Varanasi K K, Mabry J M, McKinley G H and Cohen R E 2010 Relationships between water wettability and ice adhesion *ACS Appl. Mater. Interf.* **2** 3100–10
[96] Bengaluru Subramanyam S, Kondrashov V, Rühe J r and Varanasi K K 2016 Low ice adhesion on nano-textured superhydrophobic surfaces under supersaturated conditions *ACS Appl. Mater. Interf.* **8** 12583–7

[97] Maitra T, Jung S, Giger M E, Kandrical V, Ruesch T and Poulikakos D 2015 Superhydrophobicity vs ice adhesion: the quandary of robust icephobic surface design *Adv. Mater Interfaces* **2** 1500330
[98] Tarquini S, Antonini C, Amirfazli A, Marengo M and Palacios J 2014 Investigation of ice shedding properties of superhydrophobic coatings on helicopter blades *Cold Reg. Sci. Technol.* **100** 50–8
[99] Chanda J, Ionov L, Kirillova A and Synytska A 2015 New insight into icing and de-icing properties of hydrophobic and hydrophilic structured surfaces based on core–shell particles *Soft Matter* **11** 9126–34
[100] Thomas S K, Cassoni R P and MacArthur C D 1996 Aircraft anti-icing and de-icing techniques and modeling *J. Aircr.* **33** 841–54
[101] Cancilla D A, Holtkamp A, Matassa L and Fang X 1997 Isolation and characterization of Microtox®-active components from aircraft de-icing/anti-icing fluids *Environ. Toxicol. Chem.* **16** 430–4
[102] Golovin K, Kobaku S P, Lee D H, DiLoreto E T, Mabry J M and Tuteja A 2016 Designing durable icephobic surfaces *Sci. Adv.* **2** e1501496
[103] Fletcher N 1958 Size effect in heterogeneous nucleation *J. Chem. Phys.* **29** 572–6
[104] Adamson A and Gast A 1997 *Physical Chemistry of Surfaces* 6th edn (New York: Wiley)
[105] Hoke J L, Georgiadis J G and Jacobi A M 2004 Effect of substrate wettability on frost properties *J. Thermophys. Heat Tr.* **18** 228–35
[106] Na B and Webb R L 2003 A fundamental understanding of factors affecting frost nucleation *Int. J. Heat Mass Transfer* **46** 3797–808
[107] Alizadeh A *et al* 2012 Dynamics of ice nucleation on water repellent surfaces *Langmuir* **28** 3180–6
[108] Tourkine P, Le Merrer M and Quere D 2009 Delayed freezing on water repellent materials *Langmuir* **25** 7214–6
[109] Wang H, Tang L M, Wu X M, Dai W T and Qiu Y P 2007 Fabrication and anti-frosting performance of super hydrophobic coating based on modified nano-sized calcium carbonate and ordinary polyacrylate *Appl. Surf. Sci.* **253** 8818–24
[110] Hermes C J L, Piucco R O, Barbosa J R and Melo C 2009 A study of frost growth and densification on flat surfaces *Exp. Therm. Fluid Sci.* **33** 371–9
[111] Zhang Q L, He M, Zeng X P, Li K Y, Cui D P, Chen J, Wang J J, Song Y L and Jiang L 2012 Condensation mode determines the freezing of condensed water on solid surfaces *Soft Matter* **8** 8285–8
[112] Guadarrama-Cetina J, Mongruel A, Gonzalez-Vinas W and Beysens D 2013 Percolation-induced frost formation *Europhys. Lett.* **101** 16009
[113] Meuler A J, McKinley G H and Cohen R E 2010 Exploiting topographical texture to impart icephobicity *ACS Nano* **4** 7048–52
[114] Kulinich S A and Farzaneh M 2011 On ice-releasing properties of rough hydrophobic coatings *Cold Reg. Sci. Technol.* **65** 60–4
[115] Farhadi S, Farzaneh M and Kulinich S A 2011 Anti-icing performance of superhydrophobic surfaces *Appl. Surf. Sci.* **257** 6264–9
[116] Nosonovsky M and Hejazi V 2012 Why superhydrophobic surfaces are not always icephobic *ACS Nano* **6** 8488–91
[117] Na B and Webb R L 2004 New model for frost growth rate *Int. J. Heat Mass Transfer* **47** 925–36

[118] Narhe R D, Khandkar M D, Shelke P B, Limaye A V and Beysens D A 2009 Condensation-induced jumping water drops *Phys. Rev.* E **80** 031604
[119] Kim M-K, Cha H, Birbarah P, Chavan S, Zhong C, Xu Y and Miljkovic N 2015 Enhanced jumping-droplet departure *Langmuir* **31** 13452–66
[120] Liu H, Zhang P, Liu M, Wang S and Jiang L 2013 Organogel-based thin films for self-cleaning on various surfaces *Adv. Mater.* **25** 4477–81
[121] Enright R, Miljkovic N, Al-Obeidi A, Thompson C V and Wang E N 2012 Condensation on superhydrophobic surfaces: the role of local energy barriers and structure length scale *Langmuir* **40** 14424–32
[122] Enright R, Miljkovic N, Alvarado J L, Kim K J and Rose J W 2014 Dropwise condensation on micro- and nanostructured surfaces *Nanosc. Microsc. Therm. Eng.* **18** 223–50
[123] Enright R, Miljkovic N, Dou N, Nam Y and Wang E N 2013 Condensation on superhydrophobic copper oxide nanostructures *J. Heat Transfer* **135** 091304
[124] Enright R, Miljkovic N, Sprittles J, Mitchell R, Nolan K, Thompson C V and Wang E N 2014 How coalescing droplets jump *ACS Nano* **8** 10352–62
[125] Miljkovic N, Enright R, Nam Y, Lopez K, Dou N, Sack J and Wang E N 2013 Jumping-droplet-enhanced condensation on scalable superhydrophobic nanostructured surfaces *Nano Lett.* **13** 179–87
[126] Miljkovic N, Enright R and Wang E N 2013 Modeling and optimization of super-hydrophobic condensation *J. Heat Trans.* **135** 111004
[127] Miljkovic N, Preston D J, Enright R, Adera S, Nam Y and Wang E N 2013 Jumping droplet dynamics on scalable nanostructured superhydrophobic surfaces *J. Heat Transfer* **135** 080907
[128] Miljkovic N, Preston D J, Enright R and Wang E N 2013 Electric-field-enhanced condensation on superhydrophobic nanostructured surfaces *ACS Nano* **7** 11043–54
[129] Miljkovic N, Xiao R, Preston D J, Enright R, McKay I and Wang E N 2013 Condensation on hydrophilic, hydrophobic, nanostructured superhydrophobic and oil-infused surfaces *J. Heat Transfer* **135** 080906–7
[130] Miljkovic N, Preston D J, Enright R and Wang E N 2013 Electrostatic charging of jumping droplets *Nat. Commun.* **4** 2517
[131] Chavan S, Cha H, Orejon D, Nawaz K, Singla N, Yeung Y F, Park D, Kang D H, Chang Y and Takata Y 2016 Heat transfer through a condensate droplet on hydrophobic and nanostructured superhydrophobic surfaces *Langmuir* **32** 7774–87
[132] Boinovich L B and Emelyanenko A M 2013 Anti-icing potential of superhydrophobic coatings *Mendeleev Commun.* **23** 3–10
[133] Eberle P, Tiwari M K, Maitra T and Poulikakos D 2014 Rational nanostructuring of surfaces for extraordinary icephobicity *Nanoscale* **6** 4874–81
[134] Quéré D 2005 Non-sticking drops *Rep. Prog. Phys.* **68** 2495
[135] Chaudhary G and Li R 2014 Freezing of water droplets on solid surfaces: an experimental and numerical study *Exp. Therm. Fluid Sci.* **57** 86–93
[136] Aliotta F, Giaquinta P V, Ponterio R C, Prestipino S, Saija F, Salvato G and Vasi C 2014 Supercooled water escaping from metastability *Sci. Rep.* **4** 7230
[137] Feuillebois F, Lasek A, Creismeas P, Pigeonneau F and Szaniawski A 1995 Freezing of a subcooled liquid droplet *J. Colloid. Interf. Sci.* **169** 90–102

[138] Chavan S, Park D, Singla N, Sokalski P, Boyina K and Miljkovic N 2018 Effect of latent heat released by freezing droplets during frost wave propagation *Langmuir* **34** 6636–44

[139] Cheng R J 1970 Water drop freezing: ejection of microdroplets *Science* **170** 1395–6

[140] Jung S, Tiwari M K and Poulikakos D 2012 Frost halos from supercooled water droplets *Proc. Natl Acad. Sci.* **109** 16073–8

[141] Esmeryan K D, Castano C E, Mohammadi R, Lazarov Y and Radeva E I 2018 Delayed condensation and frost formation on superhydrophobic carbon soot coatings by controlling the presence of hydrophilic active sites *J. Phys. D: Appl. Phys.* **51** 055302

[142] Rose J and Glicksman L 1973 Dropwise condensation—the distribution of drop sizes *Int. J. Heat Mass Transfer* **16** 411–25

[143] Kim A, Lee C, Kim H and Kim J 2015 Simple approach to superhydrophobic nano-structured AI for practical antifrosting application based on enhanced self-propelled jumping droplets *ACS Appl. Mater. Interf.* **7** 7206–13

[144] Sanders C T 1974 The influence of frost formation and defrosting on the performance of air coolers *PhD Thesis* Technische Hogeschool, Delft

[145] Na B and Webb R L 2004 Mass transfer on and within a frost layer *Int. J. Heat Mass Transfer* **47** 899–911

[146] El Cheikh A and Jacobi A 2014 A mathematical model for frost growth and densification on flat surfaces *Int. J. Heat Mass Transfer* **77** 604–11

[147] Lee K-S, Jhee S and Yang D-K 2003 Prediction of the frost formation on a cold flat surface *Int. J. Heat Mass Transfer* **46** 3789–96

[148] Le Gall R, Grillot J and Jallut C 1997 Modelling of frost growth and densification *Int. J. Heat Mass Transfer* **40** 3177–87

[149] Brian P T, Reid R C and Shah Y T 1970 Frost deposition on cold surfaces *Industr. Eng. Chem. Fundam.* **9** 375–80

[150] Hermes C J, Loyola F R and Nascimento V S Jr 2014 A semi-empirical correlation for the frost density *Int. J. Refrig.* **46** 100–4

[151] Hermes C J 2012 An analytical solution to the problem of frost growth and densification on flat surfaces *Int. J. Heat Mass Transfer* **55** 7346–51

[152] Storey B and Jacobi A 1999 The effect of streamwise vortices on the frost growth rate in developing laminar channel flows *Int. J. Heat Mass Transfer* **42** 3787–802

[153] Schneider H 1978 Equation of the growth rate of frost forming on cooled surfaces *Int. J. Heat Mass Transfer* **21** 1019–24

[154] Lee Y and Ro S 2005 Analysis of the frost growth on a flat plate by simple models of saturation and supersaturation *Exp. Therm. Fluid Sci.* **29** 685–96

[155] Jones B and Parker J 1975 Frost formation with varying environmental parameters *J. Heat Transfer* **97** 255–9

[156] Hejazi V, Sobolev K and Nosonovsky M 2013 From superhydrophobicity to icephobicity: forces and interaction analysis *Sci. Rep.* **3** 2194

[157] Menini R and Farzaneh M 2009 Elaboration of Al_2O_3/PTFE icephobic coatings for protecting aluminum surfaces *Surf. Coat. Technol.* **203** 1941–6

[158] Alizadeh A, Bahadur V, Shang W, Zhu Y, Buckley D, Dhinojwala A and Sohal M 2013 Influence of substrate elasticity on droplet impact dynamics *Langmuir* **29** 4520–4

[159] Sokuler M, Auernhammer G K, Roth M, Liu C, Bonacurrso E and Butt H-J 2009 The softer the better: fast condensation on soft surfaces *Langmuir* **26** 1544–7

[160] Wong T-S, Kang S H, Tang S K, Smythe E J, Hatton B D, Grinthal A and Aizenberg J 2011 Bioinspired self-repairing slippery surfaces with pressure-stable omniphobicity *Nature* **477** 443

[161] Boreyko J B, Srijanto B R, Nguyen T D, Vega C, Fuentes-Cabrera M and Collier C P 2013 Dynamic defrosting on nanostructured superhydrophobic surfaces *Langmuir* **29** 9516–24

[162] Darmanin T and Guittard F 2014 Recent advances in the potential applications of bioinspired superhydrophobic materials *J. Mater. Chem.* A **2** 16319–59

[163] Li Y, Li L and Sun J 2010 Bioinspired self-healing superhydrophobic coatings *Angewandte Chemie.* **122** 6265–9

[164] Wang S and Jiang L 2007 Definition of superhydrophobic states *Adv. Mater.* **19** 3423–4

[165] Quéré D 2008 Wetting and roughness *Annu. Rev. Mater. Res.* **38** 71–99

[166] Ulman A 1996 Formation and structure of self-assembled monolayers *Chem. Rev.* **96** 1533–54

[167] Enright R, Miljkovic N, Al-Obeidi A, Thompson C V and Wang E N 2012 Condensation on superhydrophobic surfaces: the role of local energy barriers and structure length scale *Langmuir* **28** 14424–32

[168] Wier K A and McCarthy T J 2006 Condensation on ultrahydrophobic surfaces and its effect on droplet mobility: ultrahydrophobic surfaces are not always water repellant *Langmuir* **22** 2433–6

[169] Miljkovic N, Enright R and Wang E N 2013 Modeling and optimization of superhydrophobic condensation *J. Heat Transfer* **135** 111004

[170] Cho H, Kim D, Lee C and Hwang W 2013 A simple fabrication method for mechanically robust superhydrophobic surface by hierarchical aluminum hydroxide structures *Curr. Appl. Phys.* **13** 762–7

[171] Yue X, Liu W and Wang Y 2016 Effects of black silicon surface structures on wetting behaviors, single water droplet icing and frosting under natural convection conditions *Surf. Coat. Technol.* **307** 278–86

[172] Mishchenko L, Hatton B, Bahadur V, Taylor J A, Krupenkin T and Aizenberg J 2010 Design of ice-free nanostructured surfaces based on repulsion of impacting water droplets *ACS Nano* **4** 7699–707

[173] Liao R, Zuo Z, Guo C, Yuan Y and Zhuang A 2014 Fabrication of superhydrophobic surface on aluminum by continuous chemical etching and its anti-icing property *Appl. Surf. Sci.* **317** 701–9

[174] Maitra T, Tiwari M K, Antonini C, Schoch P, Jung S, Eberle P and Poulikakos D 2013 On the nanoengineering of superhydrophobic and impalement resistant surface textures below the freezing temperature *Nano Lett.* **14** 172–82

[175] Song J, Xu W and Lu Y 2012 One-step electrochemical machining of superhydrophobic surfaces on aluminum substrates *J. Mater. Sci.* **47** 162–8

[176] Miwa M, Nakajima A, Fujishima A, Hashimoto K and Watanabe T 2000 Effects of the surface roughness on sliding angles of water droplets on superhydrophobic surfaces *Langmuir* **16** 5754–60

[177] Hao Q, Pang Y, Zhao Y, Zhang J, Feng J and Yao S 2014 Mechanism of delayed frost growth on superhydrophobic surfaces with jumping condensates: more than interdrop freezing *Langmuir* **30** 15416–22

[178] Boreyko J B, Zhao Y and Chen C-H 2011 Planar jumping-drop thermal diodes *Appl. Phys. Lett.* **99** 234105

[179] Boreyko J B and Chen C-H 2013 Vapor chambers with jumping-drop liquid return from superhydrophobic condensers *Int. J. Heat Mass Transfer* **61** 409–18
[180] Zhao Y, Preston D J, Lu Z, Zhang L, Queeney J and Wang E N 2018 Effects of millimetric geometric features on dropwise condensation under different vapor conditions *Int. J. Heat Mass Transfer* **119** 931–8
[181] Ma X-H, Zhou X-D, Lan Z, Yi-Ming L and Zhang Y 2008 Condensation heat transfer enhancement in the presence of non-condensable gas using the interfacial effect of dropwise condensation *Int. J. Heat Mass Transfer* **51** 1728–37
[182] Deng X, Mammen L, Butt H-J and Vollmer D 2012 Candle soot as a template for a transparent robust superamphiphobic coating *Science* **335** 67–70
[183] Wang N, Xiong D, Deng Y, Shi Y and Wang K 2015 Mechanically robust superhydrophobic steel surface with anti-icing, UV-durability, and corrosion resistance properties *ACS Appl. Mater. Interf.* **7** 6260–72
[184] Goetz L A, Jalvo B, Rosal R and Mathew A P 2016 Superhydrophilic anti-fouling electrospun cellulose acetate membranes coated with chitin nanocrystals for water filtration *J. Membr. Sci.* **510** 238–48
[185] de Leon A C C, Pernites R B and Advincula R C 2012 Superhydrophobic colloidally textured polythiophene film as superior anticorrosion coating *Appl. Mater. Interf.* **4** 3169–76
[186] Zhang X, Wang L and Levänen E 2013 Superhydrophobic surfaces for the reduction of bacterial adhesion *RSC Adv.* **3** 12003–20
[187] Banerjee I, Pangule R C and Kane R S 2011 Antifouling coatings: recent developments in the design of surfaces that prevent fouling by proteins, bacteria, and marine organisms *Adv. Mater.* **23** 690–718
[188] Wang Y, Xue J, Wang Q, Chen Q and Ding J 2013 Verification of icephobic/anti-icing properties of a superhydrophobic surface *Appl. Mater. Interf.* **5** 3370–81
[189] Zhang Y, Ge D and Yang S 2014 Spray-coating of superhydrophobic aluminum alloys with enhanced mechanical robustness *J. Colloid. Interf. Sci.* **423** 101–7
[190] Barthwal S, Kim Y S and Lim S-H 2013 Mechanically robust superamphiphobic aluminum surface with nanopore-embedded microtexture *Langmuir* **29** 11966–74
[191] Budunoglu H, Yildirim A and Bayindir M 2012 Flexible and mechanically stable antireflective coatings from nanoporous organically modified silica colloids *J. Mater. Chem.* **22** 9671–7
[192] Chen X, Ma R, Zhou H, Zhou X, Che L, Yao S and Wang Z 2013 Activating the microscale edge effect in a hierarchical surface for frosting suppression and defrosting promotion *Sci. Rep.* **3** 2515
[193] Susoff M, Siegmann K, Pfaffenroth C and Hirayama M 2013 Evaluation of icephobic coatings—screening of different coatings and influence of roughness *Appl. Surf. Sci.* **282** 870–9
[194] Yang S, Xia Q, Zhu L, Xue J, Wang Q and Chen Q-m 2011 Research on the icephobic properties of fluoropolymer-based materials *Appl. Surf. Sci.* **257** 4956–62
[195] Zhu L, Xue J, Wang Y, Chen Q, Ding J and Wang Q 2013 Ice-phobic coatings based on silicon-oil-infused polydimethylsiloxane *Appl. Mater. Interf.* **5** 4053–62
[196] Chen J, Liu J, He M, Li K, Cui D, Zhang Q, Zeng X, Zhang Y, Wang J and Song Y 2012 Superhydrophobic surfaces cannot reduce ice adhesion *Appl. Phys. Lett.* **101** 111603

[197] Sojoudi H, McKinley G H and Gleason K K 2015 Linker-free grafting of fluorinated polymeric cross-linked network bilayers for durable reduction of ice adhesion *Mater. Horizons* **2** 91–9

[198] Chen K, Zhou S, Yang S and Wu L 2015 Fabrication of all-water-based self-repairing superhydrophobic coatings based on UV-responsive microcapsules *Adv. Funct. Mater.* **25** 1035–41

[199] Isimjan T T, Wang T and Rohani S 2012 A novel method to prepare superhydrophobic, UV resistance and anti-corrosion steel surface *Chem. Eng. J.* **210** 182–7

[200] Gao Y, Gereige I, El Labban A, Cha D, Isimjan T T and Beaujuge P M 2014 Highly transparent and UV-resistant superhydrophobic SiO_2-coated ZnO nanorod arrays *Appl. Mater. Interf.* **6** 2219–23

[201] Zhi D, Lu Y, Sathasivam S, Parkin I P and Zhang X 2017 Large-scale fabrication of translucent and repairable superhydrophobic spray coatings with remarkable mechanical, chemical durability and UV resistance *J. Mater. Chem.* A **5** 10622–31

[202] Boinovich L B, Emelyanenko A M, Ivanov V K and Pashinin A S 2013 Durable icephobic coating for stainless steel *Appl. Mater. Interf.* **5** 2549–54

[203] Lee E J, Kim J J and Cho S O 2010 Fabrication of porous hierarchical polymer/ceramic composites by electron irradiation of organic/inorganic polymers: route to a highly durable, large-area superhydrophobic coating *Langmuir* **26** 3024–30

[204] Lee S-M, Kim K-S, Pippel E, Kim S, Kim J-H and Lee H-J 2012 Facile route toward mechanically stable superhydrophobic copper using oxidation–reduction induced morphology changes *J. Phys. Chem.* C **116** 2781–90

[205] Chang Y S 2011 Performance analysis of frostless heat exchanger by spreading antifreeze solution on heat exchanger surface *J. Therm. Sci. Tech.-Jpn* **6** 123–31

[206] Stoecker W, Lux J Jr and Kooy R 1983 Energy considerations in hot-gas defrosting of industrial refrigeration coils *ASHRAE Trans.* **89** 549–73

[207] Niederer D H 1976 Frosting and defrosting effects on coil heat transfer *ASHRAE Trans.* **82** 467–73

[208] Ganesan P, Vanaki S M, Thoo K K and Chin W M 2016 Air-side heat transfer characteristics of hydrophobic and super-hydrophobic fin surfaces in heat exchangers: a review *Int. Commun. Heat Mass Transf.* **74** 27–35

[209] Nath S, Ahmadi S F and Boreyko J B 2016 A review of condensation frosting *Nanosc. Microsc. Therm. Eng.* **21** 81–101

[210] Jin M, Feng X, Xi J, Zhai J, Cho K, Feng L and Jiang L 2005 Super-hydrophobic PDMS surface with ultra-low adhesive force *Macromol. Rapid Commun.* **26** 1805–9

[211] Song X, Zhai J, Wang Y and Jiang L 2005 Fabrication of superhydrophobic surfaces by self-assembly and their water-adhesion properties *J. Phys. Chem.* B **109** 4048–52

[212] Ma M and Hill R M 2006 Superhydrophobic surfaces *Curr. Opin. Coll. Interf. Sci.* **11** 193–202

[213] Yin L, Wang Y, Ding J, Wang Q and Chen Q 2012 Water condensation on superhydrophobic aluminum surfaces with different low-surface-energy coatings *Appl. Surf. Sci.* **258** 4063–8

[214] Xu L, Chen W, Mulchandani A and Yan Y 2005 Reversible conversion of conducting polymer films from superhydrophobic to superhydrophilic *Angew. Chem. Int. Ed.* **44** 6009–12

[215] Ma X, Ding G, Zhang Y and Wang K 2007 Airside heat transfer and friction characteristics for enhanced fin-and-tube heat exchanger with hydrophilic coating under wet conditions *Int. J. Refrig.* **30** 1153–67
[216] Motlagh N V, Birjandi F C, Sargolzaei J and Shahtahmassebi N 2013 Durable, super-hydrophobic, superoleophobic and corrosion resistant coating on the stainless steel surface using a scalable method *Appl. Surf. Sci.* **283** 636–47
[217] Rungraeng N, Cho Y-C, Yoon S H and Jun S 2012 Carbon nanotube-polytetrafluoro-ethylene nanocomposite coating for milk fouling reduction in plate heat exchanger *J. Food Eng.* **111** 218–24
[218] Zhang J, Li J and Han Y 2004 Superhydrophobic PTFE surfaces by extension *Macromol. Rapid Commun.* **25** 1105–8
[219] Wang C-C, Lee W-S, Sheu W-J and Chang Y-J 2002 A comparison of the airside performance of the fin-and-tube heat exchangers in wet conditions; with and without hydrophilic coating *Appl. Therm. Eng.* **22** 267–78
[220] Zhou Y, Yi-Zhi W, Yi-Fan Y, Mao-Gang G and Xiao-Liang X 2012 A simple way to fabricate an aluminum sheet with superhydrophobic and self-cleaning properties *Chin. Phys. Soc.* **21**
[221] Hong K and Webb R 2000 Wetting coatings for dehumidifying heat exchangers *HVAC&R Research* **6** 229–42
[222] Wang C-C and Chang C-T 1998 Heat and mass transfer for plate fin-and-tube heat exchangers, with and without hydrophilic coating *Int. J. Heat Mass Transfer* **41** 3109–20
[223] Tadanaga K, Kitamuro K, Matsuda A and Minami T 2003 Formation of superhydro-phobic alumina coating films with high transparency on polymer substrates by the sol–gel method *J. Sol–Gel Sci. Technol.* **26** 705–8
[224] Thulukkanam K 2013 *Heat Exchanger Design Handbook* (Boca Raton, FL: CRC Press)
[225] Shin J, Tikhonov A V and Kim C 2003 Experimental study on frost structure on surfaces with different hydrophilicity: density and thermal conductivity *J. Heat Transfer* **125** 84–94
[226] Hosoda T 1967 Effects of frost on the heat transfer coefficient *Hitachi Hyoron* **49** 647–51
[227] Kondepudi S and O'Neal D 1989 Effect of frost growth on the performance of louvered finned tube heat exchangers *Int. J. Refrig.* **12** 151–8
[228] Song S, Bullard C and Hrnjak P 2002 Frost deposition and refrigerant distribution in microchannel heat exchangers/Discussion *ASHRAE Trans.* **108** 944
[229] Verma P, Carlson D, Wu Y, Hrnjak P and Bullard C 2002 Experimentally validated model for frosting of plain fin-round-tube heat exchangers *Proc. of IIF–IIR Commission D* p 1
[230] Moallem E, Cremaschi L and Fisher D E 2010 Experimental investigation of frost growth on microchannel heat exchangers *13th Int. Refrigeration Conf.* (*Purdue, West Lafayette, IN*) Paper 2416
[231] Xia Y, Hrnjak P S and Jacobi A M 2005 Air-side thermal-hydraulic performance of louvered-fin, flat-tube heat exchangers with sequential frost-growth cycles *ASHRAE Trans.* **111** 487–95
[232] Yonko J D 1967 An investigation of the thermal conductivity of frost while forming on a flat horizontal plate *ASHRAE Trans.* **73** 1.1–1.11
[233] Liu Y and Kulacki F 2018 An experimental study of defrost on treated surfaces: effect of frost slumping *Int. J. Heat Mass Transfer* **119** 880–90

[234] Shi Y, Tang G and Xia H 2015 Investigation of coalescence-induced droplet jumping on superhydrophobic surfaces and liquid condensate adhesion on slit and plain fins *Int. J. Heat Mass Transfer* **88** 445–55

[235] Padhmanabhan S, Fisher D, Cremaschi L and Moallem E 2011 Modeling non-uniform frost growth on a fin-and-tube heat exchanger *Int. J. Refrig.* **34** 2018–30

[236] Kondepudi S N and O'Neal D L 1993 Performance of finned-tube heat exchangers under frosting conditions: I. Simulation model *Int. J. Refrig.* **16** 175–80

[237] Kondepudi S N and O'Neal D L 1993 Performance of finned-tube heat exchangers under frosting conditions: II. Comparison of experimental data with model *Int. J. Refrig.* **16** 181–4

[238] Seker D, Karatas H and Egrican N 2004 Frost formation on fin-and-tube heat exchangers. Part I—Modeling of frost formation on fin-and-tube heat exchangers *Int. J. Refrig.* **27** 367–74

[239] Seker D, Karatas H and Egrican N 2004 Frost formation on fin-and-tube heat exchangers. Part II—Experimental investigation of frost formation on fin-and-tube heat exchangers *Int. J. Refrig.* **27** 375–7

[240] Tso C, Cheng Y and Lai A 2006 An improved model for predicting performance of finned tube heat exchanger under frosting condition, with frost thickness variation along fin *Appl. Therm. Eng.* **26** 111–20

[241] Korte C and Jacobi A 2001 Condensate retention effects on the performance of plain-fin-and-tube heat exchangers: retention data and modeling *J. Heat Transfer* **123** 926–36

[242] Wang F, Liang C, Yang M, Fan C and Zhang X 2015 Effects of surface characteristic on frosting and defrosting behaviors of fin-tube heat exchangers *Appl. Therm. Eng.* **75** 1126–32

[243] Biguria G and Wenzel L A 1970 Measurement and correlation of water frost thermal conductivity and density *Industr. Eng. Chem. Fundam.* **9** 129–38

[244] Breque F and Nemer M 2016 Frosting modeling on a cold flat plate: comparison of the different assumptions and impacts on frost growth predictions *Int. J. Refrig.* **69** 340–60

[245] Şahin A Z 2000 Effective thermal conductivity of frost during the crystal growth period *Int. J. Heat Mass Transfer* **43** 539–53

[246] Kim K and Lee K-S 2013 Frosting and defrosting characteristics of surface-treated louvered-fin heat exchangers: effects of fin pitch and experimental conditions *Int. J. Heat Mass Transfer* **60** 505–11

[247] Xia Y P, Hrnjak P S and Jacobi A M 2005 Air-side thermal-hydraulic performance of louvered-fin, flat-tube heat exchangers with sequential frost-growth cycles *ASHRAE Trans.* **111** 487–95

[248] Amer M and Wang C-C 2017 Review of defrosting methods *Renew. Sustain. Energy Rev.* **73** 53–74

[249] Ding Y, Ma G, Chai Q and Jiang Y 2004 Experiment investigation of reverse cycle defrosting methods on air source heat pump with TXV as the throttle regulator *Int. J. Refrig.* **27** 671–8

[250] Qu M, Xia L, Deng S and Jiang Y 2012 An experimental investigation on reverse-cycle defrosting performance for an air source heat pump using an electronic expansion valve *Appl. Energy* **97** 327–33

[251] Stoecker W F, Lux J J and Kooy R J 1983 Energy considerations in hot-gas defrosting of industrial refrigeration coils *ASHRAE J.* **25** 66

[252] Niederer D H 1976 Frosting and defrosting effects on coil heat transfer *ASHRAE Trans.* **82** 467–73

[253] Tso C, Wong Y, Jolly P and Ng S 2001 A comparison of hot-gas by-pass and suction modulation method for partial load control in refrigerated shipping containers *Int. J. Refrig.* **24** 544–53
[254] Reindl D, Jekel T and Elleson J 2005 *Industrial Refrigeration Energy Efficiency Guidebook* (Madison, WI: Industrial Refrigeration Consortium, The University of Wisconsin)
[255] Liu Z, Tang G and Zhao F 2003 Dynamic simulation of air-source heat pump during hot-gas defrost *Appl. Therm. Eng.* **23** 675–85
[256] Byun J-S, Lee J and Jeon C-D 2008 Frost retardation of an air-source heat pump by the hot gas bypass method *Int. J. Refrig.* **31** 328–34
[257] Dart D M 1959 Effect of fin bond on heat transfer *ASHRAE Trans.* **1** 67–71
[258] Mader G and Thybo C 2012 A new method of defrosting evaporator coils *Appl. Therm. Eng.* **39** 78–85
[259] Mei V C, Domitrovic R E, Chen F C and Kilpatrick J K 2002 A frost-less heat pump/ discussion *ASHRAE Trans.* **108** 452
[260] Kwak K and Bai C 2010 A study on the performance enhancement of heat pump using electric heater under the frosting condition: heat pump under frosting condition *Appl. Therm. Eng.* **30** 539–43
[261] Yang C, Hong S, Gao C, Zhang D and Li Y 2011 An improvement test approach of look-up table in SRAM-based FPGAs *Adv. Mater. Res.* **159** 116–23
[262] Petrenko V F, Sullivan C R, Kozlyuk V, Petrenko F V and Veerasamy V 2011 Pulse electro-thermal de-icer (PETD) *Cold Reg. Sci. Technol.* **65** 70–8
[263] Elsharkawy M, Tortorella D, Kapatral S and Megaridis C M 2016 Combating frosting with Joule-heated liquid-infused superhydrophobic coatings *Langmuir* **32** 4278–88
[264] Li D and Chen Z 2014 Experimental study on instantaneously shedding frozen water droplets from cold vertical surface by ultrasonic vibration *Exp. Therm. Fluid Sci.* **53** 17–25
[265] Schaefer V J and Langmuir I 1953 *Project cirrus Gen. Electric Res. Lab.* Final Report 52
[266] Marshall J and Gunn K 1957 A first experiment on snow crystal growth *Artificial Stimulation of Rain* (London: Pergamon), pp 340–5
[267] Ma H and Peterson G 1995 Thermodynamic analysis of the influence of electric fields on frost formation *J. Thermophys. Heat Tr.* **9** 562–4
[268] Molki M, Ohadi M and Bloshteyn M 2000 Frost reduction under intermittent electric field *National Heat Transfer Conf. (Pittsburgh, PA)*
[269] Yang S 1995 Variation of heat transfer coefficient on an oscillating circular cylinder during temperature-rising process with water vapor condensation in humid air flow *MS Thesis* Taipei: Tatung University
[270] Cheng C-H and Shiu C-C 2003 Oscillation effects on frost formation and liquid droplet solidification on a cold plate in atmospheric air flow *Int. J. Refrig.* **26** 69–78
[271] Wu X and Webb R L 2001 Investigation of the possibility of frost release from a cold surface *Exp. Therm. Fluid Sci.* **24** 151–6
[272] Adachi K, Saiki K and Sato H 1998 Suppression of frosting on a metal surface using ultrasonic vibrations *Ultrasonics Symposium Proceedings* (Piscataway, NJ: IEEE), pp 759–62
[273] Adachi K, Saiki K, Sato H and Ito T 2003 Ultrasonic frost suppression *Japan J. Appl. Phys.* **42** 682
[274] Ghosh A, Beaini S, Zhang B J, Ganguly R and Megaridis C M 2014 Enhancing dropwise condensation through bioinspired wettability patterning *Langmuir* **30** 13103–15

[275] Petrenko V 2009 Pulse electrothermal and heat-storage ice detachment apparatus and methods *Google patents*
[276] Petrenko V F 2010 System and method for icemaker and aircraft wing with combined electromechanical and electrothermal pulse deicing *Google patents*
[277] Petrenko V F, Higa M, Starostin M and Deresh L 2003 Pulse electrothermal de-icing *The 13th Int. Offshore and Polar Engineering Conf. (25–30 May, Honolulu, HI)* (Int. Society of Offshore and Polar Engineers)

IOP Publishing

Nanoscale Energy Transport
Emerging phenomena, methods and applications
Bolin Liao

Chapter 16

Reliably measuring the efficiency of thermoelectric materials

Qing Zhu and Zhifeng Ren

While most research on thermoelectric materials has been focused on improving the figure of merit (ZT), less attention has been paid to reliable characterizations of their efficiency from prediction to measurement. In this chapter, recent progress on the prediction and measurement of efficiency is summarized. The numerical calculation of efficiency based on the finite difference method by considering the temperature-dependent properties is regarded as a benchmark to evaluate the prediction results from the equations and the measurement results. A comparison of prediction results from the equations shows that $(ZT)_{\text{eng}}$ is a more reliable indicator than $(ZT)_{\text{avg}}$ for efficiency prediction. Since an accurate measurement of heat flow is required for efficiency measurements but is very challenging, we introduce two methods to measure the heat flow. One is to accurately control the input heat from the heater, and the other is to measure the heat flow using a standard sample. Methods to eliminate errors in the measurements are discussed in detail. Finally, we also provide a new method to measure the efficiency without considering the contact problem.

16.1 Introduction

The figure of merit, $ZT = (\frac{S^2}{\rho\kappa})T$, is used as the crucial parameter for determining the performance of thermoelectric (TE) materials, where S, ρ, κ and T are the Seebeck coefficient, electrical resistivity, thermal conductivity and absolute temperature, respectively. The efficiency is determined by the Carnot efficiency (η_c) and ZT according to

$$\eta_{\max} = \eta_c \frac{\sqrt{1 + (ZT)_{\text{avg}}} - 1}{\sqrt{1 + (ZT)_{\text{avg}}} + \frac{T_c}{T_h}}, \tag{16.1}$$

where $(ZT)_{avg}$ is the average ZT between the hot-side temperature (T_h) and the cold-side temperature (T_c), and η_c equals $(T_h - T_c)/T_h$ [1]. It is important to note that this equation is derived by assuming that the TE properties are temperature-independent, which is rarely applicable to real materials and leads to an unreliable prediction of efficiency. Additionally, due to the intrinsic errors of the measurement equipment in the lab, the measured ZT of a material can be different from its actual ZT. Therefore, the reliable prediction and accurate measurement of the efficiency of a single leg are equally important in evaluating the performance of any TE material [2]. Since TE materials are fabricated into TE modules for real applications, the evaluation of the heat conversion capability of any TE module will also be discussed in this chapter [3].

16.2 Prediction of efficiency from mathematical methods

16.2.1 Prediction of efficiency from the FDM

Under steady-state conditions, the first law of thermodynamics for one-dimensional heat transfer in a single-leg device yields

$$\frac{\mathrm{d}}{\mathrm{d}x}\left(\kappa(T(x))\frac{\mathrm{d}T(x)}{\mathrm{d}x}\right) + J^2\rho(T(x)) - JT(x)\frac{\mathrm{d}S(T(x))}{\mathrm{d}x} \\ -\varepsilon\sigma\frac{W}{A}\left[T^4(x) - T_r^4\right] = 0, \tag{16.2}$$

where $T(x)$ is the temperature distribution through the TE leg as a function of position, T_r is the temperature of the surroundings, J is electric current density, ε is the emissivity of the TE leg, σ is the Stefan–Boltzmann constant and $S(T(x))$, $\rho(T(x))$ and $\kappa(T(x))$ are the temperature-dependent Seebeck coefficient, electrical resistivity and thermal conductivity, respectively [2]. The single-leg device efficiency model takes into account the temperature-dependent properties of the TE material and the radiative heat transfer between the TE leg and its surroundings. In the iterative method, the TE leg is mathematically subdivided into a large number of increments ($\mathrm{d}x$), permitting the assumption of constant material properties in each segment when determining the energy balance. The energy associated with heat conduction, joule heating, the Thomson effect and radiation can then be extracted separately. In this way, output power, efficiency and radiation loss can be obtained. By taking the temperature-dependent properties into consideration, efficiency can be precisely obtained from the finite difference method (FDM). The FDM has been incorporated into some commercial software packages such as MATLAB, COMSOL Multiphysics and ANSYS to provide accurate prediction of efficiency.

16.2.2 Prediction of efficiency from equations

The formula (equation (16.1)) for η_{max} incorporating the dimensionless ZT has been conventionally used to evaluate TE materials. The properties of a realistic TE material are temperature-dependent. Due to its assumption of temperature-independence, equation (16.1) only correctly predicts the maximum efficiency at a small temperature difference between the cold and hot sides or for the limited number of TE materials that have a constant Z value over their operating temperature range. Additionally, there are several ways to define $(ZT)_{avg}$. Two of them are: (i) $Z_{int} = \frac{\int_{T_c}^{T_h} Z(T)\mathrm{d}T}{\Delta T}$, an integration with respect to temperature and (ii) $Z_{T_{avg}}$ corresponding to the Z value at the average temperature (T_{avg}). Therefore, the $(ZT)_{avg}$ is insufficient because of its variation depending on how it is averaged.

The numerical simulation considering the temperature-dependent properties is a reliable method to predict efficiency, but it is complex and requires good knowledge of the FDM. An engineering dimensionless figure of merit $(ZT)_{eng}$ as a function of the temperature difference between the cold and hot sides was proposed to reliably and accurately predict the practical efficiency, overcoming the reporting of unrealistic efficiency when using $(ZT)_{avg}$ [4, 5].

The maximum efficiency considering the temperature-dependent properties is expressed as

$$\eta_{max} = \eta_c \frac{\sqrt{1 + (ZT)_{eng}\alpha_1\eta_c^{-1}} - 1}{\alpha_0\sqrt{1 + (ZT)_{eng}\alpha_1\eta_c^{-1}} + \alpha_2}, \tag{16.3}$$

where $(ZT)_{eng}$ is the engineering figure of merit, which is dependent on ΔT and is defined as

$$(ZT)_{eng} = \frac{\left(\int_{T_c}^{T_h} S(T)\mathrm{d}T\right)^2}{\int_{T_c}^{T_h} \rho(T)\mathrm{d}T \int_{T_c}^{T_h} \kappa(T)\mathrm{d}T}\Delta T, \tag{16.4}$$

$$\alpha_i = \frac{S(T_h)\Delta T}{\int_{T_c}^{T_h} S(T)\mathrm{d}T} - \frac{\int_{T_c}^{T_h} \tau(T)\mathrm{d}T}{\int_{T_c}^{T_h} S(T)\mathrm{d}T} W_T\eta_c - \mathrm{i}W_J\eta_c, \tag{16.5}$$

$$W_J = \frac{\int_{T_c}^{T_h}\int_{T}^{T_h} \rho(T)\mathrm{d}T\mathrm{d}T}{\Delta T \int_{T_c}^{T_h} \rho(T)\mathrm{d}T}, \tag{16.6}$$

$$W_T = \frac{\int_{T_c}^{T_h}\int_{T}^{T_h} \tau(T)\mathrm{d}T\mathrm{d}T}{\Delta T \int_{T_c}^{T_h} \tau(T)\mathrm{d}T}. \tag{16.7}$$

To obtain a reliable prediction for the maximum efficiency from equation (16.3), calculation of W_J and W_T is essential.

Figure 16.1(a) shows the temperature-dependent ZT and average ZT of the p-type Ni-doped MgAgSb [2], and their corresponding efficiency calculations using equation (16.1) are shown in figure 16.1(b) and are matched with the numerical simulation of an ideal case because the Z versus T is not much different to the constant. In figure 16.1(c), however, n-type In_4Se_{3-x} [6] is shown to have larger variations in its average ZT depending on the method used due to the strong temperature dependence that results in very different efficiency predictions (figure 16.1(d)). Even though the peak ZT of In_4Se_{3-x} is higher than that of Ni-doped MgAgSb, its average ZT and predicted efficiency are much lower, which means that the peak ZT value is not the right indicator for a TE material's efficiency. Clearly, average ZT values vary with different averaging techniques (figure 16.1(a) and (c)) and the analytical prediction of some materials based on the average ZT is far different from the numerical analysis (figure 16.1(d)). The maximum efficiency values of Ni-doped MgAgSb and In_4Se_{3-x} based on $(ZT)_{\text{eng}}$ and $(ZT)_{\text{avg}}$ are compared to the numerical analysis in figure 16.1(b) and (d), respectively, showing that the efficiency of In_4Se_{3-x} based on $(ZT)_{\text{eng}}$ is much lower but is in good agreement with the result from numerical simulations. Employing

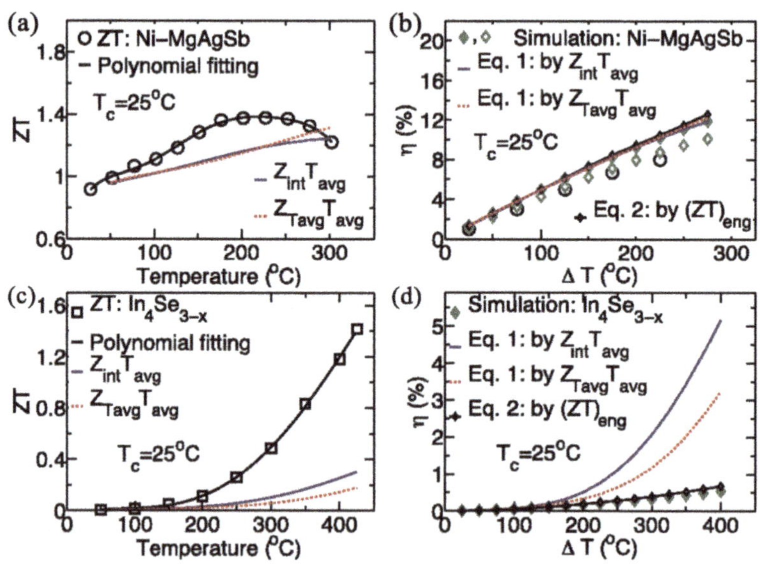

Figure 16.1. (a) ZT versus T for Ni-doped MgAgSb. Circles are measured data and the solid black line is the fitted curve. Solid blue and dashed red lines are the average ZT values using $Z_{\text{int}}T_{\text{avg}}$ and $Z_{T_{\text{avg}}}T_{\text{avg}}$, respectively. (b) Efficiency versus ΔT at T_c = 25 °C for Ni-doped MgAgSb by experimental measurement (circles), by numerical simulation of ideal (solid diamonds) and actual (open diamonds) conditions, by the conventional formula using integration (solid blue line) and average temperature (dashed red line) for Z_{avg}, and by the new formula (open squares on the solid black line). The ideal and actual conditions of the numerical predictions denote the cases excluding and including electric contact/parasitic resistance (a total of 36 $\mu\Omega$ cm^2 from two ends), respectively. (c) ZT versus T and (d) efficiency versus ΔT at T_c = 25 °C for In_4Se_{3-x}. The x-axis in (a) and (c) indicates the hot-side temperature for the average ZT and the measurement temperature for ZT data and their polynomial fittings. Reproduced with permission from [4]. Copyright 2015 National Academy of Sciences USA.

$(ZT)_{\text{eng}}$ enables the analytical prediction of maximum efficiency more accurately without the need to carry out numerical simulations or an experimental set-up and measurements. $(ZT)_{\text{eng}}$, rather than ZT, predicts the practical performance of a TE material at any given temperature difference.

To further show the advantages of $(ZT)_{\text{eng}}$, effective ZT ($(ZT)_{\text{eff}}$), which is more reliable than $(ZT)_{\text{avg}}$, is introduced and defined as

$$(ZT)_{\text{eff}} = \frac{\left(\int_{T_c}^{T_h} S(T)\mathrm{d}T\right)^2}{\Delta T \int_{T_c}^{T_h} \rho(T)\kappa(T)\mathrm{d}T} T_{\text{avg}}. \tag{16.8}$$

The scattered pattern in figure 16.2(a) shows that $(ZT)_{\text{avg}}$ is not a direct indicator of a material's maximum efficiency. Figure 16.2(b) shows the efficiency corresponding to the $(ZT)_{\text{eff}}$ of the materials. The relationship of η versus $(ZT)_{\text{eff}}$ still shows a scattered pattern, so $(ZT)_{\text{eff}}$ is also not effective for the reliable prediction of a material's efficiency under operable temperature conditions. Figure 16.2(c) plots the efficiency values of the materials according to $(ZT)_{\text{eng}}$. Based on the correlation between η versus $(ZT)_{\text{eng}}$, the efficiency of a material can be easily recognized by

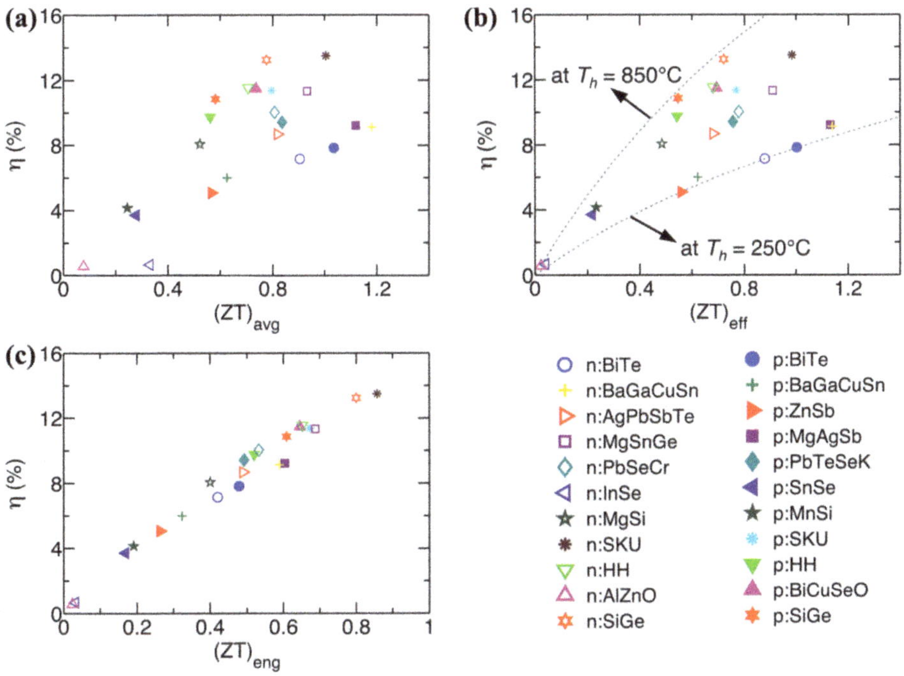

Figure 16.2. The predicted efficiency of materials as a function of (a) $(ZT)_{\text{avg}}$, (b) $(ZT)_{\text{eff}}$ and (c) $(ZT)_{\text{eng}}$ with all efficiency values evaluated by a numerical simulation accounting for the temperature dependence of each property. T_c is fixed at 50 °C for all efficiency values. Reproduced with permission from [5]. Copyright 2017 Royal Society of Chemistry.

evaluating $(ZT)_{eng}$. Thus, the efficiency of any material can be evaluated by $(ZT)_{eng}$ without the need for tedious efficiency calculations. For example, any type of material with a $(ZT)_{eng}$ of 0.6 exhibits an efficiency of ~10%. By the simple calculation of $(ZT)_{eng}$, the performance of different materials can be reasonably compared. Clearly, this relationship shows the benefit of taking into account each accumulated property based on ΔT rather than T_{avg}. The TE materials in figure 16.2 are listed in table 16.1.

16.3 Efficiency measurement

16.3.1 Challenges of efficiency measurement

The efficiency of TE materials, η, is defined as the maximum ratio of the output power, P, to the total heat flux into the module from the hot side, Q_{in}:

$$\eta = \frac{P}{Q_{in}} \times 100\%. \tag{16.9}$$

Thus, accurate measurements of P and Q_{in} are required to ensure an accurate measurement of η.

Output power can be obtained from the following:

$$P = IV_L = I(V_{oc} - IR), \tag{16.10}$$

where P is the output power, V_L is the load voltage, V_{oc} is the open circuit voltage, I is the current in the circuit and R is the internal resistance including the electrical contact resistance. By tuning the current in the circuit and measuring the corresponding load voltage, the maximum output power can be obtained. V_{oc} can be accurately obtained by determining the temperature difference and the Seebeck coefficient of the TE material. Additionally, I can be precisely controlled by an

Table 16.1. The maximum operable temperature for p-type and n-type materials at the hot side. Data from [5].

p-type			n-type		
Material	$T_{h,max}$ (°C)	Ref.	Material	$T_{h,max}$ (°C)	Ref.
$Bi_{0.4}Sb_{1.6}Te_3$	250	—	$Bi_2Te_{2.7}Se_{0.3}S_{0.015}$	250	—
β-Zn_4Sb_3	250	[7]	$Ba_8Ga_{14}Cu_2Sn_{30}$	275	[8]
$Ba_8Ga_{15.75}Cu_{0.25}Sn_{30}$	275	[8]	$AgPb_mSbTe_{m+2}$	400	[9]
MgAgSb	295	[10]	In_4Se_{3-x}	400	[6]
$Pb_{0.98}Te_{0.75}Se_{0.25}K_{0.02}$	450	[11]	$Mg_2Sn_{0.75}Ge_{0.25}$	450	[12]
SnSe	450	[13]	$Pb_{0.995}SeCr_{0.005}$	450	[14]
$MnSi_{1.78}$	500	[15]	$Ba_{0.08}La_{0.05}Yb_{0.04}Co_4Sb_{12}$	550	[16]
$Ce_{0.45}Nd_{0.45}Fe_{3.5}Co_{0.5}Sb_{12}$	550	[17]	Mg_2Si-0.5 at% Sb/1.0 at% Zn	550	[18]
$Hf_{0.19}Zr_{0.76}Ti_{0.048}CoSb_{0.8}Sn_{0.2}$	650	[19]	$Hf_{0.25}Zr_{0.75}NiSn_{0.99}Sb_{0.01}$	650	[20]
$Bi_{0.875}Ba_{0.125}CuSeO$	650	[21]	ZnO-0.25 at% Al	700	[22]
$Si_{80}Ge_{20}B_5$	850	[23]	$Si_{80}Ge_{20}P_2$	850	[24]

external instrument. According to equation (16.10), the electrical contact resistance should be as small as possible to generate the maximum output power ($\frac{V_{oc}^2}{4R}$), which is the fundamental step toward reliable efficiency measurement.

For most good TE materials, either direct soldering onto the electrode is difficult, or the solder reacts with the TE material and degrades quickly at the working temperature. A metallized contact layer is thus needed between the TE elements and electrodes. There are several ways to prepare a contact layer on the TE legs. For low-temperature applications, sputtering and electroplating/chemical vapor deposition methods allow the preparation of a very thin layer ($<5\ \mu$m) of contact material on a properly prepared TE leg. For high-temperature applications, the contact layer can be joined to the TE material by hot pressing. The latter method is usually capable of making relatively thick contact layers at low cost and works better to create a diffusion barrier at high temperatures. In the cases where these two methods are not applicable, conducting paste can be used to bond the TE legs to the electrode for a limited time. Other key difficulties such as thermal stresses, parasitic thermal resistance and chemical stability should also be addressed for real applications [25].

Unlike P, which can be accurately measured if the contact problem is carefully addressed, Q_{in} is extremely difficult to accurately determine during efficiency measurement. There are three ways that heat loss occurs: heat conduction, heat convection and heat radiation, all of which need to be eliminated by all means. The heat convection loss can be avoided by performing the measurement in the vacuum chamber. However, the methods to reduce heat conduction and radiation loss vary in different measurement systems. The heat flow is so difficult to accurately measure that few reports exist on the experimental demonstration of efficiency for TE modules, unicouples and single-leg devices.

16.3.2 Methods of efficiency measurement

The efficiency measurement systems vary in different labs, but the fundamentals of their design are quite similar. The maximum output power can be obtained by tuning the current in the circuit as mentioned above, or matching the load resistance with the internal resistance. However, the load resistance including the resistance in the lead wires may be significantly larger than the internal resistance. Thus the second method is limited to the TE module which has a much larger internal resistance than the lead wires. For a single-leg device, usually the first method is adopted. There are two common ways to measure Q_{in}. One is to precisely control the power supply of the heater in order to know how much heat generated from the heater will flow into the hot side of the TE sample; the other way is to use a heat flow meter to measure heat flow, which is based on Fourier's law. According to the difference in Q_{in} measurement, two different measurement systems have been developed. Mathematical methods are also applied to verify the measurement results.

16.3.2.1 Power supply

The heat input is equal to the electrical heater input power if the parasitic heat losses such as radiative heat transfer and wire heat conduction from the heater assembly to the surroundings are negligible. This method has been successfully employed to accurately measure the efficiency of different TE materials [2, 26, 27].

MgAgSb-based materials have been developed as a promising p-type candidate for power generation at near-room temperatures since they possess an exciting efficiency of about 8.5% between 20 °C and 245 °C [2]. The sample is fabricated with silver contact pads using a one-step hot-pressing technique, eliminating the typically required sample-metallization process (figure 16.3). This significantly simplifies the fabrication of TE elements with low electrical and thermal-contact resistance. The parasitic heat losses from the heater assembly are reduced by surrounding the single-leg device with a heated radiation shield that is maintained close to the hot-junction temperature and by thermally grounding all wires attached to the heater assembly to this shield (figure 16.4). This ensures that most of the supplied electrical heater power corresponds to the hot-junction heat input of the single-leg device. However, the TE leg with an imposed temperature difference is to some extent heated by the surrounding radiation shield at the hot-junction temperature, which affects the

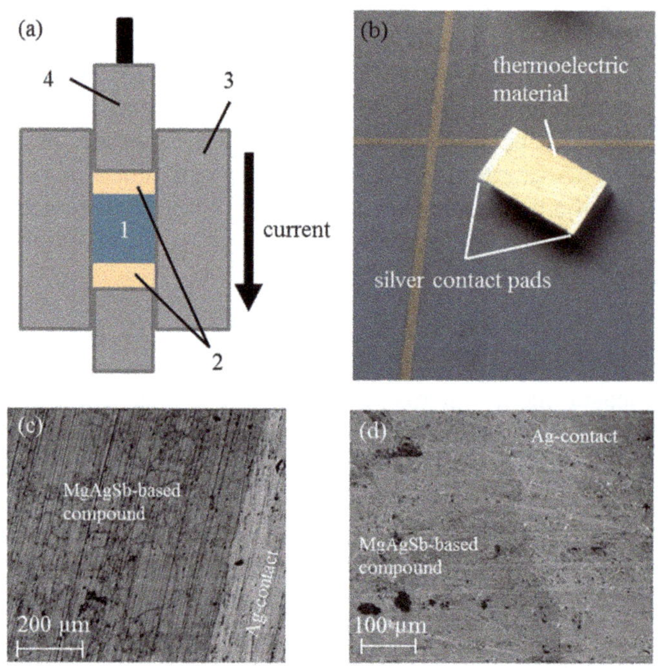

Figure 16.3. TE sample fabrication with electrical contact pads using the hot-pressing technique. (a) TE material powder (1) is sandwiched between electrical contact pad material (2) inside the graphite die (3) of the hot-press. While the graphite piston (4) applies pressure, the current flowing through the piston, die and materials increases the hot-press temperature. (b) A p-type $MgAg_{0.965}Ni_{0.005}Sb_{0.99}$ sample (3 mm × 3 mm × 5 mm) as fabricated using a three-step process (ball milling, hot pressing and cutting) with silver contact pads with a thickness of ~0.25–0.35 mm on each end. SEM images with magnifications of (c) 100× and (d) 395× show a clean and well-defined interface between the MgAgSb-based compound and the Ag contact pad after the hot-pressing and annealing process. Reproduced with permission from [2]. Copyright 2015 Royal Society of Chemistry.

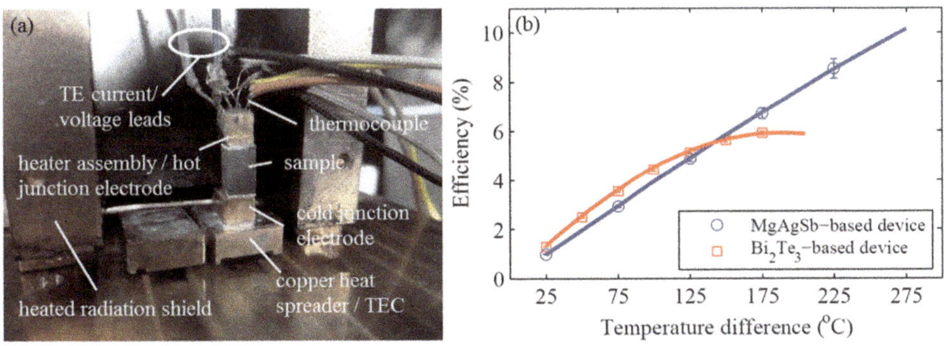

Figure 16.4. (a) Fabricated single TE leg device with the sample soldered to the copper heater assembly (also acting as the hot-junction electrode) and the copper cold-junction electrode. The device is soldered onto a TE cooler (TEC), which is mounted onto a liquid-cooled cold plate and surrounded by a heated radiation shield. (b) Efficiency results of a single TE leg device based on the p-type nickel-doped MgAgSb compound (blue circles) compared to those of a device based on a commercial p-type doped bismuth telluride sample (red squares). Solid lines correspond to simulation results obtained with FDM. Reproduced with permission from [2]. Copyright 2015 Royal Society of Chemistry.

efficiency. The heat radiation transfer is corrected by the simulation based on FDM. Another major challenge of the single-leg efficiency measurement is that the large TE currents required for the efficiency measurement can result in significant Joule heating in the current-carrying lead, which affects the hot-junction heat input power measurement. Thus, the diameter of the current-carrying wire is chosen to be large enough to minimize this effect. With all of the possible heat loss reduced, the Q_{in} can be calculated by the current (I_{E}) and voltage (V_{E}) supplied to the heater.

Instead of using a complex design to eliminate the heat loss, the heat loss in the efficiency measurement of skutterudites can be estimated [26]. The n-type ($Yb_{0.35}Co_4Sb_{12}$) and p-type ($NdFe_{3.5}Co_{0.5}Sb_{12}$) skutterudites are fabricated into a unicouple device for efficiency testing. $CoSi_2$ for the n-type electrodes and Co_2Si for the p-type electrodes are directly bonded onto the TE powder during the hot pressing to produce the device. An in-line unicouple configuration, in which the legs are on the same axis, replaces the conventional π-configuration (figure 16.5). Heat is generated in the middle at the heater assembly to temperatures of up to 600 °C while the ends are maintained near room temperature. There are five K-type thermocouples in total to measure the temperatures of the four electrodes and the heater. To improve thermal contact, Ag epoxy was used at the interface of the hot electrode and the Cu heater, and In–Ga eutectic alloy was used at the interface of the cold electrode and the Cu heat sink. The Ag epoxy is baked off at high temperature, leaving a strong bond with low electrical and thermal resistance. Radiation at the heater and at the sides of the legs is the dominant mode of heat loss, which is carefully calibrated. The Q_{Loss} term is evaluated by suspending the heater assembly in a vacuum and measuring the steady-state electric power needed to keep it at a certain temperature. In this case, the Joule heating in the heater will be balanced by the radiation and conduction losses. Additionally, Q_{LegLoss} has also been corrected by FDM.

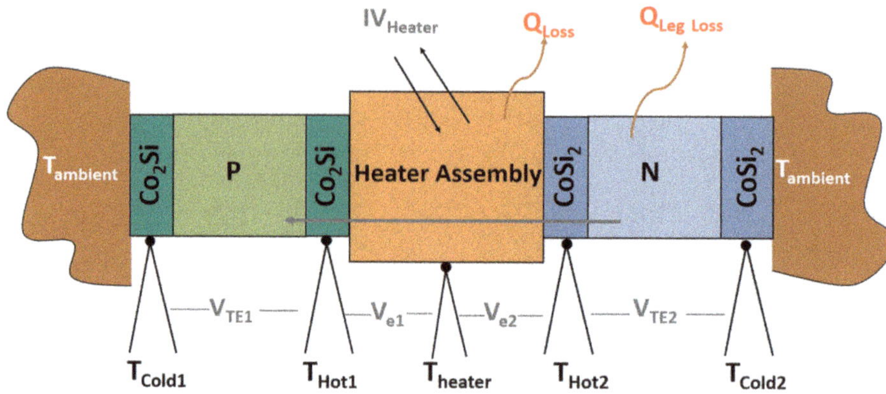

Figure 16.5. Unicouple with the in-line configuration. The schematic of the p–n unicouple during the efficiency measurement. The TE voltage is measured at the alumel wires between K-type thermocouples. T_{Ambient} and Q_{LegLoss} represent the ambient temperature and the heat transfer due to leg loss, respectively. Reproduced with permission from [26]. Copyright 2013 Wiley.

16.3.2.2 Heat flow meter

A standard sample (constantan), for which the thermal conductivity is already known, is used to measure the heat flow according to the one-dimensional Fourier's law:

$$Q_{\mathrm{in}} = \kappa \frac{A}{L} \Delta T, \tag{16.11}$$

where A, L, ΔT and κ represent the cross-sectional area, length, temperature difference and thermal conductivity of the standard sample, respectively.

It is not suitable to place the constantan at the hot-side end due to the heavy heat loss at high temperature, but the input heat can be obtained from the following under a steady state:

$$Q_{\mathrm{in}} = P + Q_{\mathrm{out}} + Q_{\mathrm{rad}}, \tag{16.12}$$

where Q_{out} represents the heat flow out of the TE sample, by placing the constantan at the cold side (figure 16.6). The basic strategy to eliminate the impact of the heat loss is to increase the total heat flow and reduce the heat loss. The TE material $Mg_{3.1}Co_{0.1}Sb_{1.5}Bi_{0.49}Te_{0.01}$ with a high peak ZT was used for efficiency studies [28]. Good contact layers on the TE material were formed by a one-step hot-pressing method with negligible electrical contact resistance, small thermal-contact resistance and good bonding strength. For a single leg, the sample was prepared with a large cross-sectional area of 3.6 mm × 3.6 mm and a small length of 4.35 mm to allow more heat flowing through the TE leg. The whole experiment is conducted in a bell jar with air pressure below 10^{-6} mbar to largely avoid the heat convection with air. Heat loss through conduction from the copper plates can be minimized by selecting thermocouples and lead wires with small diameters (0.002 inch). To roughly estimate

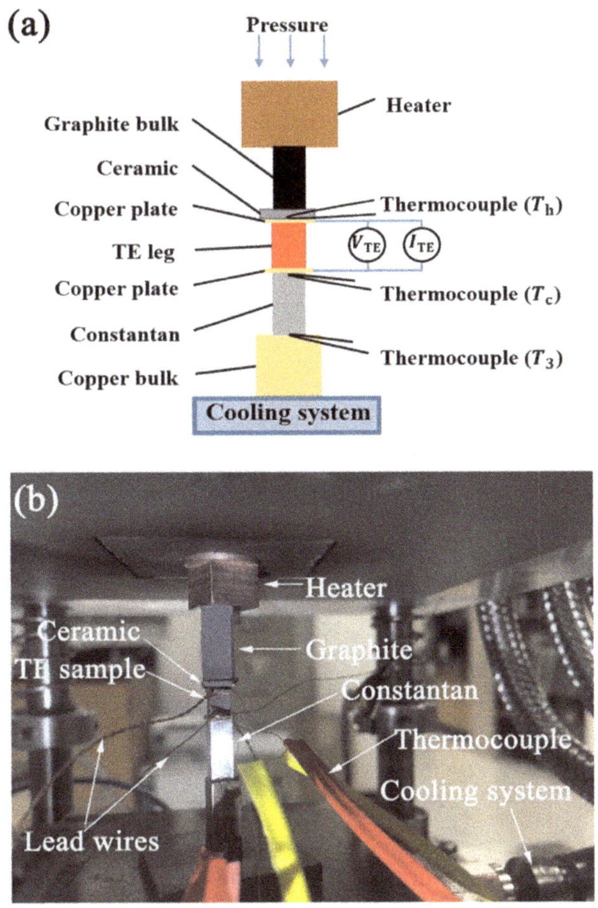

Figure 16.6. Experimental set-up to measure single TE leg efficiency under a large temperature difference: (a) schematic and (b) photograph of the set-up. Reproduced with permission from [43]. Copyright 2019 Elsevier.

the heat radiation loss from the side wall of the TE leg, the emissivity of the TE leg is assumed to be 0.5 [2]. Based on FDM, heat radiation loss is as much as 5% of the total heat flow, which can be ignored. As the constantan bulk is placed below the TE sample with a relatively low temperature during the measurement, the heat radiation from the constantan is also quite small. Furthermore, to reduce the impact of radiation from the heater to the TE leg, a piece of graphite with the same cross-sectional area is placed between the heater and the TE leg. The resistance of one lead wire is measured to be 10 mΩ and the current at the maximum efficiency (10.6%) is 3 A. The generated joule heat in the lead wire is ~0.09 W, which can be conducted either to the TE leg or to the surrounding environment and is significantly smaller than the total heat flow (~2 W) through the TE leg. Additionally, the heat conduction loss through the lead wires and the thermocouples from the bottom copper plate can also be estimated from Fourier's law and ignored when compared to the total heat flow.

This method is much simpler than the method described in the previous section, as the heat loss from the heater is not a concern. However, the heat flow measured by the temperature readings from the thermocouples is not as accurate as expected due to the intrinsic temperature-reading errors from the thermocouples. Thus, a relatively large temperature difference is needed to eliminate the impact of the temperature-reading errors.

16.3.3 TE module

The efficiency of a single TE leg can be obtained using the equations discussed above or FDM. However, the prediction of efficiency for a three-dimensional TE module is more complicated. Commercial software packages such as ANSYS and COMSOL, which incorporate the FDM, provide a convenient way to obtain the efficiency of a TE module. The three-dimensional TE effect in a steady state can be expressed by a set of differential equations (equations (16.13)–(16.16)) with temperature (T) and electric potential (V) as two unknown variables; electrical current density (J) and heat flux (q), as intermediate variables, are vectors in three spatial directions, respectively:

$$J = \sigma(-\nabla V - S\nabla T), \tag{16.13}$$

$$q = -k_{\text{total}}\nabla T, \tag{16.14}$$

$$\nabla J = 0, \tag{16.15}$$

$$\nabla(k_{\text{total}}\nabla T) + J^2/\sigma - TJ\nabla S = 0, \tag{16.16}$$

where σ denotes the electrical conductivity, equivalent to the inverse of resistivity. The three-dimensional distribution of J, q, V and T is numerically solved under given electrical and thermal boundary conditions. Important values that characterize the module, such as electrical current I, load voltage V_L, heat dissipated from the cold side of the module Q_{out} and electrical power output P, are calculated from J, q, V and T.

The measured heat flow is usually larger than the prediction results. The larger heat flow in the experiment is attributed to the parasitic radiation in the open space in the module from the hot-side substrate. In the fabricated module, the TE legs occupy only a small area of the module substrate. The rest of the open area provides a radiation-transfer path. Thus, the heat flow measured by the flow meter placed at the cold side of the TE module is overestimated due to the radiation from the hot-side substrate.

In some cases, the TE module is designed to work in the air, where heat loss is severe. In order to diminish heat losses by radiation and convection within the real TE module, insulating materials have been used to fill the gaps between the TE legs (figure 16.7). It is thus beneficial to reduce the gap as much as possible by employing insulating materials that possess as low a thermal conductivity as possible. Although such fillers generate thermal bypasses that consume part of the input heat flow and

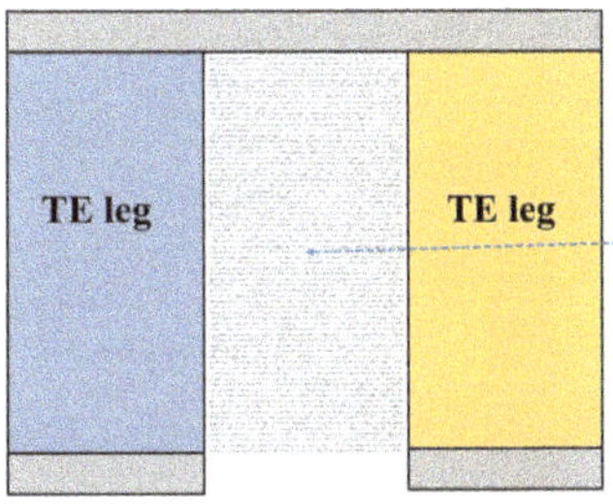

Figure 16.7. Schematic diagram of the glass fibers used as filling in the TE module.

therefore degrade the generation performance, the glass fibers that have been added into the gap evidently improve the efficiency of the TE module, confirming the effectiveness of fillers in minimizing heat losses by radiation and convection [29].

16.4 Double four-point probe method

Despite the methods used to eliminate heat loss, making a good contact for any TE material requires great effort. Efficiency measurement based on the four-point probe method provides a way to rapidly screen the efficiency without considering the contact problem. The four-point probe method has been widely studied as a technique for avoiding the consideration of electrical contact resistance [30–32]. A double four-point probe method was successfully implemented to measure the efficiency of TE materials regardless of the electrical contact resistance between the probes and the TE leg. The schematic and photograph of the set-up is shown in figure 16.8. Its name refers to the eight probes in total, including four probes in contact with the TE leg and four wires from two thermocouples for temperature readings. Since the properties of the TE material are temperature-dependent and current flows in three dimensions inside the TE leg, COMSOL Multiphysics was applied to predict the output power and efficiency in order to demonstrate the reliability of this method [33–37].

An n-type Mg_3Sb_2-based TE material ($Mg_{3.1}Co_{0.1}Sb_{1.5}Bi_{0.49}Te_{0.01}$) with a high peak ZT of ~1.7 [28] was polished to dimensions of ~3.8 mm × 3.8 mm × 6 mm for the experiment. Attempts were made with different braze materials to connect a copper plate with one end of the TE leg working as the hot-side end, but without much success. The unsuccessful brazing is attributed to the degradation of the contact layer or the reaction between the braze material and the TE material at high temperatures. However, the other end of the TE material, working as the cold-side end, was successfully connected to a copper plate by low-temperature solder ($In_{52}Sn_{48}$, melting point 118 °C) after being electroplated with a nickel layer. Electrical contact-resistance scanning showed that there is negligible electrical contact resistance between the copper plate and the TE material, which is important for output power and efficiency measurement. For further proof, other TE materials such as skutterudites [38], Zintl materials [39], half-Heuslers [40], GeTe [41], Bi_2Te_3 [42], etc, have been tested, all showing negligible electrical contact resistance. Therefore, the set-up to measure efficiency was modified according to the schematic

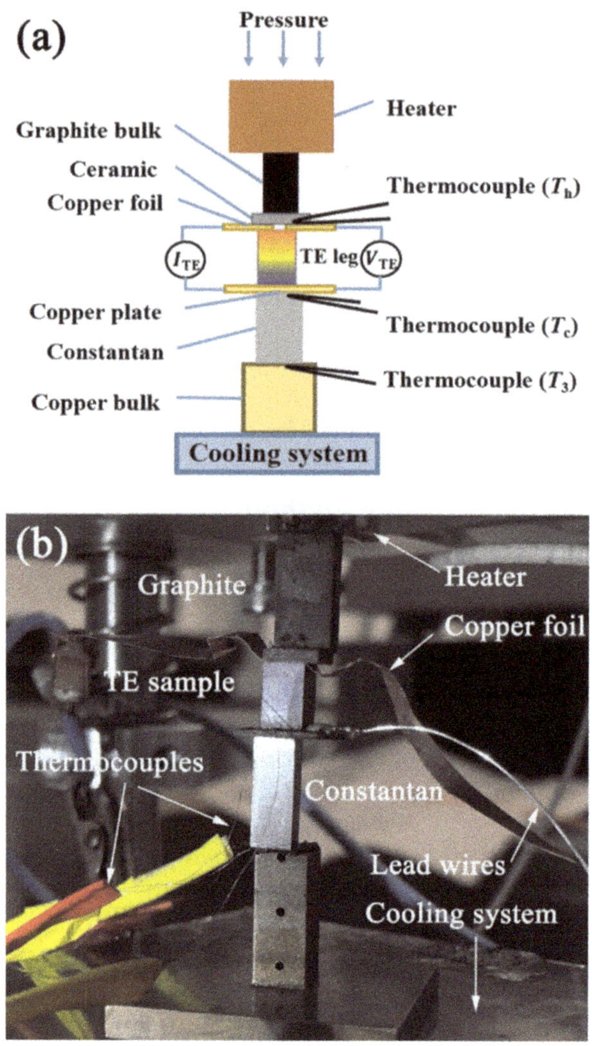

Figure 16.8. (a) Schematic and (b) photograph of the efficiency measurement set-up.

diagram shown in figure 16.8(a). While the TE cold-side end was connected to a copper plate by low-temperature solder, two copper foils were mechanically contacted with the TE hot-side end as in the set-up for the output-power measurement.

Constantan, as a standard sample with its thermal conductivity already known, was placed at the bottom of the TE leg to measure heat flow into the TE leg. The joule heat generated from the hot-side interface will increase the hot-side temperature, thus giving rise to more heat flow. However, this does not influence the heat flow measurement since the constantan at the bottom can be used to measure the heat flow regardless of the hot-side temperature. The heat conduction through the copper foils at the hot side does not affect heat flow measurement either. Meanwhile,

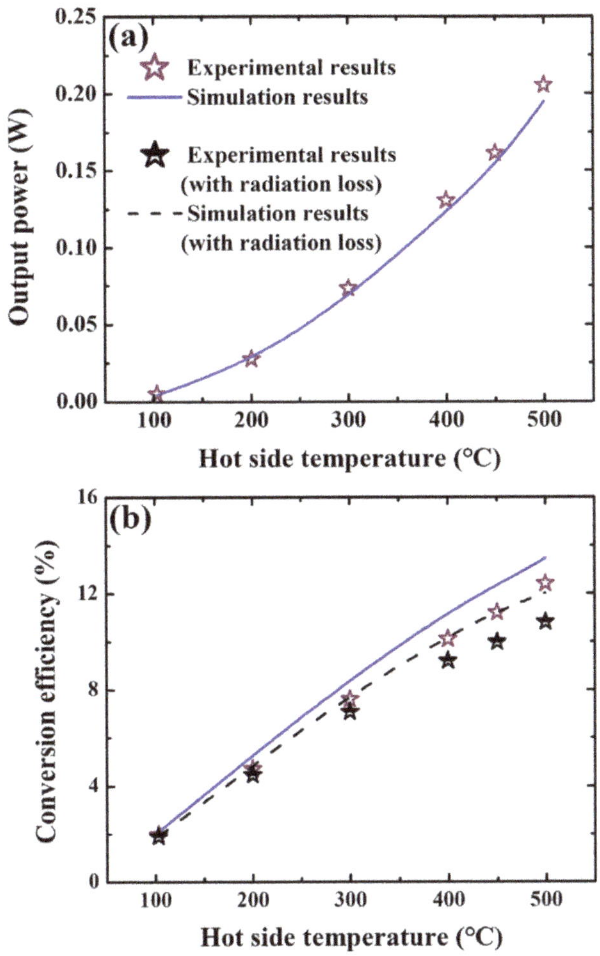

Figure 16.9. Comparison of (a) output power and (b) efficiency between the measurement and 1D model prediction at different temperature differences (T_h: 100, 200, 300, 400, 450 and 500 °C; and T_c: 28, 44, 57, 69, 74 and 83 °C, respectively.).

generation of joule heat at the cold-side interface can be avoided. Thus, connecting the cold-side end with a copper plate not only eliminates the generated joule heat at the cold side, but also ensures an accurate measurement of output power. The efficiency measured from the double four-point probe should be compared with that predicted from the one-dimensional model where current flows through the entire cross-sectional area.

Despite the joule heat at the interface, T_h can remain at a certain value by proportional integral derivative (PID) control. By increasing T_h from 100 °C to 500 °C, a series of output power and efficiency values were measured at different temperatures, agreeing well with the 1D model prediction results (figure 16.9). Output power, determined by the electrical properties of the TE material, is almost unaffected by the heat radiation. However, heat radiation has a little influence on the efficiency measurement. As shown in figure 16.9(b), the efficiency will be slightly lowered if corrected for heat radiation loss. This method could be more accurate for TE materials with a relatively large thermal conductivity, since the influence of radiation would be comparatively smaller.

16.5 Conclusions

In this chapter, methods to accurately predict and measure the efficiency of TE materials from single-leg device to a TE module are introduced. While FDM is used to accurately predict efficiency, efficiency calculated from $(ZT)_{\mathrm{eng}}$ is close to the results from FDM, demonstrating that $(ZT)_{\mathrm{eng}}$ is a better indicator than $(ZT)_{\mathrm{avg}}$ for efficiency prediction. For efficiency measurement, we introduce the challenges in the measurement and the methods to overcome the challenges first. Then different methods of efficiency measurement are presented.

References

[1] Liu W S, Hu J Z, Zhang S M, Deng M J, Han C G and Liu Y 2017 *Mater. Today Phys.* **1** 50–60

[2] Kraemer D, Sui J, McEnaney K, Zhao H, Jie Q, Ren Z F and Chen G 2015 *Energ. Environ. Sci.* **8** 1299–308

[3] Wang H, McCarty R, Salvador J R, Yamamoto A and Konig J 2014 *J. Electron. Mater.* **43** 2274–86

[4] Kim H S, Liu W S, Chen G, Chua C W and Ren Z F 2015 *P. Natl Acad. Sci. USA* **112** 8205–10

[5] Kim H S, Liu W S and Ren Z F 2017 *Energ. Environ. Sci.* **10** 69–85

[6] Rhyee J S, Lee K H, Lee S M, Cho E, Kim S I, Lee E, Kwon Y S, Shim J H and Kotliar G 2009 *Nature* **459** 965–8

[7] Toberer E S, Rauwel P, Gariel S, Tafto J and Snyder G J 2010 *J. Mater. Chem.* **20** 9877–85

[8] Saiga Y, Du B, Deng S K, Kajisa K and Takabatake T 2012 *J. Alloy Compd.* **537** 303–7

[9] Zhou M, Li J F and Kita T 2008 *J. Am. Chem. Soc.* **130** 4527–32

[10] Zhao H Z, Sui J E, Tang Z J, Lan Y C, Jie Q G, Kraemer D, McEnaney K N, Guloy A, Chen G and Ren Z F 2014 *Nano. Energy* **7** 97–103

[11] Zhang Q, Cao F, Liu W S, Lukas K, Yu B, Chen S, Opeil C, Broido D, Chen G and Ren Z F 2012 *J. Am. Chem. Soc.* **134** 10031–8

[12] Liu W S *et al* 2015 *P. Natl Acad. Sci. USA* **112** 3269–74

[13] Zhao L D, Lo S H, Zhang Y S, Sun H, Tan G J, Uher C, Wolverton C, Dravid V P and Kanatzidis M G 2014 *Nature* **508** 373

[14] Zhang Q, Chere E K, McEnaney K, Yao M L, Cao F, Ni Y Z, Chen S, Opeil C, Chen G and Ren Z F 2015 *Adv. Energy Mater.* **5** 1401977

[15] Chen X, Shi L, Zhou J S and Goodenough J B 2015 *J. Alloy Compd.* **641** 30–6

[16] Shi X, Yang J, Salvador J R, Chi M F, Cho J Y, Wang H, Bai S Q, Yang J H, Zhang W Q and Chen L D 2011 *J. Am. Chem. Soc.* **133** 7837–46

[17] Jie Q, Wang H Z, Liu W S, Wang H, Chen G and Ren Z F 2013 *Phys. Chem. Chem. Phys.* **15** 6809–16

[18] Oto Y, Iida T, Sakamoto T, Miyahara R, Natsui A, Nishio K, Kogo Y, Hirayama N and Takanashi Y 2013 *Phys. Status Solidi* C **10** 1857–61

[19] He R, Kim H S, Lan Y C, Wang D Z, Chen S and Ren Z F 2014 *RSC Adv.* **4** 64711–6
[20] Chen S, Lukas K C, Liu W S, Opeil C P, Chen G and Ren Z F 2013 *Adv. Energy Mater.* **3** 1210–4
[21] Sui J H, Li J, He J Q, Pei Y L, Berardan D, Wu H J, Dragoe N, Cai W and Zhao L D 2013 *Energ. Environ. Sci.* **6** 2916–20
[22] Jood P, Mehta R J, Zhang Y L, Peleckis G, Wang X L, Siegel R W, Borca-Tasciuc T, Dou S X and Ramanath G 2011 *Nano Lett.* **11** 4337–42
[23] Joshi G *et al* 2008 *Nano Lett.* **8** 4670–4
[24] Wang X W *et al* 2008 *Appl. Phys. Lett.* **93** 193121
[25] Liu W S, Jie Q, Kim H S and Ren Z F 2015 *Acta Mater.* **87** 357–76
[26] Muto A, Yang J, Poudel B, Ren Z F and Chen G 2013 *Adv. Energy Mater.* **3** 245–51
[27] Muto A, Kraemer D, Hao Q, Ren Z F and Chen G 2009 *Rev. Sci. Instrum.* **80** 093901
[28] Mao J *et al* 2017 *P. Natl Acad. Sci. USA* **114** 10548–53
[29] Zhang Q H, Liao J C, Tang Y S, Gu M, Ming C, Qiu P F, Bai S Q, Shi X, Uher C and Chen L D 2017 *Energ. Environ. Sci.* **10** 956–63
[30] Shi J S and Sun Y C 1997 *Rev. Sci. Instrum.* **68** 1814–7
[31] Yoshimoto S *et al* 2007 *Nano Lett.* **7** 956–9
[32] Saini S, Mele P, Miyazaki K and Tiwari A 2016 *Energ. Convers. Manag.* **114** 251–7
[33] Hu X K, Jood P, Ohta M, Kunii M, Nagase K, Nishiate H, Kanatzidis M G and Yamamoto A 2016 *Energ. Environ. Sci.* **9** 517–29
[34] Wu G X and Yu X 2014 *Energ. Convers. Manag.* **86** 99–110
[35] Skomedal G *et al* 2016 *Energ. Convers. Manag.* **110** 13–21
[36] Fabian-Mijarigos A, Min G and Alvarez-Quintana J 2017 *Energ. Convers. Manag.* **148** 1372–81
[37] Donoso-Garcia P and Henriquez-Vargas L 2015 *Energy* **93** 1189–98
[38] Rogl G and Rogl P 2017 *Mater. Today Phys.* **3** 48–69
[39] Mao J, Wu Y X, Song S W, Shuai J, Liu Z H, Pei Y Z and Ren Z F 2017 *Mater. Today Phys.* **3** 1–6
[40] He R, Zhu H T, Sun J Y, Mao J, Reith H, Chen S, Schierning G, Nielsch K and Ren Z F 2017 *Mater. Today Phys.* **1** 24–30
[41] Liu Z H *et al* 2018 *P. Natl Acad. Sci. USA* **115** 5332–7
[42] Poudel B *et al* 2008 *Science* **320** 634–8
[43] Zhu Q, Song S, Zhu H and Ren Z 2019 *J. Power Sources* **414** 393

Chapter 17

Thermophotovoltaic energy conversion: materials and device engineering

Tobias Burger, Caroline Sempere and Andrej Lenert

This chapter reviews the basic principles of thermophotovoltaic (TPV) conversion and examines how candidate materials perform across various factors that contribute to overall conversion efficiency. The best approaches within each factor are contextualized by theoretical limits. Ongoing efforts to improve upon reported efficiency factors are discussed, along with the practical challenges that face the translation of lab-scale demonstrations to realistic generators, such as thermal stability and parasitic losses. Overall, this chapter reveals that the largest losses relative to theoretical limits are due to spectral inefficiencies. Improvements to out-of-band reflectivity are needed to reduce net radiative exchange between the emitter and cell. In addition to spectral inefficiencies, the uniquely high photocurrents of TPV generators make the system vulnerable to ohmic losses. Ultimately, addressing these challenges may enable TPV generation to offer superior performance and lower cost than conventional power cycles.

17.1 Introduction

Thermophotovoltaic (TPV) generators are solid-state heat engines that utilize the photovoltaic effect to convert thermal radiation into power. A primary energy source, such as solar [1–8], nuclear [9], [10], electrical [11], [12], or chemical [13], is used to sustain a temperature difference between the thermal emitter and the photovoltaic (PV) cell, and, in turn, drive a net flow of radiative energy (figure 17.1). Conversion of radiation with sufficiently high photon energy, above the PV cell's bandgap, generates electrical power.

Interest in TPV conversion has been motivated by the technology's solid-state design, dispatchable nature and potential for high power densities. These characteristics are desirable for remote power generation and several emerging generation and storage approaches, as discussed below.

doi:10.1088/978-0-7503-1738-2ch17

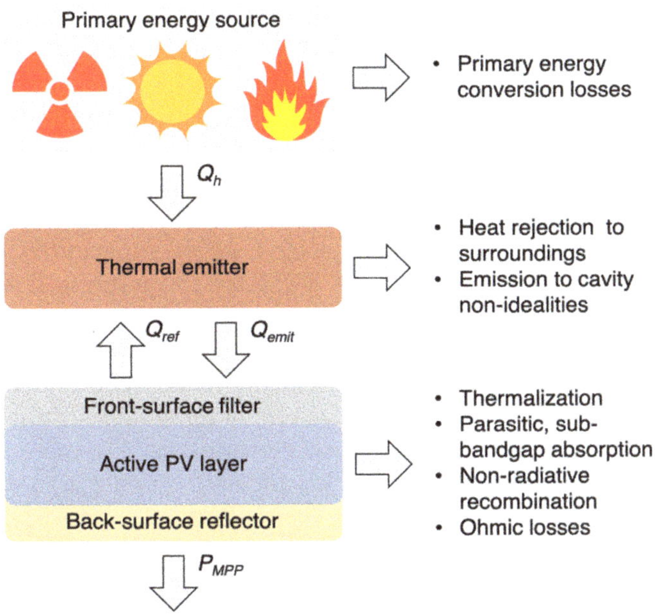

Figure 17.1. Heat flows and energy conversion in a TPV generator.

The solid-state design of TPV generators makes the technology suitable for small-scale power generation near the point of use (i.e., kW scale). This approach may eliminate losses inherent to MW scale energy production, far from the point of use, by enabling the functional utilization of waste heat for space and water heating. The feasibility of using TPVs for residential co-generation has been evaluated through detailed system design and simulation [14], economic analysis [15], and development of a compact furnace-generator [16]. In addition to being suitable for small scales, recent advances in semiconductor manufacturing have enabled fabrication of high-quality, thin-film (<10 μm) cells, characterized by the removal of the parent wafer after growth [17–21]. These thin-film cells may offer high-density power generation with the added advantage of a flexible and lightweight architecture, which is desirable for remote and portable power generation applications.

TPV generators may also offer several advantages over conventional cycles owing to their ability to respond quickly to sudden electricity demands on the electrical grid [11, 12]. Grid-scale storage is increasingly necessary to facilitate a transition from fossil fuels to intermittent renewable energy sources, such as solar and wind. Beyond reducing carbon emissions, diversification of primary energy sources and implementation of rapidly dispatchable power can help improve grid resilience. Studies suggest that TPVs may offer competitive modular costs, lower maintenance costs and faster ramp rates than combined cycle systems [11, 22]. Further, solid-state engines such as TPVs are well-suited for operation with constant-temperature heat, whereas turbines suffer from entropic losses associated with heating of the working fluid [22]. In one envisioned application, a high-temperature thermal energy storage material, such as Si, could offer energy storage for up to several months [12].

Charging would be accomplished via resistive heating, while TPV generators would be used to discharge the stored thermal energy during periods of high demand. Despite the appeal of TPVs for a range of applications, current TPV generators operate at conversion efficiencies well below their thermodynamic limits.

This chapter provides a critical discussion of high-performance TPV generators, from both a materials development and a device engineering perspective. Prior TPV literature reviews have emphasized system components (i.e., devices [23] and emitters [24, 25]), prototype systems [26], and target applications, such as direct solar energy conversion [7], radioisotope decay for space propulsion [10], combustion-driven systems [13], and waste heat recovery [27]. This review identifies leading approaches and specific areas where improvement is needed. First, a general framework, consisting of four discrete, physically meaningful performance metrics, is used to evaluate TPV performance. The roles of material selection, as well as cell and emitter architecture for enhancing cell performance, are then discussed in the context of each performance metric. Specifically, optimization of wavelength-selective radiative transport between the thermal emitter and PV cell is discussed in section 17.2.1. Losses in the PV cell, resulting from non-radiative recombination processes and ohmic losses, are considered in sections 17.2.2–17.2.4. Within each section, this chapter provides a summary of ongoing challenges along with suggestions for improved generator performance, as informed by the analysis of trends across material systems and architectures throughout the literature.

17.2 Framework for analyzing the performance of TPV systems

The conversion efficiency of TPV generators (η_{TPV}) can be expressed as the product of several factors, describing spectral management across the optical cavity and carrier management in the PV cell:

$$\eta_{\mathrm{TPV}} = \mathrm{SE} \cdot \mathrm{IQE} \cdot \mathrm{VF} \cdot \mathrm{FF}, \tag{17.1}$$

where SE is the spectral efficiency, IQE is the internal quantum efficiency, VF is the voltage factor and FF is the fill factor.

This chapter is organized according to the structure of the governing efficiency (equation (17.1)). Analysis of each factor includes a theory-based discussion, a review of the best factors reported across material systems in the literature and a summary of ongoing efforts to improve these factors.

Spectral efficiency describes the usable spectral overlap of emission from the thermal emitter and absorption within the cell. Discussion of the importance of spectral management and a summary of SE achieved by the best thermal emitters and PV cells across material systems is provided in section 17.2.1. Although SE is not solely a property of the emitter or the PV cell, a comparison of spectral efficiency across reported emitters and cells can be performed if the opposing component is assumed to be a blackbody.

Internal quantum efficiency is a spectrally dependent quantity that measures the efficiency of photoexcitation and charge carrier collection in the PV cell. High IQE is indicative of a high-quality material, where a large fraction of absorbed radiation contributes to the photocurrent. Section 17.2.2 evaluates quantum efficiency, both internal and external, reported for state-of-the-art PV cells in the literature.

Charge carrier management is described jointly by the voltage factor and the fill factor. Variables affecting the voltage factor, VF, are discussed in section 17.2.3. The voltage factor can be related to the dark current density, J_0, to analyze the quality of charge carrier management while normalizing for the effects of illumination intensity. Finally, fill factor, FF, is a measure of the maximum power generation of a PV cell relative to the product of the open-circuit voltage, V_{oc}, and the short-circuit current density, J_{sc}. The importance of FF, particularly at the high current densities characteristic of TPV generators, is discussed in section 17.2.4.

17.2.1 Spectral efficiency

The performance of TPV generators is highly dependent on the radiative exchange between the thermal emitter and the PV cell. Although emission and subsequent absorption of above-bandgap photons are necessary for power generation, suppression and/or reflection of sub-bandgap photons are, concurrently, critical for achieving high conversion efficiency [28]. Spectral efficiency captures the importance of both considerations:

$$\mathrm{SE} = \frac{E_g \cdot \int_{E_g}^{\infty} \varepsilon_{\mathrm{eff}}(E) \cdot b(E, T_h)\mathrm{d}E}{\int_0^{\infty} \varepsilon_{\mathrm{eff}}(E) \cdot E \cdot b(E, T_h)\mathrm{d}E}. \tag{17.2}$$

Here, E_g is the bandgap energy, b is the blackbody spectral photon flux and T_h is the hot side (i.e., thermal emitter) temperature. The effective spectral emissivity of the emitter-cell pair, $\varepsilon_{\mathrm{eff}}$, is given by

$$\varepsilon_{\mathrm{eff}} = \frac{\varepsilon_e \varepsilon_c}{\varepsilon_e + \varepsilon_c - \varepsilon_e \varepsilon_c}, \tag{17.3}$$

where ε_e is the emitter emissivity and ε_c is the cell emissivity. Equivalently, the effective spectral emissivity can be written in terms of the spectral reflectance of the emitter, R_e, and of the cell, R_c [20]:

$$\varepsilon_{\mathrm{eff}} = \frac{(1 - R_e)(1 - R_c)}{1 - R_e R_c}. \tag{17.4}$$

The spectral efficiency discussed here shares similarities with the photon over-excitation factor [23] and the ultimate efficiency [28]. However, SE considers the net radiative exchange between the emitter and the PV cell, thereby considering the effects of multiple reflections across the cavity. Notably, spectral efficiency is a function of both the spectral emissivity of the thermal emitter and the PV cell.

SE shares the same upper bound as the ultimate efficiency, which is shown in figure 17.2 for a cell or an emitter with a single cutoff energy, E_{cutoff}, that matches the bandgap. For a finite, out-of-band emissivity, $\varepsilon_{\mathrm{out}}$, there is an optimal bandgap which maximizes the spectral efficiency. This optimum arises because of the tradeoff

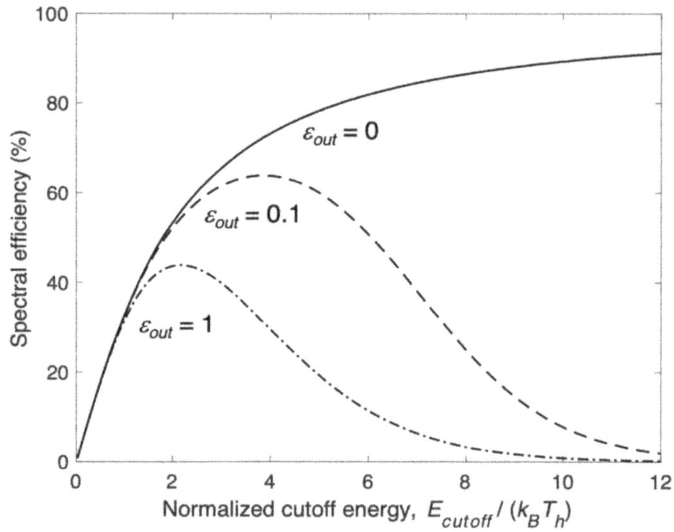

Figure 17.2. Spectral efficiency limits. Spectral efficiency shown as a function of normalized cutoff energy for perfect in-band (i.e., above-bandgap) absorption ($\varepsilon_{in} = 1$). Limits are calculated for three out-of-band (i.e., sub-bandgap) conditions: total suppression ($\varepsilon_{out} = 0$), intermediate suppression ($\varepsilon_{out} = 0.1$) and no suppression ($\varepsilon_{out} = 1$).

between photocurrent and bandgap energy—i.e., for a given emission temperature, the photocurrent decreases as the bandgap increases.

This section examines the design of both thermal emitters and PV cells to identify the most effective and robust methods for improving spectral utilization. Control of the radiative exchange may be engineered at the hot side, described here as emissive spectral control, and at the cold side, described here as absorptive spectral control.

While intrinsic material properties alone offer limited spectral control for TPV generators, targeted engineering of components has been shown to promote above-bandgap photon transport and impede sub-bandgap transport. Photonic design of emitter micro-scale geometry is a common strategy to promote above-bandgap emission while simultaneously suppressing sub-bandgap emission at the hot side. Strategies for absorptive spectral control at the PV cell typically consist of the use of front-surface filters (FSFs) and/or back-surface reflectors (BSRs).

17.2.1.1 Emissive spectral control

Selective emitters can improve TPV generator performance through preferential emission of above-bandgap photons, thereby simultaneously promoting power generation and inhibiting parasitic absorption in the PV cell. Utilization of emissive spectral control is advantageous for high-performance TPV generators, as it may relax the requirement for ultrahigh emitter temperatures and minimize the sensitivity to cavity non-idealities. The development of thermal emitters requires both spectral engineering to achieve selective emission and the design of thermally stable architectures to withstand operation at elevated temperatures (figure 17.3).

This chapter examines the spectral efficiency of thermal emitters when paired with a blackbody PV cell based on their measured spectral properties and the highest

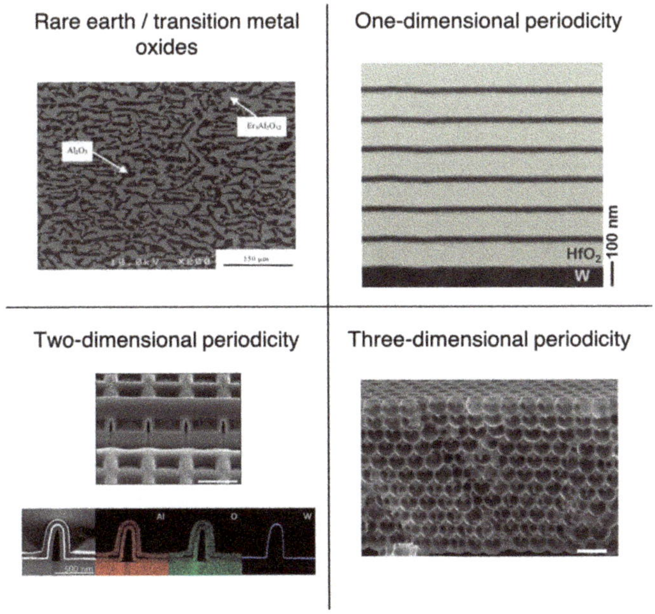

Figure 17.3. Examples of thermal emitter structures. Top left: SEM image of $Al_2O_3/Er_3Al_5O_{12}$ eutectic ceramic emitter microstructure. Reproduced with permission from [29]. Copyright 2005 Elsevier. Top right: cross-sectional STEM image of a W/HfO_2 stack emitter (false color adapted from elemental mapping in [30]). Reproduced with permission from [30]. Copyright Springer Nature. Bottom left: cross-sectional SEM image of a $Al_2O_3/W/Al_2O_3$ cavity array and TEM and EDS elemental image of the cavity wall. Reproduced with permission from [31]. Copyright 2019 American Chemical Society. Bottom right: cross-sectional SEM image of a HfB_2 inverse colloidal crystal emitter. Reproduced with permission from [32]. Copyright 2013 Nature.

temperature at which the emitters were tested. While the optimal T_h is a function of E_{cutoff} and the emitter's spectral properties (figure 17.2), calculating SE using the highest tested temperature acknowledges the difficulty of developing emitters with high-temperature stability. This analysis only includes thermal emitters that were aged for at least 1 h at or above 1023 K (750 °C). SE data for the thermally stable emitters are compiled in the appendix and are summarized visually in figure 17.4. Although not as common, spectral data measured at elevated/operating temperatures are preferred because optical properties are temperature dependent. To broaden the data set, however, this chapter also includes emitters that meet the basic stability criterion, but were characterized at room temperature.

Emitters have been categorized based on the extent of structural engineering, ranging from intrinsic material properties to three-dimensional structuring. Structurally tunable designs include photonic crystals, characterized by periodicity at a length-scale on the order of the wavelength of interest, and metamaterials, characterized by periodicity at a length-scale shorter than the wavelength of interest. Photonic crystals may be further characterized by the number of dimensions exhibiting periodicity. For example, periodic emitter designs exhibiting one-, two- and three-dimensional periodicity include alternating layer stacks, wire or cavity arrays, and inverse opals, respectively (figure 17.3).

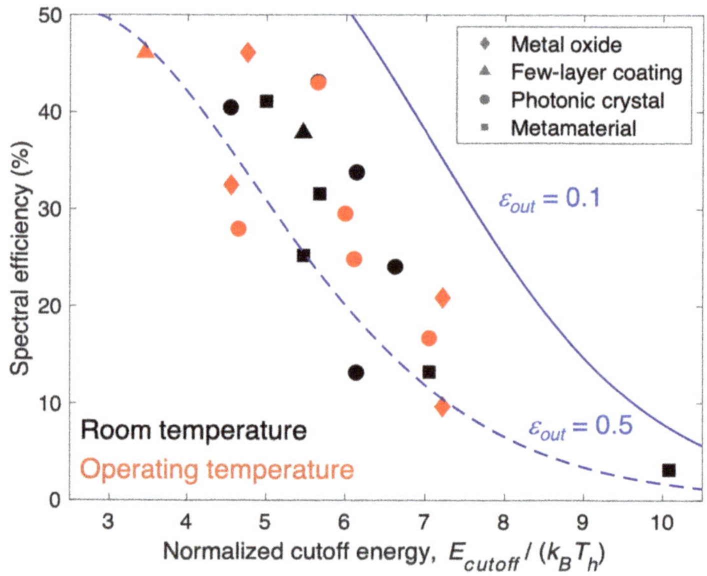

Figure 17.4. Spectral efficiency of demonstrated thermal emitters. The upper bounds of spectral efficiency for 0.1 and 0.5 sub-bandgap emissivity are provided to illustrate the trend as a function of normalized cutoff energy. Black symbols indicate spectral characterization at room temperature, while red symbols distinguish studies that have characterized spectral properties at elevated temperatures.

The best emissive spectral efficiency identified in this chapter is 46%, which is achieved by a few-layer coating consisting of HfO_2/W/HfO_2 [33] and a MgO emitter with NiO loading [34]. These emitters are followed closely by 2D photonic crystals with SE of 43% [35, 36]. Other emitters with SE exceeding 35% include structures containing protective HfO_2 layers [30, 37, 38]. Since SE calculations are sensitive to both T_h and E_{cutoff}, high SE demonstrations benefit from achieving high-temperature operation, typically enabled by use of a metal oxide or HfO_2 protective layer [30, 34, 37, 38], or spectral engineering of a relatively low E_{cutoff}, such as the few-layer coating of HfO_2/W/HfO_2 [33] and polycrystalline Ta 2D photonic crystal with a HfO_2 coating [37]. Overall, however, the spectral efficiency remains significantly below the theoretical limit.

Among intrinsically selective materials, several rare-earth and transition-metal oxides have been identified for use as thermal emitters in TPV generators because of their high-temperature stability and infrared emission peaks. Specifically, Er_2O_3 and NiO exhibit emissive properties appropriate for use with PV materials such as InGaAs, Ge and GaSb, while Yb_2O_3 exhibits high emissivity near the band edge of Si [39]. At comparable normalized cutoffs, the SE of metal oxide emitters is among the highest reported. The high operating temperature of these emitters benefits SE. However, these approaches cannot tailor the thermal emission properties to the desired bandgap.

Alternatively, a variety of tunable thermal emitters have been developed through the geometrical design of the emitter architecture. Although metamaterials generally have lower SE than photonic crystals at comparable normalized cutoffs, the SE data do not reveal a clear dependence on the emitter type.

To offer a measure of spectral selectivity that normalizes for the effects of T_h and E_{cutoff}, an alternative performance metric is considered. Here, spectral selectivity, S, is defined as the ratio of the SE of a thermal emitter to the SE of a blackbody emitter, SE_{BB}, at the corresponding E_{cutoff} and T_h (equation (17.5)):

$$S = \frac{SE(E_{cutoff}, T_h)}{SE_{BB}(E_{cutoff}, T_h)}. \tag{17.5}$$

By this definition, a blackbody emitter is characterized by an S of 1, while selective emitters have S exceeding 1. S has been calculated for each of the thermally tested emitters identified in this chapter (figure 17.5). Generally, selectivity scales with the ratio of in-band to out-of-band emissivity (i.e., $\varepsilon_{in}/\varepsilon_{out}$). This emissivity contrast ratio serves as an intuitive measure of selectivity but does not directly correspond to spectral efficiency. Deviations from this trend are a result of the metric's dependence on the absolute magnitude of the emissivities and the distribution of spectral power.

The highest S identified among thermally stable emitters considered here is 4, achieved by a Ni nanopyramid 2D metamaterial emitter [40]. If only spectral properties measured at operating temperatures are considered, an S of 3.77 represents the highest value [41]. This selectivity was achieved through suppression of sub-bandgap emission from an Yb_2O_3 emitter, utilizing a mantle structure. Among emitters with S exceeding 2.2, all but one case utilized a photonic or

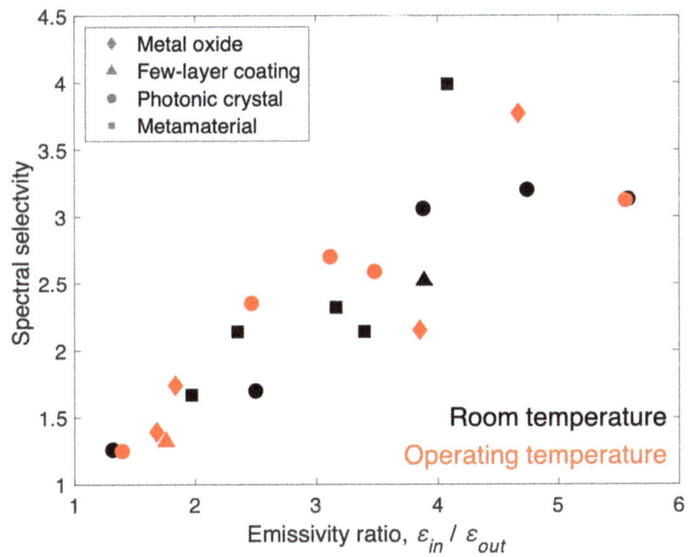

Figure 17.5. Spectral selectivity of demonstrated thermal emitters. Selectivity generally increases with the ratio of in-band to out-of-band emissivity. Black symbols indicate spectral characterization at room temperature, while red symbols distinguish studies that have characterized spectral properties at elevated temperatures.

metamaterial design, highlighting the value of tunable spectral properties for achieving high spectral selectivity.

In addition to high SE, the development of practical, high-performance emitters will require improved thermal stability. Thermal emitters exhibiting selectivity at room temperature can experience micro-scale morphological changes and phase transitions, resulting in degradation of their selective properties [30, 32, 38, 42–44]. The remainder of this section summarizes efforts to demonstrate thermal stability for tunable thermal emitters at high temperatures.

Thermal stability testing across tunable thermal emitters can be described in terms of aging temperature and duration. Stability measurements for thermal emitters that have been spectrally characterized at elevated temperatures are sparse (figure 17.6). Notably, among studies reporting spectral properties measured at elevated temperatures, all but one considers photonic crystal structures. Heightened interest in developing photonic crystal thermal emitters may be motivated by the architecture's relatively high SE.

While several structurally tunable thermal emitters have exhibited stability at temperatures exceeding 1600 K, none have been aged for longer than 24 h [30, 32, 34]. Generally, the highest temperature stability testing has been conducted over short timespans. Tunable emitters with reported thermal stability beyond 1600 K have one common characteristic—a HfO_2 coating. In fact, HfO_2 is present in all but one thermal emitter exhibiting stability above 1400 K. To date, the thermal stability of structured emitters at temperatures optimized for PV cells with bandgaps exceeding 1 eV, such as Si, remains to be demonstrated.

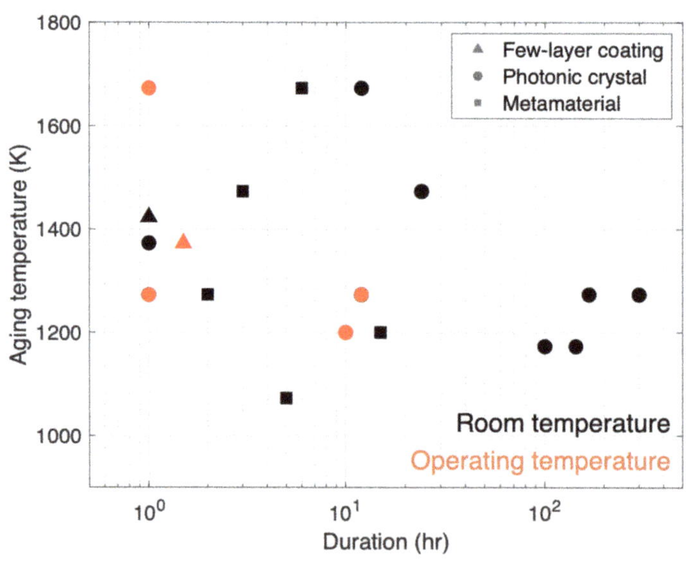

Figure 17.6. Duration and temperature of thermal emitter aging conditions. Black symbols indicate spectral characterization at room temperature, while red symbols distinguish studies that have characterized spectral properties at elevated temperatures.

Emitters that have exhibited thermal stability for relatively long aging times, more than 24 h, have been limited to lower temperatures, under 1300 K [36, 37, 45–47]. The longest duration for a high-temperature test considered here is 300 h [47]. Notably, three of the four emitters with the longest aging duration (>100 h) also utilize HfO_2 coatings to improve the emitter lifetime [36, 46, 47]. Overall, this analysis suggests that HfO_2 coating is an effective strategy for enhancing thermal stability at high operating temperatures. A HfO_2 coating improves thermal stability by acting as a diffusion inhibitor, preventing surface reactions and structural degradation [36, 42, 48]. Conformal HfO_2 coatings may be achieved through atomic layer deposition techniques [36, 37].

In addition to high temperatures, practical implementation of high-performance, tunable thermal emitters may be limited by practical difficulties associated with maintaining inert or vacuum conditions in a TPV generator. Thermal stability testing above 1023 K conducted throughout the literature has been performed almost exclusively in inert environments, consisting of Ar or He, or under vacuum. One study reports thermal stability for Pt array 2D metamaterial at 1023 K in air, which represents the highest thermal stability of a tunable thermal emitter in air identified in this chapter; however, no details regarding aging duration were provided [49].

Despite the progress in reported thermal stability, further verification of emitter stability is necessary, as TPV generators require lifetimes on the order of years. It is apparent from this analysis that a technological gap continues to exist between the reported spectral properties of tunable thermal emitters at room temperature and the development of high-temperature stable, high-SE emitters for TPV generators. In the absence of such emitters, alternative techniques for enhancing spectral efficiency are necessary for improving TPV generator performance, particularly for wider bandgap PV cells.

17.2.1.2 Absorptive spectral control

The primary goal of absorptive spectral control is to design PV cells with high reflectance at wavelengths longer than the absorbing material's bandgap, such that low-energy (i.e., sub-bandgap) radiation is returned to the thermal emitter, rather than lost by heat dissipation in the cell. Through this *external* photon recycling technique, selectively absorptive PV cells enable selective, radiative transport across the TPV cavity, even when operated in tandem with broadband emitters. Relaxing the selectivity constraint otherwise placed on the thermal emitter is beneficial for the practical design of TPV generators, as the optical properties of thermal emitters degrade with increasing temperature [24, 39, 50, 51].

Absorptive spectral control may be achieved through the use of a back-surface reflector (BSR) behind the active layer of the PV cell and/or a front-surface filter (FSF) between the active layer and the thermal emitter. In the case of cells with BSRs, sub-bandgap radiation passes through the cell, reflects at the backside of the active layer and returns to the thermal emitter. This configuration is advantageous, as it allows above-bandgap radiation to reach the cell un-impeded. However, it also makes the system vulnerable to parasitic sub-bandgap absorption within the cell.

Alternatively, FSFs selectively allow transmission of above-bandgap radiation into the cell, while reflecting sub-bandgap radiation before it interacts with the active layers. This configuration eliminates sub-bandgap parasitic loss pathways in the cell but may reduce power generation if above-bandgap radiation is reflected or absorbed in the filter.

This chapter compares the spectral efficiency of cells, from various material systems, based on their reported spectral properties, assuming a blackbody emitter at the T_h provided in the reference (table 17.1). The spectral range used to calculate SE for each cell, and what fraction of the overall emissive power at the relevant T_h is captured by that range, is also provided. This fraction provides a measure of uncertainty associated with the reported SE. Where spectral data are not reported, weighted averages reported in the reference were used to calculate SE.

It should be noted that a BSR is present in all of the cells with the highest SE. BSRs are generally metallic, allowing them to serve as broadband reflectors in addition to serving as the back electrical contact. In some cases, the addition of a low-index spacer between the metallic contact and semiconductor material has helped to improve out-of-band reflectance, R_{out} [19, 21, 53]. This approach, however, can require additional backside patterning to enable an electrical connection.

In theory, concurrent use of an ideal BSR and ideal FSF would prove redundant. In practice, however, the use of both strategies may help to overcome the spectral imperfections of a single strategy. This is exemplified by the 0.6 eV InGaAs cell, where the addition of an FSF improved R_{out} and η_{TPV} relative to the same cell with only a BSR [52, 56], and enabled the highest overall SE [52]. It should be noted that these cells were not removed from the semi-insulating InP growth substrate.

Removal of the growth substrate following cell fabrication can, in principle, decrease parasitic absorption. The first substrate removal demonstration, however, is believed to have induced damage to the epi-film, thereby reducing the cell's R_{out} [19]. Recently, growth substrate removal was performed without damaging an $In_{0.53}Ga_{0.47}As$ cell, enabling high R_{out} [20]. Although this $In_{0.53}Ga_{0.47}As$ example

Table 17.1. Spectral efficiency of TPV cells.

Material	E_g [eV]	T_h [K]	E_g/k_BT_h	R_{out} [%][a]	A_{in} [%][a]	SE [%]	S	E range [eV] BB fraction	Control feature	Ref.
InGaAsSb	0.54	1273	4.9	76	95	44	2.3	0.31–1.24/64%	BSR thin-film	[19]
InGaAs	0.60	1312	5.3	94	92	64	3.9	0.06–1.48/99%	BSR, FSF	[52]
Ge	0.67	1373	5.7	44	85	19	1.4	0.46–4.58/43%	BSR	[53]
GaSb	0.73	1573	5.4	—	72	—	—	—	BSR	[54]
$In_{0.53}Ga_{0.47}As$	0.75	1480	5.9	94	68	53	4.3	0.08–0.99/95%	BSR thin-film	[20]
Si	1.12	2300	5.65	95	—	—	—	—	BSR	[55]

[a] Weighted averages R_{out} and A_{in} have been calculated using spectral properties collected from graphical data. ±0.5% (absolute) error can be expected as a result of the data collection technique.

has a comparable R_{out} to the 0.6 eV InGaAs cell, it has a lower SE. This is partially due to the lack of an ARC, leading to sub-optimal in-band absorptance, A_{in}.

Analysis of spectral selectivity, *S*, across these high-performance cells reveals that the lattice-matched $In_{0.53}Ga_{0.47}As$ cell [20] exhibits the best spectral efficiency relative to its blackbody counterpart with the same T_h and E_g, despite its lower absolute SE in comparison to the 0.6 eV InGaAs cell [52]. While utilization of highly reflective BSRs and FSFs has enabled absorptive spectral control to surpass the selectivity achieved through emissive spectral control, as indicated by a comparison of SE and *S* between the two strategies, additional improvements in cold-side spectral efficiencies are needed. Simulations suggest that the use of dielectric layers, cladding the absorber layer and optimization of layer thicknesses to leverage interference effects can enable SE as high as 74% [21]. This level of spectral control, however, has not been demonstrated for a PV cell at this time.

17.2.2 Quantum efficiency

Internal quantum efficiency, IQE, is a spectrally dependent measurement of photo-excitation and charge carrier collection. IQE may be expressed quantitatively as the ratio of charge carriers collected to the number of photons *absorbed* by the cell.

Poor IQE may indicate the presence of parasitic loss channels for above-bandgap photons. Specifically, absorption of convertible photons in metallic or heavily doped layers, such as BSRs and contacts, can lower IQE. Additionally, IQE losses can be indicative of a low-quality absorber material or unoptimized geometry in which the charge carrier diffusion length is shorter than the cell's characteristic collection path.

A summary of representative, spectral IQE reported for high-performance PV cells across the literature is provided in figure 17.7. Average, weighted IQE is reported in table 17.2. In the absence of reported average IQE values, weighted

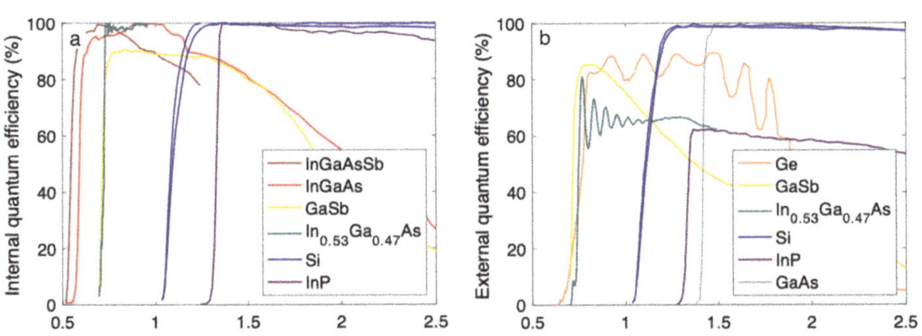

Figure 17.7. Spectral quantum efficiency. Reported (a) IQE and (b) EQE shown for high-performance cells in TPV generator and non-concentrating solar PV configurations. Spectral IQE data for Ge and $In_{0.53}Ga_{0.47}As$ have been calculated using spectral EQE and in-band reflectance, R_{in}, properties collected from graphical data.

Table 17.2. Summary of weighted average EQE and IQE for various TPV and solar PV cell materials reported in the literature.

Material	E_g [eV]	T_h [K]	EQE [%][a]	IQE [%][b]	E range [eV]/BB fraction	Ref.
InGaAsSb	0.53	1223	—	89.77	0.52–1.2/98.1%	[57]
InGaAs	0.6	2050	—	93.9	0.52–2.7/99.8%	[58]
Ge	0.67	1373	55.5	—	0.63–4.1/99.4%	[53]
GaSb	0.73	1573	77.9	85.9	0.71–2.5/100%	[54]
$In_{0.53}Ga_{0.47}As$	0.75	1480	65	98	0.69–1.5/98.4%	[20]
Si	1.12	5777	96.5	99.25	1.0–4.5/100%	[59]
Si	1.12	5777	96.9	97.4	1.0–4.1/98.7%	[60]
InP	1.344	5777	56	94.6	1.2–3.5/95.0%	[61]
GaAs	1.424	5777	92.8	—	1.3–4.1/100%	[62, 63]

[a] Weighted average EQE values have been calculated using spectral properties collected from graphical data for all materials except lattice-matched $In_{0.53}Ga_{0.47}As$. ±0.5% (absolute) error can be expected as a result of the data collection technique.

[b] Weighted average IQE values for the InGaAs, InP and both Si cells have been calculated using spectral properties collected from graphical data. ±0.5% (absolute) error can be expected as a result of the data collection technique.

estimates have been calculated based on blackbody emitters with T_h, selected from the relevant publication.

Figure 17.7(a) shows that mature PV technologies tend to exhibit the highest IQE. In particular, the $In_{0.53}Ga_{0.47}As$ PV cell and both Si PV cells considered here, representative of the highest efficiency TPV cells and highest efficiency Si solar PV, respectively, are characterized by IQE approaching unity. Other high-performance cells, such as InGaAs and InP, exhibit high IQE, in excess of 93%.

This chapter further provides spectral (figure 17.7(b)) and average *external* quantum efficiency, EQE, for high-performance cells. EQE is the ratio of collected charge carriers to photons *incident* on the cell and is related to IQE by the cell's emissivity:

$$\mathrm{EQE}(E) = \varepsilon_c(E) \cdot \mathrm{IQE}(E). \tag{17.6}$$

EQE, thus, represents a joint measurement that cannot decouple reflection losses from poor photoexcitation and charge collection.

17.2.3 Bandgap utilization

Effective utilization of the bandgap is an important measure of cell performance. Improving voltage also benefits the fill factor (as discussed in section 17.2.4). Voltage factor represents one of several factors contributing to TPV efficiency, as previously identified (equation (17.1)). It is defined as the ratio of open-circuit voltage V_{oc} to the bandgap voltage V_g:

$$\text{VF} = \frac{V_{oc}}{V_g} = \frac{q \cdot V_{oc}}{E_g}, \tag{17.7}$$

where q is the elementary charge of an electron. A summary of VF for high-performance PV cells across various material systems is provided in table 17.3.

By rearranging the ideal diode equation (equation (17.8)), it is apparent that the open-circuit voltage of a cell (at constant temperature) may be improved by increasing the photocurrent or decreasing the dark current (equation (17.9)):

$$J = J_{ph} - J_0\left[\exp\left(\frac{qV}{k_B T_c}\right) - 1\right], \tag{17.8}$$

$$V_{oc} = V_{th} \cdot \ln\left(\frac{J_{ph}}{J_0} + 1\right). \tag{17.9}$$

Here, J is the current density, k_B is the Boltzmann constant, T_c is the cold-side (i.e., cell) temperature, V_{th} is the thermal voltage, J_0 is the dark current density and J_{ph} is the photocurrent density. In the case of solar PV, photocurrent may be increased through optical concentration of incident sunlight. For TPV generators, photocurrent may be increased by improving the view factor between the emitter and cell, increasing the emitter temperature, or increasing the emissivity of the emitter and/or the above-bandgap absorptivity of the cell.

The voltage factor is a convenient metric for evaluating bandgap utilization across material systems and can guide material selection. However, it offers limited utility in comparing the quality of carrier management, as it varies considerably with

Table 17.3. Summary of fill factor, voltage factor and dark current density calculated for high-performance cells of various material systems in the literature.

Material	E_g [eV]	J_{sc} [mA cm^{-2}]	V_{oc} [V]	FF [%]	VF [%]	J_0 [A cm^{-2}]	Ref.
InGaAsSb	0.53	2900	0.306	67	57.7	2.36×10^{-5}	[57]
InGaAs	0.6	2865	0.417	66.2	69.4	2.59×10^{-7}	[52]
InGaAs	0.6	2260	0.355	66.5	59.2	2.24×10^{-6}	[58]
Ge	0.67	1650	0.353	70.8	62.7	1.78×10^{-6}	[53]
GaSb	0.726	147	0.377	56.8	51.9	1.57×10^{-7}	[66]
GaSb	0.726	3670	0.43	69	58.9	1.47×10^{-7}	[54]
$In_{0.53}Ga_{0.47}As$	0.75	918	0.529	73	70.5	7.38×10^{-10}	[20]
Si	1.12	8760	0.8	72.3	71.4	1.52×10^{-13}	[55]
Si	1.12	42.6	0.727	84.3	64.9	2.22×10^{-14}	[59]
Si	1.12	42.7	0.738	84.9	65.9	1.43×10^{-14}	[60]
InP	1.344	31.2	0.939	82.6	69.8	4.18×10^{-18}	[61]
GaAs	1.424	29.8	1.127	86.7	79.2	2.63×10^{-21}	[62, 63]

the illumination conditions used to test the cell. Alternatively, dark current effectively quantifies the quality of carrier management for a given cell by normalizing for the effects of variable illumination. Dark current density may be decreased by decreasing the recombination rate through improvements in manufacturing, materials and cell design. The following discussion considers dark current as a metric for evaluating charge carrier management among cells in the literature.

Dark current densities for the best cells within each considered material system are presented with respect to bandgap energy in figure 17.8. Dark current density has been calculated based on the reported open-circuit voltage, photocurrent and cell temperature using equation (17.10). It should be noted that this analysis assumes negligible shunting losses.

$$J_0 = J_{\text{ph}}\left[\exp\left(\frac{qV_{\text{oc}}}{k_{\text{B}}T_{\text{c}}}\right) - 1\right]^{-1}. \tag{17.10}$$

A lower limit of dark current density exists in the absence of non-radiative recombination, described here as the radiative recombination limit. Dark current density in the radiative recombination limit is described by equation (17.11).

$$J_0 = \frac{2 \cdot q}{h^3 \cdot c^2}\pi(n^2 + 1)k_{\text{B}}T_{\text{c}}(2k_{\text{B}}^2T_{\text{c}}^2 + k_{\text{B}}T_{\text{c}}E_{\text{g}} + E_{\text{g}}^2)\exp\left(\frac{-E_{\text{g}}}{k_{\text{B}}T_{\text{c}}}\right), \tag{17.11}$$

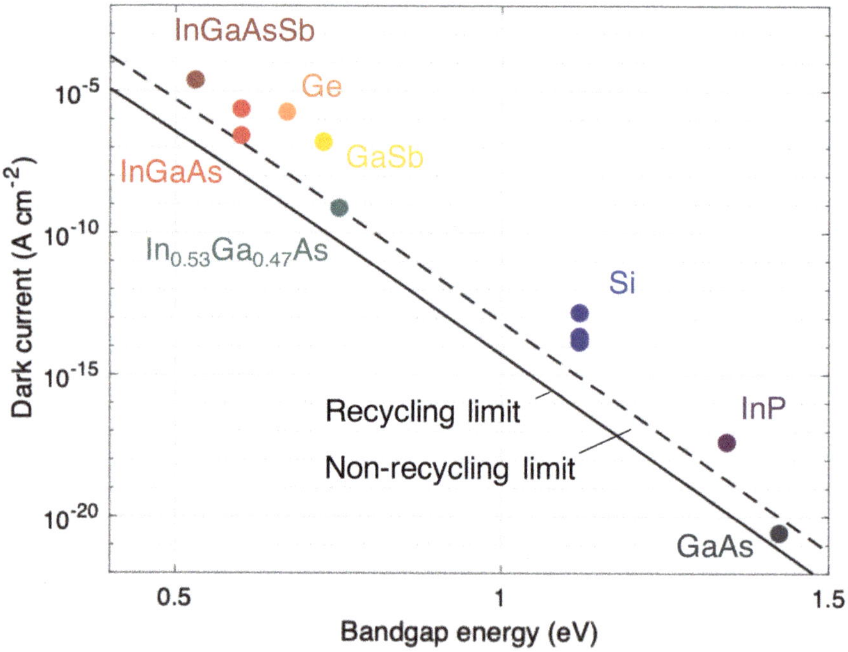

Figure 17.8. Dark current density. J_0 calculated for high-performance cells in TPV generator and non-concentrating solar PV configurations from reported V_{oc} and T_{c} data, compared to the radiative recombination limit as a function of bandgap energy. Limits for dark current density, with (solid) and without (dashed) internal photon recycling, are also shown.

where h is Planck's constant, c is the speed of light in a vacuum and n is the real part of the refractive index [64]. In the limit of perfect photon recycling, radiative emission is limited to the front surface of the converter, thereby reducing bulk recombination losses [65]. Equation (17.11) captures the radiative limit in the absence of internal photon recycling by setting $n = 3.6$ for a typical III–V semiconductor and in the presence of photon recycling by setting $n = 0$.

Low bandgap cells exhibit higher dark current since a larger fraction of the room-temperature emission spectrum lies above the bandgap, resulting in increased radiative recombination. Generally, the best dark currents observed for more mature material systems (e.g., $J_0 \sim 1.4 \times 10^{-14}$ for Si [60]) are approaching the limit at their corresponding bandgap ($J_{0,\ \mathrm{rad\ limit}} = \sim 1.1 \times 10^{-15}$ at $E_g = 1.12$ eV). Among material systems traditionally considered for TPV application (<1.1 eV), cells have yet to surpass the non-recycling limit. The cells that are closest to surpassing this limit are high-quality thin III–V materials with highly reflective back surfaces that can enhance internal photon recycling. Notably, the highest performance $In_{0.53}Ga_{0.47}As$ TPV cell reports dark current of the same order of magnitude as the non-photon-recycling radiative limit.

Additionally, the highest performance GaAs solar PV cell reports dark current lower than the non-photon-recycling radiative limit for its bandgap. Thin-film crystalline cells are excellent candidates for TPV applications because they can synergistically reflect low-energy photons to the emitter, achieving high spectral efficiency (as discussed in section 17.2.1.2) [20, 21, 52, 55]. The best dark currents for less-mature low bandgap cells (e.g., $J_0 = \sim 1.8 \times 10^{-6}$ for Ge [53]) are relatively high compared to the radiative limit at their bandgap ($J_{0,\ \mathrm{rad\ limit}} = \sim 1.8 \times 10^{-6}$ at $E_g = 0.67$ eV).

17.2.4 Fill factor

This section examines the fill factors for the best PV cells demonstrated to date within each material system relevant for TPV generators. Evaluation of this data reveals a dependence of FF on V_{oc}, and indirectly on E_g, in good agreement with past theoretical models describing the low-illumination regime. Further, this section examines the effect of high photocurrents on the fill factor and highlights the importance of low series resistance for TPV generators.

The maximum power point (MPP) describes the operating condition that results in peak power generation from a PV cell. Peak power generation occurs at forward bias V_{MPP} and current density J_{MPP}. The maximum power itself, P_{MPP}, is then the product of V_{MPP} and J_{MPP}. Fill factor (FF) is defined as the ratio of P_{MPP} to the product of V_{oc} and the short-circuit current density, J_{sc}:

$$\mathrm{FF} = \frac{P_{\mathrm{MPP}}}{V_{oc} \cdot J_{sc}} = \frac{J_{\mathrm{MPP}} \cdot V_{\mathrm{MPP}}}{J_{sc} \cdot V_{oc}}. \tag{17.12}$$

Fill factor is primarily affected by the cell's series and shunt resistances, R_s and R_{sh}, respectively.

In the presence of series resistance, a portion of the generated power is dissipated as heat along the current's path. Contributions to R_s include bulk resistance along the longitudinal depth of the cell, sheet resistance along the lateral direction, interfacial resistance at the contacts, and line losses along the length of the contact fingers and busbar. Proper design of highly conductive, selective contacts at the edges of the cell and optimized design of the contact grid geometry can help to minimize resistive losses. However, care should be taken in selecting the appropriate objective function when optimizing the grid design for TPV cells. In solar PV, an opaque top contact grid will shade the active layer and decrease output power, directly affecting the efficiency. For TPV cells, however, a reflective top grid could have a substantially smaller effect on conversion efficiency, as photons reflected by the grid can be reabsorbed by the emitter or active regions of the cell.

Low shunt resistance can sharply affect a cell's fill factor in the case of low-illumination conditions. However, shunting effects are not as important for most TPV cells because of their characteristically high current densities. Nevertheless, fabrication defects and film irregularities, such as pinholes or cracks, can provide a lower resistance path for current flow that bypasses the cell junction, thereby reducing the operating voltage.

In the case of negligible ohmic and shunting losses (i.e., R_s approaches 0 and R_{sh} approaches ∞), the upper limit of a cell's fill factor may be expressed as a function of V_{oc} and cell temperature, T_c, as previously described by Shockley and Queisser [67] (equation (17.13)). Shockley and Queisser's theoretical limit for FF is given by

$$\mathrm{FF_{SQ}} = \frac{\left(\frac{V_{\mathrm{MPP}}}{V_{\mathrm{th}}}\right)^2}{\left(1 + \frac{V_{\mathrm{MPP}}}{V_{\mathrm{th}}} - \exp\left(\frac{-V_{\mathrm{MPP}}}{V_{\mathrm{th}}}\right)\right) \cdot \frac{V_{\mathrm{oc}}}{V_{\mathrm{th}}}}, \tag{17.13}$$

where the ratio of open-circuit voltage to thermal voltage is described by

$$\frac{V_{\mathrm{oc}}}{V_{\mathrm{th}}} = \frac{V_{\mathrm{MPP}}}{V_{\mathrm{th}}} + \ln\left(1 + \frac{V_{\mathrm{MPP}}}{V_{\mathrm{th}}}\right). \tag{17.14}$$

A simpler, empirical model for the upper FF limit as a function of V_{oc} and T_c was developed by Green [68] (equation (17.15)). This model agrees well with Shockley and Queisser's theoretical limit model for $V_{oc}/V_{th} > 5$ ($E_g > 0.13$ eV) [28]. This approximation is therefore suitable for all TPV material systems considered here:

$$\mathrm{FF_{Green}} = \frac{\frac{V_{\mathrm{oc}}}{V_{\mathrm{th}}} - \ln\left(\frac{V_{\mathrm{oc}}}{V_{\mathrm{th}}} + 0.72\right)}{\frac{V_{\mathrm{oc}}}{V_{\mathrm{th}}} + 1}. \tag{17.15}$$

FFs reported in the literature for TPV-relevant cells are presented here as a function of reduced voltage (figure 17.9). These data are compared to the theoretical limit proposed by Shockley and Queisser and the empirical model developed by Green. In the case of low-illumination conditions ($J_{sc} < 50$ mA cm^{-2}), high-quality cells have exhibited fill factors approaching the upper bound at the corresponding bandgap (see Si and GaAs). However, there exists a gap between the best FFs reported for low-illumination conditions, characteristic of non-concentrated solar PV cells, and those reported at higher photocurrents ($J_{sc} \geqslant 1$ A cm^{-2}), characteristic of TPV generators. This gap is apparent from the discrepancy between the best FF reported for a Si solar PV cell [60], 84.9%, and the best FF for a Si TPV cell, 72.3% [55].

The FF gap between cells across illumination conditions is a result of the increasingly stringent requirements for low R_s with increasing photocurrent. Power lost to series resistance increases quadratically as a function of the operating current density:

$$P_{\text{loss,series}} = R_s \cdot J_{\text{MPP}}^2. \tag{17.16}$$

In the case of unconcentrated solar PV ($C = 1$), low photocurrent loosens the requirements on R_s. With increasing photocurrents, however, this loss pathway can outweigh the performance gains associated with improved V_{oc}.

Estimates of ohmic power loss for representative high-performance cells from the literature are provided in table 17.4. The best $In_{0.53}Ga_{0.47}As$ PV cell exhibits low R_s

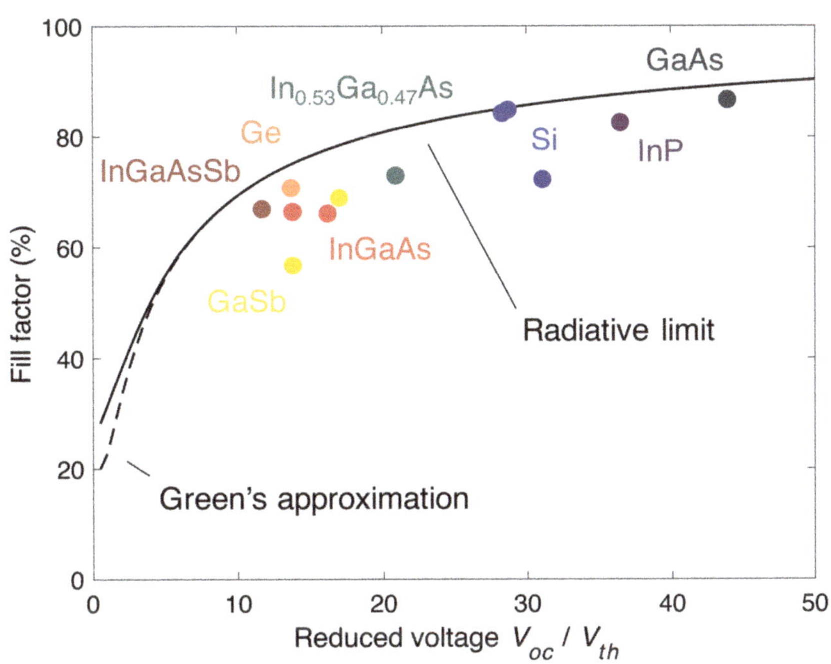

Figure 17.9. Fill factor, FF, reported for high-performance cells in TPV generator and non-concentrating solar PV configurations. Comparison to the radiative recombination limit validates the experimentally observed trend with respect to reduced voltage. Green's approximation, a compact expression of the radiative limit, is valid for high values of reduced voltage, characteristic of the TPV cell materials considered here.

Table 17.4. Evaluation of ohmic power losses for several representative TPV and PV cells.

Material	J_{MPP} [mA cm^{-2}]	R_s[Ω cm^{-2}]	P_{MPP} [mW cm^{-2}]	P_{loss} [mW cm^{-2}]	P_{loss}/P_{MPP} [%]	Ref.
$In_{0.53}Ga_{0.47}As$	834.1	0.044	355	30.6	8.6	[20]
Si	40.6	0.1	26.1	0.164	0.63	[59]
Si	40.2	0.32	26.8	0.517	1.93	[60]

relative to the Si cells considered. However, due to the high photocurrents produced in TPV generators, the ohmic losses in the $In_{0.53}Ga_{0.47}As$ cell amount to over 8% of the power generation. For comparison, the highest estimate of ohmic losses for the Si PV cells considered is under 2%. Since one of the primary advantages of TPV generators, from the perspective of cost per power ($/W), is high-density power generation, innovations in materials and cell design are needed to minimize series resistance without adversely affecting other factors.

17.3 Discussion and summary

This chapter assesses the state of TPV materials by examining, independently, the various efficiency factors that contribute to overall conversion efficiency. The best approaches within each efficiency factor are contextualized by theoretical limits. Ongoing efforts to improve upon reported efficiency factors are discussed, along with practical challenges that face translation of lab-scale demonstrations to realistic generators, such as thermal stability and parasitic losses.

Analysis of high-performance cells across material systems reveals relatively good charge carrier collection and management, which, in at least one material system, are approaching the theoretical limits of the relevant efficiency factors, albeit at substantially lower power densities than theoretically possible. This has been observed separately for internal quantum efficiency (IQE), voltage factor (VF) and fill factor (FF). However, photon management, as described by the spectral efficiency, remains significantly below its theoretical limits.

Internal quantum efficiency data were compiled from several cells, representing the best reported for InGaAs, GaSb, $In_{0.53}Ga_{0.47}As$, Si and InP. In each case, weighted average IQE exceeds 85%, with several materials, Si and $In_{0.53}Ga_{0.47}As$, approaching unity. Generally, high IQE values, reported throughout the literature, suggest that only incremental improvements to quantum efficiency can be expected.

Comparison of the dark current estimates, a measure of bandgap utilization by a PV cell, to the radiative recombination limit has revealed that mature cell technologies are operating at or near the non-recycling radiative limit. In fact, the best GaAs solar PV cell has surpassed the non-recycling limit and is approaching the photon recycling radiative limit. The best $In_{0.53}Ga_{0.47}As$ TPV cell, on the other hand, is operating near the non-recycling limit, despite having a configuration that would, in theory, enable photon recycling. This indicates that non-radiative recombination is still important within the lattice-matched $In_{0.53}Ga_{0.47}As$ material system. Several other mature technologies are approaching the non-recycling radiative limit, as well. Low dark currents also manifest as high voltage factors, exceeding 70% in the case of mature technologies, such as GaAs, Si and $In_{0.53}Ga_{0.47}As$. Less-mature, narrow-bandgap materials are more likely to benefit from further improvements to dark current and voltage factor, as the gap between experimental demonstration and the radiative recombination limit appears to widen for these materials.

As was the case for dark current, fill factors of relatively mature technologies, such as Si and GaAs, are approaching the detailed-balance limit. This is the case for operation under low-illumination conditions (e.g., non-concentrating solar). However, substantially higher current densities can be achieved with high-efficiency TPV generators. In this regime, TPV generators suffer from ohmic losses, sometimes exceeding 8% of power generation. To overcome ohmic losses in TPV generators, it will be necessary to minimize series resistance through contact optimization and cell design. Reduction of ohmic losses, without lowering the other efficiency factors, represents an important challenge that will need to be addressed to achieve high efficiency and low cost per power.

In contrast to carrier management in state-of-the-art cells, spectral utilization remains well below theoretical limits. The upper bound of spectral efficiency is set by the emitter temperature and the cutoff energy (i.e., bandgap). Examination of thermally stable (>1023 K) selective emitters reveals that most demonstrations to date exhibit spectral efficiency below 50%, which is significantly lower than the other efficiency factors. Further, the spectral efficiency of structurally tunable thermal emitters is currently limited by their operating temperatures, quantified here as the highest temperature at which the emitter was aged for at least one hour. In terms of absorptive spectral control at the cold side, the best InGaAs and $In_{0.53}Ga_{0.47}As$ cells exhibit spectral efficiency exceeding that offered by any thermally stable emitter. However, the highest SE observed, 64%, falls short of the theoretical limit of 85% at the relevant T_h and E_g.

Overall, this chapter suggests that the most apparent pathway to improved TPV conversion efficiency, beyond the state-of-the-art, is through advances in spectral efficiency. Spectral inefficiencies represent the biggest losses among the efficiency factors considered in this chapter. Improvements to out-of-band reflectivity, specifically, will help enable external photon recycling, thereby reducing net radiative exchange between the emitter and cell. In addition to spectral inefficiencies, the uniquely high photocurrents of TPV generators leave the system vulnerable to substantial ohmic losses. Optimized materials for selective contacts and grid design can help to alleviate this loss pathway and improve power densities. Ultimately, addressing these challenges may enable TPV generation to offer superior performance and lower cost than conventional, mechanical power cycles for applications in distributed power generation and dispatchable grid-scale storage.

Appendix: Emitter data

Table A1. Summary of reported data for thermally tested emitters considered in section 17.2.1.1.

Material	Class[b]	E_{cutoff} [eV]	T_h [K]	$E_{cutoff}/k_B T_h$	ε_{out}[a]	ε_{in}[a]	SE [%]	S	E range [eV]/BB fraction	Aging conditions	Ref.
HfO_2/Mo/HfO_2	FLC	0.41	1373	3.47	0.42	0.81	46.1	1.32	0.25–3.1 /79.8%	1.5 h at 1373 K under vacuum	[33]
Ta3%W alloy with HfO_2 coating	2D PhC	0.57	1473	4.55	0.36	0.89	40.5	1.70	0.41–0.89/47.4%	300 h at 1273 K, 24 h at 1473 K in Ar	[37, 47]
Pt puck array on Al_2O_3/Pt stack	2D MM	0.6	1273	5.47	0.47	0.94	25.2	1.67	0.25–1.4/76.4%	2 h at 1273 K in Ar	[69]
Polycrystalline Ta with HfO_2 coating	2D PhC	0.62	1273	5.65	0.14	0.82	43.1	3.13	0.41–0.89/40.9%	144 h at 1173 K, 1 h at 1273 K in Ar	[36]
W cavity array	2D PhC	0.62	1200	6.00	0.25	0.86	29.5	2.59	0.16–1.3/90.8%	10 1 h cycles at 1200 K under vacuum	[70]
Ta cavity array with HfO_2 coating	2D PhC	0.62	1273	5.65	0.14	0.80	43.0	3.12	0.16–9.6/92.1%	1 h at 1273 K under vacuum	[35]
Polycrystalline Ta with HfO_2 coating	2D PhC	0.62	1173	6.13	0.18	0.83	33.8	3.20	0.41–1.2/38.4%	100 h at 1173 K under vacuum	[46]
HfO_2 coated W inverse colloidal crystal	3D PhC	0.67	1673	4.65	0.66	0.92	28.0	1.25	0.25–0.98/78.2%	1 h at 1673 K in Ar	[32]
W/HfO_2	FLC	0.67	1423	5.46	0.07	0.27	37.9	2.52	0.25–3.1/81.1%	1 h at 1423 K under vacuum	[38]
W coated colloidal crystal	3D PhC	0.67	1273	6.11	0.10	0.24	24.8	2.35	0.25–0.99/74.4%	12 h at 1273 K in Ar	[32]
MgO with 2wt% NiO loading	TMO	0.69	1677	4.76	0.19	0.73	46.1	2.15	0.14–1.14/92.6%	1677 K	[34]
W/Al_2O_3	1D MM	0.72	1473	5.67	0.28	0.88	31.6	2.32	0.38–1.5/60.6%	3 h at 1473 K under vacuum	[43]

(*Continued*)

Table A1. *(Continued)*

Material	Class[b]	E_{cutoff} [eV]	T_h [K]	$E_{cutoff}/k_B T_h$	ε_{out}[a]	ε_{in}[a]	SE [%]	S	E range [eV]/BB fraction	Aging conditions	Ref.
W/HfO_2 layered metamaterial	1D MM	0.72	1673	4.99	0.23	0.77	41.1	2.14	0.36–2.4/71.7%	6 h at 1673 K under vacuum	[30]
Ni inverse opal with MgO coating	3D PhC	0.726	1373	6.14	0.57	0.75	13.2	1.26	0.50–3.1/36.3%	1 h 1373 K in Ar	[71]
W/CNT	2D PhC	0.726	1273	6.62	0.21	0.83	24.1	3.06	0.25–2.4/76.3%	168 h at 1273 K in He	[45]
$Al_2O_3/Er_3Al_5O_{12}$ eutectic	REO	0.726	1850	4.55	0.26	0.44	32.5	1.39	0.62–1.4/40.1%	1850 K	[29]
$Al_2O_3/W/Al_2O_3$ submicron cavity	2D MM	0.729	1200	7.05	0.38	0.88	13.2	2.14	0.25–3.1/73.4%	Three 5 h cycles at 1200 K under vacuum	[31]
W cavity array	2D PhC	0.729	1200	7.05	0.30	0.95	16.7	2.70	0.16–1.3/89.7%	Ten 1 h cycles at 1200 K under vacuum	[70]
Ni nanopyramid array	2D MM	0.93	1073	10.08	0.19	0.76	3.2	3.99	0.084–4.2/97.3%	5 h at 1073 K under vacuum	[30]
Yb_2O_3 foam	REO	1.12	1800	7.22	0.29	0.53	9.6	1.74	0.075–1.6/98.8%	200 cycles at 1800 K, duration omitted	[41]
Yb_2O_3 mantle	REO	1.12	1800	7.22	0.13	0.61	20.8	3.77	0.024–1.6/99.2%	200 cycles at 1800 K, duration omitted	[41]
Al_2O_3 coated W	3D PhC	—	—	—	—	—	N/A	N/A	—	12 h at 1273 K in forming gas	[42]
HfO_2 coated W	3D PhC	—	—	—	—	—	N/A	N/A	—	12 h at 1673 K in forming gas	[42]

[a] Weighted average ε_{out} and ε_{in} have been calculated using spectral properties collected from graphical data. Error may have resulted from the data extraction process.
[b] REO = rare-earth oxide; TMO = transition-metal oxide; FLC = few-layer coating; PhC = photonic crystal; MM = metamaterial.

References

[1] Bhatt R, Kravchenko I and Gupta M 2020 High-efficiency solar thermophotovoltaic system using a nanostructure-based selective emitter *Sol. Energy* **197** 538–45

[2] Bierman D M *et al* 2016 Enhanced photovoltaic energy conversion using thermally based spectral shaping *Nat. Energy* **1** 16068

[3] Ungaro C, Gray S K and Gupta M C 2015 Solar thermophotovoltaic system using nanostructures *Opt. Express* **23** A1149–56

[4] Lenert A *et al* 2014 A nanophotonic solar thermophotovoltaic device *Nat. Nanotechnol.* **9** 126–30

[5] Rephaeli E and Fan S 2009 Absorber and emitter for solar thermo-photovoltaic systems to achieve efficiency exceeding the Shockley-Queisser limit *Opt. Express* **17** 15145–59

[6] Wang Y, Liu H and Zhu J 2019 Solar thermophotovoltaics: Progress, challenges, and opportunities *APL Mater.* **7** 080906

[7] Zhou Z, Sakr E, Sun Y and Bermel P 2016 Solar thermophotovoltaics: reshaping the solar spectrum *Nanophotonics* **5** 1–21

[8] Harder N-P and Würfel P 2003 Theoretical limits of thermophotovoltaic solar energy conversion *Semicond. Sci. Technol.* **18** S151–7

[9] Crowley C J 2005 Thermophotovoltaic converter performance for radioisotope power systems *AIP Conf. Proc.* **746** 601–14

[10] Datas A and Martí A 2016 Thermophotovoltaic energy in space applications: review and future potential *Sol. Energy Mater. Sol. Cells* **161** 285–96

[11] Datas A, Ramos A, Martí A, del Cañizo C and Luque A 2016 Ultra high temperature latent heat energy storage and thermophotovoltaic energy conversion *Energy* **107** 542–9

[12] Amy C, Seyf H R, Steiner M A, Friedman D J and Henry A 2019 Thermal energy grid storage using multi-junction photovoltaics *Energy Environ. Sci.* **12** 334–43

[13] Daneshvar H, Prinja R and Kherani N P 2015 Thermophotovoltaics: fundamentals, challenges and prospects *Appl. Energy* **159** 560–75

[14] Durisch W and Bitnar B 2010 Novel thin film thermophotovoltaic system *Sol. Energy Mater. Sol. Cells* **94** 960–5

[15] Bianchi M, Ferrari C, Melino F and Peretto A 2012 Feasibility study of a thermo-photovoltaic system for CHP application in residential buildings *Appl. Energy* **97** 704–13

[16] Fraas L M, Avery J E and Huang H X 2003 Thermophotovoltaic furnace–generator for the home using low bandgap GaSb cells *Semicond. Sci. Technol.* **18** S247–53

[17] Lee K, Zimmerman J D, Xiao X, Sun and Forrest S R 2012 Reuse of GaAs substrates for epitaxial lift-off by employing protection layers *J. Appl. Phys.* **111** 033527

[18] Lee K, Zimmerman J D, Hughes T W and Forrest S R 2014 Non-destructive wafer recycling for low-cost thin-film flexible optoelectronics *Adv. Funct. Mater.* **24** 4284–91

[19] Wang C A *et al* 2004 Wafer bonding and epitaxial transfer of GaSb-based epitaxy to GaAs for monolithic interconnection of thermophotovoltaic devices *J. Electron. Mater.* **33** 213–7

[20] Omair Z *et al* 2019 Ultraefficient thermophotovoltaic power conversion by band-edge spectral filtering *Proc. Natl Acad. Sci.* **116** 15 356–61

[21] Burger T, Fan D, Lee K, Forrest S R and Lenert A 2018 Thin-film architectures with high spectral selectivity for thermophotovoltaic cells *ACS Photonics* **5** 2748–54

[22] Seyf H R and Henry A 2016 Thermophotovoltaics: a potential pathway to high efficiency concentrated solar power *Energy Environ. Sci.* **9** 2654–65

[23] Mauk M G 2006 Survey of thermophotovoltaic (TPV) devices *Mid-infrared Semiconductor Optoelectronics* vol 118 (London: Springer), pp 673–738

[24] Sakakibara R *et al* 2019 Practical emitters for thermophotovoltaics: a review *J. Photonics Energy* **9** 032713

[25] Gupta M C, Ungaro C, Foley J J and Gray S K 2017 Optical nanostructures design, fabrication, and applications for solar/thermal energy conversion *Sol. Energy* **165** 100–14

[26] Ferrari C and Melino F 2014 Overview and status of thermophotovoltaic systems *Energy Proc.* **45** 160–9

[27] Licht A, Pfiester N, DeMeo D, Chivers J and Vandervelde T E 2019 A review of advances in thermophotovoltaics for power generation and waste heat harvesting *MRS Adv.* **4** 2271–82

[28] Bauer T 2011 *Thermophotovoltaics: Basic Principles and Critical Aspects of System Design* (Berlin: Springer)

[29] Nakagawa N, Ohtsubo H, Waku Y and Yugami H 2005 Thermal emission properties of $Al_2O_3/Er_3Al_5O_{12}$ eutectic ceramics *J. Eur. Ceram. Soc.* **25** 1285–91

[30] Chirumamilla M *et al* 2019 Metamaterial emitter for thermophotovoltaics stable up to 1400 °C *Sci. Rep.* **9** 7241

[31] Kim S *et al* 2019 Optical tunneling mediated sub-skin-depth high emissivity tungsten radiators *Nano Lett.* **19** 7093

[32] Arpin K A *et al* 2013 Three-dimensional self-assembled photonic crystals with high temperature stability for thermal emission modification *Nat. Commun.* **4** 2630

[33] Wang Y *et al* 2018 Hybrid solar absorber–emitter by coherence-enhanced absorption for improved solar thermophotovoltaic conversion *Adv. Opt. Mater.* **6** 1800813

[34] Ferguson L G and Dogan F 2001 A highly efficient NiO-Doped MgO matched emitter for thermophotovoltaic energy conversion *Mater. Sci. Eng.* B **83** 35–41

[35] Rinnerbauer V, Yeng Y X, Senkevich J J, Joannopoulos J D, Soljačić M and Celanovic I 2013 Large area selective emitters/absorbers based on 2D tantalum photonic crystals for high-temperature energy applications *Proc. SPIE* **8632** 863207

[36] Rinnerbauer V *et al* 2013 High-temperature stability and selective thermal emission of polycrystalline tantalum photonic crystals *Opt. Express* **21** 11482

[37] Stelmakh V *et al* 2013 High-temperature tantalum tungsten alloy photonic crystals: stability, optical properties, and fabrication *Appl. Phys. Lett.* **103** 123903

[38] Shimizu M, Kohiyama A and Yugami H 2018 Evaluation of thermal stability in spectrally selective few-layer metallo-dielectric structures for solar thermophotovoltaics *J. Quant. Spectrosc. Radiat. Transf.* **212** 45–9

[39] Pfiester N A and Vandervelde T E 2017 Selective emitters for thermophotovoltaic applications *Phys. Status Solidi* **214** 1600410

[40] Li P *et al* 2015 Large-scale nanophotonic solar selective absorbers for high-efficiency solar thermal energy conversion *Adv. Mater.* **27** 4585–91

[41] Bitnar S *et al* 2004 Practical thermophotovoltaic generators *Semiconductors* **38** 941–5

[42] Arpin K A, Losego M D and Braun P V 2011 Electrodeposited 3D tungsten photonic crystals with enhanced thermal stability *Chem. Mater.* **23** 4783–88

[43] Kim J H, Jung S M and Shin M W 2019 Thermal degradation of refractory layered metamaterial for thermophotovoltaic emitter under high vacuum condition *Opt. Express* **27** 3039

[44] Rinnerbauer V *et al* 2012 Recent developments in high-temperature photonic crystals for energy conversion *Energy Environ. Sci.* **5** 8815–23

[45] Cui K, Lemaire P, Zhao H, Savas T, Parsons G and Hart A J 2018 Tungsten-carbon nanotube composite photonic crystals as thermally stable spectral-selective absorbers and emitters for thermophotovoltaics *Adv. Energy Mater.* **8** 1801471

[46] Chan W R, Stelmakh V, Ghebrebrhan M, Soljačić M, Joannopoulos J D and Čelanović I 2017 Enabling efficient heat-to-electricity generation at the mesoscale *Energy Environ. Sci.* **10** 1367–71

[47] Stelmakh V 2017 A practical high temperature photonic crystal for high performance thermophotovoltaics *PhD Thesis* (Massachusetts Institute of Technology)

[48] Nagpal P, Josephson D P, Denny N R, DeWilde J, Norris D J and Stein A 2011 Fabrication of carbon/refractory metal nanocomposites as thermally stable metallic photonic crystals *J. Mater. Chem.* **21** 10836

[49] Pfiester N A, Naka N and Vandervelde T E 2015 Platinum metmamaterials for thermophotovoltaic selective emitters *31st EU-PVSEC* pp 55–8

[50] Doiron C F and Naik G V 2018 Semiconductors for high selectivity thermal emitters *J. Opt.* **20** 084001

[51] Baranov D G, Xiao Y, Nechepurenko I A, Krasnok A, Alù A and Kats M A 2019 Nanophotonic engineering of far-field thermal emitters *Nat. Mater.* **18** 920–30

[52] Wernsman B *et al* 2004 Greater than 20% radiant heat conversion efficiency of a thermophotovoltaic radiator/module system using reflective spectral control *IEEE Trans. Electron Devices* **51** 512–5

[53] Fernández J, Dimroth F, Oliva E, Hermle M and Bett A W 2007 Back-surface optimization of germanium TPV cells *AIP Conf. Proc.* 890 190–97

[54] Tang L, Fraas L M, Liu Z, Zhang Y, Duan H and Xu C 2019 N-type vapor diffusion for the fabrication of GaSb thermophotovoltaic cells to increase the quantum efficiency in the long wavelength range *Sol. Energy Mater. Sol. Cells* **194** 137–41

[55] Swanson R M 1980 Recent developments in thermophotovoltaic conversion *1980 Int. Electron Devices Meeting* pp 186–9

[56] Siergiej R R *et al* 2003 20% efficient InGaAs/InPAs thermophotovoltaic cells *AIP Conf. Proc.* 653 414–23

[57] Dashiell M W *et al* 2006 Quaternary InGaAsSb thermophotovoltaic diodes *IEEE Trans. Electron Devices* **53** 2879–91

[58] Hudait M K, Brenner M and Ringel S A 2009 Metamorphic $In_{0.7}Al_{0.3}As/In_{0.69}Ga_{0.31}As$ thermophotovoltaic devices grown on graded $InAs_yP_{1-y}$ buffers by molecular beam epitaxy *Solid. State. Electron.* **53** 102–6

[59] Haase F *et al* 2018 Laser contact openings for local poly-Si-metal contacts enabling 26.1%-efficient POLO-IBC solar cells *Sol. Energy Mater. Sol. Cells* **186** 184–93

[60] Yoshikawa K *et al* 2017 Silicon heterojunction solar cell with interdigitated back contacts for a photoconversion efficiency over 26% *Nat. Energy* **2** 17032

[61] Wanlass M W 2017 Systems and methods for advanced ultra–high–performance InP solar cells *US Patent* https://patents.google.com/patent/US9590131B2/en

[62] Kayes B M *et al* 2011 27.6% conversion efficiency, a new record for single-junction solar cells under 1 Sun illumination in *2011 37th IEEE Photovoltaic Specialists Conf.* 000 004–8

[63] Green M A *et al* 2019 Solar cell efficiency tables (version 53) *Prog. Photovoltaics Res. Appl.* **27** 3–12

[64] Henry C H 1980 Limiting efficiencies of ideal single and multiple energy gap terrestrial solar cells *J. Appl. Phys.* **51** 4494–500

[65] Miller O D, Yablonovitch E and Kurtz S R 2012 Strong internal and external luminescence as solar cells approach the Shockley–Queisser limit *IEEE J. Photovoltaics* **2** 303–11

[66] Shimizu M, Kohiyama A and Yugami H 2015 High-efficiency solar-thermophotovoltaic system equipped with a monolithic planar selective absorber/emitter *J. Photonics Energy* **5** 053099

[67] Shockley W and Queisser H J 1961 Detailed balance limit of efficiency of p–n junction solar cells *J. Appl. Phys.* **32** 510–9

[68] Green M A 1982 *Solar Cells: Operating Principles, Technology, and System Applications* (Englewood Cliffs, NJ: Prentice-Hall)

[69] Woolf D N *et al* 2018 High-efficiency thermophotovoltaic energy conversion enabled by a metamaterial selective emitter *Optica* **5** 213

[70] Yeng Y X *et al* 2012 Enabling high-temperature nanophotonics for energy applications *Proc. Natl Acad. Sci.* **109** 2280–5

[71] Kim Y, Kim M-J, Kim Y, Lee H and Lee S 2019 Nanostructured radiation emitters: design rules for high-performance thermophotovoltaic systems *ACS Photonics* **6** 2260–7

www.ingramcontent.com/pod-product-compliance
Ingram Content Group UK Ltd.
Pitfield, Milton Keynes, MK11 3LW, UK
UKHW051909100926
8556IPUK00005B/820

9 780750 317368